$1\ lb_f = 32.174\ ft\cdot lb/s^2$
$1\ lb_f = 4.4482\ N$
$1\ lb_f = 4.4482\ kg\cdot m/s^2$
$1\ lb_f = 32.1739\ poundals$
$1\ lb_f/in^2 \equiv (1\ psi) = 6894.76\ N/m^2$
$1\ lb_f/ft^2 = 47.880\ N/m^2$
$1\ bar = 10^5\ N/m^2$
$1\ atm = 14.696\ lb_f/in^2$
$1\ atm = 2116.2\ lb_f/ft^2$
$1\ atm = 1.0132 \times 10^5\ N/m^2$
$1\ atm = 1.0132\ bar$

14. Specific heat
   $1\ Btu/lb\cdot{}^\circ F = 1\ kcal/kg\cdot{}^\circ C = 1\ cal/g\cdot{}^\circ C$
   $1\ Btu/lb\cdot{}^\circ F = 4186.69\ J/kg\cdot{}^\circ C$
   (or $W\cdot s/kg\cdot{}^\circ C$)
   $1\ Btu/lb\cdot{}^\circ F = 4.18669\ J/g\cdot{}^\circ K$
   (or $W\cdot s/g\cdot{}^\circ C$)
   $1\ J/g\cdot{}^\circ C = 0.23885\ Btu/lb\cdot{}^\circ F$
   ($cal/g\cdot{}^\circ C$ or $kcal/kg\cdot{}^\circ C$)

15. Speed
   $1\ ft/s = 0.3048\ m/s$
   $1\ m/s = 3.2808\ ft/s$
   $1\ mile/h = 1.4667\ ft/s$
   $1\ mile/h = 0.44704\ m/s$

16. Surface tension
   $1\ lb_f/ft = 14.5937\ N/m$
   $1\ N/m = 0.068529\ lb_f/ft$

17. Temperature
   $1\ K = 1.8{}^\circ R$
   $T({}^\circ F) = 1.8(K - 273) + 32$
   $$T(K) = \frac{1}{1.8}\ ({}^\circ F - 32) + 273$$

$$T({}^\circ C) = \frac{1}{1.8}\ ({}^\circ R - 492)$$
$$\Delta T({}^\circ C) = 1.8\ \Delta T({}^\circ F)$$

18. Thermal conductivity
   $1\ Btu/h\cdot ft\cdot{}^\circ F = 1.7303\ W/m\cdot{}^\circ C$
   $1\ Btu/h\cdot ft\cdot{}^\circ F = 1.7303 \times 10^{-2}\ W/cm\cdot{}^\circ C$
   $1\ Btu/h\cdot ft\cdot{}^\circ F = 0.4132\ cal/s\cdot m\cdot{}^\circ C$
   $1\ W/m\cdot{}^\circ C = 0.5779\ Btu/h\cdot ft\cdot{}^\circ F$
   $1\ W/cm\cdot{}^\circ C = 57.79\ Btu/h\cdot ft\cdot{}^\circ F$

19. Thermal resistance
   $1\ h\cdot{}^\circ F/Btu = 1.896{}^\circ C/W$
   $1{}^\circ C/W = 0.528\ h\cdot{}^\circ F/Btu$

20. Viscosity
   $1\ poise = 1\ g/cm\cdot s$
   $1\ poise = 10^{-2}\ centipoise$
   $1\ poise = 241.9\ lb/ft\cdot h$
   $1\ centipoise = 2.419\ lb/ft\cdot h$
   $1\ lb/ft\cdot s = 1.4882\ kg/m\cdot s$
   $1\ lb/ft\cdot s = 14.882\ poises$
   $1\ lb/ft\cdot s = 1488.2\ centipoises$
   $1\ lb/ft\cdot h = 0.4134 \times 10^{-3}\ kg/m\cdot s$
   $1\ lb/ft\cdot h = 0.4134 \times 10^{-2}\ poise$
   $1\ lb/ft\cdot h = 0.4134\ centipoise$

21. Volume
   $1\ in^3 = 16.387\ cm^3$
   $1\ cm^3 = 0.06102\ in^3$
   $1\ oz\ (U.S.\ fluid) = 29.573\ cm^3$
   $1\ ft^3 = 0.0283168\ m^3$
   $1\ ft^3 = 28.3168\ liters$
   $1\ ft^3 = 7.4805\ gal\ (U.S.)$
   $1\ m^3 = 35.315\ ft^3$
   $1\ gal\ (U.S.) = 3.7854\ liters$
   $1\ gal\ (U.S.) = 3.7854 \times 10^{-3}\ m^3$
   $1\ gal\ (U.S.) = 0.13368\ ft^3$

CONSTANTS

g_c = gravitational acceleration = $32.1739\ ft\cdot lb/lb_f\cdot s^2$
   conversion factor
   = $4.1697 \times 10^8\ ft\cdot lb/lb_f\cdot h^2$
   = $1\ g\cdot cm/dyn\cdot s^2$
   = $1\ kg\cdot m/N\cdot s^2$
   = $1\ lb\cdot ft/poundal\cdot s^2$
   = $1\ slug\cdot ft/lb_f\cdot s^2$

J = mechanical equivalent of heat = $778.16\ ft\cdot lb_f/Btu$

$\mathscr{R}$ = gas constant = $1544\ ft\cdot lb_f/lb\ mol\cdot{}^\circ R$
   = $0.730\ ft^3\cdot atm/lb\ mol\cdot{}^\circ R$
   = $0.08205\ m^3\cdot atm/kg\ mol\cdot K$
   = $8.314\ J/g\ mol\cdot K$
   = $8.314\ N\cdot m/g\ mol\cdot K$
   = $1.987\ cal/g\ mol\cdot K$

σ = Stefan-Boltzmann constant = $0.1714 \times 10^{-8}\ Btu/h\cdot ft^2\cdot{}^\circ R^4$
   = $0.56697 \times 10^{-8}\ W/m^2\cdot K^4$
   5.67×10^{-8}

BASIC HEAT TRANSFER

BASIC HEAT TRANSFER

M. Necati Özışık

Professor of Mechanical
and Aerospace Engineering
North Carolina State University

McGraw-Hill Book Company

New York ∎ St. Louis ∎ San Francisco
Auckland ∎ Bogotá ∎ Düsseldorf
Johannesburg ∎ London ∎ Madrid
Mexico ∎ Montreal ∎ New Delhi
Panama ∎ Paris ∎ São Paulo
Singapore ∎ Sydney ∎ Tokyo ∎ Toronto

BASIC HEAT TRANSFER

1234567890   DODO   7832109876

This book was set in Times Roman.   The editors were
B. J. Clark, M. E. Margolies, and Barbara Tokay; the designer
was Joseph Gillians; the production supervisor was Charles Hess.
The drawings were done by J & R Services, Inc.
R. R. Donnelly & Sons Company was printer and binder.

Library of Congress Cataloging in Publication Data

Özışık, M.   Necati.
    Basic heat transfer.

    Includes bibliographies and index.
    1. Heat—Transmission. I. Title.
QC320.093   536′.2   76-7999
ISBN 0-07-047980-1

TO
GÜL and
HAKAN

Contents

CHAPTER 16   MASS TRANSFER   467

APPENDIXES

PROBLEMS   525

INDEX   563

Preface

The material which must be covered in one semester undergraduate level heat transfer course is so extensive that a systematic and unified presentation of the subject matter is essential for effective teaching. Most textbooks available in the market follow a pedagogy in which the reader is introduced to the subject with the solution of particular problems without first being exposed to the fundamentals. This approach appears to be an easy way to start the instruction, but has the disadvantage that, within the limited time available, it becomes almost impossible to return to the fundamentals. As a result, the student's knowledge on the subject does not extend beyond the specific problems covered during the course.

In teaching heat transfer over the past several years, the author has observed that teaching effectiveness is improved and the student's capacity for dealing with the analysis of heat transfer problems is significantly increased if the fundamentals are presented first and close attention is paid to the proper posing of the physical problems before proceeding to the solutions. At the undergraduate level, the fundamentals should be presented with the minimum amount of mathematical complexity and with careful description of the physical significance of various quantities in the mathematical expressions.

It is in this philosophy of approach that the present text differs fundamentally from the existing undergraduate level heat transfer texts. There is sufficient material in this book, systematically arranged at different levels, for the spectrum of its possible uses to include: a first course in heat transfer at the junior level, a basic heat transfer course at a higher level, or a two-quarter undergraduate heat transfer sequence. The text can also serve as a reference volume for engineering graduates and industry. A background in ordinary differential equations at the sophomore level is sufficient to follow the material in this book.

In Chapter 1 the basic concepts in the area of heat transfer are discussed. Chapter 2 is devoted to the derivation of the heat conduction equation and a discussion of dimensionless parameters, the boundary conditions, and the mathematical formulation of physical problems with a unified approach. The aim of this chapter is to provide a good understanding of the physical significance of the heat conduction equation and the proper formulation of heat conduction problems.

The three chapters that follow are devoted to the methods of solution of heat conduction problems at three distinct levels. In particular, Chapter 3 deals with the analysis of one-dimensional, steady-state heat conduction in slabs, cylinders, spheres and through fins. In Chapter 4, a unified approach for the solution of two physically different heat conduction problems is presented. The solution of two-dimensional steady-state and one-dimensional unsteady heat conduction problems are brought together in this chapter because of their common mathematical base. This aim is achieved by a systematic tabulation, presented in Table 4-1, of the fundamental solutions common to these two different class of problems. Once the reader becomes familiar with the use of this table, the analysis of heat conduction problems discussed in Chapter 4 becomes a relatively easy matter. In Chapter 5, the finite difference technique and its application to the solution of heat conduction problems are presented with a concise and rigorous approach.

In the teaching of convection heat transfer, the physical significance of various quantities in the energy equation should be tied to the fluid mechanics aspects of the problem. To this end, Chapter 6 is devoted to the derivation of the equations of motion and energy. The aim of this chapter is to provide a good appreciation of the physical significance of various terms and the dimensionless groups in the resulting expressions and to serve as a ready reference for the equations needed in the four subsequent chapters on the analysis of convective heat transfer. If individual course objectives do not require it, the detailed derivations of this chapter may be omitted without effecting the continuity of the subject. Emphasis may be placed, instead, on the understanding of the physical significance of these equations.

Chapters 7 and 8 deal with heat transfer in internal and external flow respectively. The aim of these two chapters is to illustrate the mathematical formulation of typical convective heat transfer problems by utilizing the equations given in Chapter 6, to present typical methods of solution, and to provide a good understanding of the physical significance of various heat transfer results. Chapter 9 is devoted to a discussion of heat transfer in internal and external turbulent flow. In order to provide some insight to the implications of turbulent flow, basic concepts of the mechanism of turbulence are discussed and various analogies between momentum and heat transfer are described before a wealth of empirical correlations of heat transfer in turbulent flow is presented. If it is not required by the course objective, the analysis of turbulent flow may be omitted, and emphasis can be placed on the application of the empirical relations. Heat transfer in free convection is presented in Chapter 10.

Chapters 11 through 13 provide a systematic analysis of radiative heat transfer in nonparticipating and participating media. Chapter 11 gives the background information on the emission, the absorption, and the reflection of radiation by the matter that are needed in the two succeeding chapters on the analysis of radiative heat exchange. The subject of radiative heat transfer between surfaces in a nonparticipating medium is presented in Chapter 12 using an approach different from those followed in most undergraduate heat transfer texts. The method of analysis is more straightforward and possesses computational advantages. The radiative heat transfer in participating media is considered in Chapter 13. Chapter 13 may be omitted, if not required by the goals of the course, without affecting the continuity of the subject. The empirical results given in this chapter and their applications may be emphasized in these cases.

In Chapters 14 and 15, a comprehensive treatment of the subjects of heat transfer in condensation, boiling and the heat exchangers is given.

Finally, in Chapter 16, the analysis of mass transfer is closely tied to the analysis of heat transfer. The systematic, simple, and rigorous approach followed in this chapter in developing the basic relations will make the teaching of this complicated subject a relatively easy matter.

Heat transfer calculations are commonly performed in engineering by using the English system and the SI (Systéme Internationale) system of units. The SI system has been adopted in a number of countries and a changeover into the SI system in the engineering field is expected to take place in the countries which are currently using the English system. In the transition period it will be necessary for the student and the engineer to be familiar with both systems of units. Therefore, both the English system and the SI system of units are simultaneously used throughout the main body of the text, in the solution of examples, and in the physical property tables.

I would like to thank Dr. J. R. Biddle, California State Polytechnic University at Pomona, Dr. D. K. Warinner, Argonne National Laboratory, and Dr. J. P. Holman, Southern Methodist University for reading the entire manuscript and making valuable suggestions.

M. Necati Özışık

One

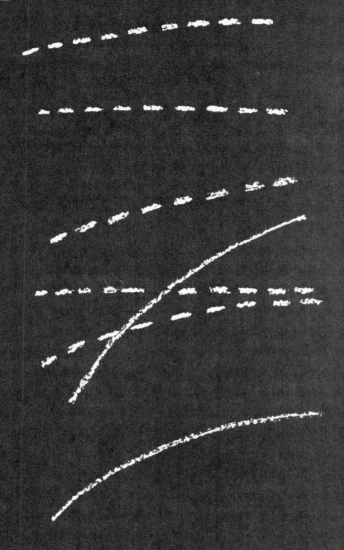

Introduction and Concepts

The concept of *energy* is used in thermodynamics to specify the state of a system. It is a well-known fact that energy is neither created nor destroyed but only changed from one form to another.   The science of *thermodynamics* deals with the relation between heat and other forms of energy, but the science of *heat transfer* is concerned with the analysis of the rate of heat transfer taking place in a system. The energy transfer by heat flow cannot be measured directly, but the concept has physical meaning because it is related to the measurable quantity called *temperature*. It has long been established by observations that, when there is temperature difference in a system, heat flows from the region of high temperature to the region of low temperature. Since heat flow takes place whenever there is a temperature gradient in a system, a knowledge of the temperature distribution in a system is essential in heat-transfer studies. Once the temperature distribution is known, a quantity of practical interest, the *heat flux*, which is the amount of heat transfer per unit area, per unit time, is readily determined from the law relating the heat flux to the temperature gradient.

The problem of determining temperature distribution and heat flow is of interest in many branches of science and engineering. In the design of heat exchangers such as boilers, condensers, radiators, etc., for example, heat-transfer analysis is essential for sizing such equipment.   In the design of nuclear-reactor cores, a thorough heat-transfer analysis of fuel elements is important for proper sizing of fuel elements to prevent burnout.   In aerospace technology, the temperature-distribution and heat-transfer problems are crucial because of weight limitations and safety considerations.   In heating and air-conditioning applications for buildings, a proper heat-transfer analysis is necessary to estimate the amount of insulation needed to prevent excessive heat losses or gains.

In the studies of heat transfer it is customary to consider three distinct modes of heat transfer: *conduction, convection,* and *radiation.*   In reality, temperature distribution in a medium is controlled by the combined effects of these three modes of heat transfer; therefore it is not actually possible to isolate entirely one mode from interactions with the other modes.   However, for simplicity in the analysis, one can consider, for example, conduction separately whenever heat transfer by convection and radiation is negligible.   With this qualification, we present below a brief qualitative description of these three distinct modes of heat transfer; they will be studied in greater detail in the following chapters.

1-1   CONDUCTION

Conduction is the mode of heat transfer in which energy exchange takes place from the region of high temperature to the region of low temperature by the kinetic motion or direct impact of molecules, as in the case of fluid at rest, and by the drift of electrons, as in the case of metals.   In a solid which is a good electric conductor, a large number of free electrons move about in the lattice; hence materials that are good electric conductors are generally good heat conductors (i.e., copper, silver, etc.).

The basic law of heat conduction based on experimental observations originates from Biot but is generally named after the French mathematical physicist Joseph Fourier [1][1] who used it in his analytic theory of heat. This law states that the rate of heat flow by conduction in a given direction is proportional to the area normal to the direction of heat flow and to the gradient of temperature in that direction. For heat flow in the x direction, for example, the Fourier law is given as

$$Q_x = -kA \frac{\partial T}{\partial x} \qquad \text{Btu/h} \quad \text{or} \quad \text{W} \tag{1-1a}$$

or

$$q_x = \frac{Q_x}{A} = -k \frac{\partial T}{\partial x} \qquad \text{Btu/h·ft}^2 \quad \text{or} \quad \text{W/m}^2 \tag{1.1b}$$

where Q_x is the rate of heat flow through area A in the positive x direction, and q_x is called the *heat flux* in the positive x direction. The proportionality constant k is called the *thermal conductivity* of the material and is a positive quantity. The minus sign is included in Eqs. (1-1) to ensure that q_x (or Q_x) is a positive quantity when the heat flow is in the positive x direction. This is apparent from the fact that the temperature should decrease in the positive x direction if the heat should flow in that direction; then $\partial T/\partial x$ is negative, and the inclusion of the negative sign in the above equations ensures that q_x (or Q_x) is a positive quantity.

The thermal conductivity k in Eqs. (1-1) has units Btu/h·ft·°F (or W/m·°C) if heat flux q_x is in Btu/h·ft^2 (or W/m^2), and the temperature gradient $\partial T/\partial x$ is in °F/ft (or °C/m). There is a wide difference in the thermal conductivities of various engineering materials, as shown in Fig. 1-1. The highest value is given by pure metals and the lowest value by gases and vapors; the amorphous insulating materials and inorganic liquids have thermal conductivities that lie in between. To give some idea of the order of magnitude of thermal conductivity for various materials we list below some typical values of k:

Metals: 30 to 240 Btu/h·ft·°F (or 52 to 415 W/m·°C)
Alloys: 7 to 70 Btu/h·ft·°F (or 12 to 120 W/m·°C)
Nonmetallic liquids: 0.1 to 0.4 Btu/h·ft·°F (or 0.173 to 0.69 W/m·°C)
Insulating materials: 0.02 to 0.1 Btu/h·ft·°F (or 0.035 to 0.173 W/m·°C)
Gases at atmospheric pressure: 0.004 to 0.1 Btu/h·ft·°F (or 0.0069 to 0.173 W/m·°C)

Thermal conductivity also varies with temperature. For most pure metals it decreases with temperature, whereas for gases and insulating materials it increases with temperature. At very low temperatures thermal conductivity varies very rapidly with temperature, as shown in Fig. 1-2. A comprehensive compilation of thermal conductivities of materials may be found in Refs. [2, 3, 4].

---

[1] Bracketed numbers indicate references at the end of the chapter.

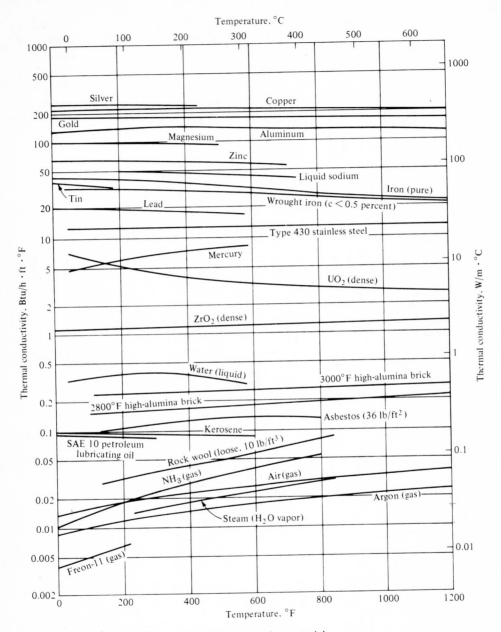

FIG. 1-1 Thermal conductivity of typical engineering materials.

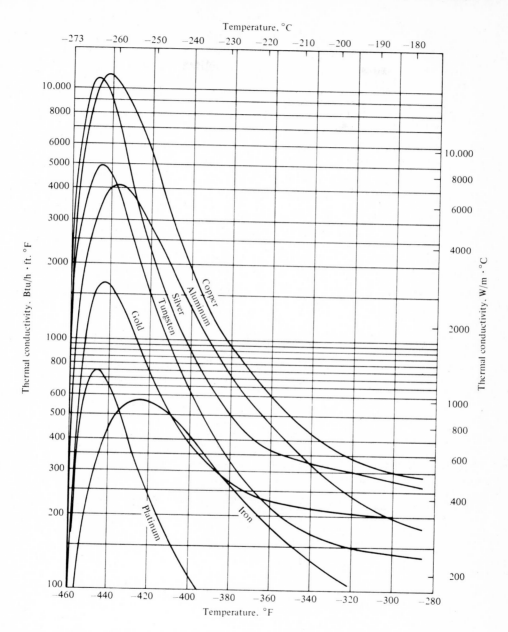

FIG. 1-2 Thermal conductivity of metals at low temperatures. (*From Powell et al.* [2].)

1-2   CONVECTION

When fluid flows over a solid body or inside a channel while temperatures of the fluid and the solid surface are different, heat transfer between the fluid and the solid surface takes place as a consequence of the motion of fluid relative to the surface; this mechanism of heat transfer is called *convection*. If the fluid motion is artificially induced, say with a pump or a fan that forces the fluid flow over the surface, the heat transfer is said to be by *forced convection*. If the fluid motion is set up by buoyancy effects resulting from density difference caused by temperature difference in the fluid, the heat transfer is said to be by *free* (or *natural*) *convection*. For example, a hot plate vertically suspended in stagnant cool air causes a motion in the air layer adjacent to the plate surface because the temperature gradient in the air gives rise to a density gradient which in turn sets up the air motion. As the temperature field in the fluid is influenced by the fluid motion, the determination of temperature distribution and of heat transfer in convection for most practical situations is a complicated matter. In engineering applications, to simplify the heat-transfer calculations between a

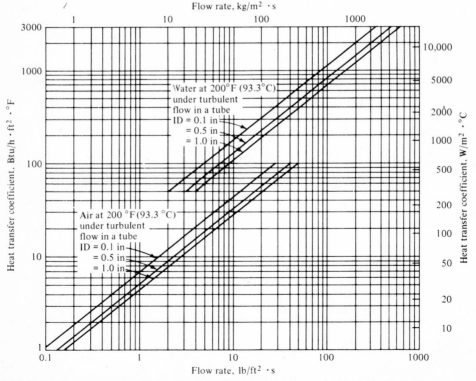

FIG. 1-3   Heat-transfer coefficient for turbulent flow of water and air at 200°F in tubes.

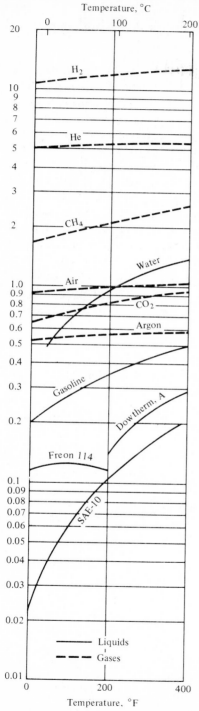

FIG. 1-4   Correction factor for Fig. 1-3 to illustrate the variation of *h* with the type of fluid and temperature.

surface at a temperature T_w and a fluid flowing over it at a mean temperature T_f, a heat-transfer coefficient h is defined as

$$q = h(T_f - T_w) \tag{1-2}$$

where q is the heat flux at the wall. This relation is sometimes called heat transfer according to "Newton's law of cooling." If the heat flux is given in the units of Btu/h·ft^2 (or W/m^2) and temperature of °F (or °C), then the heat-transfer coefficient h has a unit of Btu/h·ft^2·°F (or W/m^2·°C).

Although h can be computed analytically for laminar flow over bodies having simple geometries, an experimental approach is the only means to determine it for flow over bodies having complex configuration. Table 1-1 gives typical values of h for forced and free convection encountered in typical engineering applications. Figures 1-3 and 1-4 are intended to give some idea of the effects of flow velocity, tube size, types of fluid, and temperature on the heat-transfer coefficient for turbulent flow inside tubes. Figure 1-3 gives the heat-transfer coefficient for water and air at 200°F as a function of flow velocity for three different tube diameters. In order to illustrate the variation of the convective heat-transfer coefficient with temperature and the type of fluid, we present in Fig. 1-4 a chart to adjust the data in Fig. 1-3 for other types of gases and liquids for temperatures from 0 to 400°F. For example, h for hydrogen at 200°F is almost 12 times higher than that of air, and for SAE-10 oil it is almost one-tenth that of water. Clearly, the heat-transfer coefficient varies significantly with the type of fluid and temperature.

TABLE 1-1   Typical Values of Convective Heat-Transfer Coefficient h in Engineering Applications

Type of Flow	h, Heat-Transfer Coefficient	
	Btu/h·ft^2·°F	W/m^2·°C
Free convection	1–5	6–28
Turbulent forced convection inside pipes		
Air	1–100	6–570
Water	50–3000	284–17,000
Boiling of water	500–10,000	2840–57,000
Condensation of steam	1000–20,000	5680–113,500

1-3   RADIATION

When two bodies at different temperatures are separated by a perfect vacuum, heat transfer between them by conduction or convection is not possible; in such situations heat transfer between the bodies takes place by *thermal radiation*. That is, the radiative energy emitted by a body because of its temperature is transmitted in the space in the form of electromagnetic waves according to Maxwell's classic electromagnetic-wave theory [5] or in the form of discrete photons according to Planck's [6] hypothesis. Both of these concepts have been utilized in the investigation of radiative-heat transfer. The emission or absorption of radiation energy by a body is a bulk process, that is, radiation originating from the interior of the body is emitted through the surface; conversely, radiation incident on the surface of a body penetrates into the depths of the medium where it is attenuated. In situations where a large proportion of the incident radiation is attenuated within a very short distance from the surface, we may speak of radiation as being absorbed or emitted by the surface. For example, in many engineering applications the absorption or emission of radiation by a metal is considered a surface process because radiation is attenuated within a distance of a few hundred angstroms from the surface; but the emission or absorption of radiation by gases is always treated as a bulk process. Radiation energy emitted by a body is proportional to the fourth power of its absolute temperature. Consider, for example, that a *black-body* (i.e., a perfect emitter and perfect absorber) of surface area A and at an *absolute temperature* T_1 is contained in an enclosure at an *absolute temperature* T_2. The body will emit radiative energy of amount $A\sigma T_1{}^4$ Btu/h (or W) and absorb radiative energy of amount $A\sigma T_2{}^4$ Btu/h (or W) so that the net radiative energy *leaving* the body becomes

$$Q = A\sigma(T_1{}^4 - T_2{}^4) \qquad \text{Btu/h (or W)} \tag{1-3}$$

where σ is the *Stefan-Boltzmann constant* which has a value of 0.1714×10^{-8} Btu/h·ft²·°R⁴ (or 5.6697×10^{-8} W/m²·K⁴) in engineering units.

If the two bodies are not perfectly black and the surface A is not completely enclosed by the other body, then the above relation may be modified as

$$Q = FA\sigma(T_1{}^4 - T_2{}^4) \qquad \text{Btu/h (or W)} \tag{1-4}$$

where the factor F is a quantity which is less than unity and accounts for the effects of geometrical arrangement of the surfaces and for bodies not being perfect emitters and absorbers.

When the difference between the temperatures T_1 and T_2 is sufficiently small in comparison with T_1, Eq. (1-4) can be linearized as

$$Q = FA\sigma(T_1 - T_2)(T_1 + T_2)(T_1{}^2 + T_2{}^2) \cong FA\sigma 4T_1{}^3(T_1 - T_2)$$

or

$$q \equiv \frac{Q}{A} = (F4\sigma T_1{}^3)(T_1 - T_2) \tag{1-5}$$

For such situations a *radiative heat-transfer coefficient* h_r may be defined:

$$h_r \equiv F4\sigma T_1{}^3 \tag{1-6}$$

Then Eq. (1-5) is written in the form

$$q = h_r(T_1 - T_2) \tag{1-7}$$

which is analogous to Eq. (1-2) for convective heat transfer. This approximate, simple expression for radiative-heat flux given by Eq. (1-7) is applicable only if $|T_1 - T_2|/T_1 \ll 1$.

1-4   COMBINED CONVECTION AND RADIATION

When heat transfer by convection and by radiation are of the same order of magnitude and occur simultaneously, a proper analysis of heat transfer by taking into consideration the interaction between the two modes of heat transfer is a very complicated matter. On the other hand, under very restrictive conditions the heat transfer by simultaneous convection and radiation can be determined approximately by linear superposition of heat fluxes due to these two different modes of heat transfer. Consider, for example, the flow of hot combustion products at temperature T_g through a cooled duct whose walls are kept at temperature T_w. Combustion products such as CO_2, CO, and H_2O absorb and emit radiation. Therefore, the heat transfer from the gas to the channel walls is by both convection and radiation, and a proper analysis of this heat-transfer problem requires a simultaneous solution of convection and radiation problems; but this is a very complicated matter. If the radiative component of the heat flux is not very strong, the total heat flux q from the gas to the wall surface may be computed approximately by taking the sum of the convective heat flux q_c and the radiative heat flux q_r as

$$q = q_c + q_r \tag{1-8}$$

When the relations for the convective and radiative heat flux given by Eqs. (1-2) and (1-7) are introduced into Eq. (1-8) we find

$$q = h_c(T_g - T_w) + h_r(T_g - T_w) = (h_c + h_r)(T_g - T_w)$$

or

$$q = h_{cr}(T_g - T_w) \tag{1-9a}$$

where the *combined convection and radiation heat-transfer coefficient* h_{cr} is defined as

$$h_{cr} = h_c + h_r \tag{1-9b}$$

1-5   HEAT TRANSFER WITH CHANGE OF PHASE

There are numerous heat-transfer processes which involve change of phase.   For example, a vapor coming into contact with a cold surface condenses, and the latent heat of vapor released during the condensation process is to be removed. Conversely, in the boiling of a liquid, the latent heat absorbed by the vapor is to be supplied to the liquid during the process of phase change from liquid to vapor.   In the melting of a solid, heat is supplied during the phase change from solid to liquid because latent heat is absorbed by the liquid.   During reentry of space vehicles at very high speeds into the earth's atmosphere, the large quantity of heat generated at the surface as a result of air friction is dissipated very rapidly by a heat-transfer process known as ablation in which part of the solid body exposed to hot, high-speed air is allowed to melt or evaporate away. Such heat-transfer processes which involve a change of phase are very important in engineering applications but their analysis is extremely complicated.   A discussion of heat transfer in the boiling of liquids and the condensation of vapors will be given later in this book.

Example 1-1   A constant temperature difference of 300°F (166.7°C) is maintained across the surfaces of a slab of 0.1-ft (0.0305-m) thickness.   Determine the rate of heat transfer per unit area across the slab for each of the following cases: The slab material is copper $(k = 220$ Btu/h$\cdot$ft$\cdot$°F or 380.7 W/m$\cdot$°C), aluminum $(k = 130$ Btu/h$\cdot$ft$\cdot$°F or 225 W/m$\cdot$°C), carbon steel $(k = 10$ Btu/h$\cdot$ft$\cdot$°F or 17.3 W/m$\cdot$°C), brick $(k = 0.5$ or 0.865), and asbestos $(k = 0.1$ or 0.173).

Solution   The Fourier law for one-dimensional heat conduction is given by Eq. (1-1b):

$$q = -k \frac{dT}{dx}$$

For the problem considered here q should be constant everywhere in the medium since there are no heat sources or heat sinks in the slab.   The integration of this equation across the slab for constant q and k gives

$$qx \Big|_0^L = -kT \Big|_{T_1}^{T_2}$$

or

$$q = k \frac{T_1 - T_2}{L} \qquad \text{Btu/h}\cdot\text{ft}^2 \text{ (or W/m}^2)$$

In the present problem $T_1 - T_2 = 300$°F, $L = 0.1$ ft, and k is specified for each material considered.   Then, the heat fluxes, for copper, aluminum, carbon steel, brick, and asbestos, respectively, are given as 6.6×10^5, 3.9×10^5, 3×10^4, 1.5×10^3, and 3×10^2 Btu/h$\cdot$ft^2 (or 20.8×10^5, 12.3×10^5, 9.5×10^4, 4.7×10^3, and 9.5×10^2 W/m^2).   We note that the heat-transfer rate is higher with higher thermal conductivity.

Example 1-2   A fluid at 500°F (260°C) flows over a flat plate which is kept at a uniform temperature of 100°F (82.2°C).   If the heat-transfer coefficient h for convection is 20 Btu/h·ft²·°F (113.5 W/m²·°C), determine the heat-transfer rate per unit area of the plate from the fluid into the plate.

Solution   Heat transfer by convection between a fluid and a solid surface is given by Eq. (1-2):

$$q = h(T_f - T_w) \qquad \text{Btu/h·ft}^2 \text{ (or W/m}^2)$$

Taking $h = 20$ Btu/h·ft²·°F (or 113.5 W/m²·°C) and $T_f - T_w = 500 - 100 = 400°F$ (or 222.2°C), the heat flux at the wall becomes

$$q = 20 \times 400 = 8 \times 10^3 \text{ Btu/h·ft}^2 \text{ (or 25.2 kW/m}^2)$$

1-6   UNITS, DIMENSIONS, AND CONVERSION FACTORS

In the field of heat transfer the physical quantities such as specific heat, thermal conductivity, heat-transfer coefficient, heat flux, etc., are expressed in terms of a few fundamental *dimensions* which include length, time, mass, and temperature, and each of these dimensions is associated with a *unit* when it is to be expressed numerically.   For example, length is the dimension of a distance and to express it numerically one may use units of feet or meters or centimeters, etc. Time may be measured in units of hours or seconds, mass in units of pounds or kilograms, temperature in units of degrees of Fahrenheit or Celsius, energy in British thermal units or joules, and so on.   When the dimensions of a physical quantity are to be expressed numerically, a consistent system of units is generally preferred.   In engineering the two most commonly used systems of units include (1) the SI system (*Systéme International d'Unités*) which is also referred to as the MKSA system and (2) the English engineering system (ft·lb·lb$_f$·s).   The basic units for length, mass, time, and temperature for each of these systems are listed in Table 1-2.   Here the symbol lb$_f$ is used for *pound force* to distinguish it from the symbol lb commonly used for *pound mass*,

TABLE 1-2   Systems of Units

Quantity	SI (MKSA) System	English Engineering System
Length	m	ft
Mass	kg	lb
Time	s	s
Temperature	K	°R
Force	newton	lb$_f$
Energy	joule	Btu
	or	or
	newton-meter (N·m)	ft·lb$_f$

but there is no such misunderstanding in the SI system because the *kilogram* is the unit of *mass* and the *newton* is the unit of *force*. The physical significance of the force units, newton and lb_f, is better envisioned by considering Newton's second law of motion written as

$$\text{Force} = \frac{1}{g_c} \times \text{mass} \times \text{acceleration} \qquad (1\text{-}10)$$

where g_c is the *gravitational conversion-factor* constant. The pound force, lb_f, is defined as the force that acts on the mass of one pound at a point on the earth where the magnitude of the gravitational acceleration is $g = 32.174 \text{ ft/s}^2$. Then, in the English engineering system, Eq. (1-10) becomes

$$1 \text{ lb}_f = \frac{1}{g_c} \times 1 \text{ lb} \times 32.174 \text{ ft/s}^2 \qquad (1\text{-}11)$$

According to this relation, one pound of force (that is, 1 lb_f) will accelerate one pound of mass (that is, 1 lb) 32.174 ft/s^2; or 1 lb_f is equal to $32.174 \text{ ft} \cdot \text{lb/s}^2$. The conversion factor g_c in the English engineering system is obtained from this relation as

$$g_c = 32.174 \text{ lb} \cdot \text{ft/lb}_f \cdot \text{s}^2 \qquad (1\text{-}12)$$

It is to be noted that the gravitational acceleration g and the gravitational conversion factor g_c are not similar quantities; g_c is constant, but g depends on the location and on the altitude.

In the SI system Eq. (1-10) becomes

$$1 \text{ newton} = \frac{1}{g_c} \times 1 \text{ kg} \times 1 \text{ m/s}^2 \qquad (1\text{-}13)$$

Clearly, in the SI system, 1 newton (that is, 1 N) is a force that will accelerate 1 kg mass, 1 m/s^2; or 1 newton force is equal to $1 \text{ kg} \cdot \text{m/s}^2$. The conversion factor g_c in the SI system becomes

$$g_c = 1 \text{ kg} \cdot \text{m/N} \cdot \text{s}^2 \qquad (1\text{-}14)$$

Energy is measured in Btu or ft·lb$_f$ in the English engineering system whereas it is measured in joules (that is, J) or newton-meters (that is, N·m) in the SI system. It is to be noted that $1 \text{ J} = 1 \text{ N} \cdot \text{m}$ and $1 \text{ J} = 1 \text{ kg} \cdot \text{m}^2/\text{s}^2$ since $1 \text{ N} = 1 \text{ kg} \cdot \text{m/s}^2$.

Power is measured in Btu/h or ft·lb$_f$/s in the English engineering system, and it is measured in watts (that is, W) or kilowatts (kW) or J/s in the SI system. It is to be noted that

$$1 \text{ kW} = 1000 \text{ W} \qquad 1 \text{ W} = 1 \text{ J/s} = 1 \text{ N} \cdot \text{m/s} = 1 \text{ kg} \cdot \text{m}^2/\text{s}^3$$

Force is measured in lb_f in the English engineering system but is measured in newtons (N) in the SI system. We note that

$$1 \ lb_f = 32.174 \ ft \cdot lb/s^2 \quad and \quad 1 \ N = 1 \ kg \cdot m/s^2$$

Pressure is measured in lb_f/in^2 in the English engineering system whereas it is measured in bars in the SI system.   It is to be noted that

$$1 \ bar = 10^5 \ N/m^2 = 10^5 \ kg/m \cdot s^2$$

and

$$1 \ atm = 0.98066 \ bar$$

In the SI system, when the size of units becomes too large or too small, multiples in powers of 10 are formed with certain prefixes.   The important ones include

10^{-12} = pico (p)

10^{-9} = nano (n)

10^{-6} = micro (μ)

10^{-3} = milli (m)

10^{-2} = centi (c)

10^{-1} = deci (d)

$10 \quad$ = deca (da)

$10^2 \quad$ = hecto (h)

$10^3 \quad$ = kilo (k)

$10^6 \quad$ = mega (M)

$10^9 \quad$ = giga (G)

10^{-12} = tera (T)

For example,

1000 W = 1 kW (kilowatt)

1,000,000 W = 1 MW (megawatt)

1,000,000 N = 1 MN (meganewton)

1000 m = 1 km (kilometer)

10^{-2} m = 1 cm (centimeter)

and so forth.

The SI system has many advantages and has been adopted in a number of countries; a changeover into the SI system in the engineering field is expected to take place in the countries where the English system of units are currently used. In the transition period it will be necessary for the student and engineer to have the ability to work the problems in both systems of units. With this thought in mind, in the main body of this book the examples are worked and the principal tables are prepared in both English engineering and SI systems of units.

To serve as a ready reference for converting from the English engineering system of units to the SI system of units, or vice versa, we present in Table 1-3 a comprehensive list of conversion factors which are useful in heat-transfer calculations. To illustrate the use of Table 1-3, we consider the conversion of density

$$\rho = 10 \ \text{lb/ft}^3 \tag{1-15}$$

from lb/ft^3 units into kg/m^3 units. In this table the conversion from lb to kg is given as 1 kg = 2.2046 lb, which is now written in the form of a conversion factor:

$$\frac{1}{2.2046} \ \text{kg/lb} \tag{1-16a}$$

The conversion from ft^3 to m^3 is given as 1 m^3 = 35.315 ft^3, which is written in the form of a conversion factor as

$$35.315 \ \text{ft}^3/\text{m}^3 \tag{1-16b}$$

Now, the right-hand side of Eq. (1-15) is multiplied by the conversion factors Eqs. (1-16a) and (1-16b) in order to eliminate lb/ft^3. We find

$$\rho = 10 \ \frac{\text{lb}}{\text{ft}^3} = \left(10 \ \frac{\text{lb}}{\text{ft}^3}\right)\left(\frac{1}{2.2046} \ \frac{\text{kg}}{\text{lb}}\right)\left(35.315 \ \frac{\text{ft}^3}{\text{m}^3}\right)$$

$$= 10 \times \frac{1}{2.2046} \times 35.315 \ \frac{\text{kg}}{\text{m}^3} \cong 160 \ \frac{\text{kg}}{\text{m}^3}$$

Example 1-3 Convert the heat-transfer coefficient $h = 20$ Btu/h·ft²·°F into the units of J/s·m²·°C or W/m²·°C.

Solution For illustration purposes we perform this conversion in two different ways: (a) by using the conversion factor for the heat-transfer coefficient directly available in Table 1-3, and (b) by deriving the appropriate conversion factor by converting each basic unit separately.

TABLE 1-3 Conversion Factors

1. Acceleration
   $1 \text{ ft/s}^2 = 0.3048 \text{ m/s}^2$
   $1 \text{ m/s}^2 = 3.2808 \text{ ft/s}^2$

2. Area
   $1 \text{ in}^2 = 6.4516 \text{ cm}^2$
   $1 \text{ in}^2 = 6.4516 \times 10^{-4} \text{ m}^2$
   $1 \text{ ft}^2 = 929 \text{ cm}^2$
   $1 \text{ ft}^2 = 0.0929 \text{ m}^2$
   $1 \text{ m}^2 = 10.764 \text{ ft}^2$

3. Density
   $1 \text{ lb/in}^3 = 27.680 \text{ g/cm}^3$
   $1 \text{ lb/in}^3 = 27.680 \times 10^3 \text{ kg/m}^3$
   $1 \text{ lb/ft}^3 = 16.019 \text{ kg/m}^3$
   $1 \text{ kg/m}^3 = 0.06243 \text{ lb/ft}^3$
   $1 \text{ slug/ft}^3 = 515.38 \text{ kg/m}^3$
   $1 \text{ lb mol/ft}^3 = 16.019 \text{ kg mol/m}^3$
   $1 \text{ kg mol/m}^3 = 0.06243 \text{ lb mol/ft}^3$

4. Diffusivity (heat, mass, momentum)
   $1 \text{ ft}^2/\text{s} = 0.0929 \text{ m}^2/\text{s}$
   $1 \text{ ft}^2/\text{h} = 0.2581 \text{ cm}^2/\text{s}$
   $1 \text{ ft}^2/\text{h} = 0.2581 \times 10^{-4} \text{ m}^2/\text{s}$
   $1 \text{ m}^2/\text{s} = 10.7639 \text{ ft}^2/\text{s}$
   $1 \text{ cm}^2/\text{s} = 3.8745 \text{ ft}^2/\text{h}$

5. Energy, heat, power
   $1 \text{ J} = 1 \text{ W}\cdot\text{s} = 1 \text{ N}\cdot\text{m}$
   $1 \text{ J} = 10^7 \text{ erg}$
   $1 \text{ Btu} = 1055.04 \text{ J}$
   $1 \text{ Btu} = 1055.04 \text{ W}\cdot\text{s}$
   $1 \text{ Btu} = 1055.04 \text{ N}\cdot\text{m}$
   $1 \text{ Btu} = 252 \text{ cal}$
   $1 \text{ Btu} = 0.252 \text{ kcal}$
   $1 \text{ Btu} = 778.161 \text{ ft}\cdot\text{lb}_f$
   $1 \text{ Btu/h} = 0.2931 \text{ W}$
   $1 \text{ Btu/h} = 0.2931 \times 10^{-3} \text{ kW}$
   $1 \text{ Btu/h} = 3.93 \times 10^{-4} \text{ hp}$
   $1 \text{ cal} = 4.1868 \text{ J (or W}\cdot\text{s or N}\cdot\text{m)}$
   $1 \text{ cal} = 3.968 \times 10^{-3} \text{ Btu}$
   $1 \text{ kcal} = 3.968 \text{ Btu}$
   $1 \text{ hp} = 550 \text{ ft}\cdot\text{lb}_f/\text{s}$
   $1 \text{ hp} = 745.7 \text{ W}$
   $1 \text{ Wh} = 3.413 \text{ Btu}$
   $1 \text{ kWh} = 3413 \text{ Btu}$

6. Heat capacity, heat per unit mass, specific heat
   $1 \text{ Btu/h}\cdot{}^\circ\text{F} = 0.5274 \text{ W/}^\circ\text{C}$
   $1 \text{ W/}^\circ\text{C} = 1.8961 \text{ Btu/h}\cdot{}^\circ\text{F}$
   $1 \text{ Btu/lb} = 2325.9 \text{ J/kg}$
   $1 \text{ Btu/lb} = 2.3259 \text{ kJ/kg}$
   $1 \text{ Btu/lb}\cdot{}^\circ\text{F} = 4186.69 \text{ J/kg}\cdot{}^\circ\text{C}$
   $1 \text{ Btu/lb}\cdot{}^\circ\text{F} = 4.18669 \text{ kJ/kg}\cdot{}^\circ\text{C}$
   (or $\text{J/g}\cdot{}^\circ\text{C}$)
   $1 \text{ Btu/lb}\cdot{}^\circ\text{F} = 1 \text{ cal/g}\cdot{}^\circ\text{C} = 1 \text{ kcal/kg}\cdot{}^\circ\text{C}$

7. Heat flux
   $1 \text{ Btu/h}\cdot\text{ft}^2 = 3.1537 \text{ W/m}^2$
   $1 \text{ Btu/h}\cdot\text{ft}^2 = 3.1537 \times 10^{-3} \text{ kW/m}^2$
   $1 \text{ W/m}^2 = 0.31709 \text{ Btu/h}\cdot\text{ft}^2$

8. Heat-generation rate
   $1 \text{ Btu/h}\cdot\text{ft}^3 = 10.35 \text{ W/m}^3$
   $1 \text{ Btu/h}\cdot\text{ft}^3 = 8.9 \text{ kcal/h}\cdot\text{m}^3$
   $1 \text{ W/m}^3 = 0.0966 \text{ Btu/h}\cdot\text{ft}^3$

9. Heat-transfer coefficient
   $1 \text{ Btu/h}\cdot\text{ft}^2\cdot{}^\circ\text{F} = 5.677 \text{ W/m}^2\cdot{}^\circ\text{C}$
   $1 \text{ Btu/h}\cdot\text{ft}^2\cdot{}^\circ\text{F} = 5.677 \times 10^{-4} \text{ W/cm}^2\cdot{}^\circ\text{C}$
   $1 \text{ W/m}^2\cdot{}^\circ\text{C} = 0.1761 \text{ Btu/h}\cdot\text{ft}^2\cdot{}^\circ\text{F}$
   $1 \text{ Btu/h}\cdot\text{ft}^2\cdot{}^\circ\text{F} = 4.882 \text{ kcal/h}\cdot\text{m}^2\cdot{}^\circ\text{C}$

10. Length
    $1 \text{ Å} = 10^{-8} \text{ cm}$
    $1 \text{ Å} = 10^{-10} \text{ m}$
    $1 \text{ μm} = 10^{-3} \text{ mm}$
    $1 \text{ μm} = 10^{-4} \text{ cm}$
    $1 \text{ μm} = 10^{-6} \text{ m}$
    $1 \text{ in} = 2.54 \text{ cm}$
    $1 \text{ in} = 2.54 \times 10^{-2} \text{ m}$
    $1 \text{ ft} = 0.3048 \text{ m}$
    $1 \text{ m} = 3.2808 \text{ ft}$
    $1 \text{ mile} = 1609.34 \text{ m}$
    $1 \text{ mile} = 5280 \text{ ft}$
    $1 \text{ light year} = 9.46 \times 10^{15} \text{ m}$

11. Mass
    $1 \text{ oz} = 28.35 \text{ g}$
    $1 \text{ lb} = 16 \text{ oz}$
    $1 \text{ lb} = 453.6 \text{ g}$
    $1 \text{ lb} = 0.4536 \text{ kg}$
    $1 \text{ kg} = 2.2046 \text{ lb}$
    $1 \text{ g} = 15.432 \text{ grains}$
    $1 \text{ slug} = 32.1739 \text{ lb}$
    $1 \text{ ton (metric)} = 1000 \text{ kg}$
    $1 \text{ ton (metric)} = 2205 \text{ lb}$
    $1 \text{ ton (short)} = 2000 \text{ lb}$
    $1 \text{ ton (long)} = 2240 \text{ lb}$

12. Mass flux
    $1 \text{ lb mol/ft}^2\cdot\text{h}$
    $\quad = 1.3563 \times 10^{-3} \text{ kg mol/m}^2\cdot\text{s}$
    $1 \text{ kg mol/m}^2\cdot\text{s} = 737.3 \text{ lb mol/ft}^2\cdot\text{h}$
    $1 \text{ lb/ft}^2\cdot\text{h} = 1.3563 \times 10^{-3} \text{ kg/m}^2\cdot\text{s}$
    $1 \text{ lb/ft}^2\cdot\text{s} = 4.882 \text{ kg/m}^2\cdot\text{s}$
    $1 \text{ kg/m}^2\cdot\text{s} = 737.3 \text{ lb/ft}^2\cdot\text{h}$
    $1 \text{ kg/m}^2\cdot\text{s} = 0.2048 \text{ lb/ft}^2\cdot\text{s}$

13. Pressure, force
    $1 \text{ N} = 1 \text{ kg}\cdot\text{m/s}^2$
    $1 \text{ N} = 0.22481 \text{ lb}_f$
    $1 \text{ N} = 7.2333 \text{ poundals}$
    $1 \text{ N} = 10^5 \text{ dyn}$

$1 \text{ lb}_f = 32.174 \text{ ft} \cdot \text{lb/s}^2$
$1 \text{ lb}_f = 4.4482 \text{ N}$
$1 \text{ lb}_f = 4.4482 \text{ kg} \cdot \text{m/s}^2$
$1 \text{ lb}_f = 32.1739 \text{ poundals}$
$1 \text{ lb}_f/\text{in}^2 \equiv (1 \text{ psi}) = 6894.76 \text{ N/m}^2$
$1 \text{ lb}_f/\text{ft}^2 = 47.880 \text{ N/m}^2$
$1 \text{ bar} = 10^5 \text{ N/m}^2$
$1 \text{ atm} = 14.696 \text{ lb}_f/\text{in}^2$
$1 \text{ atm} = 2116.2 \text{ lb}_f/\text{ft}^2$
$1 \text{ atm} = 1.0132 \times 10^5 \text{ N/m}^2$
$1 \text{ atm} = 1.0132 \text{ bar}$

14. Specific heat

$1 \text{ Btu/lb} \cdot °\text{F} = 1 \text{ kcal/kg} \cdot °\text{C} = 1 \text{ cal/g} \cdot °\text{C}$
$1 \text{ Btu/lb} \cdot °\text{F} = 4186.69 \text{ J/kg} \cdot °\text{C}$
  (or $\text{W} \cdot \text{s/kg} \cdot °\text{C}$)
$1 \text{ Btu/lb} \cdot °\text{F} = 4.18669 \text{ J/g} \cdot °\text{K}$
  (or $\text{W} \cdot \text{s/g} \cdot °\text{C}$)
$1 \text{ J/g} \cdot °\text{C} = 0.23885 \text{ Btu/lb} \cdot °\text{F}$
  (cal/g $\cdot °\text{C}$ or kcal/kg $\cdot °\text{C}$)

15. Speed

$1 \text{ ft/s} = 0.3048 \text{ m/s}$
$1 \text{ m/s} = 3.2808 \text{ ft/s}$
$1 \text{ mile/h} = 1.4667 \text{ ft/s}$
$1 \text{ mile/h} = 0.44704 \text{ m/s}$

16. Surface tension

$1 \text{ lb}_f/\text{ft} = 14.5937 \text{ N/m}$
$1 \text{ N/m} = 0.068529 \text{ lb}_f/\text{ft}$

17. Temperature

$1 \text{ K} = 1.8°\text{R}$
$T(°\text{F}) = 1.8(\text{K} - 273) + 32$

$T(\text{K}) = \dfrac{1}{1.8} (°\text{F} - 32) + 273$

$T(°\text{C}) = \dfrac{1}{1.8} (°\text{R} - 492)$

$\Delta T(°\text{C}) = 1.8 \, \Delta T(°\text{F})$

18. Thermal conductivity

$1 \text{ Btu/h} \cdot \text{ft} \cdot °\text{F} = 1.7303 \text{ W/m} \cdot °\text{C}$
$1 \text{ Btu/h} \cdot \text{ft} \cdot °\text{F} = 1.7303 \times 10^{-2} \text{ W/cm} \cdot °\text{C}$
$1 \text{ Btu/h} \cdot \text{ft} \cdot °\text{F} = 0.4132 \text{ cal/s} \cdot \text{m} \cdot °\text{C}$
$1 \text{ W/m} \cdot °\text{C} = 0.5779 \text{ Btu/h} \cdot \text{ft} \cdot °\text{F}$
$1 \text{ W/cm} \cdot °\text{C} = 57.79 \text{ Btu/h} \cdot \text{ft} \cdot °\text{F}$

19. Thermal resistance

$1 \text{ h} \cdot °\text{F/Btu} = 1.896 °\text{C/W}$
$1 °\text{C/W} = 0.528 \text{ h} \cdot °\text{F/Btu}$

20. Viscosity

$1 \text{ poise} = 1 \text{ g/cm} \cdot \text{s}$
$1 \text{ poise} = 10^{-2} \text{ centipoise}$
$1 \text{ poise} = 241.9 \text{ lb/ft} \cdot \text{h}$
$1 \text{ centipoise} = 2.419 \text{ lb/ft} \cdot \text{h}$
$1 \text{ lb/ft} \cdot \text{s} = 1.4882 \text{ kg/m} \cdot \text{s}$
$1 \text{ lb/ft} \cdot \text{s} = 14.882 \text{ poises}$
$1 \text{ lb/ft} \cdot \text{s} = 1488.2 \text{ centipoises}$
$1 \text{ lb/ft} \cdot \text{h} = 0.4134 \times 10^{-3} \text{ kg/m} \cdot \text{s}$
$1 \text{ lb/ft} \cdot \text{h} = 0.4134 \times 10^{-2} \text{ poise}$
$1 \text{ lb/ft} \cdot \text{h} = 0.4134 \text{ centipoise}$

21. Volume

$1 \text{ in}^3 = 16.387 \text{ cm}^3$
$1 \text{ cm}^3 = 0.06102 \text{ in}^3$
$1 \text{ oz (U.S. fluid)} = 29.573 \text{ cm}^3$
$1 \text{ ft}^3 = 0.0283168 \text{ m}^3$
$1 \text{ ft}^3 = 28.3168 \text{ liters}$
$1 \text{ ft}^3 = 7.4805 \text{ gal (U.S.)}$
$1 \text{ m}^3 = 35.315 \text{ ft}^3$
$1 \text{ gal (U.S.)} = 3.7854 \text{ liters}$
$1 \text{ gal (U.S.)} = 3.7854 \times 10^{-3} \text{ m}^3$
$1 \text{ gal (U.S.)} = 0.13368 \text{ ft}^3$

Constants

g_c = gravitational acceleration conversion factor	$= 32.1739 \text{ ft} \cdot \text{lb/lb}_f \cdot \text{s}^2$
	$= 4.1697 \times 10^8 \text{ ft} \cdot \text{lb/lb}_f \cdot \text{h}^2$
	$= 1 \text{ g} \cdot \text{cm/dyn} \cdot \text{s}^2$
	$= 1 \text{ kg} \cdot \text{m/N} \cdot \text{s}^2$
	$= 1 \text{ lb} \cdot \text{ft/poundal} \cdot \text{s}^2$
	$= 1 \text{ slug} \cdot \text{ft/lb}_f \cdot \text{s}^2$
J = mechanical equivalent of heat	$= 778.16 \text{ ft} \cdot \text{lb}_f/\text{Btu}$
$\mathcal{R}$ = gas constant	$= 1544 \text{ ft} \cdot \text{lb}_f/\text{lb mol} \cdot °\text{R}$
	$= 0.730 \text{ ft}^3 \cdot \text{atm/lb mol} \cdot °\text{R}$
	$= 0.08205 \text{ m}^3 \cdot \text{atm/kg mol} \cdot \text{K}$
	$= 8.314 \text{ J/g mol} \cdot \text{K}$
	$= 8.314 \text{ N} \cdot \text{m/g mol} \cdot \text{K}$
	$= 1.987 \text{ cal/g mol} \cdot \text{K}$
σ = Stefan-Boltzmann constant	$= 0.1714 \times 10^{-8} \text{ Btu/h} \cdot \text{ft}^2 \cdot °\text{R}^4$
	$= 0.56697 \times 10^{-8} \text{ W/m}^2 \cdot \text{K}^4$

(a) From Table 1-3 we have 1 Btu/h·ft²·°F = 5.677 W/m²·°C which is written in the form of a conversion factor as

$$5.677 \frac{W/m^2 \cdot {}^\circ C}{Btu/h \cdot ft^2 \cdot {}^\circ F}$$

Then the conversion is performed as

$$h = 20 \ Btu/h \cdot ft^2 \cdot {}^\circ F = (20 \ Btu/h \cdot ft^2 \cdot {}^\circ F)\left(5.677 \frac{W/m^2 \cdot {}^\circ C}{Btu/h \cdot ft^2 \cdot {}^\circ F}\right)$$

$$\cong 113.6 \ W/m^2 \cdot {}^\circ C = 113.6 \ J/s \cdot m^2 \cdot {}^\circ C$$

since
$$1 \ W = 1 \ J/s$$

(b) The conversion factors for the basic units are obtained from Table 1-3 as

1 Btu = 1055.04 J   or   1055.04 J/Btu

1 m² = 10.764 ft²   or   10.764 ft²/m²

$\Delta T({}^\circ C) = 1.8 \ \Delta T({}^\circ F)$ or   1.8 °F/°C

and the conversion of hour to seconds is given by

1 h = 3600 s   or $\frac{1}{3600}$ h/s

Now, by using these conversion factors the conversion of h is performed:

$$h = 20 \frac{Btu}{h \cdot ft^2 \cdot {}^\circ F} = \left(20 \frac{Btu}{h \cdot ft^2 \cdot {}^\circ F}\right)\left(1055.04 \frac{J}{Btu}\right)\left(10.764 \frac{ft^2}{m^2}\right)\left(\frac{1}{3600} \frac{h}{s}\right)\left(1.8 \frac{{}^\circ F}{{}^\circ C}\right)$$

$$= 20 \frac{1055.04 \times 10.764 \times 1.8}{3600} \frac{J}{s \cdot m^2 \cdot {}^\circ C} \cong 113.6 \frac{J}{s \cdot m^2 \cdot {}^\circ C} = 113.6 \frac{W}{m^2 \cdot {}^\circ C}$$

which is the same as that obtained in part a.

REFERENCES

1   Fourier, J. B.: "Theorie analytique de la chaleur," Paris, 1822.   (English translation by A. Freeman, Dover Publications Inc., New York, 1955.)
2   Powell, R. W., C. Y. Ho, and P. E. Liley: Thermal Conductivity of Selected Materials, NSRDS-NBS 8, U.S. Department of Commerce, National Bureau of Standards, 1966.
3   "Thermophysical Properties of Matter," vols. 1–3, 1F1/Plenum Data Corporation, New York, 1969.
4   Ho, C. Y., R. W. Powell, and P. E. Liley: Thermal Conductivity of Elements, vol. 1, First supplement to *Journal of Physical and Chemical Reference Data* (1972), American Chemical Society, Washington, D.C.
5   Stratton, J. A.: "Electromagnetic Theory," McGraw-Hill Book Company, New York, 1941.
6   Planck, M.: "The Theory of Heat Radiation," Dover Publications, Inc., New York, 1959.

Two

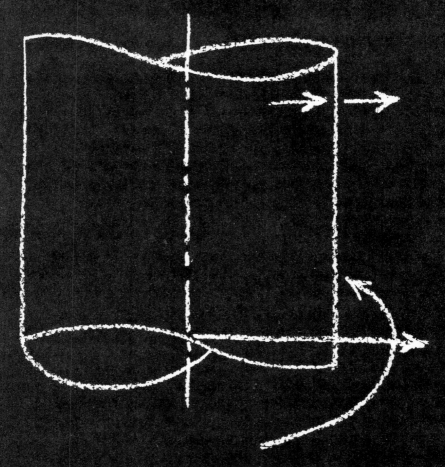

Conduction—
Basic Equations

In this chapter the one-dimensional heat-conduction law is generalized into the three-dimensional form, the equation of heat conduction is derived, different types of boundary conditions are discussed, and dimensionless heat-conduction parameters are presented. The purpose of this chapter is to provide a good understanding of the physical significance of various terms in the heat-conduction equation and its boundary conditions, hence to enable the reader to formulate mathematically simple heat-conduction problems subject to different types of boundary conditions. It is intended that, in the application of heat conduction in the subsequent chapters, the equations needed will be obtained from those given in this chapter by appropriate simplification.

2-1   HEAT-FLUX COMPONENTS

In Chap. 1 the temperature was assumed to vary in the x direction only, and the heat flux q_x in the x direction was related to the temperature gradient dT/dx by the Fourier law. In general, the temperature varies in the x, y, and z directions; hence there is heat flow in those directions. To determine the relations for the heat-flux components q_x, q_y, and q_z in the x, y, and z directions, respectively, we consider an *isotropic medium*, that is, a medium in which the thermal conductivity k at any given location does not vary "at uniform temperature" with the direction at that point. This is true for most materials encountered in engineering applications; exceptions include, for example, laminated sheets in which the thermal conductivity along laminations is not the same as that in the direction perpendicular to the laminations; crystals in which the thermal conductivity varies with direction; and wood in which the thermal conductivity is different along the grain, across the grain, and circumferentially. Then, the three heat-flux components q_x, q_y, and q_z in the x, y, and z directions, respectively, are related to the temperature gradients in the x, y, and z directions by the Fourier law in the following manner:

$$q_x = -k \frac{\partial T}{\partial x} \tag{2-1}$$

$$q_y = -k \frac{\partial T}{\partial y} \tag{2-2}$$

$$q_z = -k \frac{\partial T}{\partial z} \tag{2-3}$$

Figure 2-1 illustrates the three components of the heat flux. Clearly, for no temperature variation, say in the y and z directions, the temperature gradients $\partial T/\partial y$ and $\partial T/\partial z$ vanish, and we recover the one-dimensional Fourier law, Eq. (2-1), which was discussed in Chap. 1.

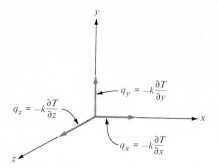

FIG. 2-1   Heat-flux components q_x, q_y, and q_z.

2-2   DIFFERENTIAL EQUATION OF HEAT CONDUCTION

The heat-flux relations given above imply that the rate of heat flow in the x, y, and z directions can be computed if the temperature gradients in those directions are known, and the temperature gradient in any direction can be determined if the temperature distribution in the medium is available. The temperature distribution in a medium can be determined from the solution of the *differential equation of heat conduction* subject to appropriate boundary conditions. We now derive the differential equation of heat conduction in the rectangular coordinate system for the general three-dimensional case, that is, when the temperature varies in the x, y, and z directions.

Consider an infinitesimal volume element $\Delta x\, \Delta y\, \Delta z$ whose surfaces are chosen parallel to the coordinate axes as illustrated in Fig. 2-2.   The energy-balance equation for this volume element may be stated as

$$
\begin{pmatrix}
\text{Net rate of heat} \\
\text{entering by conduction} \\
\text{into element} \\
\Delta x\, \Delta y\, \Delta z \\
\text{I}
\end{pmatrix}
+
\begin{pmatrix}
\text{rate of energy} \\
\text{generated in} \\
\text{element} \\
\Delta x\, \Delta y\, \Delta z \\
\text{II}
\end{pmatrix}
=
\begin{pmatrix}
\text{rate of increase} \\
\text{of internal energy} \\
\text{of element} \\
\Delta x\, \Delta y\, \Delta z \\
\text{III}
\end{pmatrix}
\tag{2-4}
$$

The three terms I, II, and III in this equation are evaluated below.

Net Rate of Heat Entering by Conduction

The net rate of heat entering the volume element by conduction is determined by summing up the net conduction-heat gains in the x, y, and z directions.   If q_x is the heat flux in the x direction at x, the rate of heat flow into the volume element in the x direction through the surface at x is (see Fig. 2-2)

$$
q_x\, \Delta y\, \Delta z \equiv Q_x
\tag{2-5a}
$$

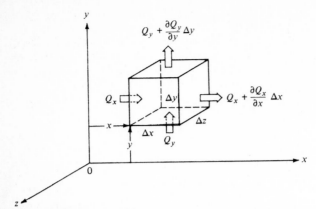

FIG. 2-2   Symbols for the derivation of the heat-conduction equation.

and the rate of heat flow out of the volume element in the x direction through the surface at $x + \Delta x$ is

$$Q_x + \frac{\partial Q_x}{\partial x} \Delta x \qquad\qquad (2\text{-}5b)$$

Then, the net rate of heat entering the volume element by conduction in the x direction is the difference between the entering and leaving heat-flow rates. We obtain

$$\text{Net heat-flow rate entering} \atop \text{element in } x \text{ direction} = -\frac{\partial Q_x}{\partial x} \Delta x = -\frac{\partial q_x}{\partial x} \Delta x\, \Delta y\, \Delta z \qquad (2\text{-}6a)$$

Similarly, the net rate of heat entering the volume element by conduction in the y and z directions are given, respectively, as

$$-\frac{\partial q_y}{\partial y} \Delta x\, \Delta y\, \Delta z \qquad\qquad (2\text{-}6b)$$

$$-\frac{\partial q_z}{\partial z} \Delta x\, \Delta y\, \Delta z \qquad\qquad (2\text{-}6c)$$

Then, the net rate of heat entering the volume element by conduction is obtained by summing up these three components:

$$I \equiv -\left(\frac{\partial q_x}{\partial x} + \frac{\partial q_y}{\partial y} + \frac{\partial q_z}{\partial z}\right) \Delta x\, \Delta y\, \Delta z \qquad (2\text{-}7)$$

Rate of Energy Generation

If there are distributed energy sources in the medium, generating heat at a rate of $g(x, y, z, t)$ per unit time, per unit volume (that is, $Btu/h \cdot ft^3$), the rate of energy generation in the element is given by

$$II \equiv g \, \Delta x \, \Delta y \, \Delta z \tag{2-8}$$

Rate of Increase of Internal Energy

In the case of solids and liquids the specific heats under constant pressure and under constant volume are equal, that is, $c_p \cong c_v \equiv c$.   Then, the rate of increase of internal energy is reflected in the rate of energy storage in the volume element and is given by

$$III \equiv \rho c_p \frac{\partial T}{\partial t} \Delta x \, \Delta y \, \Delta z \tag{2-9}$$

where ρ and c_p do not vary with time.

Now, substituting Eqs. (2-7) to (2-9) into Eq. (2-4) and canceling out $\Delta x \, \Delta y \, \Delta z$, we obtain:

$$-\left(\frac{\partial q_x}{\partial x} + \frac{\partial q_y}{\partial y} + \frac{\partial q_z}{\partial z}\right) + g = \rho c_p \frac{\partial T}{\partial t} \tag{2-10}$$

When the relations for q_x, q_y, and q_z as given by Eqs. (2-1) to (2-3) are introduced into Eq. (2-10), the energy equation becomes

$$\frac{\partial}{\partial x}\left(k \frac{\partial T}{\partial x}\right) + \frac{\partial}{\partial y}\left(k \frac{\partial T}{\partial y}\right) + \frac{\partial}{\partial z}\left(k \frac{\partial T}{\partial z}\right) + g = \rho c_p \frac{\partial T}{\partial t} \tag{2-11}$$

where

$$T \equiv T(x, y, z, t) \quad \text{and} \quad g \equiv g(x, y, z, t)$$

Equation (2-11) is called the *partial differential equation of heat conduction.*

The physical significance of various terms in this equation is better understood if we now examine the special cases of Eq. (2-11).

Uniform Thermal Conductivity (i.e., Independent of Position and Temperature)   Equation (2-11) becomes

$$\nabla^2 T + \frac{g}{k} = \frac{1}{\alpha}\frac{\partial T}{\partial t} \tag{2-12}$$

where the laplacian operator $\nabla^2 T$ is defined as

$$\nabla^2 T \equiv \frac{\partial^2 T}{\partial x^2} + \frac{\partial^2 T}{\partial y^2} + \frac{\partial^2 T}{\partial z^2} \qquad (2\text{-}13)$$

and the thermal diffusivity α as

$$\alpha \equiv \frac{k}{\rho c_p} \qquad (2\text{-}14)$$

The first and second terms on the left-hand side of Eq. (2-12) represent the heat gains by *conduction* and *generation*, respectively, and the right-hand side represents *the rate of change of temperature with time* in the solid.

We note that the thermal diffusivity combines three physical properties ρ, c_p, and k into a single constant; it has a dimension of length^2/time, (that is, ft^2/h or m^2/s).

Uniform Thermal Conductivity, No Heat Sources   For no heat sources we set $g = 0$, and Eq. (2-12) simplifies to

$$\nabla^2 T = \frac{1}{\alpha}\frac{\partial T}{\partial t} \qquad (2\text{-}15)$$

which is sometimes called the *Fourier equation* of heat conduction or the *diffusion equation*.

Uniform Thermal Conductivity, Steady State   The steady-state condition implies that the temperature does not vary with time, hence $\partial T/\partial t$ vanishes; Eq. (2-12) simplifies to

$$\nabla^2 T + \frac{g}{k} = 0 \qquad (2\text{-}16a)$$

which is called *Poisson's equation*.   If it is further assumed that the temperature varies only in the x direction (i.e., no variation in the y and z directions) Eq. (2-16a) reduces to

$$\frac{d^2 T}{dx^2} + \frac{g}{k} = 0 \qquad (2\text{-}16b)$$

Uniform Thermal Conductivity, Steady State and No Heat Generation   In this case $\partial T/\partial t$ vanishes and we set $g = 0$; Eq. (2-12) simplifies to

$$\nabla^2 T = 0 \qquad (2\text{-}17a)$$

which is called the *Laplace equation*. If it is further assumed that the temperature varies only in the x direction, Eq. (2-17a) reduces to

$$\frac{d^2 T}{dx^2} = 0 \tag{2-17b}$$

Thermal Diffusivity

It is instructive to discuss the physical significance of thermal diffusivity α as defined above. Table 2-1 gives the magnitude of thermal diffusivity of typical engineering materials; it is apparent from this table that there are wide differences between the thermal diffusivities of materials. The physical significance of thermal diffusivity is associated with the speed of propagation of heat into the solid during changes of temperature with time. The higher the thermal diffusivity, the faster is the propagation of heat in the medium. This statement is better understood by referring to the following specific heat-conduction problem. A semi-infinite medium in the region from $x = 0$ to $x \to \infty$ is initially at a uniform temperature

TABLE 2-1 Thermal Diffusivity of Typical Materials

| | Average Temperature | | Thermal α | Diffusivity $\alpha \times 10^6$ |
	°F	°C	ft²/hr	m²/s
Metals				
Aluminum	32	0	3.33	85.9
Copper	32	0	4.42	114.1
Gold	68	20	4.68	120.8
Iron, pure	32	0	0.70	18.1
Cast iron ($c \simeq 4\%$)	68	20	0.66	17.0
Lead	70	21.1	0.95	25.5
Mercury	32	0	0.172	4.44
Nickel	32	0	0.60	15.5
Silver	32	0	6.60	170.4
Steel, mild	32	0	0.48	12.4
Tungsten	32	0	2.39	61.7
Zinc	32	0	1.60	41.3
Nonmetals				
Asbestos	32	0	0.010	0.258
Brick, fire clay	400	204.4	0.020	0.516
Cork, ground	100	37.8	0.006	0.155
Glass, Pyrex			0.023	0.594
Granite	32	0	0.050	1.291
Ice	32	0	0.046	1.187
Oak, across grain	85	29.4	0.0062	0.160
Pine, across grain	85	29.4	0.0059	0.152
Quartz sand, dry			0.008	0.206
Rubber, soft			0.003	0.077
Water	32		0.005	0.129

T_0.   At the time $t = 0$ a uniform temperature $T = 0$ is applied at the boundary
surface $x = 0$, and this boundary is kept at zero temperature for all times $t > 0$.
Clearly, for times $t > 0$, the temperature in the medium will vary with the position
and time.   Suppose we are interested in the time required for the temperature
to decrease from its initial value T_0 to half of this value, $\frac{1}{2}T_0$, at a position 1 ft
from the boundary surface; Table 2-2 gives the time required for several different
materials.   It is apparent from these results that the larger the thermal diffusivity,
the shorter is the time required for the applied heat to penetrate into the depth
of the solid.

TABLE 2-2   Effect of Thermal Diffusivity on the Rate of Heat Propagation

Material		Silver	Copper	Steel	Glass	Cork
α	ft^2/h	6.6	4.0	0.5	0.023	0.006
	$10^6 \times$ m^2/s	170	103	12.9	0.59	0.155
Time		9.5 min	16.5 min	2.2 h	2.00 days	77 days

2-3   HEAT-CONDUCTION EQUATION IN OTHER COORDINATE SYSTEMS

In the foregoing analysis we derived the heat-conduction equation in the
rectangular coordinate system, and this equation is useful in the analysis of heat
conduction in solids having shapes such as a slab, semi-infinite medium, rectangle,
or parallelepiped.   On the other hand, to analyze heat conduction for bodies
having shapes such as a cylinder and a sphere, the heat-conduction equation
should be given in the cylindrical and the spherical coordinate system, respectively;
the purpose of using different coordinate systems is to ensure that the coordinate
surfaces coincide with the boundary surfaces of the region.   Thus, the equation
in the rectangular coordinate system is used for rectangular bodies, in the
cylindrical coordinate system for cylindrical bodies, and in the spherical coordinate
system for spherical bodies.   Other orthogonal coordinate systems are also available for solving the heat-conduction equation for bodies having other shapes.
For example, the *conical coordinate system* may be used for conical bodies, the
ellipsoidal coordinate system for ellipsoidal bodies, the *parabolic coordinate system*
for parabolic bodies, and so forth.   However, the solution of the heat-conduction
equation in such coordinate systems is beyond the scope of this book; we shall
consider heat conduction only in the rectangular, cylindrical, and spherical
coordinate systems.   The heat-conduction equation can be derived for the
cylindrical and spherical coordinate systems by writing an energy-balance equation and following a procedure similar to that described previously.   However,
there is no need for such derivations, because the desired results are readily
obtainable by the standard coordinate-transformation technique. That is, by

referring to the heat-conduction equation (2-12) we note that the only term that is affected by the coordinate transformation is the laplacian operator $\nabla^2 T$. This term in the rectangular coordinate system was given previously as

$$\nabla^2 T \equiv \frac{\partial^2 T}{\partial x^2} + \frac{\partial^2 T}{\partial y^2} + \frac{\partial^2 T}{\partial z^2} \tag{2-18}$$

where $T \equiv T(x, y, z, t)$. Therefore, the heat-conduction equation (2-12) can be written in other coordinate systems merely by transforming the laplacian operator. Here we consider the transformation into the cylindrical (r, ϕ, z) and the spherical (r, ϕ, ψ) coordinate systems as shown in Figs. 2-3 and 2-4, respectively. We present only the results, since such transformations are available in any standard text on mathematics.

The heat-conduction equation (2-12) in the *cylindrical coordinate system* is given as

$$\frac{1}{r} \frac{\partial}{\partial r} \left(r \frac{\partial T}{\partial r} \right) + \frac{1}{r^2} \frac{\partial^2 T}{\partial \phi^2} + \frac{\partial^2 T}{\partial z^2} + \frac{g}{k} = \frac{1}{\alpha} \frac{\partial T}{\partial t} \tag{2-19}$$

where $T \equiv T(r, \phi, z, t)$. In the *spherical coordinate system* it is given as

$$\frac{1}{r^2} \frac{\partial}{\partial r} \left(r^2 \frac{\partial T}{\partial r} \right) + \frac{1}{r^2 \sin \psi} \frac{\partial}{\partial \psi} \left(\sin \psi \frac{\partial T}{\partial \psi} \right) + \frac{1}{r^2 \sin^2 \psi} \frac{\partial^2 T}{\partial \phi^2} + \frac{g}{k} = \frac{1}{\alpha} \frac{\partial T}{\partial t} \tag{2-20}$$

where $T \equiv T(r, \phi, \psi, t)$.

Problems of heat conduction, in general, involve finding solutions to the partial differential equations (2-12), (2-19), or (2-20), subject to a given set of boundary and initial conditions; but these equations are too complicated to solve in their general form. Therefore, only the simple equations derived from these general equations will be considered in this book.

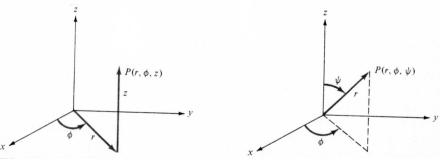

FIG. 2-3   Cylindrical coordinate system (r, ϕ, z). FIG. 2-4   Spherical coordinate system (r, ϕ, ψ).

We now examine the simplification of the above heat-conduction equations in the rectangular, cylindrical, and spherical coordinates for the special case of *one-dimensional, time-dependent situations with constant thermal conductivity.*

1   *The rectangular coordinate system.*   Temperature varies in the x direction and with time only.   Equation (2-12) simplifies to

$$\frac{\partial^2 T}{\partial x^2} + \frac{g}{k} = \frac{1}{\alpha}\frac{\partial T}{\partial t} \tag{2-21}$$

where $T \equiv T(x, t)$.

2   *The cylindrical coordinate system.*   Temperature varies in the r direction and with time only.   Equation (2-19) reduces to

$$\frac{1}{r}\frac{\partial}{\partial r}\left(r\frac{\partial T}{\partial r}\right) + \frac{g}{k} = \frac{1}{\alpha}\frac{\partial T}{\partial t} \tag{2-22}$$

where $T \equiv T(r, t)$.

3   *The spherical coordinate system.*   Temperature varies in the r direction and with time only.   Equation (2-20) becomes

$$\frac{1}{r^2}\frac{\partial}{\partial r}\left(r^2\frac{\partial T}{\partial r}\right) + \frac{g}{k} = \frac{1}{\alpha}\frac{\partial T}{\partial t} \tag{2-23a}$$

or in the alternative form is

$$\frac{1}{r}\frac{\partial^2}{\partial r^2}(rT) + \frac{g}{k} = \frac{1}{\alpha}\frac{\partial T}{\partial t} \tag{2-23b}$$

If it is further assumed that the steady-state conditions prevail, the time derivative term $\partial T/\partial t$ should be set equal to zero in the above equations.

2-4   BOUNDARY CONDITIONS

To solve the differential equation of heat conduction for the purpose of determining the temperature distribution in a medium a set of *boundary conditions* and an *initial condition* are needed.   The initial condition specifies the temperature distribution in the medium at the origin of the time coordinate (that is, $t = 0$), and it is needed only for time-dependent problems.   The boundary conditions specify the temperature or the heat-flow situation at the boundaries of the region.   For example, at a given boundary surface, the temperature distribution may be prescribed, or the heat-flux distribution may be prescribed, or there may be heat exchange by convection with a medium at a prescribed temperature.   With this consideration, we use the terminology the boundary condition of the *first kind*, the *second kind*, and the *third kind*, respectively, to characterize these three different situations at the boundaries.   To formalize the mathematical representation of these boundary conditions we present below a discussion of these three types of boundary conditions.

Boundary Condition of the First Kind

When the distribution or the value of temperature is prescribed at a boundary surface, the boundary condition is said to be of the *first kind* at that boundary. A typical illustration of the boundary condition of the first kind for a slab and a two-dimensional region is shown in Fig. 2-5.   In the case of the slab geometry, for example, the boundary conditions

$$T(x, t)\Big|_{x=0} = T_0 \quad \text{and} \quad T(x, t)\Big|_{x=L} = T_L \qquad (2\text{-}24)$$

denote that the temperatures of the boundary surfaces $x = 0$ and $x = L$ are specified as T_0 and T_L, respectively.   In the case of the two-dimensional geometry shown in Fig. 2-5, the boundary condition

$$T(x, y, t)\Big|_{x=0} = f_1(y) \qquad (2\text{-}25)$$

denotes that the temperature of the boundary surface at $x = 0$ is prescribed as $f_1(y)$ which is a function of y.   If the temperature at the boundary is zero, i.e.,

$$T\Big|_{x=0} = 0 \qquad (2\text{-}26)$$

the boundary condition is called the *homogeneous boundary condition of the first kind*.

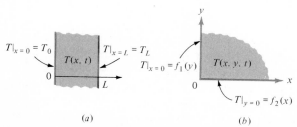

(a) (b)

FIG. 2-5   Illustration of boundary conditions of the first kind.

Boundary Condition of the Second Kind

When the heat flux at a boundary is prescribed, it is called the boundary condition of the *second kind*. Since the heat flux, by definition, is equal to the product of the thermal conductivity k of the material and the derivative of temperature normal to the boundary surface [that is, $q = -k(\partial T/\partial x)$, etc.], then the boundary condition of the second kind is equivalent to prescribing the derivative of temperature normal to the boundary surface. In Fig. 2-6 we illustrate the boundary condition of the second kind for a one-dimensional and a two-dimensional region. In the case of the one-dimensional region in Fig. 2-6a, suppose the heat flux q_0 *entering* the medium through the boundary surface at $x = 0$ is prescribed. This condition can be written

$$-k\frac{\partial T(x, t)}{\partial x}\bigg|_{x=0} = q_0 \tag{2-27a}$$

or

$$-\frac{\partial T}{\partial x}\bigg|_{x=0} = \frac{q_0}{k} \equiv f_0 \tag{2-27b}$$

which is the boundary condition shown in Fig. 2-6a. In the case of the two-dimensional region in Fig. 2-6b, let heat flux $q_1(y)$ *entering* the region through the boundary surface at $x = 0$ be prescribed. This condition is written

$$-k\frac{\partial T(x, y, t)}{\partial x}\bigg|_{x=0} = q_1(y) \tag{2-28a}$$

Rearranging this relation we obtain

$$-\frac{\partial T}{\partial x}\bigg|_{x=0} = \frac{q_1(y)}{k} \equiv f_1(y) \tag{2-28b}$$

which is the boundary condition shown in Fig. 2-6b.

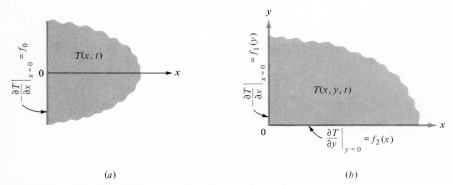

(a) (b)

FIG. 2-6   Illustration of boundary conditions of the second kind.

If the derivative of temperature normal to the boundary surface is zero, i.e.,

$$\left. \frac{\partial T}{\partial x} \right|_{x=0} = 0 \tag{2-29}$$

the boundary condition is called the *homogeneous boundary condition of the second kind*. This type of boundary condition indicates a *thermally insulated* or an *adiabatic boundary*, or a *symmetry condition*.

Boundary Condition of the Third Kind

When the boundary surface is subjected to a convective heat transfer into a medium at a prescribed temperature, the boundary condition is said to be of the *third kind*. To demonstrate the physical significance of this boundary condition we consider a slab geometry in which convection takes place from both boundaries into an environment at a temperature T_∞ as illustrated in Fig. 2-7a. Let the heat-transfer coefficients at the boundary surfaces $x=0$ and $x=L$ be h_1 and h_2, respectively. To derive the boundary condition at the surface $x=0$, we write the energy balance at the boundary surface as

$$\begin{pmatrix} \text{Heat entering} \\ \text{by convection} \end{pmatrix} = \begin{pmatrix} \text{heat leaving} \\ \text{by conduction} \end{pmatrix} \tag{2-30}$$

For a unit area on the boundary surface, various terms in this relation are written

$$h_1 \left(T_\infty - T \bigg|_{x=0} \right) = -k \left. \frac{\partial T}{\partial x} \right|_{x=0} \tag{2-31a}$$

Rearranging, we find

$$\left[-k \frac{\partial T}{\partial x} + h_1 T \right]_{x=0} = h_1 T_\infty \equiv f_1 \tag{2-31b}$$

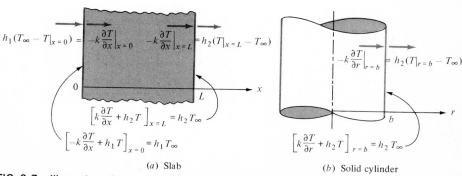

(a) Slab   (b) Solid cylinder

FIG. 2-7   Illustration of boundary conditions of the third kind.

Similarly, by writing the energy balance for the boundary surface at $x = L$,

$$\begin{pmatrix} \text{Heat entering} \\ \text{by conduction} \end{pmatrix} = \begin{pmatrix} \text{heat leaving} \\ \text{by convection} \end{pmatrix} \tag{2-32}$$

we obtain

$$-k \frac{\partial T}{\partial x}\bigg|_{x=L} = h_2 \left(T \bigg|_{x=L} - T_\infty \right) \tag{2-33a}$$

or

$$\left[k \frac{\partial T}{\partial x} + h_2 T \right]_{x=L} = h_2 T_\infty \equiv f_2 \tag{2-33b}$$

In the case of a solid cylinder of radius $r = b$ as shown in Fig. 2-7b, consider convection from the boundary surface at $r = b$ into an environment at a temperature T_∞ with a heat-transfer coefficient h_2. We write the energy-balance equation at the boundary surface $r = b$ as

$$\begin{pmatrix} \text{Heat entering} \\ \text{by conduction} \end{pmatrix} = \begin{pmatrix} \text{heat leaving} \\ \text{by convection} \end{pmatrix}$$

and obtain

$$-k \frac{\partial T}{\partial r}\bigg|_{r=b} = h_2 \left(T \bigg|_{r=b} - T_\infty \right) \tag{2.34a}$$

or

$$\left[k \frac{\partial T}{\partial r} + h_2 T \right]_{r=b} = h_2 T_\infty \equiv f_2 \tag{2.34b}$$

The boundary condition of the form given by Eq. (2.31b), (2.33b), or (2-34b) is called the *boundary condition of the third kind*, which can be expressed more compactly with a single expression in the form

$$\left[k \frac{\partial T}{\partial n_i} + h_i T \right]_{\text{boundary } i} = f_i \tag{2-35}$$

where h_i is a specified coefficient, f_i is a prescribed function, and $\partial T / \partial n_i$ denotes the derivative of temperature normal to the boundary surface i in the *outward* direction. To illustrate the interpretation of the general relation given by Eq. (2-35), we consider the boundary condition at the surface $x = L$ of the slab shown in Fig. 2-7a. At this surface the *outward*-drawn normal to the boundary surface is in the *positive x direction*; hence we replace $\partial T / \partial n_i$ by $\partial T / \partial x$ and from Eq. (2-35) obtain

$$\left[k \frac{\partial T}{\partial x} + h_2 T \right]_{x=L} = f_2 \tag{2-36}$$

where we set $h_i = h_2$ and $f_i = f_2$. In the case of the boundary surface $x = 0$ of the slab, the *outward*-drawn normal is in the *negative x direction*; hence we replace $\partial T/\partial n_i$ by $-\partial T/\partial x$ and obtain

$$\left[-k\frac{\partial T}{\partial x} + h_1 T\right]_{x=0} = f_1 \tag{2-37}$$

where we set $h_i = h_1$ and $f_i = f_1$. In the case of a solid cylinder, the *outward*-drawn normal at the boundary surface $r = b$ is in the *positive r direction*; hence we replace $\partial T/\partial n_i$ by $\partial T/\partial r$, and so forth.

If the right-hand side of Eq. (2-35) is equal to zero, i.e.,

$$\left[k\frac{\partial T}{\partial n_i} + h_i T\right]_{\text{boundary } i} = 0 \tag{2-38}$$

the boundary condition is said to be the *homogeneous boundary condition of the third kind*, which characterizes *convection into a medium at zero temperature*, that is, $T_\infty = 0$.

We now illustrate with the following example the mathematical formulation of simple heat-conduction problems subject to different types of boundary conditions.

Example 2-1 Consider one-dimensional, steady-state heat conduction in a slab with constant thermal conductivity in a region $0 \le x \le L$. Heat is generated in the slab at a rate of $g_0 e^{-\beta x}$ Btu/h·ft^3 (or W/m^3), while the boundary surfaces at $x = 0$ are kept insulated and at $x = L$ dissipate heat by convection into a medium at temperature T_∞ with a heat-transfer coefficient h Btu/h·ft^2·°F (or W/m^2·°C). Write the mathematical formulation of this heat-conduction problem.

Solution The heat-conduction equation is immediately available from Eq. (2-16b) by setting in that equation $g = g_0 e^{-\beta x}$. The boundary condition at $x = 0$ is a homogeneous boundary condition of the second kind, and that at $x = L$ is of the third kind. Then the mathematical formulation is given as

$$\frac{d^2 T(x)}{dx^2} + \frac{1}{k} g_0 e^{-\beta x} = 0 \qquad \text{in } 0 \le x \le L$$

$$\frac{dT(x)}{dx} = 0 \qquad \text{at } x = 0$$

$$k\frac{dT(x)}{dx} + hT(x) = hT_\infty \qquad \text{at } x = L$$

Example 2-2 Consider one-dimensional, steady-state heat conduction in a hollow cylinder with constant thermal conductivity in the region $a \le r \le b$. Heat is generated in the cylinder at a rate of g_0 Btu/h·ft^3 (or W/m^3), while heat is dissipated by convection into fluids flowing inside and outside the cylindrical tube. Heat-transfer coefficients for the inside and outside fluids are h_a and h_b, and temperatures of the inside and outside fluids are T_a and T_b, respectively. Write the mathematical formulation of this heat-conduction problem.

Solution   The heat-conduction equation is obtainable from Eq. (2-22) by setting $\partial T/\partial t = 0$ for the steady state and $g = g_0$. The boundary conditions at $r = a$ and $r = b$ are both of the third kind. Then the mathematical formulation is given as

$$\frac{1}{r}\frac{d}{dr}\left(r\frac{dT}{dr}\right) + \frac{g_0}{k} = 0 \qquad \text{in } a \leq r \leq b$$

$$-k\frac{dT}{dr} + h_a T = h_a T_a \qquad \text{at } r = a$$

$$k\frac{dT}{dr} + h_b T = h_b T_b \qquad \text{at } r = b$$

where $T \equiv T(r)$.

Example 2-3   Write the mathematical formulation of one-dimensional, steady-state heat conduction for a hollow sphere with constant thermal conductivity in the region $a \leq r \leq b$, when heat is supplied into the sphere at a rate of q_0 Btu/h·ft² (or W/m²) from the boundary surface at $r = a$ and dissipated by convection from the boundary surface at $r = b$ into a medium at zero temperature with a heat-transfer coefficient h.

Solution   The heat-conduction equation is obtained from Eq. (2-23) by setting in that equation $\partial T/\partial t = 0$ for the steady state and $g = 0$ for no heat generation. The mathematical formulation becomes

$$\frac{d}{dr}\left(r^2\frac{dT}{dr}\right) = 0 \qquad \text{in } a \leq r \leq b$$

$$-k\frac{dT}{dr} = q_0 \qquad \text{at } r = a$$

$$k\frac{dT}{dr} + hT = 0 \qquad \text{at } r = b$$

where $T \equiv T(r)$. We note that the boundary condition at $r = a$ is of the second kind and at $r = b$ is a homogeneous boundary condition of the third kind.

Nonlinear Boundary Conditions

The boundary conditions for heat-transfer problems involving the fourth-power radiation law, free convection, melting or solidification, and ablation are called *nonlinear boundary conditions* because they involve a power of temperature. The analysis of heat-conduction problems subject to such nonlinear boundary conditions is a complicated matter and beyond the scope of this work. Therefore, in this book we shall be concerned with the solution of simple heat-conduction problems subject to boundary conditions of the first, the second, or the third kind as discussed above.

2-5 DIMENSIONLESS HEAT-CONDUCTION PARAMETERS

The number of variables in a heat-conduction problem can be reduced by introducing dimensionless variables. To illustrate typical dimensionless variables in heat conduction and their physical significance, we consider the following one-dimensional, time-dependent heat-conduction problem for a slab and transform the equation into *nondimensional* form.

A slab in the region $0 \leq x \leq L$ with constant thermal properties is initially (i.e., for $t = 0$) at a uniform temperature T_0. For times $t > 0$, the boundary surface at $x = 0$ is kept insulated, the boundary surface at $x = L$ dissipates heat by convection into a medium at temperature T_∞ with a heat-transfer coefficient h, and heat is generated within the slab at a rate of g Btu/h·ft^3 (or W/m^3). The mathematical formulation of this heat-conduction problem is given as

$$\frac{\partial^2 T}{\partial x^2} + \frac{g}{k} = \frac{1}{\alpha} \frac{\partial T}{\partial t} \qquad \text{in } 0 \leq x \leq L, \text{ for } t > 0 \tag{2-39a}$$

subject to the boundary conditions

$$\frac{\partial T}{\partial x} = 0 \qquad\qquad \text{at } x = 0, \text{ for } t > 0 \tag{2-39b}$$

$$k \frac{\partial T}{\partial x} + hT = hT_\infty \qquad \text{at } x = L, \text{ for } t > 0 \tag{2-39c}$$

and the initial condition

$$T = T_0 \qquad \text{in } 0 \leq x \leq L, \text{ for } t = 0 \tag{2-39d}$$

where $T \equiv T(x, t)$.

To nondimensionalize Eqs. (2-39) we choose the slab thickness L as a reference length, the environment temperature T_∞ as a reference temperature, and define the following dimensionless variables:

$$X = \frac{x}{L} = \text{dimensionless space coordinate} \tag{2-40a}$$

$$\theta = \frac{T - T_\infty}{T_0 - T_\infty} = \text{dimensionless temperature} \tag{2-40b}$$

When these dimensionless variables are introduced into Eqs. (2-39) we obtain

$$\frac{\partial^2 \theta}{\partial X^2} + \frac{gL^2}{(T_0 - T_\infty)k} = \frac{\partial \theta}{\partial(\alpha t/L^2)} \qquad \text{in } 0 \leq X \leq 1, \text{ for } t > 0 \tag{2-41a}$$

$$\frac{\partial \theta}{\partial X} = 0 \qquad\qquad \text{at } X = 0, \text{ for } t > 0 \qquad\qquad (2\text{-}41b)$$

$$\frac{\partial \theta}{\partial X} + \frac{hL}{k}\,\theta = 0 \qquad\qquad \text{at } X = 1, \text{ for } t > 0 \qquad\qquad (2\text{-}41c)$$

$$\theta = 1 \qquad\qquad \text{in } 0 \le X \le 1, \text{ for } t = 0 \qquad\qquad (2\text{-}41d)$$

Equations (2-41) can be written more compactly as

$$\frac{\partial^2 \theta}{\partial X^2} + G = \frac{\partial \theta}{\partial \tau} \qquad \text{in } 0 \le X \le 1, \text{ for } \tau > 0 \qquad\qquad (2\text{-}42a)$$

$$\frac{\partial \theta}{\partial X} = 0 \qquad\qquad \text{at } X = 0, \text{ for } \tau > 0 \qquad\qquad (2\text{-}42b)$$

$$\frac{\partial \theta}{\partial X} + \text{Bi } \theta = 0 \qquad \text{at } X = 1, \text{ for } \tau > 0 \qquad\qquad (2\text{-}42c)$$

$$\theta = 1 \qquad\qquad \text{in } 0 \le X \le 1, \text{ for } \tau = 0 \qquad\qquad (2\text{-}42d)$$

where we have defined the new dimensionless parameters as

$$\text{Bi} \equiv \frac{hL}{k} = \text{Biot number} \qquad\qquad (2\text{-}43a)$$

$$G \equiv \frac{gL^2}{k(T_0 - T_\infty)} = \text{dimensionless heat generation} \qquad\qquad (2\text{-}43b)$$

$$\tau \equiv \frac{\alpha t}{L^2} = \text{Fourier number} \equiv \text{Fo} \qquad\qquad (2\text{-}43c)$$

The *Fourier number* and the *Biot number* are important dimensionless parameters which are frequently used in heat-conduction problems. The physical significance of the Fourier number is better envisioned if it is rearranged in the form

$$\tau = \frac{\alpha t}{L^2} = \frac{k(1/L)L^2}{\rho c_p L^3/t} = \frac{\substack{\text{rate of heat conduction} \\ \text{across } L \text{ in volume } L^3, \\ \text{Btu/h} \cdot {}^\circ\text{F}}}{\substack{\text{rate of heat storage} \\ \text{in volume } L^3, \\ \text{Btu/h} \cdot {}^\circ\text{F}}} \qquad\qquad (2\text{-}44)$$

Thus, the Fourier number is a measure of the rate of heat conduction in comparison with the rate of heat storage in a given volume element. Therefore, the larger the Fourier number, the deeper is the penetration of heat into a solid over a given period of time.

The physical significance of the Biot number is better understood if it is rearranged in the form

$$\mathrm{Bi} = \frac{hL}{k} = \frac{h}{k/L} = \frac{\text{heat-transfer coefficient at the surface of solid}}{\text{internal conductance of solid across length } L} \qquad (2\text{-}45)$$

That is, the Biot number is the ratio of the heat-transfer coefficient to the unit conductance of a solid over the characteristic dimension.

REFERENCES

1 Schneider, P. J.: "Conduction Heat Transfer," Addison-Wesley Publishing Company, Inc., Reading, Mass., 1955.
2 Carslaw, H. S., and J. C. Jaeger: "Conduction of Heat in Solids," 2d ed., Oxford University Press, London, 1959.
3 Özişik, M. N.: "Boundary Value Problems of Heat Conduction," International Textbook Company, Scranton, Pa., 1968.
4 Myers, G. M.: "Analytical Methods in Conduction Heat Transfer," McGraw-Hill Book Company, New York, 1971.
5 Holman, J. P.: "Heat Transfer," 3d ed., McGraw-Hill Book Company, New York, 1972.
6 Kreith, F.: "Principles of Heat Transfer," 3d ed., Intext Educational Publishers, New York, 1973.

`Three

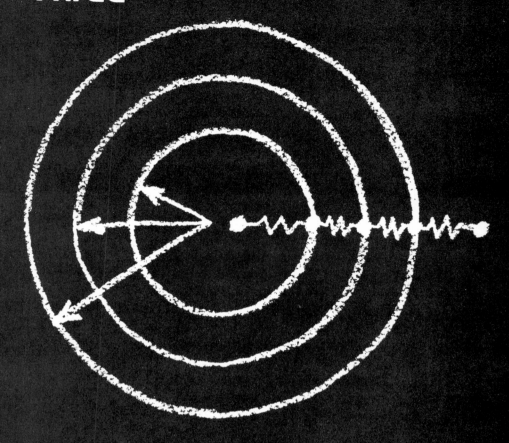

Conduction—
One-dimensional Steady State

In this chapter we present the application of one-dimensional, steady-state heat conduction in a slab, cylinder, and sphere, subject to different types of boundary conditions; discuss the determination of heat flow through a slab having a temperature-dependent thermal conductivity; introduce the concept of *thermal resistance* in analogy with the electric resistance for the analysis of heat transfer in composite parallel layers; and derive the one-dimensional fin equation and illustrate its application to the prediction of heat transfer from finned surfaces.

3-1 THE SLAB

Consider a large slab of thickness L as illustrated in Fig. 3-1. It is assumed that the steady-state condition is established (i.e., the temperature does not vary with time) and the temperature varies in the x direction only. The heat-conduction equation for this case is obtained from Eqs. (2-12) by setting the derivatives with respect to the t, y, and z variables equal to zero; we find

$$\frac{d^2 T(x)}{dx^2} + \frac{g(x)}{k} = 0 \qquad \text{in } 0 \le x \le L \tag{3-1}$$

The generation term $g(x)$ in this equation represents the energy generation in the medium per unit time and per unit volume (that is, Btu/h·ft³ or W/m³), and it is encountered in applications such as the generation of energy by the fission of neutrons in nuclear-reactor fuel elements or by the passage of electric current through an electric-resistance heater, and so forth.

The temperature distribution $T(x)$ in the slab is obtained by the integration of Eq. (3-1) subject to appropriate boundary conditions at $x = 0$ and $x = L$. As discussed in Chap. 2, the boundary condition at any of these boundary surfaces may be of the first, second, or third kind, thus resulting in nine different combinations of these boundary conditions for the slab problem considered here. Although the problem of proving the existence and uniqueness of the solution of the differential equation (3-1) subject to a given set of these boundary conditions is a matter that belongs to pure mathematical analysis, it is important to know if there is any combination of these boundary conditions for which the solution is

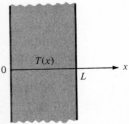

FIG. 3-1 Coordinates for one-dimensional heat conduction in a slab.

not unique. It can be proved that when both boundary conditions are of the second kind the solution is not defined uniquely. The implications of this situation is envisioned by considering a slab in which heat is generated continuously while both boundary surfaces at $x = 0$ and $x = L$ are kept insulated (that is, $\partial T/\partial x = 0$ at $x = 0$ and $x = L$). In this problem the steady-state solution can never be established because the heat generated in the medium cannot escape from the slab because both boundaries are insulated. Therefore care must be exercised for such situations.

Once Eq. (3-1) is integrated subject to appropriate boundary conditions at $x = 0$ and $x = L$ and the temperature distribution $T(x)$ in the slab is established, the heat flux $q(x)$ (i.e., the heat flow per unit area per unit time in the x direction) anywhere in the medium is determined from the definition of the heat flux:

$$q(x) = -k\,\frac{dT(x)}{dx} \tag{3-2}$$

If T is in °F (or °C), x is in ft (or m), k is in Btu/h·ft·°F (or W/m·°C), then the heat flux $q(x)$ is in Btu/h·ft^2 (or W/m^2). By definition, *the heat flow is in the positive x direction if $q(x)$ is a positive quantity, and vice versa.*

We now illustrate with simple examples the determination of the temperature distribution $T(x)$ and the heat flux $q(x)$ in a slab by the integration of differential equation (3-1) subject to different combinations of boundary conditions at the two boundary surfaces.

Example 3-1 Determine the steady-state temperature distribution and the heat flux in a slab in the region $0 \le x \le L$ for uniform thermal conductivity and no heat generation when the boundary surfaces at $x = 0$ and $x = L$ are kept at uniform temperature T_0 and T_1, respectively.

Solution The mathematical formulation of this heat-conduction problem is given as

$$\frac{d^2 T(x)}{dx^2} = 0 \qquad \text{in } 0 \le x \le L \tag{3-3a}$$

$$T(x) = T_0 \qquad \text{at } x = 0 \tag{3-3b}$$

$$T(x) = T_1 \qquad \text{at } x = L \tag{3-3c}$$

The integration of Eq. (3-3a) yields

$$T(x) = C_1 x + C_2 \tag{3-4}$$

The integration constants C_1 and C_2 are determined by the application of the boundary conditions (3-3b) and (3-3c); the resulting temperature distribution becomes

$$\frac{T(x) - T_0}{T_1 - T_0} = \frac{x}{L} \tag{3-5}$$

We note that for the considered problem the temperature in the slab varies linearly with the distance. The heat flux q is obtained by its definition in Eq. (3-2) as

$$q = -k \frac{dT(x)}{dx} = k \frac{T_0 - T_1}{L} \tag{3-6a}$$

and the heat-flow rate Q through an area A of the slab becomes

$$Q = qA = Ak \frac{T_0 - T_1}{L} = \frac{T_0 - T_1}{L/kA} \tag{3-6b}$$

We note that for this particular case both q and Q are independent of position.

Example 3-2   Determine the steady-state temperature distribution and the heat flux in a slab in the region $0 \le x \le L$ for uniform thermal conductivity k and a uniform heat generation in the medium at a rate of g_0 Btu/h·ft^3 (or W/m^3) when the boundary surface at $x = 0$ is kept at a uniform temperature T_0 and the boundary surface at $x = L$ dissipates heat by convection into an environment at a constant temperature T_∞ with a heat-transfer coefficient h.

Solution   The mathematical formulation of this heat-conduction problem is given as

$$\frac{d^2 T(x)}{dx^2} + \frac{g_0}{k} = 0 \qquad \text{in } 0 \le x \le L \tag{3-7a}$$

$$T(x) = T_0 \qquad \text{at } x = 0 \tag{3-7b}$$

$$k \frac{dT(x)}{dx} + hT(x) = hT_\infty \qquad \text{at } x = L \tag{3-7c}$$

The first and the second integration of Eq. (3-7a) give, respectively,

$$\frac{dT(x)}{dx} = -\frac{g_0}{k} x + C_1 \tag{3-8a}$$

$$T(x) = -\frac{1}{2} \frac{g_0}{k} x^2 + C_1 x + C_2 \tag{3-8b}$$

The application of the boundary condition Eq. (3-7b) to Eq. (3-8b) immediately gives the integration constant C_2 as $C_2 = T_0$. The integration constant C_1 is determined by introducing Eqs. (3-8) into the boundary condition equation (3-7c), that is,

$$k\left(-\frac{g_0}{k} L + C_1\right) + h\left(-\frac{g_0}{2k} L^2 + C_1 L + T_0\right) = hT_0 \tag{3-9a}$$

or

$$C_1 = \frac{T_\infty - T_0}{(1 + k/hL)L} + \frac{g_0 L}{2k} \frac{1 + 2(k/hL)}{1 + k/hL} \tag{3-9b}$$

The substitution of these coefficients into Eq. (3-8b) gives the temperature distribution in the slab:

$$T(x) - T_0 = \frac{T_\infty - T_0}{1 + 1/\text{Bi}} \frac{x}{L} + \frac{g_0 L^2}{2k} \left[\frac{1 + 2/\text{Bi}}{1 + 1/\text{Bi}} \frac{x}{L} - \left(\frac{x}{L}\right)^2 \right] \tag{3-10a}$$

where the *Biot number*, Bi, is defined as

$$\text{Bi} = \frac{hL}{k} \tag{3-10b}$$

As discussed in Chap. 2, the Biot number compares the relative magnitudes of the surface heat-transfer coefficient and the internal conductance of the slab to heat transfer. The heat flux $q(x)$ anywhere in the medium is determined from

$$q(x) = -k\frac{dT(x)}{dx} = g_0 x - C_1 k \tag{3-11}$$

where the constant C_1 is as given by Eq. (3-9b). We note that when there is heat generation in the medium the heat flux varies with the position. For most practical problems the amount of heat flow at the boundaries is of interest. For example, the heat flux at the boundary surface $x = 0$ becomes

$$q(x)\bigg|_{x=0} = -C_1 k = -k\frac{T_\infty - T_0}{(1 + 1/\text{Bi})L} - \frac{g_0 L}{2}\frac{1 + 2/\text{Bi}}{1 + 1/\text{Bi}} \tag{3-12}$$

We now examine some special cases of this problem:

1 When the Biot number becomes infinite (that is, $\text{Bi} \to \infty$) the heat-transfer coefficient h becomes infinite (that is, $h \to \infty$) and the boundary condition (3-7c) reduces to

$$T(x) = T_\infty \qquad \text{at } x = L \tag{3-13}$$

The solution for this particular case is immediately obtainable from Eqs. (3-10) by setting $\text{Bi} \to \infty$. We find

$$T(x) - T_0 = (T_\infty - T_0)\frac{x}{L} + \frac{g_0 L^2}{2k}\left[\frac{x}{L} - \left(\frac{x}{L}\right)^2\right] \tag{3-14}$$

For the special case of no heat generation, that is, $g_0 = 0$, Eq. (3-14) simplifies to the solution given previously by Eq. (3-5) for the boundary condition of the first kind at both boundaries.

2 When $\text{Bi} \to 0$, we have $h \to 0$; then the boundary condition (3-7c) simplifies to

$$\frac{dT(x)}{dx} = 0 \qquad \text{at } x = 0 \tag{3-15}$$

and the solution for this particular case is obtainable from Eqs. (3-10) by setting $\text{Bi} \to 0$.

$$T(x) - T_0 = \frac{g_0 L^2}{2k}\left[2\frac{x}{L} - \left(\frac{x}{L}\right)^2\right] \tag{3-16}$$

3-2 THE CYLINDER

We now consider the one-dimensional, steady-state heat conduction in a cylinder. When the temperature $T(r)$ is assumed to be a function of the r variable only, the derivatives with respect to z, ϕ, and t variables vanish and the heat-conduction equation (2-19) in the cylindrical coordinate system simplifies to

$$\frac{1}{r}\frac{d}{dr}\left(r\frac{dT}{dr}\right) + \frac{g(r)}{k} = 0 \tag{3-17}$$

where $T \equiv T(r)$ and $g(r)$ represents the energy generation in the medium in the dimensions of energy per unit time per unit volume (that is, $Btu/h \cdot ft^3$). The solution of Eq. (3-17) subject to appropriate boundary conditions gives the temperature distribution in the cylinder. Once the temperature distribution $T(r)$ is established, the heat flux $q(r)$ anywhere in the medium is determined from the definition

$$q(r) = -k\frac{dT(r)}{dr} \tag{3-18}$$

When $q(r)$ is a positive quantity, the heat flow is in the positive r direction, and vice versa.

Two boundary conditions are needed for the solution of Eq. (3-17). The *solid cylinder* in a region $0 \le r \le b$ and the *hollow cylinder* in a region $a \le r \le b$, as illustrated in Fig. 3-2a and b, respectively, are the two cases of interest in practical applications. We give a brief discussion of appropriate boundary conditions for these two cases.

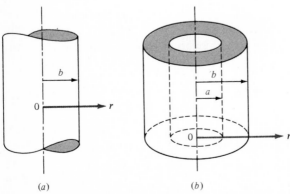

(a) (b)

FIG. 3-2   Coordinates for one-dimensional heat conduction in a cylinder. (a) Solid cylinder; (b) hollow cylinder.

Solid Cylinder

The heat-conduction problem for a solid cylinder in the region $0 \leq r \leq b$ requires two boundary conditions, one at $r = 0$ and the other at $r = b$. The boundary condition at $r = 0$ is not generally stated explicitly in the formulation of the problem; it is understood from the physical consideration that the *temperature should remain finite at $r = 0$*. This condition implies that any elementary solution of the differential equation that leads to the result $T \to \infty$ as $r \to 0$ should be excluded from the solution. In the present problem we have symmetry about the center point $r = 0$; therefore an alternative condition at the center that leads to the same result may be taken as $dT/dr = 0$ at $r = 0$. The boundary condition at $r = b$ may be chosen as the boundary condition of the first or the third kind; however, care must be exercised when the boundary condition at $r = b$ is of the second kind. For example, if the boundary at $r = b$ is insulated (that is, $dT/dr = 0$) while heat is being generated in the medium, the steady-state condition cannot be established because heat generated in the medium cannot escape through the boundaries and the temperature will continue to increase.

Hollow Cylinder

The solution of the heat-conduction equation (3-17) for a hollow cylinder in the region $a \leq r \leq b$, as illustrated in Fig. 3-2b, requires two boundary conditions, one at $r = a$ and the other at $r = b$. The boundary condition at any of these boundary surfaces may be of the first, the second, or the third kind, thus resulting in nine different combinations; but the case when both boundary conditions are of the second kind requires special consideration.

We now illustrate with examples the determination of one-dimensional, steady-state temperature distribution and heat flux in solid and hollow cylinders subject to different types of boundary conditions.

Example 3-3   Determine the steady-state temperature distribution $T(r)$ in a solid cylinder in the region $0 \leq r \leq b$ in which heat is generated at a constant rate of g_0 Btu/h·ft³ (or W/m³) while the boundary surface at $r = b$ is kept at a uniform temperature T_1. Determine the magnitude of the temperature difference between the cylinder-center and the boundary-surface temperatures for $g_0 = 144{,}000$ Btu/h·ft³ (1.49×10^6 W/m³), $k = 10$ Btu/h·ft·°F (17.3 W/m·°C), and $b = 1$ in (2.54×10^{-2} m).

Solution   The mathematical formulation of this heat-conduction problem is given as

$$\frac{1}{r} \frac{d}{dr}\left[r \frac{dT(r)}{dr}\right] + \frac{g_0}{k} = 0 \qquad \text{in } 0 \leq r \leq b \tag{3-19a}$$

$$T(r) = T_1 \qquad \text{at } r = b \tag{3-19b}$$

The integration of Eq. (3-19a) gives

$$\frac{dT(r)}{dt} = -\frac{g_0}{2k} r + \frac{C_1}{r} \tag{3-20a}$$

$$T(r) = -\frac{g_0}{4k} r^2 + C_1 \ln r + C_2 \tag{3-20b}$$

The integration constant C_1 is determined from the requirement that the temperature should remain finite (or $dT/dr = 0$) at $r = 0$. We find

$$C_1 = 0 \tag{3-21a}$$

The integration constant C_2 is determined by the application of the boundary condition at $r = b$ as

$$C_2 = T_1 + \frac{g_0 b^2}{4k} \tag{3-21b}$$

Then the solutions for the heat flux $q(r)$ and the temperature distribution $T(r)$ in the medium become

$$q(r) = -k \frac{dT(r)}{dr} = \frac{g_0 r}{2} \tag{3-22a}$$

$$T(r) - T_1 = \frac{g_0 b^2}{4k} \left[1 - \left(\frac{r}{b}\right)^2 \right] \tag{3-22b}$$

The temperature T_c at the center of the cylinder is obtained from Eq. (3-22b) by setting $r = 0$ to yield

$$T_c - T_1 = \frac{g_0 b^2}{4k} \tag{3-23}$$

An alternative representation of temperature distribution $T(r)$ in the cylinder is obtained by combining Eqs. (3-22b) and (3-23).

$$\frac{T(r) - T_1}{T_c - T_1} = 1 - \left(\frac{r}{b}\right)^2 \tag{3-24}$$

which shows that the temperature distribution in the solid is parabolic.

The heat flux at the boundary surface $r = b$ is obtained by setting $r = b$ in Eq. (3-22a):

$$q(r)\bigg|_{r=b} = \frac{g_0 b}{2} \tag{3-25}$$

In this equation, both g_0 and b being positive quantities, the heat flux at the boundary surface is a positive quantity; hence the heat flow is in the *positive r direction*, or out of the cylinder surface. This result is also expected by physical considerations.

For the numerical part of the problem, the difference between the center and the boundary-surface temperatures is immediately obtainable from Eq. (3-23) as follows: If the English engineering system of units is used, we find

$$T_c - T_1 = \frac{g_0 b^2}{4k} = \frac{144,000}{4 \times 10} \left(\frac{1}{12}\right)^2 = 25°F \ (13.89°C) \tag{3.26a}$$

If the SI system of units is used, we have

$$T_c - T_1 = \frac{(1.49 \times 10^6) \times (2.54 \times 10^{-2})^2}{4 \times 17.3} = 13.89°C \tag{3-26b}$$

Example 3-4 Determine the steady-state temperature distribution $T(r)$ and the radial heat-flow rate Q for a length H in a hollow cylinder in a region $a \le r \le b$ in which heat is generated at a constant rate of g_0 Btu/h·ft³ (or W/m³) while the boundary surfaces at $r = a$ and $r = b$ are kept at uniform temperatures T_1 and T_2, respectively.

Solution The mathematical formulation of this problem is given as

$$\frac{1}{r}\frac{d}{dr}\left[r\frac{dT(r)}{dr}\right] + \frac{g_0}{k} = 0 \qquad \text{in } a \le r \le b \tag{3-27a}$$

$$T(r) = T_0 \qquad\qquad \text{at } r = a \tag{3-27b}$$

$$T(r) = T_1 \qquad\qquad \text{at } r = b \tag{3-27c}$$

Equation (3-27a) is integrated as

$$\frac{dT(r)}{dr} = -\frac{g_0}{2k}r + \frac{C_1}{r} \tag{3-28a}$$

$$T(r) = -\frac{g_0}{4k}r^2 + C_1 \ln r + C_2 \tag{3-28b}$$

and the boundary conditions (3-27b) and (3-27c) are applied to Eq. (3-28b), to give, respectively,

$$T_0 = -\frac{g_0}{4k}a^2 + C_1 \ln a + C_2 \tag{3-29a}$$

$$T_1 = -\frac{g_0}{4k}b^2 + C_1 \ln b + C_2 \tag{3-29b}$$

Equations (3-29) provide two relations for the determination of C_1 and C_2; a simultaneous solution gives

$$C_1 = \frac{(T_1 - T_0) + (g_0/4k)(b^2 - a^2)}{\ln (b/a)} \tag{3-30a}$$

$$C_2 = \left(T_0 + \frac{g_0 a^2}{4k}\right) - \left[(T_1 - T_0) + \frac{g_0}{4k}(b^2 - a^2)\right]\frac{\ln a}{\ln (b/a)} \tag{3-30b}$$

Then the temperature distribution $T(r)$ in the cylinder becomes

$$\frac{T(r) - T_0}{T_1 - T_0} = \frac{\ln (r/a)}{\ln (b/a)} + \frac{g_0(b^2 - a^2)}{4k(T_1 - T_0)}\left[\frac{\ln (r/a)}{\ln (b/a)} - \frac{(r/a)^2 - 1}{(b/a)^2 - 1}\right] \tag{3-31}$$

The radial heat-flow rate at any position r through the cylinder for a length H of the cylinder is determined from

$$Q(r) = q(r) \text{ area} = -k\frac{dT(r)}{dr}2\pi rH \tag{3-32a}$$

When $dT(r)/dr$ is substituted from Eq. (3-28a) we find

$$Q(r) = 2\pi H\left(\frac{g_0}{2}r^2 - C_1 k\right) \tag{3-32b}$$

where C_1 is given by Eq. (3-30a).

For the case of no heat generation, by setting $g_0 = 0$, the temperature distribution $T(r)$ and the heat-flow rate $Q(r)$ are obtained, respectively, from Eqs. (3-31) and (3-32b) as

$$\frac{T(r) - T_0}{T_1 - T_0} = \frac{\ln (r/a)}{\ln (b/a)} \tag{3-33}$$

and

$$Q = \frac{2\pi kH}{\ln (b/a)} (T_0 - T_1) = \frac{T_0 - T_1}{\ln (b/a)/2\pi kH} \tag{3-34}$$

We note that the total heat-flow rate Q is independent of the radial position for the case of no heat generation in the cylinder.

3-3   THE SPHERE

When the temperature $T(r)$ is a function of the r variable only, the heat-conduction equation (2-20) in the spherical coordinate system simplifies to

$$\frac{1}{r^2} \frac{d}{dr} \left[r^2 \frac{dT(r)}{dr} \right] + \frac{g(r)}{k} = 0 \tag{3-35}$$

Once the temperature distribution $T(r)$ is established from the solution of this equation, subject to appropriate boundary conditions, the heat flux $q(r)$ anywhere in the medium is determined from

$$q(r) = -k \frac{dT(r)}{dr} \tag{3-36}$$

A brief discussion of the boundary conditions appropriate for a *solid sphere* in the region $0 \le r \le b$ and a *hollow sphere* in the region $a \le r \le b$, as illustrated in Fig. 3-3, is now given.

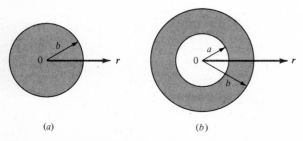

(a)   (b)

FIG. 3-3   Coordinates for one-dimensional heat conduction in a sphere. (*a*) Solid sphere; (*b*) hollow sphere.

Solid Sphere

The solution of Eq. (3-35) requires two boundary conditions at $r = 0$ and $r = b$ in the case of a solid sphere in the region $0 \leq r \leq b$. The boundary condition at $r = 0$ is chosen by the physical consideration that the *temperature* $T(r)$ *should remain finite at the center* $r = 0$. This condition requires that any elementary solution of the differential equation leading to $T(r) \rightarrow \infty$ at $r \rightarrow 0$ should be excluded from the solution. The present problem possesses symmetry about $r = 0$; therefore an alternative condition that leads to the same result may be taken as $dT/dr = 0$ at $r = 0$. The boundary condition at the outer surface $r = b$ can be chosen as of the first or the third kind, but the boundary condition of the second kind gives rise to the difficulties discussed previously.

Hollow Sphere

In the case of a hollow sphere in the region $a \leq r \leq b$, as illustrated in Fig. 3-3b, the boundary conditions at any of these surfaces may be taken as of the first, the second, or the third kind. When the boundary conditions are both of the second kind for the inner and outer boundaries the difficulty arises for the reasons discussed previously.

We now illustrate with representative examples the determination of temperature distribution and heat flow in a sphere.

Example 3-5   Determine the steady-state temperature distribution $T(r)$ and the total radial heat flow Q in a hollow sphere in a region $a \leq r \leq b$ when the boundary surfaces at $r = a$ and $r = b$ are kept at uniform temperatures T_0 and T_1, respectively.

Solution   The mathematical formulation of the problem is given as

$$\frac{d}{dr}\left[r^2 \frac{dT(r)}{dr}\right] = 0 \qquad \text{in } a \leq r \leq b \tag{3-37a}$$

$$T(r) = T_0 \qquad \text{at } r = a \tag{3-37b}$$

$$T(r) = T_1 \qquad \text{at } r = b \tag{3-37c}$$

The integration of Eq. (3-37a) yields

$$\frac{dT(r)}{dr} = \frac{C_1}{r^2} \tag{3-38}$$

$$T(r) = -\frac{C_1}{r} + C_2 \tag{3-39}$$

The application of the boundary conditions (3-37b) and (3-37c) to Eq. (3-39) gives, respectively,

$$T_0 = -\frac{C_1}{a} + C_2 \tag{3-40a}$$

$$T_1 = -\frac{C_1}{b} + C_2 \tag{3-40b}$$

and the constants C_1 and C_2 are determined from the solution of Eqs. (3-40) as

$$C_1 = \frac{ab}{a-b}(T_0 - T_1) \tag{3-41a}$$

$$C_2 = \frac{bT_1 - aT_0}{b-a} \tag{3-41b}$$

Then the temperature distribution $T(r)$ in the sphere becomes

$$T(r) = \frac{1}{b-a}\left[aT_0\left(\frac{b}{r} - 1\right) + bT_1\left(1 - \frac{a}{r}\right)\right] \tag{3-42}$$

The total radial heat flow Q through the sphere at any position r is determined from

$$Q = q(r) \text{ area} = -k\frac{dT(r)}{dr}4\pi r^2 = -4\pi kC_1 \tag{3-43a}$$

and when C_1 is substituted from Eq. (3-41a) we obtain

$$Q = 4\pi k\frac{ab}{b-a}(T_0 - T_1) = \frac{T_0 - T_1}{(b-a)/4\pi kab} \tag{3-43b}$$

We note that for no heat generation the total heat-transfer rate Q is independent of position.

3-4   THE CONCEPT OF THERMAL RESISTANCE

In the problems of one-dimensional, steady-state heat conduction in finite regions for the special case of no heat generation, constant thermal conductivity, and prescribed temperature at the two boundaries, the total heat-transfer rate Q through the solid can be related to the *thermal resistance R* of the solid in the form

$$Q = \frac{\Delta T}{R} \tag{3-44}$$

where Q = total heat-transfer rate through solid, Btu/h (or W)
   ΔT = difference between temperatures of the two boundary surfaces of the region, °F (or °C)
   R = thermal resistance of solid, h·°F/Btu (or °C/W)

The thermal-resistance concept is analogous to the electric-resistance concept defined by the relation

$$\text{Current} = \frac{\text{potential difference}}{\text{electric resistance}} \tag{3-45}$$

Clearly, the total heat flow Q is analogous to the electric current and the temperature difference to voltage difference. The thermal-resistance concept is used in many engineering applications. We now examine the determination of the thermal resistances of a slab, a hollow cylinder, and a hollow sphere.

Slab

Consider the one-dimensional, steady-state heat conduction through a slab in the region $0 \leq x \leq L$, having a constant thermal conductivity k and boundaries at $x = 0$ and $x = L$ kept at uniform temperatures T_0 and T_1, respectively. The solution of this problem was considered previously in Example 3-1, and the heat flux q was given by $q = k(T_0 - T_1)/L$. Then the total heat-transfer rate Q through an area A of the slab is given by

$$Q = Aq = Ak \frac{T_0 - T_1}{L} \equiv \frac{T_0 - T_1}{R_{\text{slab}}} \tag{3-46a}$$

where the *thermal resistance of the slab* R_{slab} is defined as

$$R_{\text{slab}} = \frac{L}{Ak} \tag{3-46b}$$

Hollow Cylinder

We now consider one-dimensional, steady-state heat conduction through a hollow cylinder in the region $a \leq r \leq b$, having a constant thermal conductivity k and boundaries at $r = a$ and $r = b$ kept at uniform temperatures T_0 and T_1, respectively. The total heat-transfer rate Q through the cylinder over a length H of the cylinder can be obtained from the solution of the same problem given by Eq. (3-34) as

$$Q = \frac{2\pi k H}{\ln (b/a)} (T_0 - T_1) \equiv \frac{T_0 - T_1}{R_{\text{cyl}}} \tag{3-47a}$$

where the *thermal resistance of the cylinder* R_{cyl} is defined as

$$R_{\text{cyl}} \equiv \frac{\ln (b/a)}{2\pi H k} \tag{3-47b}$$

The thermal resistance given by Eq. (3-47b) is now rearranged in a form similar to that for a slab:

$$R_{\text{cyl}} = \frac{\ln (b/a)}{2\pi k H} = \frac{(b - a) \ln (2\pi b H/2\pi a H)}{(b - a)2\pi H k} = \frac{L_{\text{cyl}} \ln (A_1/A_0)}{(A_1 - A_0)k} \equiv \frac{L_{\text{cyl}}}{A_{\text{cyl}} k} \tag{3-47c}$$

where $A_0 = 2\pi aH$ = area of inner surface of cylinder
$A_1 = 2\pi bH$ = area of outer surface of cylinder
$L_{cyl} = b - a$ = thickness of cylinder

$$A_{cyl} = \frac{A_1 - A_0}{\ln (A_1/A_0)} = \text{logarithmic mean area (i.e., logarithmic mean of inner and outer areas)}$$

Thus the thermal resistance for a hollow cylinder given by Eq. (3-47c) is of exactly the same form as that for a slab given by Eq. (3-46b), except that the *logarithmic mean area* is used for the cylinder.

Hollow Sphere

The one-dimensional, steady-state heat conduction through a hollow sphere in the region $a \leq r \leq b$ having a constant thermal conductivity k and the boundaries at $r = a$ and $r = b$ kept at uniform temperatures T_0 and T_1, respectively, was solved in Example 3-5, and the total heat-transfer rate Q through the sphere given by Eq. (3-43b).   The *thermal resistance for the sphere* R_{sph} is now determined from Eq. (3-43b) as

$$Q = 4\pi K \frac{ab}{b - a}(T_0 - T_1) = \frac{T_0 - T_1}{(1/4\pi k)[(b - a)/ab]} \equiv \frac{T_0 - T_1}{R_{sph}} \tag{3-48a}$$

where the *thermal resistance of the sphere,* R_{sph}, is defined as

$$R_{sph} \equiv \frac{1}{4\pi k}\frac{b - a}{ab} \tag{3-48b}$$

This relation is rearranged in a form similar to that for a slab as follows:

$$R_{sph} = \frac{1}{4\pi k}\frac{b - a}{ab} = \frac{b - a}{k\sqrt{(4\pi a^2)(4\pi b^2)}} = \frac{L_{sph}}{k\sqrt{A_0 A_1}} = \frac{L_{sph}}{A_g k} \tag{3-48c}$$

where $A_0 = 4\pi a^2$ = area of inner surface area of sphere
$A_1 = 4\pi b^2$ = area of outer surface area of sphere
$L_{sph} = b - a$ = thickness of sphere
$A_g = \sqrt{A_0 A_1}$ = geometric mean area (i.e., geometric mean of inner and outer areas)

The thermal resistance for a hollow sphere given by Eq. (3-48c) is of a form similar to that for a slab except that the area is replaced by the geometric mean area.

3-5 COMPOSITE MEDIUM

There are many engineering applications in which heat transfer takes place through a medium composed of several different parallel layers each having different thermal conductivity. Consider, for example, a hot fluid flowing inside a tube covered with a uniform layer of thermal insulation. The thermal conductivities of the tube metal and of insulation are different; hence the heat-transfer problem from the hot fluid to the colder outer environment involves conduction through a composite medium consisting of two parallel concentric cylinders. The thermal-resistance concept discussed above is now applied to the prediction of one-dimensional, steady-state heat-transfer rate by conduction through a composite structure consisting of parallel plates, coaxial cylinders, or concentric spheres. Figures 3-4 and 3-5 show typical configurations of such composite layers for slabs, cylinders, and spheres. For generality it is assumed that the interior and exterior surfaces of these composite structures are subjected to convective heat transfer to fluids at constant mean temperatures T_a and T_b, and with heat-transfer coefficients h_a and h_b, respectively. It is also assumed that the parallel layers in the composite structure are in *perfect thermal contact*, that is, there is no temperature drop at the interfaces, i.e., the temperature is continuous at the interface of the two layers in contact. We now examine the determination of the heat flow through such a layer for the cases of slabs, cylinders, and spheres.

Parallel Slabs

Consider a composite wall consisting of three parallel layers in perfect thermal contact as shown in Fig. 3-4. The heat is transferred from the hot fluid at a mean temperature T_a through this composite layer to the cold fluid of a mean

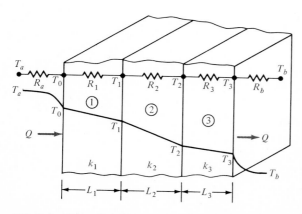

FIG. 3-4 Thermal resistances for one-dimensional heat flow through parallel walls in perfect thermal contact.

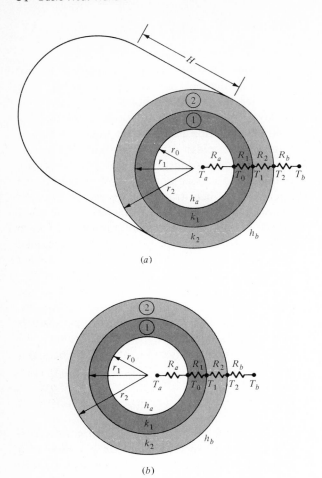

(a)

(b)

FIG. 3-5 Thermal resistances for radial heat flow through a hollow composite cylinder and sphere with layers in perfect thermal contact. (a) Coaxial cylinder; (b) concentric spheres.

temperature T_b; the heat-transfer rate Q through an area A of this composite structure is the same through each layer and is given by

$$Q = Ah_a(T_a - T_0) = Ak_1 \frac{T_0 - T_1}{L_1} = Ak_2 \frac{T_1 - T_2}{L_2} = Ak_3 \frac{T_2 - T_3}{L_3} = Ah_b(T_3 - T_b)$$

$$(3\text{-}49)$$

where the terms like $Ah \, \Delta T$ represent heat transfer by convection and the terms like $Ak(\Delta T/L)$ represent heat transfer by conduction through various layers.

Equation (3-49) is now written in terms of the thermal resistances of the layers:

$$Q = \frac{T_a - T_0}{R_a} = \frac{T_0 - T_1}{R_1} = \frac{T_1 - T_2}{R_2} = \frac{T_2 - T_3}{R_3} = \frac{T_3 - T_b}{R_b} \qquad (3\text{-}50a)$$

where various thermal resistances are defined as

$$R_a = \frac{1}{Ah_a} \qquad R_1 = \frac{L_1}{Ak_1} \qquad R_2 = \frac{L_2}{Ak_2} \qquad R_3 = \frac{L_3}{Ak_3} \qquad \text{and} \qquad R_b = \frac{1}{Ah_b}$$
$$(3\text{-}50b)$$

When the interface temperatures T_0, T_1, T_2, and T_3 are eliminated from Eqs. (3-50a) by summing up the numerators and denominators of the individual ratios, we obtain the total heat-transfer rate Q through the composite layer given by the following simple expression

$$Q = \frac{T_a - T_b}{R} \qquad \text{Btu/h (or W)} \qquad (3\text{-}51a)$$

where $\quad R = R_a + R_1 + R_2 + R_3 + R_b \qquad (3\text{-}51b)$

We note that Eq. (3-51a) is of exactly the same form as Eq. (3-44), but in this case the *total thermal resistance* R in the path of heat flow from temperature T_a to temperature T_b consists of the sum of several different thermal resistances as given by Eq. (3-51b). The thermal resistance R has the units h·°F/Btu (or °C/W).

Thus, when the temperatures T_a and T_b and the magnitudes of the individual thermal resistances in the path of heat flow are known, the total heat-flow rate Q through an area A of a composite structure is determined by means of the simple relation given by Eq. (3-51a).

The temperature at any one of the interfaces of this composite structure is readily determined by means of Eqs. (3-50a) and (3-51a). For example, the surface temperature T_a is given by

$$T_a - T_0 = (T_a - T_b) \frac{R_a}{R} \qquad (3\text{-}52a)$$

or the interface temperature T_1 by

$$T_a - T_1 = (T_a - T_b) \frac{R_a + R_1}{R} \qquad (3\text{-}52b)$$

and so forth.

The concept of *overall heat-transfer coefficient U* is frequently used in heat-transfer applications to characterize the unit conductance of a composite layer.

It is related to the total thermal resistance R of the composite layer by

$$UA = \frac{1}{R} \tag{3-53}$$

where U has the units $\text{Btu/h} \cdot \text{ft}^2 \cdot {}^\circ\text{F}$ (or $\text{W/m}^2 \cdot {}^\circ\text{C}$). The total heat-transfer rate Q through an area A of a composite structure from temperature T_a to T_b is evaluated by using the unit conductance U from the relation

$$Q = AU(T_a - T_b) \qquad \text{Btu/h (or W)} \tag{3-54}$$

Coaxial Cylinders

Consider a composite cylindrical structure consisting of two coaxial layers in perfect thermal contact, as illustrated in Fig. 3-5a. A hot fluid at a temperature T_a flows inside the tube, and heat is transferred to the tube wall with a heat-transfer coefficient h_a. On the outside, heat transfer takes place from the exterior surface of the tube to a cold fluid at temperature T_b with a heat-transfer coefficient h_b. The total heat-transfer rate Q from the hot to the cold fluid over the length H of the cylindrical structure is the same through each layer and is given by

$$Q = \frac{T_a - T_0}{R_a} = \frac{T_0 - T_1}{R_1} = \frac{T_1 - T_2}{R_2} = \frac{T_2 - T_b}{R_b} \tag{3-55a}$$

where various thermal resistances are defined as

$$R_a = \frac{1}{2\pi r_0 H h_a} \qquad R_1 = \frac{1}{2\pi H k_1} \ln \frac{r_1}{r_0}$$

$$R_2 = \frac{1}{2\pi H k_2} \ln \frac{r_2}{r_1} \qquad \text{and} \qquad R_b = \frac{1}{2\pi r_2 H h_b} \tag{3-55b}$$

Here the thermal resistances R_1 and R_2 for conduction through a cylindrical layer are written in accordance with the thermal-resistance expression given by Eq. (3-47b).

When the interface temperatures are eliminated from Eq. (3-55a), we obtain the following simple expression for the total heat-flow rate Q:

$$Q = \frac{T_a - T_b}{R} \qquad \text{Btu/h (or W)} \tag{3-56a}$$

where $\quad R = R_a + R_1 + R_2 + R_b \tag{3-56b}$

Thus, the total heat-transfer rate through a composite cylindrical structure is equal to the overall temperature difference $T_a - T_b$ divided by the sum of the thermal resistances in the path of the heat flow.

The interface temperatures are readily determined by means of Eqs. (3-55a) and (3-56a). For example, the interface temperature T_1 is given by

$$T_a - T_1 = (T_a - T_b) \frac{R_a + R_1}{R} \tag{3-57}$$

An overall heat-transfer coefficient U can also be defined for heat transfer through a composite cylinder; in such cases *it is necessary to specify the area on which U is based* because the area of a cylinder varies in the radial direction. For example, the overall heat-transfer coefficient U_0 based on the interior surface A_0 of the cylinder is defined as

$$U_0 A_0 = \frac{1}{R} \tag{3-58a}$$

and U_2 based on the exterior surface A_2 is defined as

$$U_2 A_2 = \frac{1}{R} \tag{3-58b}$$

where R is the total thermal resistance, i.e.,

$$R = R_a + R_1 + R_2 + R_b \tag{3-58c}$$

and the areas A_0 and A_2 are given by

$$A_0 = 2\pi r_0 H \qquad A_2 = 2\pi r_2 H \tag{3-58d}$$

For most engineering applications the overall heat-transfer coefficient is based on the *external surface* of a cylinder because the outer diameter is readily measured.

For the two-layer composite cylinder shown in Fig. 3-5a, the overall heat-transfer coefficient U_2 based on the outer surface is given by Eqs. (3-58b), (3-58c), and (3-55b) as

$$U_2 = \frac{1}{\dfrac{r_2}{r_0 h_a} + \dfrac{r_2}{k_1} \ln\left(\dfrac{r_1}{r_0}\right) + \dfrac{r_2}{k_2} \ln\left(\dfrac{r_2}{r_1}\right) + \dfrac{1}{h_b}} \tag{3-59}$$

Concentric Spheres

Figure 3-5b shows a composite sphere consisting of two concentric layers. The interior and exterior surfaces are subjected to heat exchange by convection with fluids at constant temperatures T_a and T_b, with heat-transfer coefficients h_a and h_b, respectively. The total radial heat flow Q through the sphere is given by

$$Q = \frac{T_a - T_0}{R_a} = \frac{T_0 - T_1}{R_1} = \frac{T_1 - T_2}{R_2} = \frac{T_2 - T_b}{R_b} \tag{3-60a}$$

where various thermal resistances are defined as [see Eq. (3-48b)]

$$R_a = \frac{1}{4\pi r_0^2 h_a} \qquad R_1 = \frac{1}{4\pi k_1} \frac{r_1 - r_0}{r_1 r_0} \tag{3-60b}$$

$$R_2 = \frac{1}{4\pi k_2} \frac{r_2 - r_1}{r_2 r_1} \quad \text{and} \quad R_b = \frac{1}{4\pi r_0^2 h_b}$$

The total heat-flow rate Q is related to the total thermal resistance R by eliminating the interface temperatures from Eq. (3-60a); we find

$$Q = \frac{T_a - T_b}{R} \qquad \text{Btu/h (or W)} \tag{3-61a}$$

where

$$R = R_a + R_1 + R_2 + R_b \qquad \text{h·°F/Btu (or °C/W)} \tag{3-61b}$$

The overall heat-transfer coefficient U_2 based on the exterior surface area of the sphere is defined as

$$U_2 \, 4\pi r_2^2 = \frac{1}{R} \tag{3-62}$$

where R is given by Eqs. (3-61b) and (3-60b).

Example 3-6   A steel tube $(k = 25$ Btu/h·ft·°F or 43.26 W/m·°C) of 2-in $(5.08 \times 10^{-2}$-m) ID, 3-in $(7.62 \times 10^{-2}$-m) OD is covered with a 1-in $(2.54 \times 10^{-2}$-m) layer of asbestos insulation $(k = 0.12$ Btu/h·ft·°F or 0.208 W/m·°C). The inside surface of the tube receives heat by convection from a hot gas at $T_a = 600°F$ (315.6°C) with a heat-transfer coefficient $h_a = 50$ Btu/h·ft²·°F (283.8 W/m²·°C) while the outer surface of the insulation is exposed to the ambient air at $T_b = 100°F$ (37.8°C) with a heat-transfer coefficient $h_b = 3$ Btu/h·ft²·°F (17.03 W/m²·°C). Determine (a) the heat loss to the ambient air per 10-ft (3.048-m) length of the tube, and (b) the temperature drops across the tube material and the insulation layer. At the considered temperature the radiation is not significant, hence can be neglected.

Solution (a) The radial heat flow through the tube is given by

$$Q = \frac{T_a - T_b}{R_a + R_1 + R_2 + R_b}$$

where various thermal resistances are determined from Eqs. (3-55b) by using the English engineering units as

$$R_a = \frac{1}{\pi \frac{2}{12} \times 10 \times 50} = 0.382 \times 10^{-2} \qquad R_1 = \frac{1}{2\pi \times 10 \times 25} \ln \frac{3}{2} = 0.258 \times 10^{-3}$$

$$R_2 = \frac{1}{2\pi \times 10 \times 0.12} \ln \frac{5}{3} = 0.678 \times 10^{-1} \qquad R_b = \frac{1}{\pi \frac{5}{12} \times 10 \times 3} = 0.255 \times 10^{-1}$$

Then the total thermal resistance R is given by

$$R = R_a + R_1 + R_2 + R_b = 0.973 \times 10^{-1} \qquad \text{h} \cdot °\text{F/Btu}$$
$$= 1.845 \times 10^{-1} \qquad °\text{C/W}$$

and the heat loss from the tube becomes

$$Q = \frac{500}{R} = 5138 \text{ Btu/h (or 1506 W)}$$

(b) The temperature drops across the tube material and the asbestos layer are determined as

$$\Delta T_{\text{tube}} = \frac{R_1}{R} (T_a - T_b) = \frac{R_1}{R} \times 500 = 1.33°\text{F} \ (0.74°\text{C})$$

$$\Delta T_{\text{asbestos}} = \frac{R_2}{R} \times 500 = 348°\text{F} \ (193.3°\text{C})$$

We note that the temperature drop across the tube material is very small compared with that across the insulation layer.

3-6 THERMAL CONTACT RESISTANCE

The interface between two solid surfaces not metallurgically bonded together constitutes a resistance to heat flow. The reason for this resistance is better envisioned by examining an enlarged view of an interface as shown in Fig. 3-6. The direct contact between the solid surfaces takes place at a limited number of spots, and the voids between them are usually filled with air or the surrounding fluid. Heat transfer through the fluid filling the voids is mainly by conduction, because there is no convection in such a thin layer of fluid; the radiation effects are negligible at normal temperatures. Then heat transfer across the interface takes place entirely by conduction through both the thin fluid layer filling the voids and the spots in direct metal-to-metal contact. If the thermal conductivity of the fluid is less than that of the solids, the interface acts as a resistance to heat

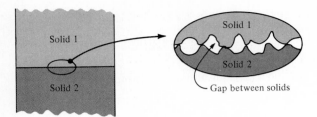

FIG. 3-6   Interface between two metal surfaces in contact.

flow; this resistance is referred to as the *thermal contact resistance*. Experimental and theoretical investigations of heat transfer through surfaces in contact are given in several references [1–6], and a comprehensive bibliography for thermal-contact-resistance studies until 1968 is given in Ref. [5].   An examination of all the work on this subject reveals that no satisfactory theory is yet available to predict the thermal contact resistance; thus, an experimental approach appears to be the only means to predict thermal contact resistance for practical purposes. Figure 3-7 illustrates the effects of surface roughness, interface temperature, and interface pressure on the *interface thermal contact conductance h* in Btu/h·ft²·°F (or W/m²·°C) determined experimentally for aluminum joints.   The reciprocal of h is called the *specific thermal contact resistance R^** (that is, $R^* = 1/h$) which is the thermal contact resistance per unit interface area and is in the units h·ft²·°F/Btu (or m²·°C/W).

In general, the interface thermal contact conductance h increases with increasing interface pressure, because the high spots are deformed under load

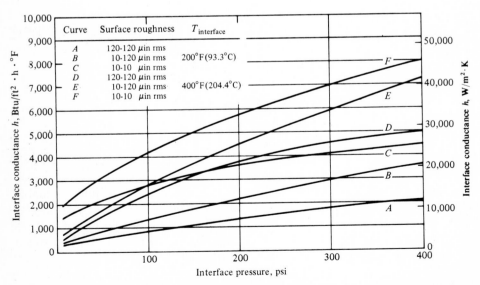

FIG. 3-7   Effect of surface roughness, interface temperature T, and pressure P on the interface-contact conductance of 7075 T6 aluminum joints. (*From Barzelay et al.* [2].)

and create greater contact area, but decreases with increasing surface roughness and waviness because the voids are enlarged and the surfaces do not come into good contact. When thin foil of good thermal conductivity is inserted between two surfaces, the contact conductance increases if the foil is softer than the interface material and decreases if the foil is harder than the interface material.

3-7 VARIABLE THERMAL CONDUCTIVITY

In many applications the thermal conductivity of the solid varies with temperature very rapidly or the temperature differences become so large that substantial variation occurs in the thermal conductivity. In such situations it may be necessary to include in the analysis the variation of thermal conductivity with temperature. In general, the solution of heat-conduction problems with variable thermal conductivity is a complicated matter; however, in the case of one-dimensional steady-state heat conduction in a solid subject to prescribed temperature boundary conditions, the effects of variable thermal conductivity on heat flow can be included in the analysis in a straightforward manner as described below.

Slab with Variable k

Consider a slab in the region $0 \leq x \leq L$ having boundary surfaces at $x = 0$ and $x = L$ kept at uniform temperatures T_0 and T_1, respectively. The thermal conductivity $k(T)$ of the material depends strongly on temperature, and the variation of $k(T)$ with temperature is assumed to be specified. The rate of heat flow through the slab is now determined by including in the analysis the temperature dependence of the thermal conductivity.

The mathematical formulation of the problem is given as

$$\frac{d}{dx}\left[k(T)\frac{dT}{dx}\right] = 0 \qquad \text{in } 0 \leq x \leq L \tag{3-63a}$$

$$T = T_0 \qquad \text{at } x = 0 \tag{3-63b}$$

$$T = T_1 \qquad \text{at } x = L \tag{3.63c}$$

The integration of Eq. (3-63a) yields

$$k(T)\frac{dT}{dx} = C \tag{3-64a}$$

or

$$k(T)\,dT = C\,dx \tag{3-64b}$$

The heat-flow rate Q through an area A of the slab is obtained from

$$Q = Aq = -Ak(T)\frac{dT}{dx} = -AC \tag{3-65}$$

in view of Eq. (3-64a). Clearly, the determination of the constant C is sufficient if only the heat-transfer rate Q through the slab is needed. The constant C is readily determined by integrating Eq. (3-64b) over the limits of T from T_0 to T_1 and of x from 0 to L as specified by the boundary conditions (3-63b) and (3-63c). We obtain

$$\int_{T_0}^{T_1} k(T)\, dT = CL$$

or

$$C = \frac{1}{L}\int_{T_0}^{T_1} k(T)\, dT = -\frac{1}{L}\int_{T_1}^{T_0} k(T)\, dT \tag{3-66}$$

The substitution of C into Eq. (3-65) gives the heat-flow rate through the slab as

$$Q = \frac{A}{L}\int_{T_1}^{T_0} k(T)\, dT \tag{3-67}$$

Once the thermal conductivity $k(T)$ is specified as a function of temperature T, the integral on the right-hand side of this equation can be evaluated, and the heat-flow rate Q through the slab is determined. To illustrate the procedure it is assumed that $k(T)$ is a linear function of temperature in the form

$$k(T) = k_0(1 + \beta T)$$

where k_0 is the thermal conductivity at $T = 0$ and the constant β is called the *temperature coefficient of thermal conductivity.* Then Eq. (3-67) becomes

$$
\begin{aligned}
Q &= \frac{Ak_0}{L}\left[(T_0 - T_1) + \frac{1}{2}\beta(T_0{}^2 - T_1{}^2)\right] \\
&= Ak_0\left(1 + \beta\,\frac{T_0 + T_1}{2}\right)\frac{T_0 - T_1}{L} \\
&= Ak_m\frac{T_0 - T_1}{L}
\end{aligned}
\tag{3-68a}
$$

where the *mean thermal conductivity* k_m is defined as

$$k_m \equiv k_0\left(1 + \beta\,\frac{T_0 + T_1}{2}\right) \tag{3-68b}$$

We note that for the linear variation of k with temperature the heat flow through the slab can be calculated by the simple relation given by Eq. (3-68a) if the thermal conductivity k_m is evaluated as the arithmetic mean of the boundary surface temperatures, that is, $(T_0 + T_1)/2$.

Hollow Cylinder with Variable k

Consider a hollow cylinder in the region $a \leq r \leq b$ with boundary surfaces at $r = a$ and $r = b$ kept at uniform temperatures T_0 and T_1, respectively. The thermal conductivity $k(T)$ of the material depends strongly on temperatures, and the variation of $k(T)$ with temperature is considered specified. The rate of heat transfer through a length H of the cylinder is now determined by including in the analysis the variation of thermal conductivity with temperature.

The mathematical formulation of the problem is given as

$$\frac{d}{dr}\left[rk(T)\frac{dT}{dr}\right] = 0 \quad \text{in } a \leq r \leq b \tag{3-69a}$$

$$T = T_0 \quad \text{at } r = a \tag{3-69b}$$

$$T = T_1 \quad \text{at } r = b \tag{3-69c}$$

The integration of Eq. (3-69a) gives

$$rk(T)\frac{dT}{dr} = C \tag{3-70a}$$

or

$$k(T)\,dT = C\,\frac{dr}{r} \tag{3-70b}$$

and the constant C is determined by integrating Eq. (3-70b) over the limits of T from T_0 to T_1 and of r from a to b as specified by the boundary conditions. We obtain

$$\int_{T_0}^{T_1} k(T)\,dT = C \ln \frac{b}{a} \tag{3-71a}$$

or

$$C = -\frac{1}{\ln(b/a)} \int_{T_1}^{T_0} k(T)\,dT \tag{3-71b}$$

The heat-flow rate Q through a length H of the cylinder is obtained from

$$Q = 2\pi r H q = -2\pi r H \left[k(T)\frac{dT}{dr}\right] = -2\pi H C \tag{3-72a}$$

where we utilized the result in Eq. (3-70a). Substituting C from Eq. (3-71b) into Eq. (3-72a), the relation for the heat flow becomes

$$Q = \frac{2\pi H}{\ln (b/a)} \int_{T_1}^{T_0} k(T)\, dT \tag{3-72b}$$

Once the thermal conductivity is specified as a function of temperature, the integral in this equation can be evaluated and the heat-transfer rate through the cylinder determined.

Example 3-7   Determine the heat flux across 0.5 ft (0.1524 m) of thick slab when one face is kept at $T_0 = 900°R$ (500 K) and the other face at $T_1 = 500°R$ (277.8 K), and the thermal conductivity varies linearly with temperature as

$$k(T) = k_0(1 + \beta T)$$

where $k_0 = 0.02$ Btu/h·ft·°F (0.0346 W/m·°C) and $\beta = 2 \times 10^{-3}$ °R^{-1}(3.6 × 10^{-3} K^{-1}).

Solution   When the thermal conductivity varies linearly with temperature, the heat-flow rate through the slab per unit area can be determined by Eqs. (3-68) by setting $A = 1$ ft^2 (or 1 m^2).   That is,

$$q = k_m \frac{T_0 - T_1}{L}$$

where

$$k_m = k_0 \left(1 + \beta \frac{T_0 + T_1}{2}\right)$$

For the considered problem, by using the English engineering units, we find

$$k_m = 0.02\left(1 + 2 \times 10^{-3} \times \frac{900 + 500}{2}\right) = 0.048 \text{ Btu/h·ft·°F (0.083 W/m·°C)}$$

$$q = 0.048 \times \frac{900 - 500}{0.5} = 38.4 \text{ Btu/h·ft}^2 \text{ (121 W/m}^2\text{)}$$

or, by using the SI system of units, we find

$$k_m = 0.0346\left(1 + 3.6 \times 10^{-3} \times \frac{500 + 277.8}{2}\right) = 0.0830 \text{ W/m·°C}$$

$$q = 0.0830 \times \frac{500 - 277.8}{0.1524} = 121 \text{ W/m}^2$$

3-8   THE ONE-DIMENSIONAL FIN EQUATION

Heat transfer by convection between a surface and the fluid surrounding it can be increased by attaching to the surface thin strips of metals called *fins*.   A large variety of fin geometries are manufactured for heat-transfer applications; Fig. 3-8 shows typical examples.   When heat transfer takes place by convection

FIG. 3-8 Types of finned tubing. (*From Brown Fintube Co.*)

from both interior and exterior surfaces of a tube or a plate, fins are generally used on the surface where the heat-transfer coefficient is low. For example, in a car radiator the outer surface of the tubes is finned because the heat-transfer coefficient for air at the outer surface is much smaller than that for water flow at the inner surface. The problem of determination of heat flow through a fin requires a knowledge of temperature distribution in the fin. In this section we derive the one-dimensional fin equation which will be used in the subsequent analysis to determine the temperature distribution and the heat flow through a fin.

Figure 3-9 shows the geometry, the coordinates, and the nomenclature for the derivation of the one-dimensional, steady-state energy equation for fins. It is

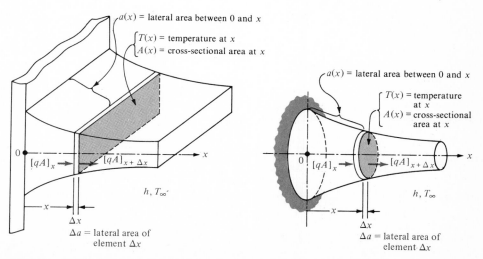

FIG. 3-9 Nomenclature for the derivation of the one-dimensional fin equation.

assumed that the temperature at any cross section of the fin is uniform so that $T(x)$ is a function of x only. This assumption is valid for most fin applications of practical interest; for a relatively thick fin the error involved in a one-dimensional analysis is less than 1 percent [7]. To derive the differential equation governing the temperature distribution in the fin, we consider a small volume element of thickness Δx as shown in Fig. 3-9 and write a statement of the steady-state energy balance for this volume element as

$$
\underbrace{\left(\begin{array}{l}\text{Net rate of heat gain}\\ \text{by conduction in}\\ x\ \text{direction into}\\ \text{volume element}\ \Delta x\end{array}\right)}_{\text{I}} + \underbrace{\left(\begin{array}{l}\text{net rate of heat gain}\\ \text{by convection through}\\ \text{lateral surfaces into}\\ \text{volume element}\ \Delta x\end{array}\right)}_{\text{II}} = 0
$$

The net heat gain by conduction is given by

$$
\text{I} = -\frac{d(qA)}{dx}\,\Delta x \tag{3-73a}
$$

and the net heat gain by convection by

$$
\text{II} = h[T_\infty - T(x)]\,\Delta a \tag{3-73b}
$$

where A = cross-sectional area at the position x
          q = conductive heat flux in the direction at x
        Δa = lateral surface area of volume element Δx
          h = the heat-transfer coefficient
        T_∞ = temperature of surrounding fluid

It is assumed that h and T_∞ are constant, but other quantities may vary with x.

The substitution of Eqs. (3-73) into the above energy equation and letting $\Delta x \rightarrow dx$ yield

$$
-\frac{d}{dx}(qA) + h[T_\infty - T(x)]\frac{da(x)}{dx} = 0 \tag{3-74}
$$

Here the conductive heat flux q is given by

$$
q = -k\frac{dT}{dx} \tag{3-75}
$$

The substitution of Eq. (3-75) into Eq. (3-74) for constant k yields

$$
\frac{d}{dx}\left[A(x)\frac{dT(x)}{dx}\right] + \frac{h}{k}[T_\infty - T(x)]\frac{da(x)}{dx} = 0 \tag{3-76}
$$

An excess temperature $\theta(x)$ is now defined as

$$\theta(x) = T(x) - T_\infty \tag{3-77}$$

Then Eq. (3-76) becomes

$$\frac{d}{dx}\left[A(x)\frac{d\theta(x)}{dx}\right] - \frac{h}{k}\frac{da(x)}{dx}\theta(x) = 0 \tag{3-78}$$

This is the *one-dimensional fin equation for fins of variable cross section*.   Here, $a(x)$ is the lateral surface of the fin between 0 and x, and $A(x)$ is the cross-sectional area of the fin at x.   For fins with variable cross section, the areas $a(x)$ and $A(x)$ should be expressed mathematically in terms of the perimeter and the slope of the fin profile.   For a fin with uniform cross section, the area A and the perimeter P are constant and the lateral area is related to P by $a(x) = Px$.

Fins of Uniform Cross Section

For fins having uniform cross section, as shown in Fig. 3-10, we have

$$A(x) = A = \text{const} \tag{3-79a}$$

$$a(x) = Px \qquad \text{or} \qquad \frac{da(x)}{dx} = P \tag{3-79b}$$

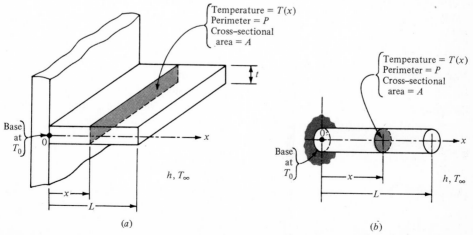

FIG. 3-10   One-dimensional fins of uniform cross section. (*a*) Plate fin; (*b*) pin fin.

where P is the perimeter of the fin, and A is the cross-sectional area. The substitution of the relations in Eqs. (3-79) into Eq. (3-78) results in

$$\frac{d^2\theta(x)}{dx^2} - m^2\theta(x) = 0 \tag{3-80a}$$

where

$$m^2 \equiv \frac{Ph}{Ak} \tag{3-80b}$$

Equation (3-80a) is called the *one-dimensional fin equation for fins of uniform cross section*. The solution of this ordinary differential equation subject to appropriate boundary conditions at the two ends of the fin gives the temperature distribution in the fin. Once the temperature distribution is known, the heat flow through the fin is readily determined. In the following section we examine the solution of Eqs. (3-80) subject to different kinds of boundary conditions at the ends of the fin.

3-9   TEMPERATURE DISTRIBUTION AND HEAT FLOW IN FINS OF UNIFORM CROSS SECTION

We now examine the application of the foregoing fin equation to the heat-transfer analysis of fins having a uniform cross section. Figure 3-10 illustrates a plate fin and a pin fin, each of uniform cross section. The temperature distribution in the fin can be established by solving the differential equation (3-80a) subject to appropriate boundary conditions. Customarily the temperature at the fin base $x = 0$ is considered known, but several different physical situations are possible at the fin tip $x = L$.   In the following analysis we consider the following three different situations at the fin tip: (1) *long fin*, (2) *negligible heat loss from the fin tip*, and (3) *convection at the fin tip*. The heat-transfer problem will be analyzed for each of these cases.

Long Fin

For a sufficiently long fin it is reasonable to assume that the temperature at the fin tip approaches the temperature T_∞ of the surrounding medium and the temperature at the fin base T_0 is considered specified. With these considerations, the mathematical formulation of the fin problem becomes

$$\frac{d^2\theta(x)}{dx^2} - m^2\theta(x) = 0 \qquad \text{in } x > 0 \tag{3-81a}$$

$$\theta(x) = T_0 - T_\infty \equiv \theta_0 \qquad \text{at } x = 0 \tag{3-81b}$$

$$\theta(x) \to 0 \qquad \text{as } x \to \infty \tag{3-81c}$$

where $m^2 \equiv Ph/Ak$.

The solution of Eqs. (3-81) is given as

$$\theta(x) = C_1 e^{-mx} + C_2 e^{mx} \tag{3-82}$$

where the two integration constants C_1 and C_2 are determined by the application of the boundary conditions. The boundary condition (3-81c) requires that $C_2 = 0$; then the application of the boundary condition (3-81b) gives $C_1 = \theta_0$, and the solution becomes

$$\frac{\theta(x)}{\theta_0} = \frac{T(x) - T_\infty}{T_0 - T_\infty} = e^{-mx} \tag{3-83}$$

Now, the temperature distribution being known, the heat flow through the fin is determined either by integrating the convective heat transfer over the entire fin surface according to the relation

$$Q = \int_{x=0}^{\infty} hP\theta(x)\, dx \qquad \text{Btu/h (or W)} \tag{3-84a}$$

or by evaluating the conductive heat flow at the fin base according to the relation

$$Q = -Ak \frac{d\theta(x)}{dx}\bigg|_{x=0} \qquad \text{Btu/h (or W)} \tag{3-84b}$$

Equations (3-84a) and (3-84b) give identical results since heat flow through the lateral surfaces by convection is equal to heat flow at the fin base by conduction. Here we prefer to use the relation given by Eq. (3-84b). Substituting $\theta(x)$ from Eq. (3-83) into Eq. (3-84b), the heat-flow rate through the fin becomes

$$Q = Ak\theta_0 m = \theta_0 \sqrt{PhkA} \qquad \text{Btu/h (or W)} \tag{3-85}$$

since $m = \sqrt{Ph/kA}$.

Fin with Negligible Heat Flow at the Tip

The heat-transfer area at the fin tip is generally small compared with the lateral area of the fin for heat transfer. For such situations the heat loss from the fin tip is negligible compared with that from the lateral surfaces, and the boundary condition at the fin tip characterizing this situation is taken as $d\theta/dx = 0$ at $x = L$. Then the mathematical formulation of the fin problem becomes

$$\frac{d^2\theta(x)}{dx^2} - m^2\theta(x) = 0 \qquad \text{in } 0 \le x \le L \tag{3-86a}$$

$$\theta(x) = T_0 - T_\infty \equiv \theta_0 \qquad \text{at } x = 0 \tag{3-86b}$$

$$\frac{d\theta(x)}{dx} = 0 \qquad \text{at } x = L \tag{3-86c}$$

The solution of differential equation (3-86a) is now taken, for convenience in the analysis, in the form (*see note 1 for a discussion of appropriate solutions*).[1]

$$\theta(x) = C_1 \cosh m(L - x) + C_2 \sinh m(L - x) \tag{3-87}$$

The boundary condition (3-86c) requires that $C_2 = 0$; then the application of the boundary condition (3-86b) gives $C_1 = \theta_0/\cosh mL$, and the complete solution becomes

$$\frac{\theta(x)}{\theta_0} = \frac{T(x) - T_\infty}{T_0 - T_\infty} = \frac{\cosh m(L - x)}{\cosh mL} \tag{3-88}$$

The heat transfer through the fin is determined by substituting the solution given by Eq. (3-88) into Eq. (3-84b); we obtain

$$Q = Ak\theta_0 m \tanh mL = \theta_0 \sqrt{PhkA} \tanh mL \tag{3-89}$$

We note that this result reduces to that given by Eq. (3-85) for the long fins, since $\tanh mL \to 1$ for sufficiently large values of mL (*see note 2*).

Fins with Convection at the Tip

A physically more realistic boundary condition at the fin tip is the one that includes heat transfer by convection between the fin tip and the surrounding fluid. Then the mathematical formulation of the heat-conduction problem becomes

$$\frac{d^2\theta(x)}{dx^2} - m^2\theta(x) = 0 \qquad \text{in } 0 \le x \le L \tag{3-90a}$$

$$\theta(x) = T_0 - T_\infty \equiv \theta_0 \qquad \text{at } x = 0 \tag{3-90b}$$

$$k\frac{d\theta(x)}{dx} + h_e\,\theta(x) = 0 \qquad \text{at } x = L \tag{3-90c}$$

where k is the thermal conductivity of the fin and h_e is the heat-transfer coefficient between the fin tip and the surrounding fluid.

The solution of the differential equation is taken as

$$\theta(x) = C_1 \cosh m(L - x) + C_2 \sinh m(L - x) \tag{3-91}$$

where the integration constants C_1 and C_2 are determined by the application of

---

[1] Notes are at the end of the chapter.

the boundary conditions.   The application of the boundary conditions (3-90b) and (3-90c), respectively, gives

$$\theta_0 = C_1 \cosh mL + C_2 \sinh mL \tag{3-92a}$$

$$-kC_2 m + h_e C_1 = 0 \tag{3-92b}$$

since

$$\frac{d\theta}{dx}\bigg|_{x=L} = \bigg[-C_1 m \sinh m(L-x) - C_2 m \cosh m(L-x)\bigg]_{x=L} = -C_2 m$$

Equations (3-92) provide two relations for the determination of the integration constants C_1 and C_2.   When C_1 and C_2 are evaluated and substituted into Eq. (3-91), the temperature distribution in the fin becomes

$$\frac{\theta(x)}{\theta_0} = \frac{T(x) - T_\infty}{T_0 - T_\infty} = \frac{\cosh m(L-x) + (h_e/mk) \sinh m(L-x)}{\cosh mL + (h_e/mk) \sinh mL} \tag{3-93}$$

For a fin with negligible heat transfer at the fin tip, we set $h_e = 0$; then Eq. (3-93) simplifies to the result given by Eq. (3-88), as expected.   It is to be noted that the heat-transfer coefficient h for the lateral surface which is hidden in the parameter m is not necessarily the same as the heat-transfer coefficient h_e at the fin tip.

3-10   FIN EFFICIENCY

In most applications fins may have variable cross-sectional areas; for such situations the determination of temperature distribution and heat flow through a fin is a complicated matter because the analysis involves the solution of the general fin equation (3-78).   Fortunately, the heat-transfer analysis of fins has been performed [7, 8] for a variety of fin geometries.   The results are presented in terms of a parameter called *fin efficiency* η, which is defined as

$$\eta = \frac{\text{actual heat transfer through fin}}{\begin{array}{l}\text{ideal heat transfer through fin} \\ \text{if entire fin surface were at} \\ \text{fin-base temperature } T_0\end{array}} = \frac{Q_{\text{fin}}}{Q_{\text{ideal}}} \tag{3-94}$$

The ideal heat-transfer rate Q_{ideal} through the fin is given by

$$Q_{\text{ideal}} = a_f h \theta_0 \tag{3-95a}$$

where a_f is the heat-transfer area of fin, and $\theta_0 = T_0 - T_\infty$.

Thus, if the fin efficiency η is known, the heat transfer Q through the fin is determined from

$$Q_{\text{fin}} = \eta Q_{\text{ideal}} = \eta a_f h \theta_0 \tag{3-95b}$$

We now illustrate the determination of fin efficiency for a fin of uniform cross section and negligible heat loss from the fin tip discussed previously. For a fin of length L and perimeter P, the lateral heat-transfer area is taken as $a_f = PL$; then the ideal heat transfer through the fin becomes

$$Q_{\text{ideal}} = PLh\theta_0 \tag{3-96a}$$

and the actual heat transfer through the fin is given by Eq. (3-89) as

$$Q_{\text{fin}} = \theta_0 \sqrt{PhkA} \tanh mL \tag{3-96b}$$

The substitution of Eqs. (3-96) into Eq. (3-94) gives the fin efficiency for this particular type of fin:

$$\eta = \frac{\theta_0 \sqrt{PhkA} \tanh mL}{\theta_0 PLh} = \frac{\tanh mL}{mL} \tag{3-97}$$

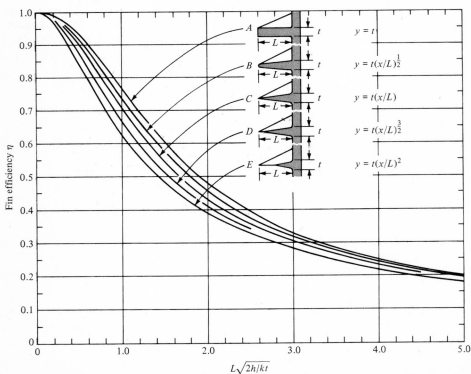

FIG. 3-11   Efficiency of axial fins where the fin thickness y varies with the distance x from the root of the fin where $y = t$. (*From Gardner* [8].)

The analytical work has shown that for fins which are sufficiently long (that is, $L \gg t$) the fin efficiency can be expressed as a function of the parameter $L\sqrt{2h/kt}$; Figs. 3-11 and 3-12 show the fin efficiency plotted against this parameter for typical fin cross sections. Figure 3-11 gives the efficiency of axial fins where the fin thickness y may vary with the distance x from the root of the fin where $y = t$. Figure 3-12 is the efficiency for circular disk fins of constant thickness. Additional charts for tapered circular and pin fins are given in the original reference [8].

In practical applications a finned heat-transfer surface is composed of the fin surfaces and the unfinned portion. Then the total heat transfer Q_{total} from such a surface is obtained by summing up the heat transfer through the fins and the unfinned portion as

$$Q_{\text{total}} = Q_{\text{fin}} + Q_{\text{unfinned}}$$
$$= \eta a_f h\theta_0 + (a - a_f)h\theta_0 \tag{3-98}$$

where a is the total heat-transfer area (i.e., fin surface + unfinned surface) and a_f is the heat-transfer area of the fins.

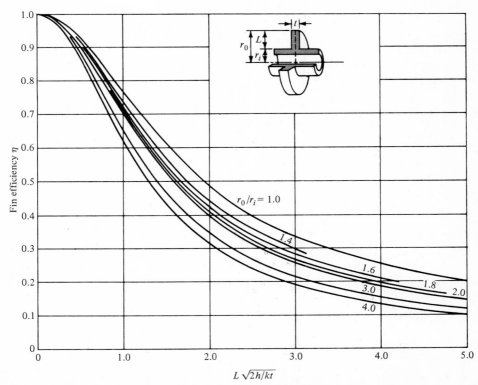

FIG. 3-12 Efficiency of circular disk fins of constant thickness. (*From Gardner [8].*)

Equation (3-98) can be written more compactly as

$$Q_{total} = [\eta\beta + (1 - \beta)]ah\theta_0 \equiv \eta'ah\theta_0 \qquad (3\text{-}99)$$

where

$$\eta' \equiv \beta\eta + (1 - \beta) = \text{area-weighted fin efficiency} \qquad (3\text{-}100a)$$

$$\beta = \frac{a_f}{a} \qquad (3\text{-}100b)$$

We note that Eqs. (3-99) and (3-95b) are of similar form except in the former the *area-weighted fin efficiency* η' is used to take into account the heat transfer through the unfinned portion. For the special case of $\beta = 1$ (i.e., fin surface only) we have $\eta' = \eta$, and Eq. (3-99) simplifies to Eq. (3-95b).

Although the addition of fins on a surface increases the surface area for heat transfer, it also increases the thermal resistance over the portion of the surface where the fins are attached. Therefore, it is expected that there may be situations in which the addition of fins does not improve heat transfer. As a practical guide, the ratio $\dfrac{Pk/A}{h}$ *should be much larger than unity to justify the use of fins.* In the case of plate fins, for example, $P/A \cong 2/t$; then $\dfrac{Pk/A}{h}$ becomes $\dfrac{2(k/t)}{h}$, which implies that internal conductance of the fin should be much greater than the heat-transfer coefficient for the fins to improve the heat-transfer rate.

Example 3-8　Compare the efficiency of a plate fin of length $L = 0.6$ in $(1.524 \times 10^{-2}$ m), thickness $t = 0.1$ in $(0.254 \times 10^{-2}$ m) for the following two cases:
  (a)　Fin material is aluminum $(k = 120$ Btu/h·ft·°F or 207.64 W/m·°C) and the heat-transfer coefficient $h = 50$ Btu/h·ft²·°F or 283.9 W/m²·°C.
  (b)　Fin material is steel $(k = 24$ Btu/h·ft·°F or 41.5 W/m·°C) and $h = 90$ Btu/h·ft²·°F or 510.9 W/m²·°C.

Solution　The fin efficiency for a plate fin with negligible heat loss from the fin tip is given by Eq. (3-97):

$$\eta = \frac{\tanh mL}{mL}$$

Then, for the two cases considered above, η is determined as follows:

  (a)　$mL = L\sqrt{\dfrac{2h}{kt}}$ since $m = \sqrt{\dfrac{Ph}{Ak}}$ and $\dfrac{P}{A} \cong \dfrac{2}{t}$

  In the English system of units, we have

$$mL = \frac{0.6}{12}\sqrt{\frac{2 \times 50}{120 \times 0.1/12}} = 0.5$$

  In the SI system of units, we have

$$mL = 1.524 \times 10^{-2}\sqrt{\frac{2 \times 283.9}{207.64 \times (0.254 \times 10^{-2})}} = 0.5$$

that is, we find the same value for mL in both systems of units since mL is a dimensionless quantity.   Then,

$$\eta = \frac{\tanh 0.5}{0.5} = \frac{0.462}{0.5} = 0.924$$

(b) $$mL = \frac{0.6}{12}\sqrt{\frac{2 \times 90}{24 \times 0.1/12}} = \frac{0.6 \times 30}{12} = 1.5$$

$$\eta = \frac{\tanh 1.5}{1.5} = \frac{0.905}{1.5} \cong 0.6$$

Example 3-9   Longitudinal thin fins are attached on the outer surface of a tube of inside radius r_a, outside radius r_b, and length H.   The hot and cold fluids flowing inside and outside the tube have mean temperatures T_i and T_o, and heat-transfer coefficients h_i and h_o, respectively.   The total heat-transfer area on the outer surface of the tube, including the surface areas of the fins and the unfinned portion of the tube, is a ft^2 and the ratio of the fin surface area a_f to the total heat-transfer area a is β. The fin efficiency η and the thermal conductivity k of the tube material are given.
(a)   Derive a relation for the heat-transfer rate Q_f through the finned tube.
(b)   Compare this heat-transfer rate Q_f with the heat-transfer rate Q_0 for the case with no fins on the tube surface.

Solution   The thermal-resistance concept can be used to determine the heat-transfer rate through the finned tube.   Figure 3-13 shows various thermal resistances in the path of heat flow between the inside and outside fluid temperatures T_i and T_o.   They include the thermal resistance R_i of inside flow, the thermal resistance R_t of the tube material, and the thermal resistance R_{of} of the outside flow including the effect of fins.   Then:

(a)   For the finned tube of length H the total heat-transfer rate Q_f through the tube is determined from

$$Q_f = \frac{T_i - T_o}{R_i + R_t + R_{of}} \qquad \text{Btu/h (or W)} \tag{3-101}$$

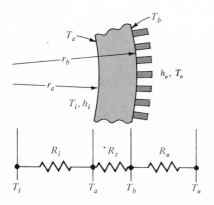

FIG. 3-13   Thermal-resistance concept for heat transfer through a finned tube.

where various thermal resistances are given as

$$R_i = \frac{1}{2\pi r_a H h_i} = \text{thermal resistance of inside flow}$$

$$R_t = \frac{1}{2\pi H k} \ln \frac{r_b}{r_a} = \text{thermal resistance of tube material}$$

$$R_{of} = \frac{1}{\eta' a h_o} = \text{thermal resistance of outside flow including effects of fins}$$
$$(\textit{see note 3 for the derivation})$$

and the area-weighted fin efficiency η' is given as

$$\eta' = \beta\eta + (1 - \beta)$$

where β and η are specified.

(b) When there are no fins at the outer surface we have $\beta = 0$. Then $\eta' = 1$ and $a = a_0 = 2\pi r_b H$. The heat-transfer rate Q_0 through the tube with no fins on the outer surface is determined from

$$Q_0 = \frac{T_i - T_o}{R_i + R_t + R_o} \qquad \text{Btu/h (or W)} \tag{3-102}$$

where R_i and R_t are as given above and R_o is the thermal resistance of the outer flow given by

$$R_o = \frac{1}{2\pi r_b H h_o}$$

The ratio of heat-transfer rate with fins to that without fins becomes

$$\frac{Q_f}{Q_0} = \frac{R_i + R_t + R_o}{R_i + R_t + R_{of}} \tag{3-103}$$

and the various thermal resistances have been defined previously.

3-11 CRITICAL RADIUS OF INSULATION

There are many practical situations in which the addition of insulation to the outside surface of a pipe does not reduce the heat loss. To establish the conditions under which this occurs, we consider an insulating layer in the form of a hollow cylinder of length H. The inside surface of the cylinder at $r = r_i$ is kept at a uniform temperature T_i while the outside surface at $r = r_o$ dissipates heat by convection into an environment at temperature T_∞ with a heat-transfer coefficient h_0. The rate of heat transfer Q through this insulation layer is determined as [see Eqs. (3-56)]

$$Q = \frac{T_i - T_\infty}{R_{\text{ins}} + R_o} \tag{3-104a}$$

where the thermal resistances R_{ins} and R_o of the insulation and of the outside surface are defined as

$$R_{\text{ins}} = \frac{1}{2\pi Hk} \ln \frac{r_o}{r_i} \quad \text{and} \quad R_o = \frac{1}{2\pi r_o H h_0} \tag{3-104b}$$

Here k is the thermal conductivity of the insulating material.

If it is assumed T_i, T_∞, k, L, h_0, and r_i remain constant while r_o varies, we note that as r_o increases the term R_o decreases but R_{ins} increases. Therefore, it is expected that Q may have a maximum for a certain value of $r_o = r_{oc}$. This *critical value of the radius* r_{oc} is determined by differentiating Eqs. (3-104) with respect to r_o and setting the resulting expression equal to zero:

$$\frac{dQ}{dr_o} = -\frac{2\pi kH(T_i - T_\infty)}{[\ln(r_o/r_i) + k/h_0 r_o]^2} \left(\frac{1}{r_o} - \frac{k}{h_0 r_o^2} \right) = 0 \tag{3-105}$$

The solution of Eq. (3-105) for r_o gives the *critical radius* r_{oc} of insulation at which the heat-transfer rate is a maximum; we find

$$r_{oc} = \frac{k}{h_0} \tag{3-106}$$

In practice the physical significance of this result is as follows: If insulation is to be added on a pipe or wire whose outer radius is less than the critical radius of the insulation and the outside surface is kept at a uniform temperature, the heat loss from the pipe will increase as the insulation is added until the outside radius of the insulation becomes equal to the critical radius r_{oc}; as the insulation thickness is increased beyond r_{oc} the heat loss from the pipe will begin to decrease.

Example 3-10   A 1-in (2.54×10^{-2}-m) OD steam pipe is to be covered with an asbestos insulation ($k = 0.12$ Btu/h·ft·°F or 0.208 W/m·°C). If the heat-transfer coefficient to the surrounding air is $h_0 = 1.0$ Btu/h·ft^2·°F (or 5.68 W/m^2·°C), examine the effect of the insulation thickness on the heat loss from the pipe.

Solution   The critical radius of the insulation is given by Eq. (3-106):

$$r_{oc} = \frac{k}{h_0} = \frac{0.12}{1.0} = 0.12 \text{ ft} = 1.44 \text{ in } (3.66 \times 10^{-2} \text{ m})$$

Since the outer radius of the pipe $r_i = 0.5$ in is less than the critical radius $r_{oc} = 1.44$ in, the addition of insulation will increase the heat loss until the outside radius of the insulation equals 1.44 in; further addition of insulation beyond $r_{oc} = 1.44$ in will result in a decrease in heat loss. The heat loss from the bare and insulated pipe is given, respectively, as

$$Q_{\text{bare}} = 2\pi r_i H h_0 \, \Delta T$$

$$Q_{\text{insulated}} = 2\pi r_o H h_0 \frac{\Delta T}{1 + (r_o h_0/k) \ln r_o/r_i}$$

and their ratio becomes

$$\frac{Q_{bare}}{Q_{insulated}} = \frac{r_i}{r_o}\left(1 + \frac{r_o h_o}{k} \ln \frac{r_o}{r_i}\right)$$

For an insulation layer with $r_o = r_{oc}$, this relation reduces to

$$\frac{Q_{bare}}{Q_{insulated}} = \frac{r_i}{r_{oc}}\left(1 + \ln \frac{r_{oc}}{r_i}\right)$$

and, setting $r_i = 0.5$ in and $r_{oc} = 1.44$ in, we find

$$\frac{Q_{bare}}{Q_{insulated}} = \frac{0.5}{1.44}(1 + \ln 2.88) = \frac{1.029}{1.44} = 0.715$$

That is, the addition of insulation until the outside radius of insulation equals $r_o = r_{oc} = 1.44$ results in increased the heat loss for this particular case.

Effects of Radiation

The result given above for the critical radius of insulation does not include the effects of thermal radiation. If the radiation effects can be linearized so that the heat-transfer coefficient h_o at the outer surface is approximated by the sum of the convective component h_c and the radiative component h_r in the form

$$h_o = h_c + h_r \tag{3-107a}$$

where

$$h_r \cong 4F\sigma T_0{}^3 \qquad \text{[See Eq. (1-6)]} \tag{3-107b}$$

then the foregoing result for the critical radius r_{oc} is given by

$$r_{oc} = \frac{k}{h_o} = \frac{k}{h_c + h_r} = \frac{k}{h_c + 4F\sigma T_0{}^3} \tag{3-108}$$

Clearly, the effect of including the thermal radiation is to reduce the resulting critical radius.

REFERENCES

1   Çetinkale, T. N., and M. Fishenden: Thermal Conductance of Metal Surfaces in Contact, General Discussion on Heat Transfer, Conference of Institution of Mechanical Engineers (London) and ASME, pp. 271–275, 1951.

2   Barzelay, M. E., K. N. Tong, and G. F. Holloway: Effect of Pressure on Thermal Conductance of Contact Joints, *NACA Tech. Note* 3295, May 1955.

3   Clausing, A. M.: Heat Transfer at the Interface of Dissimilar Metals—The Influence of Thermal Strain, *Int. J. Heat Mass Transfer,* **9**: 791–801 (1966).

4 Veziroğlu, T. N., and S. Chandra: Direction Effect in Thermal Contact Conductance, *Fourth Int. Heat Transfer Conf., Paris*, **1:** Cu 3.5 (1970).

5 Moore, C. J., Jr., H. A. Blum, and H. Atkins: Subject Classification Bibliography for Thermal Contact Resistance Studies, *ASME Paper 68-WA/HT*-18, December 1968.

6 Cooper, M. G., B. B. Mikic, and M. M. Yovanovich: Thermal Contact Conductances, *Int. J. Heat Mass Transfer*, **12:** 279–300 (1969).

7 Harper, W. P., and D. R. Brown: Mathematical Equations for Heat Conduction in the Fins of Air-cooled Engines, *NACA Rep.* 158, 1922.

8 Gardner, K. A.: Efficiency of Extended Surfaces, *Trans. ASME*, **67:** 621–631 (1945).

NOTES

1 The solution of Eq. (3-86a) can be taken in several different forms, such as

$$\theta(x) = C_2 e^{-mx} + C_3 e^{mx}$$

or

$$\theta(x) = C_4 \cosh mx + C_5 \sinh mx$$

or

$$\theta(x) = C_1 \cosh m(L - x) + C_2 \sinh m(L - x)$$

The reason we have chosen the form given by Eq. (3-87) is that it is easier to determine the integration constants for the type of boundary conditions considered in Eqs. (3-86).

2 Typical values of $\tanh mL$ as a function of mL are

mL	1	1.5	2	3	4
$\tanh mL$	0.7616	0.905	0.964	0.995	0.999

3 The thermal resistance R_{of} of outside flow including the effects of fins is obtained by the following consideration.

The total heat-transfer rate from the outer finned surface, by Eq. (3-99), is

$$Q_f = \eta' a h_0 (T_b - T_o) = \frac{T_b - T_o}{1/\eta' a h_0} \equiv \frac{T_b - T_o}{R_{of}}$$

where

$$R_{of} \equiv \frac{1}{\eta' a h_0}$$

and T_b is the temperature of the outer surface of the tube or at the fin root.

Four

Conduction—
Two-dimensional Steady and
One-dimensional Unsteady States

In the previous chapters we discussed the method of solution of heat-conduction problems in which the temperature was a function of a single space variable. There are many applications in which temperature is a function of time and/or more than one space variable.   When the boundaries of the region are regular, that is, when the boundary surfaces coincide with the coordinate surfaces of an orthogonal coordinate system, the heat-conduction problem can be solved analytically. Various mathematical techniques, such as the *separation of variables, Laplace transform, the integral transform, complex variable, variational method*, and so on, are available for analytical solution.   A discussion of such methods and their application to the solution of multidimensional, time-dependent heat-conduction problems can be found in several references [1–4].

The method of *separation of variables*, when applicable, is the most straightforward approach to solve heat-conduction problems in more than one variable. In this chapter we discuss the method of separation of variables in the solution of the heat-conduction equation in more than one variable, with emphasis on problems in the rectangular coordinate system involving (1) two-dimensional steady-state heat conduction in a rectangular region and (2) one-dimensional, time-dependent heat conduction in a slab.   Basic to the solution of a partial differential equation with the separation of variables is the *orthogonal expansion technique*.   Therefore, a brief discussion of *orthogonal functions* arising from the separation of the heat-conduction problem is presented.   Although the method of separation of variables is straightforward in principle, its application to specific problems requires a considerable amount of tedious analysis in the determination of appropriate elementary solutions for a given set of boundary conditions.   To circumvent this difficulty *a unified approach is taken for the application of the method of separation of variables*.   That is, in the first section of this chapter the auxiliary problem, called *the eigenvalue problem*, arising from the solution of the heat-conduction problem by the separation of variables is presented; the *orthogonality property* of its solutions is discussed; and the solutions of this auxiliary problem for all possible nine combinations of boundary conditions for a slab geometry are systematically tabulated for use in the subsequent applications. By using these tabulated solutions, the analysis of two-dimensional steady and the one-dimensional unsteady heat conduction in a slab geometry to be considered in this chapter becomes a straightforward matter.

Finally, an approximate method of analysis of time-dependent heat-conduction problems by the method called *lumped-system analysis* is described.   This method is applicable when the internal specific thermal resistance of the solid is small compared with the specific thermal resistance for convection at the boundaries of the solid.   For such situations the temperature distribution in the solid is substantially uniform at any instant; hence the temperature is assumed to be a function of the time variable only and the lumped-system analysis becomes applicable.

4-1 ORTHOGONAL FUNCTIONS

In the analysis of heat-conduction problems by the method of separation of variables, the separation process gives rise to a set of auxiliary differential equations.   One of these auxiliary problems is called the *eigenvalue problem*, and its solutions are called the *eigenfunctions*.   The complete solution of the heat-conduction problem is then constructed by taking a linear sum of all the appropriate elementary solutions of the auxiliary problems.   The unknown expansion coefficients associated with this summation are determined by constraining this solution to meet the *nonhomogeneous boundary condition* (or the initial condition) for the original problem.   The *orthogonality* property of the eigenfunctions plays an important role in the determination of these unknown expansion coefficients.   The subject of orthogonality of functions was investigated originally by Sturm and Liouville in 1836, and it is for this reason the eigenvalue problems are sometimes called the *Sturm-Liouville problems*.   In this section we give a brief discussion of the eigenvalue problem and the orthogonality property of the eigenfunctions and present a systematic tabulation of eigenfunctions for the nine different combinations of boundary conditions in the rectangular coordinate system.   The reader should consult Refs. [1–5] for a discussion of the determination of eigenfunctions in the cylindrical and spherical coordinate systems.

Consider a second-degree ordinary differential equation for a function $\Psi(x)$ in the region $0 \le x \le L$ given in the form

$$\frac{d^2\Psi(x)}{dx^2} + \lambda^2\Psi(x) = 0 \qquad \text{in } 0 \le x \le L \tag{4-1a}$$

subject to the boundary conditions

$$-k\frac{d\Psi(x)}{dx} + h\Psi(x) = 0 \qquad \text{at } x = 0 \tag{4-1b}$$

$$k\frac{d\Psi(x)}{dx} + h\Psi(x) = 0 \qquad \text{at } x = L \tag{4-1c}$$

where λ, h, and k are constants.   The problem defined by Eqs. (4-1) is called an *eigenvalue problem*.   We note that the boundary conditions at $x = 0$ and $x = L$ are *homogeneous boundary conditions of the third kind;* in addition, the differential equation is also homogeneous.   Then, the problem defined by Eqs. (4-1) has a solution only for certain values of the parameter $\lambda = \lambda_n$, $n = 1, 2, 3, \ldots$, where λ_n's are called the *eigenvalues* (or the *characteristic numbers*), and has trivial solutions (that is, $\Psi = 0$) when λ is not an eigenvalue.   The nontrivial solutions $\Psi(\lambda_n, x)$ are called the *eigenfunctions*.   If $\Psi(\lambda_m, x)$ and $\Psi(\lambda_n, x)$ denote the two different eigenfunctions corresponding to the eigenvalues λ_m and λ_n, respectively, the *orthogonality* property of the eigenfunctions in the region $0 \leq x \leq L$ can be stated by the following relation:

$$\int_0^L \Psi(\lambda_m, x)\Psi(\lambda_n, x)\, dx = \begin{cases} 0 & \text{for } \lambda_m \neq \lambda_n \\ N & \text{for } \lambda_m = \lambda_n \end{cases} \tag{4-2a}$$

where N is called the *normalization integral* defined as

$$\int_0^L \Psi^2(\lambda_m, x)\, dx = N \tag{4-2b}$$

The reader should consult any standard text on mathematics [1] for the proof of the above orthogonality relation.

We note that in the above eigenvalue problem we have chosen, for generality, homogeneous boundary conditions of the third kind at the boundaries $x = 0$ and $x = L$. Such eigenvalue problems arise in the solution of heat-conduction problems in the rectangular coordinate system with boundary conditions of the third kind at $x = 0$ and $x = L$. If the heat-conduction problem is subjected to boundary conditions, say, of the second kind at $x = 0$ and of the first kind at $x = L$, then the boundary conditions for the above eigenvalue problem could be a homogeneous boundary condition of the second kind at $x = 0$ and of the first kind at $x = L$, and so forth. Clearly, nine different combinations of such boundary conditions are possible for a finite region $0 \leq x \leq L$, and any of these combinations can readily be constructed from the general boundary conditions (4-1b) and (4-1c). For each of these nine combinations of boundary conditions, there are the corresponding eigenfunctions, eigenvalues, and normalization integrals; we now illustrate their determination for simple cases with the following examples. Once the physical significance of eigenfunctions, eigenvalues, and normalization integrals is understood, we shall present them in tabular form for ready reference in the analysis of heat-conduction problems in the rectangular coordinate system.

Example 4-1　Determine the eigenvalues, the eigenfunctions, and the normalization integral for the following eigenvalue problem:

$$\frac{d^2\Psi(x)}{dx^2} + \lambda^2\Psi(x) = 0 \qquad \text{in } 0 \leq x \leq L \tag{4-3a}$$

$$\Psi(x) = 0 \qquad \text{at } x = 0 \tag{4-3b}$$

$$\Psi(x) = 0 \qquad \text{at } x = L \tag{4-3c}$$

Solution　The solution of the differential equation (4-3a) is taken in the form

$$\Psi(x) = C_1 \sin \lambda x + C_2 \cos \lambda x \tag{4-4}$$

The boundary condition at $x = 0$ requires that $C_2 = 0$. Then the solution becomes

$$\Psi(x) = C_1 \sin \lambda x \tag{4-5}$$

If this solution should also satisfy the boundary condition at $x = L$, for $C_1 \neq 0$ we should have

$$\sin \lambda L = 0 \qquad \text{or} \qquad \lambda_n = n\pi \qquad \text{where } n = 1, 2, 3, \ldots \tag{4-6}$$

We now summarize our results. The *eigenfunctions* $\Psi(\lambda_n, x)$ of the eigenvalue problem of Eqs. (4-3) are (omitting the constant C_1 which is not needed)

$$\Psi(\lambda_n, x) = \sin \lambda_n x \tag{4-7a}$$

where the *eigenvalues* λ_n are the roots of

$$\sin \lambda L = 0 \tag{4-7b}$$

or they are given by

$$\lambda_n = \frac{n\pi}{L} \qquad n = 1, 2, 3, \ldots \tag{4-7c}$$

The *normalization integral* N becomes

$$N = \int_0^L \sin^2 \lambda_n x \, dx = \int_0^L \left(\sin \frac{n\pi}{L} x \right)^2 dx = \frac{L}{2} \tag{4-7d}$$

The *orthogonality property* of these eigenfunctions is written

$$\int_0^L \sin \lambda_n x \sin \lambda_m x \, dx = \begin{cases} 0 & \text{for } \lambda_n \neq \lambda_m \\ \dfrac{L}{2} & \text{for } \lambda_n = \lambda_m \end{cases} \tag{4-8}$$

where $\lambda_n = n\pi/L$, $n = 1, 2, 3, \ldots$.

Example 4-2 Determine the eigenvalues, the eigenfunctions, and the normalization integral for the following eigenvalue problem:

$$\frac{d^2 \Psi(x)}{dx^2} + \lambda^2 \Psi(x) = 0 \qquad \text{in } 0 \leq x \leq L \tag{4-9a}$$

$$\frac{d\Psi(x)}{dx} = 0 \qquad \text{at } x = 0 \tag{4-9b}$$

$$\Psi(x) = 0 \qquad \text{at } x = L \tag{4-9c}$$

Solution The solution of the differential equation (4-9a) can be taken in the form

$$\Psi(x) = C_1 \sin \lambda x + C_2 \cos \lambda x \tag{4-10a}$$

and its derivative becomes

$$\frac{d\Psi(x)}{dx} = C_1 \lambda \cos \lambda x - C_2 \lambda \sin \lambda x \tag{4-10b}$$

The boundary condition at $x = 0$ requires that $C_1 = 0$; then the solution takes the form

$$\Psi(x) = C_2 \cos \lambda x \tag{4-11}$$

If this solution should satisfy the boundary condition at $x = L$ (for $C_2 \neq 0$) we should have

$$\cos \lambda L = 0 \qquad \text{or} \qquad \lambda_n L = (2n + 1)\frac{\pi}{2} \qquad \text{where } n = 0, 1, 2, 3, \ldots \qquad (4\text{-}12)$$

We now summarize the results.   The *eigenfunctions* of the eigenvalue problem of Eqs. (4-9) are

$$\Psi(\lambda_n, x) = \cos \lambda_n x \qquad (4\text{-}13a)$$

The *eigenvalues* λ_n are the roots of

$$\cos \lambda_n L = 0 \qquad (4\text{-}13b)$$

or they are given by

$$\lambda_n = \frac{(2n + 1)\pi}{2L} \qquad \text{where } n = 0, 1, 2, 3, \ldots \qquad (4\text{-}13c)$$

The *normalization integral N* becomes

$$N = \int_0^L \cos^2 \lambda_n x \, dx = \int_0^L \left[\cos \frac{(2n + 1)\pi}{2L} \right]^2 dx = \left| \frac{L}{2} \right. \qquad \text{for } n = 0, 1, 2, 3, \ldots \qquad (4\text{-}14)$$

The *orthogonality relation* for these eigenfunctions is given as

$$\int_0^L \cos \lambda_n x \cos \lambda_m x \, dx = \begin{cases} 0 & \text{for } \lambda_n \neq \lambda_m \\ N & \text{for } \lambda_n = \lambda_m \end{cases} \qquad (4\text{-}15)$$

It will be apparent in later sections that the eigenvalue problem given by Eqs. (4-9) is needed for the solution of the heat-conduction equation by the separation of variables when the boundary condition at $x = 0$ is of the second kind and at $x = L$, of the first kind.

Example 4-3   Determine the eigenvalues and the eigenfunctions of the following eigenvalue problem:

$$\frac{d^2\Psi(x)}{dx^2} + \lambda^2\Psi(x) = 0 \qquad \text{in } 0 \leq x \leq L \qquad (4\text{-}16a)$$

$$\frac{d\Psi(x)}{dx} = 0 \qquad \text{at } x = 0 \qquad (4\text{-}16b)$$

$$k \frac{d\Psi(x)}{dx} + h\Psi(x) = 0 \qquad \text{at } x = L \qquad (4\text{-}16c)$$

Solution   The solution of differential equation (4-16a) satisfying the boundary condition at $x = 0$ is taken as (see the previous problem)

$$\Psi(x) = \cos \lambda x \qquad (4\text{-}17)$$

If this solution should also satisfy the boundary condition at $x = L$ we should have

$$\left[k \frac{d}{dx} (\cos \lambda x) + h \cos \lambda x \right]_{x = L} = 0$$

or

$$-k\lambda \sin \lambda L + h \cos \lambda L = 0$$

or

$$\lambda L \tan \lambda L = \frac{hL}{k} \tag{4-18}$$

We now summarize these results.   The *eigenfunctions* of the eigenvalue problem of Eqs. (4-16) are

$$\Psi(\lambda_n, x) = \cos \lambda_n x \tag{4-19}$$

The *eigenvalues* λ_n are the positive roots of the following transcendental equation:

$$\lambda L \tan \lambda L = \frac{hL}{k} \tag{4-20}$$

The *normalization integral* N is given by

$$N = \int_0^L (\cos \lambda_n x)^2 \, dx = \frac{L(\lambda_n{}^2 + H^2) + H}{2(\lambda_n{}^2 + H^2)} \tag{4-21}$$

where $H \equiv h/k$.   The reader should consult Refs. [2,3] for the derivation of the result of this normalization integral.   The *orthogonality property* of these eigenfunctions is given as

$$\int_0^L \cos \lambda_n x \cos \lambda_m x \, dx = \begin{cases} 0 & \text{for } \lambda_n \neq \lambda_m \\ N & \text{for } \lambda_n = \lambda_m \end{cases} \tag{4-22}$$

where λ_n's are the roots of the transcendental equation (4-20) and the normalization integral N is as given above.

We note that for the eigenvalue problem considered here the eigenvalues λ_n are not given explicitly, but they should be determined from the solution of the transcendental equation (4-20).   This equation can be solved numerically and the values of λ_n can be determined for any given values of h, L, and k.   The physical significance of the magnitudes of these eigenvalues is better envisioned if a geometrical interpretation is given for the solution of Eq. (4-20).   For this purpose we write Eq. (4-20) in the form

$$\frac{\lambda L}{\text{Bi}} = \cot \lambda L \equiv Z \tag{4-23}$$

where $\text{Bi} \equiv hL/k$ is the Biot number.   Figure 4-1 shows a plot of the curves

$$Z = \cot \lambda L \quad \text{and} \quad Z = \frac{\lambda L}{\text{Bi}} \tag{4-24}$$

as a function of the parameter λL.   The former equation gives a set of cotangent curves, and the latter results in a straight line passing through the origin.   Clearly, according to Eq. (4-23), the values of $\lambda_n L$ corresponding to the points of intersection of the straight line with the cotangent curves give the eigenvalues λ_n, $n = 1, 2, 3, \ldots$.
The transcendental equation (4-20) can be written more compactly in the form

$$\beta \tan \beta = c \tag{4-25}$$

where $\beta \equiv \lambda L$ and $c \equiv hL/k$.

TABLE 4-1 Eigenfunctions $\psi(\lambda, x)$, Eigenvalues λ, and the Normalization Integral N of the Eigenvalue Problem:

$$\frac{d^2\psi(x)}{dx^2} + \lambda^2\psi(x) = 0 \quad \text{in } 0 \leq x \leq L$$

subject to boundary conditions at $x = 0$ and $x = L$ as shown in the table

	Boundary Condition at $x = 0$	Boundary Condition at $x = L$	Eigenfunction† $\psi(\lambda, x)$	Eigenvalues† λ's Are Positive Roots of	$\dfrac{1}{N} = \dfrac{1}{\int_0^L \psi^2(\lambda, x)\,dx}$†
1	$\psi = 0$	$\psi = 0$	$\sin \lambda x$	$\sin \lambda L = 0$	$\dfrac{2}{L}$
2	$\psi = 0$	$\dfrac{d\psi}{dx} = 0$	$\sin \lambda x$	$\cos \lambda L = 0$	$\dfrac{2}{L}$
3	$\psi = 0$	$k\dfrac{d\psi}{dx} + h\psi = 0$	$\sin \lambda x$	$\lambda \cot \lambda L = -H$	$\dfrac{2(\lambda^2 + H^2)}{L(\lambda^2 + H^2) + H}$

	BC at $x=0$	BC at $x=L$	Eigenfunction	Eigenvalue equation	Normalization
4	$\dfrac{d\psi}{dx} = 0$	$\psi = 0$	$\cos \lambda x$	$\cos \lambda L = 0$	$\dfrac{2}{L}$
5	$\dfrac{d\psi}{dx} = 0$	$\dfrac{d\psi}{dx} = 0$	$\cos \lambda x$	$\sin \lambda L = 0$	$\begin{cases}\dfrac{2}{L} & \text{for } \lambda \neq 0 \\[4pt] \dfrac{1}{L} & \text{for } \lambda = 0\text{[‡‡]}\end{cases}$
6	$\dfrac{d\psi}{dx} = 0$	$k\dfrac{d\psi}{dx} + h\psi = 0$	$\cos \lambda x$	$\lambda \tan \lambda L = H$	$\dfrac{2(\lambda^2 + H^2)}{L(\lambda^2 + H^2) + H}$
7	$-k\dfrac{d\psi}{dx} + h\psi = 0$	$\psi = 0$	$\sin \lambda(L - x)$	$\lambda \cot \lambda L = -H$	$\dfrac{2(\lambda^2 + H^2)}{L(\lambda^2 + H^2) + H}$
8	$-k\dfrac{d\psi}{dx} + h\psi = 0$	$\dfrac{d\psi}{dx} = 0$	$\cos \lambda(L - x)$	$\lambda \tan \lambda L = H$	$\dfrac{2(\lambda^2 + H^2)}{L(\lambda^2 + H^2) + H}$
9	$-k\dfrac{d\psi}{dx} + h\psi = 0$	$k\dfrac{d\psi}{dx} + h\psi = 0$	$\lambda \cos \lambda x + H \sin \lambda x$	$\tan \lambda L = \dfrac{2\lambda H}{\lambda^2 - H^2}$	$\dfrac{2}{L(\lambda^2 + H^2) + 2H}$

† H is defined as $H = h/k$.

‡‡ $\lambda = 0$ is an eigenvalue for this case.

We list in Table 4-1 various eigenvalues, eigenfunctions, and the normalization integrals as discussed below.

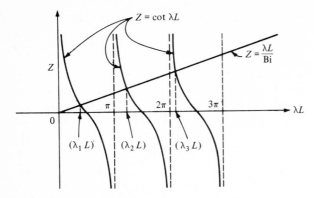

FIG. 4-1   Graphical solution of the equation $\cot \lambda L = \lambda L / \text{Bi}$.

A tabulation of the roots of this transcendental equation as well as the transcendental equation $\beta \cot \beta = -c$ is given in Appendix B. The reader should consult Ref. [5] for iterative solutions of transcendental equations.

Summary of Eigenfunctions and Eigenvalues

The foregoing examples illustrate that the solution of the differential equation

$$\frac{d^2\psi(x)}{dx^2} + \lambda^2\psi(x) = 0 \qquad \text{in } 0 \leq x \leq L \tag{4-26}$$

depends on the type of boundary conditions chosen at the boundaries $x = 0$ and $x = L$. Since the boundary condition may be of the first, second, or third kind at any one of these boundaries, nine different combinations are possible. For convenience in the subsequent applications of the method of separation of variables, we list in Table 4-1 the eigenfunctions, the eigenvalue relations, and the normalization integrals systematically for the nine different combinations of boundary conditions. In this table we included the reciprocal of the normalization integral $1/N$ rather than N because $1/N$ will be needed in the analysis. For example, case 1 in this table corresponds to the eigenvalue problem considered in Example 4-1; that is, the boundary conditions are both of the first kind at $x = 0$ and $x = L$. Then, the eigenfunctions are given as $\psi(\lambda_n, x) = \sin \lambda_n x$; the eigenvalues are the positive roots of $\sin \lambda L = 0$, or they are given by $\lambda_n = n\pi/L$, $n = 1, 2, 3, \ldots$; and the reciprocal of the normalization integral is given as $1/N = 2/L$. The eigenvalue problems considered in Examples 4-2 and 4-3 are listed in Table 4-1 as cases 4 and 6 respectively. This table will be referred to frequently in the subsequent sections on the solution of heat-conduction problems by the method of separation of variables.

4-2 STEADY-STATE HEAT CONDUCTION IN A RECTANGULAR REGION

In this section the method of separation of variables is applied to the solution of the steady-state heat-conduction equation in a rectangular region in which there is no heat generation, the thermal conductivity is constant, and *only one of the boundary conditions is nonhomogeneous* while the other three boundary conditions are homogeneous. Any combination of boundary conditions of the first, second, or third kind may be considered at the boundary surfaces except the case when all boundary conditions are of the second kind. The latter case requires special attention, hence is excluded from the analysis. *When the problem involves more than one nonhomogeneous boundary condition, it can be separated into simpler problems, each involving only one nonhomogeneous boundary condition, as discussed in the next section.*

Consider the steady-state heat conduction in a rectangular region $0 \le x \le a$, $0 \le y \le b$, subject to the boundary conditions as shown in Fig. 4-2. We note that only one of these boundary conditions, the one at $y = b$, is nonhomogeneous, and the remaining three are homogeneous. The physical significance of these boundary conditions is that the surfaces at $x = 0$, $x = a$, and $y = 0$ are maintained at zero temperature, while the boundary surface at $y = b$ is subjected to a prescribed temperature distribution $f(x)$. If it is assumed that the thermal conductivity is constant and there is no heat generation in the medium, the mathematical formulation of this heat-conduction problem is

$$\frac{\partial^2 T(x, y)}{\partial x^2} + \frac{\partial^2 T(x, y)}{\partial y^2} = 0 \qquad \text{in } 0 \le x \le a, 0 \le y \le b \tag{4-27}$$

$$T(x, y) = 0 \qquad \text{at } x = 0 \tag{4-28a}$$

$$T(x, y) = 0 \qquad \text{at } x = a \tag{4-28b}$$

$$T(x, y) = 0 \qquad \text{at } y = 0 \tag{4-28c}$$

$$T(x, y) = f(x) \qquad \text{at } y = b \tag{4-28d}$$

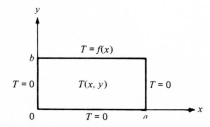

FIG. 4-2   Steady-state heat conduction in a rectangular region.

This heat-conduction problem is solved by the *method of separation of variables* as now described. It is to be noted that the method of separation of variables is applied in this section to the solution of the heat-conduction equation for a rectangle, with no heat generation in the medium and with only one of the boundary conditions being nonhomogeneous. In the next section we shall discuss the cases involving more than one nonhomogeneous boundary condition. The reader should consult Refs. [2–4] for its application to problems in the cylindrical and spherical geometries and to three-dimensional problems.

It is assumed that the temperature $T(x, y)$ can be represented as a product of two functions in the form

$$T(x, y) = X(x)Y(y) \tag{4-29}$$

where $X(x)$ is a function of x only and $Y(y)$ a function of y only. Substitution of Eq. (4-29) into Eq. (4-27) yields

$$\frac{1}{X}\frac{d^2X}{dx^2} = -\frac{1}{Y}\frac{d^2Y}{dy^2} \tag{4-30}$$

where $X \equiv X(x)$ and $Y \equiv Y(y)$. The left-hand side of Eq. (4-30) is a function of x only, and the right-hand side is a function of y only. This equality is possible only if both sides are equal to the same constant, say $-\lambda^2$. Then Eq. (4-30) becomes

$$\frac{1}{X}\frac{d^2X}{dx^2} = -\frac{1}{Y}\frac{d^2Y}{dy^2} = -\lambda^2 \tag{4-31}$$

where the sign of λ^2 is chosen to be negative with the following considerations. We look back to the boundary conditions of the above heat-conduction equation and note that the nonhomogeneous boundary-condition function $f(x)$, that is, Eq. (4-28d), is defined in the interval $0 \le x \le a$. Later in the analysis it will be necessary to represent this function in the interval $0 \le x \le a$ in terms of the eigenfunctions involving the x variable; hence the $X(\lambda, x)$ functions must be the eigenfunctions. To obtain an eigenvalue equation for $X(\lambda, x)$ in the form given by Eq. (4-1a), the sign of λ^2 must be chosen negative. Then the following eigenvalue problem is obtained for the $X(\lambda, x)$ function from Eq. (4-31) and the boundary conditions (4-28a) and (4-28b).

$$\frac{d^2X}{dx^2} + \lambda^2X = 0 \qquad \text{in } 0 \le x \le a \tag{4-32a}$$

$$X = 0 \qquad \text{at } x = 0 \tag{4-32b}$$

$$X = 0 \qquad \text{at } x = a \tag{4-32c}$$

where the boundary conditions (4-32b) and (4-32c) are obtained, respectively, from the boundary conditions (4-28a) and (4-28b) by substituting $T(x, y) = X(x)Y(y)$ and noting that the $Y(y)$ function should not vanish at $x = 0$ and $x = a$.

The differential equation for the Y separation is obtained from Eq. (4-31) and the boundary condition (4-28c) as

$$\frac{d^2 Y}{dy^2} - \lambda^2 Y = 0 \qquad \text{in } 0 \le y \le b \tag{4-33a}$$

$$Y = 0 \qquad\qquad \text{at } y = 0 \tag{4-33b}$$

where the boundary condition (4-33b) is obtained from the boundary condition (4-28c) by substituting $T(x, y) = X(x)Y(y)$ and noting that $X(x)$ should not vanish at $Y = 0$.

The three homogeneous boundary conditions, Eqs. (4-28a), (4-28b), and (4-28c), of the original heat-conduction problem are used in the above auxiliary problems for the X and Y separation functions, but the nonhomogeneous boundary condition $T = f(x)$ at $y = b$, given by Eq. (4-28d), has not yet been utilized. It will be used in the final steps of the solution.

We now examine the solutions of the auxiliary problems given by Eqs. (4-32) and (4-33). The problem given by Eqs. (4-32) is an eigenvalue problem similar to the one considered in Example 4-1. Therefore its solution can be obtained from Example 4-1 or directly from Table 4-1, case 1, by replacing L by a. We find

Eigenfunctions: $\qquad\qquad X(x) = \sin \lambda_n x \tag{4-34a}$

Eigenvalues: $\qquad\qquad \lambda_n = \dfrac{n\pi}{a} \qquad n = 1, 2, 3, \ldots \tag{4-34b}$

Normalization integral: $\quad \dfrac{1}{N} = \dfrac{2}{a} \tag{4-34c}$

and the orthogonality relation for the eigenfunctions are given as [see Eq. (4-2a)]

$$\int_0^a \sin \lambda_n x \sin \lambda_m x \, dx = \begin{cases} 0 & \text{for } \lambda_m \ne \lambda_n \\ N & \text{for } \lambda_m = \lambda_n \end{cases} \tag{4-35}$$

The solution of Eq. (4-33a) for the Y separation is taken as

$$Y(y) = C_1 \sinh \lambda_n y + C_2 \cosh \lambda_n y \tag{4-36}$$

If this solution should satisfy the boundary condition (4-33b), we would have $C_2 = 0$. Then the solution for $Y(y)$ is taken as

$$Y(y) = \sinh \lambda_n y \tag{4-37}$$

where we omitted the constant C_1.

Equation (4-29) states that the solution for $T(x, y)$ should be taken as a product of these two separation functions in the form $T(x, y) = X(x)Y(y) = \sin \lambda_n x \sinh \lambda_n y$; in this case we do not have a single solution for the separation functions but many solutions for each consecutive value λ_n, $n = 1, 2, 3, \ldots$. Therefore, the complete solution for the temperature should be taken as a linear sum of all these permissible solutions in the form

$$T(x, y) = \sum_{n=1}^{\infty} C_n \sinh \lambda_n y \sin \lambda_n x \tag{4-38}$$

where the C_n's are the unknown expansion coefficients which are yet to be determined.   We note that the solution given by Eq. (4-38) satisfies the differential equation of heat conduction (4-27) and its three homogeneous boundary conditions (4-28a), (4-28b), and (4-28c).   But it does not yet satisfy the non-homogeneous boundary condition (4-28d).   Therefore, the unknown coefficients C_n's can be determined by constraining the above solution to satisfy the non-homogeneous boundary condition (4-28d) as now described.

If Eq. (4-38) is the solution of the above heat-conduction problem it should also satisfy the nonhomogeneous boundary condition (4-28d); that is,

$$f(x) = \sum_{n=1}^{\infty} (C_n \sinh \lambda_n b) \sin \lambda_n x \qquad \text{in } 0 \leq x \leq a \tag{4-39}$$

This equation is a representation of function $f(x)$ defined in the interval $0 \leq x \leq a$ in terms of the eigenfunctions $\sin \lambda_n x$ of the eigenvalues problem given by Eqs. (4-32).   The orthogonality property of the eigenfunctions formally given by Eq. (4-2a) is utilized to determine the unknown coefficients C_n.   For this particular case the orthogonality property of $\sin \lambda_n x$ functions is written as [see Eq. (4-2a)]

$$\int_0^a \sin \lambda_m x \sin \lambda_n x \, dx = \begin{cases} 0 & \text{for } \lambda_n \neq \lambda_m \\ N & \text{for } \lambda_n = \lambda_m \end{cases} \tag{4-40}$$

where N is given by Eq. (4-34c).

We now multiply both sides of Eq. (4-39) by $\sin \lambda_m x$ and integrate it from $x = 0$ to $x = a$;

$$\int_0^a f(x') \sin \lambda_m x' \, dx' = \sum_{n=1}^{\infty} C_n \sinh \lambda_n b \int_0^a \sin \lambda_m x' \sin \lambda_n x' \, dx' \tag{4-41a}$$

where we have replaced x by x', since x' is merely a dummy integration variable. In view of the orthogonality property of the eigenfunctions given by Eq. (4-40), all the terms in the summation on the right-hand side of Eq. (4-41a) vanish except the term $\lambda_m = \lambda_n$. Then Eq. (4-41a) reduces to

$$\int_0^a f(x') \sin \lambda_n x' \, dx' = C_n \sinh \lambda_n b \, N$$

or

$$C_n = \frac{1}{N \sinh \lambda_n b} \int_0^a f(x') \sin \lambda_n x' \, dx' \qquad (4\text{-}41b)$$

The substitution of Eq. (4-41b) into Eq. (4-38) gives the desired solution of the above heat-conduction problem:

$$T(x, y) = \frac{2}{a} \sum_{n=1}^{\infty} \frac{\sinh \lambda_n y}{\sinh \lambda_n b} \sin \lambda_n x \int_0^a f(x') \sin \lambda_n x' \, dx' \qquad (4\text{-}42a)$$

since $1/N = 2/a$ by Eq. (4-34c). Here the λ_n's are given by

$$\lambda_n = \frac{n\pi}{a} \qquad n = 1, 2, 3, \ldots \qquad (4\text{-}42b)$$

Once the temperature-distribution function $f(x')$ is prescribed, the integral on the right-hand side of Eq. (4-42a) can be evaluated, and the temperature distribution anywhere in the medium is determined by summing up the terms in the series. Such series usually converge slowly; hence a large number of terms are to be evaluated for sufficient convergence. The computations can readily be performed with a high-speed digital computer. Once the temperature distribution is known, the heat flux in the medium is readily determined from the definition of heat flux, as now described.

Heat Flow

In practice, the heat flow at the boundaries of the region is of interest. For example, the heat flux at the boundary surface $x = 0$ of this rectangular region is determined from the definition of the heat flux as

$$q_{(x, y)} \bigg|_{x=0} = -k \frac{\partial T(x, y)}{\partial x} \bigg|_{x=0} \qquad \text{Btu/h·ft}^2 \text{ or W/m}^2 \qquad (4\text{-}43a)$$

If $T(x, y)$ given in Eqs. (4-42) is substituted into this equation and the indicated operations are performed, one finds

$$q(x, y) \bigg|_{x=0} = -\frac{2k}{a} \sum_{n=1}^{\infty} \frac{\sinh \lambda_n y}{\sinh \lambda_n b} \lambda_n \int_0^a f(x') \sin \lambda_n x' \, dx' \qquad (4\text{-}43b)$$

Here we note that the heat flux at the boundary surface $x = 0$ varies along the boundary in the y direction. The total heat-flow rate Q through the boundary surface at $x = 0$ is obtained by integrating Eq. (4-43b) with respect to y from $y = 0$ to $y = b$. For example, the total heat-flow rate Q through the boundary surface at $x = 0$, over the length $y = 0$ to $y = b$ for a depth W perpendicular to the plane of Fig. 4-2, is obtained as

$$Q = W \int_{y=0}^{b} q(x, y) \Big|_{x=0} \, dy \qquad \text{Btu/h or W} \qquad (4\text{-}44a)$$

and the substitution of $q(x, y)|_{x=0}$ from Eq. (4-43b) into this equation yields

$$Q = -W \frac{2k}{a} \sum_{n=1}^{\infty} \frac{\cosh \lambda_n b - 1}{\sinh \lambda_n b} \int_{0}^{a} f(x') \sin \lambda_n x' \, dx' \qquad (4\text{-}44b)$$

The heat-flow rates through other boundaries are determined in a similar manner.

In the following examples we examine the temperature distribution $T(x, y)$ given by Eqs. (4-42) for several different special cases of the boundary-surface temperature-distribution function $f(x)$.

Example 4-4   Compute the temperature distribution $T(x, y)$ given by Eqs. (4-42) for the case of boundary-surface temperature-distribution function $f(x)$ given by

$$f(x) = T_0 \sin \frac{\pi x}{a} \qquad (4\text{-}45)$$

where T_0 is a constant temperature. Also determine the heat flux of the boundary surface $x = 0$ and the total rate of heat flow Q through this boundary over the length $y = 0$ to $y = b$ and for a depth of W.

Solution   We first evaluate the integral term in $T(x, y)$ given by Eq. (4-42a) by setting $f(x) = T_0 \sin(\pi x/a)$ as specified by Eq. (4-45).

$$I_n \equiv \int_{0}^{a} f(x') \sin \lambda_n x' \, dx' = T_0 \int_{0}^{a} \sin \frac{\pi x'}{a} \sin \frac{n\pi x'}{a} \, dx' = \begin{cases} 0 & \text{for } n = 2, 3, 4, 5, \ldots \\ \dfrac{a}{2} T_0 & \text{for } n = 1 \end{cases}$$

$$(4\text{-}46)$$

To perform this integration we have utilized the orthogonality property of the eigenfunctions given by Eq. (4-40) and noted that $f(x) = T_0 \sin(\pi x/a) = T_0 \sin \lambda_1 x$, where $\lambda_1 = \pi/a$. When Eq. (4-46) is substituted into Eq. (4-42a), all the terms in the summation vanish except for the term for $n = 1$; then the temperature distribution for this special case becomes

$$T(x, y) = T_0 \frac{\sinh(\pi y/a)}{\sinh(\pi b/a)} \sin \frac{\pi x}{a} \qquad (4\text{-}47)$$

The heat flux at the boundary surface $x = 0$ is given by

$$q(x, y)\bigg|_{x=0} = -k\frac{\partial T}{\partial x}\bigg|_{x=0} = -\frac{\pi T_0 k}{a}\frac{\sinh(\pi y/a)}{\sinh(\pi b/a)} \qquad \text{Btu/h} \cdot \text{ft}^2 \text{ or W/m}^2 \qquad (4\text{-}48)$$

It is of interest to examine the direction of heat flow at this boundary. When T_0 is a positive quantity, the heat flux given by the equation is a negative quantity; this implies that heat flow at the boundary surface $x = 0$ is in the negative x direction, or heat flows out of the region. This result is also expected by physical considerations.

The total heat flow through the boundary surface $x = 0$ over the region $0 \le y \le b$ for a depth W is determined as

$$Q = W\int_{y=0}^{b} q(x, y)\bigg|_{x=0} dy = -\frac{\pi T_0 k}{a}\frac{W}{\sinh(\pi b/a)}\int_{0}^{b}\sinh\left(\frac{\pi y}{a}\right) dy$$

$$= -\frac{T_0 kW}{\sinh(\pi b/a)}\left(\cosh\frac{\pi b}{a} - 1\right) \qquad (4\text{-}49)$$

Example 4-5 Compute the temperature distribution $T(x, y)$ given by Eqs. (4-42) for the case of the boundary-surface temperature-distribution function $f(x)$ given as

$$f(x) = T_0 \qquad (4\text{-}50)$$

which characterizes a uniform temperature T_0 at the boundary surface $y = b$. Also determine for this case the heat-flux distribution at the boundary surface $y = 0$.

Solution The integral term in Eq. (4-42a) for $f(x) = T_0$ becomes

$$I_n \equiv T_0\int_{0}^{a}\sin\left(\frac{n\pi}{a}x'\right) dx' = -\frac{aT_0}{n\pi}(\cos n\pi - 1)$$

$$= -\frac{aT_0}{n\pi}[(-1)^n - 1] = \begin{cases} 0 & \text{for } n \text{ even} \\ \dfrac{2aT_0}{n\pi} & \text{for } n \text{ odd} \end{cases} \qquad (4\text{-}51)$$

When the result in Eq. (4-51) is substituted into the solution given by Eq. (4-42a), all the even terms of the summation vanish and we obtain

$$T(x, y) = \frac{4T_0}{\pi}\sum_{n=1, 3, 5, \ldots}^{\infty}\frac{1}{n}\frac{\sinh(n\pi y/a)}{\sinh(n\pi b/a)}\sin\frac{n\pi x}{a} \qquad (4\text{-}52)$$

The heat-flux distribution at the boundary $y = 0$ is determined as

$$q(x, y)\bigg|_{y=0} = -k\frac{\partial T(x, y)}{\partial y}\bigg|_{y=0} = -\frac{4kT_0}{\pi}\sum_{n=1, 3, 5, \ldots}^{\infty}\frac{1}{n}\frac{\sin(n\pi x/a)}{\sinh(n\pi b/a)}\frac{d}{dx}\left(\sinh\frac{n\pi y}{a}\right)\bigg|_{y=0}$$

$$= -\frac{4kT_0}{a}\sum_{n=1, 3, 5, \ldots}^{\infty}\frac{\sin(n\pi x/a)}{\sinh(n\pi b/a)} \qquad \text{Btu/h} \cdot \text{ft}^2 \text{ or W/m}^2 \qquad (4\text{-}53)$$

4-3   SEPARATION INTO SIMPLER PROBLEMS

In the preceding section we considered the solution of the steady-state heat-conduction problem by the method of separation of variables for a rectangular region in which *only one of the boundary conditions was nonhomogeneous.*   When more than one boundary condition is nonhomogeneous, the problem can be separated into simpler problems in which only one of the boundary conditions is nonhomogeneous.   Then the method of separation discussed in the previous section is applied to solve the resulting simpler problems.   To illustrate the procedure we consider the steady-state heat conduction in a rectangle in the region $0 \le x \le a, 0 \le y \le b$, subjected to boundary conditions two of which are nonhomogeneous and the remaining two are homogeneous.   The mathematical formulation of the problem is

$$\frac{\partial^2 T}{\partial x^2} + \frac{\partial^2 T}{\partial y^2} = 0 \qquad \text{in } 0 \le x \le a, 0 \le y \le b \qquad (4\text{-}54a)$$

$$T = 0 \qquad \text{at } x = 0 \qquad (4\text{-}54b)$$

$$T = f_1(y) \qquad \text{at } x = a \qquad (4\text{-}54c)$$

$$T = 0 \qquad \text{at } y = 0 \qquad (4\text{-}54d)$$

$$T = f_2(x) \qquad \text{at } y = b \qquad (4\text{-}54e)$$

Here the boundary conditions at $x = a$ and $y = b$ are nonhomogeneous.   We separate this problem into two simpler problems (see Fig. 4.3) by letting

$$T(x, y) = T_1(x, y) + T_2(x, y) \qquad (4\text{-}55)$$

where the temperature functions $T_1(x, y)$ and $T_2(x, y)$ are the solutions of the following two steady-state heat-conduction problems in the same rectangular region:

$$\frac{\partial^2 T_1}{\partial x^2} + \frac{\partial^2 T_1}{\partial y^2} = 0 \qquad \text{in } 0 \le x \le a, 0 \le y \le b \qquad (4\text{-}56a)$$

$$T_1 = 0 \qquad \text{at } x = 0 \qquad (4\text{-}56b)$$

$$T_1 = f_1(y) \qquad \text{at } x = a \qquad (4\text{-}56c)$$

$$T_1 = 0 \qquad \text{at } y = 0 \qquad (4\text{-}56d)$$

$$T_1 = 0 \qquad \text{at } y = b \qquad (4\text{-}56e)$$

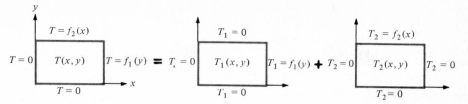

FIG. 4-3 Separation of heat conduction in a rectangle into simpler problems.

and

$$\frac{\partial^2 T_2}{\partial x^2} + \frac{\partial^2 T_2}{\partial y^2} = 0 \qquad \text{in } 0 \leq x \leq a, 0 \leq y \leq b \tag{4-57a}$$

$$T_2 = 0 \qquad \text{at } x = 0 \tag{4-57b}$$

$$T_2 = 0 \qquad \text{at } x = a \tag{4-57c}$$

$$T_2 = 0 \qquad \text{at } y = 0 \tag{4-57d}$$

$$T_2 = f_2(x) \qquad \text{at } y = b \tag{4-57e}$$

We note that only one boundary condition is nonhomogeneous in each of these problems.   The validity of the above separation into simpler problems is proved by adding the differential equations and the boundary conditions in the problems of Eqs. (4-56) and (4-57):

$$\frac{\partial^2 (T_1 + T_2)}{\partial x^2} + \frac{\partial^2 (T_1 + T_2)}{\partial y^2} = 0 \qquad \text{in } 0 \leq x \leq a, 0 \leq y \leq b \tag{4-58a}$$

$$T_1 + T_2 = 0 \qquad \text{at } x = 0 \tag{4-58b}$$

$$T_1 + T_2 = f_1(y) \qquad \text{at } x = a \tag{4-58c}$$

$$T_1 + T_2 = 0 \qquad \text{at } y = 0 \tag{4-58d}$$

$$T_1 + T_2 = f_2(x) \qquad \text{at } y = b \tag{4-58e}$$

and in view of Eq. (4-55) it is clear that the heat-conduction problem of Eqs. (4-58) is identical to the original problem given by Eqs. (4-54).   Figure 4-3 illustrates the above separation procedure into simpler problems.

The problems given by Eqs. (4-56) and (4-57) can readily be solved by the method of separation of variables; in fact, the problem of Eqs. (4-57) is exactly the same heat-conduction problem considered in the preceding section.   When the solutions of these simple problems are added according to Eq. (4-55), the solution of the original heat-conduction problem is obtained.

4-4 TRANSIENT HEAT CONDUCTION IN A SLAB

The method of separation of variables is now applied to solve one-dimensional, time-dependent heat conduction in a slab which is initially at a prescribed temperature distribution and for times $t > 0$ the boundary surfaces are subjected to *homogeneous boundary conditions*. These homogeneous boundary conditions can be any combination of the boundary conditions of the first, second, or third kind.

To illustrate the general procedure in the method of solution, we consider a slab in the region $0 \leq x \leq L$, which initially (i.e., at $t = 0$) has a temperature distribution described by the function $F(x)$, and for times $t > 0$ the boundary surface at $x = 0$ is kept insulated and that at $x = L$ dissipates heat by convection into a medium at zero temperature. The mathematical formulation of this heat-conduction problem is (see Fig. 4-4)

$$\frac{\partial^2 T(x, t)}{\partial x^2} = \frac{1}{\alpha}\frac{\partial T(x, t)}{\partial t} \qquad \text{in } 0 \leq x \leq L, t > 0 \tag{4-59}$$

subject to the boundary conditions

$$\frac{\partial T(x, t)}{\partial x} = 0 \qquad\qquad \text{at } x = 0, t > 0 \tag{4-60a}$$

$$k\frac{\partial T(x, t)}{\partial x} + hT(x, t) = 0 \qquad \text{at } x = L, t > 0 \tag{4-60b}$$

and the initial condition

$$T(x, t) = F(x) \qquad \text{for } t = 0, \text{ in } 0 \leq x \leq L \tag{4-60c}$$

Clearly, the boundary conditions at $x = 0$ and $x = L$ are the homogeneous boundary conditions of the second and third kind, respectively. It is to be noted that in this example, for convenience in mathematical analysis, we considered convection into a medium at zero temperature, which is difficult to realize or

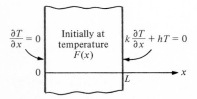

FIG. 4-4 Boundary and initial conditions for transient heat conduction in a slab.

imagine physically. However, if convection takes place into a medium at a uniform temperature T_∞, one can introduce a new temperature $\theta(x, t)$ defined as $\theta(x, t) = T(x, t) - T_\infty$. Then the mathematical formulation of this problem in terms of $\theta(x, t)$ is equivalent to convection into a medium at $\theta = 0$.

To solve this problem it is assumed that the temperature $T(x, t)$ can be represented as a product of functions in the form

$$T(x, t) = X(x)\Gamma(t) \tag{4-61}$$

where $X(x)$ is a function of x only, and $\Gamma(t)$ is a function of t only. Substitution of Eq. (4-61) into Eq. (4-59) yields

$$\frac{1}{X}\frac{d^2X}{dx^2} = \frac{1}{\alpha\Gamma}\frac{d\Gamma}{dt} \tag{4-62}$$

where $X \equiv X(x)$ and $\Gamma \equiv \Gamma(t)$. The left-hand side of this equality is a function of x only, and the right-hand side is a function of t only. This equality is possible only if both sides are equal to the same constant $-\lambda^2$; then Eq. (4-62) becomes

$$\frac{1}{X}\frac{d^2X}{dx^2} = \frac{1}{\alpha\Gamma}\frac{d\Gamma}{dt} = -\lambda^2 \tag{4-63}$$

Here, the negative sign is chosen for λ^2 in order to ensure that the temperature will decay with time. Then the following eigenvalue problem is obtained for the $X(x)$ functions from Eq. (4-63) and the boundary conditions (4-60a) and (4-60b) of the heat-conduction equation:

$$\frac{d^2X}{dx^2} + \lambda^2X = 0 \qquad \text{in } 0 \le x \le L \tag{4-64a}$$

$$\frac{dX}{dx} = 0 \qquad \text{at } x = 0 \tag{4-64b}$$

$$k\frac{dX}{dx} + hX = 0 \qquad \text{at } x = L \tag{4-64c}$$

The differential equation for the function Γ is obtained from Eq. (4-63) as

$$\frac{d\Gamma}{dt} + \alpha\lambda^2\Gamma = 0 \qquad \text{for } t > 0 \tag{4-65}$$

We note that the boundary conditions (4-60b) and (4-60c) of the heat-conduction problem have been used in the above auxiliary problems, but the initial condition $T = F(x)$ will be utilized at the final stages of the solution.

The solution of the eigenvalue problem given by Eqs. (4-64) has already been considered in Example 4-3. Therefore the eigenfunctions, the eigenvalue relation, and the normalization integral can be obtained either from Example 4-3 or directly from Table 4-1, case 6. Then, the eigenfunctions are

$$X(x) = \cos \lambda_n x \qquad (4\text{-}66a)$$

Eigenvalues λ_n are the positive roots of the following transcendental equation

$$\lambda L \tan \lambda L = \frac{hL}{k} \qquad (4\text{-}66b)$$

and the normalization integral N is given by

$$\frac{1}{N} = \frac{2(\lambda_n^2 + H^2)}{L(\lambda_n^2 + H^2) + H} \qquad (4\text{-}66c)$$

where $H = h/k$. The orthogonality property of the eigenfunctions is given as

$$\int_0^L \cos \lambda_n x \cos \lambda_m x \, dx = \begin{cases} 0 & \text{for } \lambda_n \neq \lambda_m \\ N & \text{for } \lambda_n = \lambda_m \end{cases} \qquad (4\text{-}66d)$$

The solution of Eq. (4-65) for the function $\Gamma(t)$ is taken as

$$\Gamma(t) = e^{-\alpha \lambda_n^2 t} \qquad (4\text{-}67)$$

Now, according to Eq. (4-61), the function $X(x)\Gamma(t) = e^{-\alpha \lambda_n^2 t} \cos \lambda_n x$ is a solution, and these solutions are valid for each consecutive value of λ_n, $n = 1, 2, 3, \ldots$. Therefore, the complete solution for the temperature $T(x, t)$ is taken as a linear sum of all the individual solutions and is given in the form

$$T(x, t) = \sum_{n=1}^{\infty} C_n e^{-\alpha \lambda_n^2 t} \cos \lambda_n x \qquad (4\text{-}68)$$

where the unknown expansion coefficients C_n are to be determined by applying the initial condition for the problem and by making use of the orthogonality property of the eigenfunctions as described below.

If Eq. (4-68) is the solution of the above heat-conduction problem, it should satisfy the initial condition (4-60c); that is, for $t = 0$ we have

$$F(x) = \sum_{n=1}^{\infty} C_n \cos \lambda_n x \qquad \text{in } 0 \leq x \leq L \qquad (4\text{-}69)$$

We note that this equation is a representation of the function $F(x)$ defined in the interval $0 \leq x \leq L$ in terms of the eigenfunctions $\cos \lambda_n x$ of the eigenvalue problem given by Eqs. (4-64).   The unknown expansion coefficients C_n are determined by following the procedure described previously.   That is, to determine C_n's we multiply both sides of Eq. (4-69) by $\cos \lambda_m x$, integrate it over the region from $x = 0$ to $x = L$, and obtain

$$\int_0^L F(x') \cos \lambda_m x' \, dx' = \sum_{n=1}^{\infty} C_n \int_0^L \cos \lambda_m x' \cos \lambda_n x' \, dx' \tag{4-70}$$

In view of the orthogonality property of the eigenfunctions given by Eq. (4-66d), all the terms in the summation on the right-hand side of Eq. (4-70) vanish except the term $\lambda_m = \lambda_n$.   Then the summation drops out and Eq. (4-70) simplifies to

$$\int_0^L F(x') \cos \lambda_n x' \, dx' = C_n N \tag{4-71a}$$

or

$$C_n = \frac{1}{N} \int_0^L F(x') \cos \lambda_n x' \, dx' \tag{4-71b}$$

The substitution of Eq. (4-71b) into Eq. (4-68) gives the solution of the above time-dependent heat-conduction problem as

$$T(x, t) = \sum_{n=1}^{\infty} \frac{1}{N} e^{-\alpha \lambda_n^2 t} \cos \lambda_n x \int_0^L F(x') \cos \lambda_n x' \, dx' \tag{4-72a}$$

where

$$\frac{1}{N} = \frac{2(\lambda_n^2 + H^2)}{L(\lambda_n^2 + H^2) + H} \tag{4-72b}$$

$$H = \frac{h}{k} \tag{4-72c}$$

and the eigenvalues λ_n's are the positive roots of the transcendental equation (4-66b).   The roots of this transcendental equation are available in Appendix B.
   We now consider the following special case of the above solution.

Slab Initially at a Uniform Temperature T_i

We have

$$F(x) = T_i \tag{4-73}$$

Then the integral term in Eq. (4-72a) becomes

$$I_n \equiv \int_0^L F(x') \cos \lambda_n x' \, dx' = T_i \int_0^L \cos \lambda_n x' \, dx' = \frac{T_i}{\lambda_n} \sin \lambda_n L \tag{4-74}$$

Substitution of Eq. (4-74) into Eq. (4-72a) gives the temperature distribution for this particular case:

$$T(x, t) = T_i \sum_{n=1}^{\infty} \frac{1}{N} e^{-\alpha \lambda_n^2 t} \frac{\sin \lambda_n L}{\lambda_n} \cos \lambda_n x \tag{4-75}$$

Knowing the temperature distribution $T(x, t)$, the heat flux at the boundary surface $x = L$ is determined as

$$q(x, t)\Big|_{x=L} = -k \frac{\partial T(x, t)}{\partial x}\Big|_{x=L} = kT_i \sum_{n=1}^{\infty} \frac{1}{N} e^{-\alpha \lambda_n^2 t} \sin \lambda_n L \tag{4-76}$$

Once the eigenvalues λ_n are determined from the solution of the transcendental equation (4-66b) for specified values of h, k, and L, the temperature distribution in the slab and the heat flux at the boundary surface $x = L$ are evaluated from the above equations for a given value of the thermal diffusivity α by summing up the terms in the series.   Such computations are readily performed with a high-speed digital computer.   In this case the convergence of these series is very rapid due to the presence of the exponential term; only for very small values of αt does the convergence become slow.   This matter will be illustrated in the next section with numerical examples.

4-5   TRANSIENT-TEMPERATURE CHARTS

The temperature distribution for one-dimensional transient heat conduction in simple geometries such as a slab, cylinder, and sphere has been calculated, and the results plotted as a function of time and position are available in the form of charts in several references [2–4,6,7].   In this section we present typical transient-temperature charts for slab, cylinder, and sphere, discuss their physical significance, and illustrate their application.

Slab

Figure 4-5 shows a typical transient-temperature chart for a slab in the region $-L \leq x \leq L$, which is initially at a uniform temperature T_i and for times $t > 0$ the boundaries at $x = -L$ and $x = L$ dissipate heat by convection into a medium at temperature T_e with a heat-transfer coefficient h.   Since the problem has symmetry about $x = 0$, one need consider only one half of the plate, say the region $0 \leq x \leq L$ with a symmetry boundary condition at $x = 0$.   With this considera-

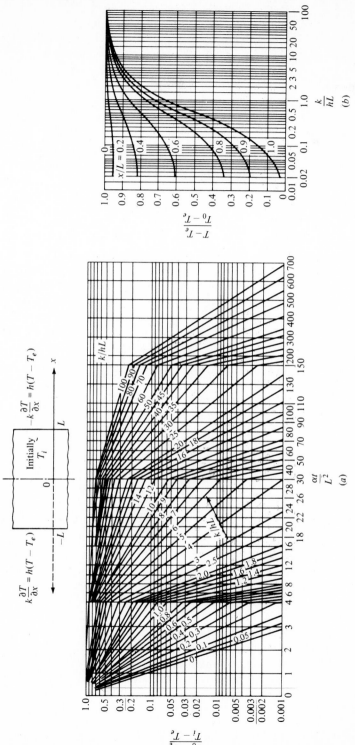

FIG. 4-5 Transient-temperature chart for a slab of thickness 2L subjected to convection at both boundary surfaces. (a) Temperature T_0 at the center plane, $x = 0$; (b) position correction for use with part a. (From Heisler [6].)

tion, the mathematical formulation of this heat-conduction problem for the case of constant thermal properties becomes

$$\frac{\partial^2 T}{\partial x^2} = \frac{1}{\alpha} \frac{\partial T}{\partial t} \qquad \text{in } 0 \leq x \leq L, \text{ for } t > 0 \qquad (4\text{-}77a)$$

$$\frac{\partial T}{\partial x} = 0 \qquad \text{at } x = 0, \text{ for } t > 0 \qquad (4\text{-}77b)$$

$$k \frac{\partial T}{\partial x} + hT = hT_e \qquad \text{at } x = L, \text{ for } t > 0 \qquad (4\text{-}77c)$$

$$T = T_i \qquad \text{for } t = 0, \text{ in } 0 \leq x \leq L \qquad (4\text{-}77d)$$

The nonhomogeneous boundary condition (4-77c) can be transformed into a homogeneous one if a dimensionless temperature $\theta(x, t)$ is defined as

$$\theta(x, t) = \frac{T(x, t) - T_e}{T_i - T_e} \qquad (4\text{-}78)$$

Then, the heat-conduction problem given by Eqs. (4-77) becomes

$$\frac{\partial^2 \theta}{\partial x^2} = \frac{1}{\alpha} \frac{\partial \theta}{\partial t} \qquad \text{in } 0 \leq x \leq L, t > 0 \qquad (4\text{-}79a)$$

$$\frac{\partial \theta}{\partial x} = 0 \qquad \text{at } x = 0, t > 0 \qquad (4\text{-}79b)$$

$$k \frac{\partial \theta}{\partial x} + h\theta = 0 \qquad \text{at } x = L \qquad t > 0 \qquad (4\text{-}79c)$$

$$\theta = 1 \qquad \text{for } t = 0, 0 \leq x \leq L \qquad (4\text{-}79d)$$

We note that this heat-conduction problem for $\theta(x, t)$ is obtainable as a special case from the more general problem considered previously by Eqs. (4-59) and (4-60) by setting in that problem the initial-condition function $F(x) = 1$. Therefore, the solution of the heat-conduction problem Eqs. (4-79) is obtainable from the previous solution (4-75) by setting in that solution $T_i = 1$ and replacing $T(x, t)$ by $\theta(x, t)$.

Figure 4-5a shows a plot of the dimensionless temperature $(T_0 - T_e)/(T_i - T_e)$ at the center plane (that is, $x = 0$) of the slab as a function of the dimensionless time $\alpha t/L^2$ for several different values of the parameter k/hL. Here T_0 is the temperature at the center plane. The temperature at other locations in the slab is determined by using the position-correction chart given in Fig. 4-5b in conjunction with Fig. 4-5a. That is, if the center-plane temperature $(T_0 - T_e)/(T_i - T_e)$ is known, the temperature at a position x/L is determined by multiplying this temperature with the correction $(T - T_e)/(T_0 - T_e)$ obtained from Fig. 4-5b for the position x/L and for the same value of k/hL.

An examination of Fig. 4-5b reveals that for values of $1/\text{Bi} \equiv k/hL$ larger than about 10, the position-correction factor is nearly unity; that is, for the cases $\text{Bi} \equiv hL/k < 0.1$ the temperature distribution within the slab may be considered uniform with an error less than about 5 percent.

Long Cylinder

Figure 4-6 shows a typical transient-temperature chart for a long solid cylinder of radius b, which is initially at a uniform temperature T_i and for times $t > 0$ the boundary at $r = b$ dissipates heat by convection into a medium at temperature T_e with a heat-transfer coefficient h. The mathematical formulation of this problem is

$$\frac{1}{r}\frac{\partial}{\partial r}\left(r\frac{\partial \theta}{\partial r}\right) = \frac{1}{\alpha}\frac{\partial \theta}{\partial t} \qquad \text{in } 0 \leq r \leq b, t > 0 \tag{4-80a}$$

$$\frac{\partial \theta}{\partial r} = 0 \qquad \text{at } r = 0, t > 0 \tag{4-80b}$$

$$k\frac{\partial \theta}{\partial r} + h\theta = 0 \qquad \text{at } r = b, t > 0 \tag{4-80c}$$

$$\theta = 1 \qquad \text{for } t = 0, 0 \leq r \leq b \tag{4-80d}$$

where the dimensionless temperature $\theta(r, t)$ is defined as

$$\theta(r, t) = \frac{T(r, t) - T_e}{T_i - T_e} \tag{4-80e}$$

Figure 4-6a gives the dimensionless center temperature $(T_0 - T_e)/(T_i - T_e)$ of the cylinder as a function of the dimensionless time $\alpha t/b^2$ for several different values of the parameter k/hb. Here T_0 is the center temperature. The temperature at other radial positions is determined by multiplying the temperature $(T_0 - T_e)/(T_i - T_e)$ from Fig. 4-6a with the position-correction factor $(T - T_e)/(T_0 - T_e)$ as obtained from Fig. 4-6b. An examination of Fig. 4-6b shows that for values of k/hb larger than about 10 the position-correction factor is nearly unity; then for such cases the temperature distribution in the cylinder is almost uniform at any instant with an error less than about 5 percent.

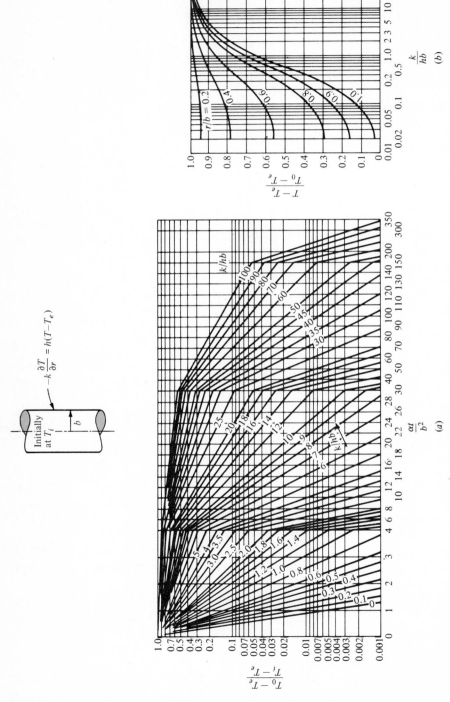

FIG. 4-6 Transient-temperature chart for a long solid cylinder of radius $r = b$ subjected to convection at the boundary surface $r = b$. (a) Temperature T_o at the axis of the cylinder; (b) position correction for use with part a. (From Heisler [6].)

Sphere

Figure 4-7 shows a typical transient-temperature chart for a solid sphere of radius b, which is initially at a uniform temperature T_i and for times $t > 0$ the boundary surface at $r = b$ dissipates heat by convection into a medium at temperature T_e with a heat-transfer coefficient h. The mathematical formulation of this problem is

$$\frac{1}{r^2} \frac{\partial}{\partial r}\left(r^2 \frac{\partial \theta}{\partial r}\right) = \frac{1}{\alpha} \frac{\partial \theta}{\partial t} \qquad \text{in } 0 \leq r \leq b, t > 0 \tag{4-81a}$$

$$\frac{\partial \theta}{\partial r} = 0 \qquad \text{at } r = 0, t > 0 \tag{4-81b}$$

$$k \frac{\partial \theta}{\partial r} + h\theta = 0 \qquad \text{at } r = b, t > 0 \tag{4-81c}$$

$$\theta = 1 \qquad \text{for } t = 0, 0 \leq r \leq b \tag{4-81d}$$

where the dimensionless temperature $\theta(r, t)$ is defined as

$$\theta(r, t) = \frac{T(r, t) - T_e}{T_i - T_e} \tag{4-81e}$$

As in the previous case Fig. 4-7a gives the dimensionless center-point temperatures $(T_0 - T_e)/(T_i - T_e)$ for the sphere; Fig. 4-7b is the position-correction factor to determine the temperature at other radial locations. Figure 4-7b shows that for values of k/hb larger than about 10 the position-correction factor is nearly unity; hence the temperature distribution in the sphere is almost uniform with an error less than about 5 percent for Bi $\equiv hb/k < 0.1$.

Example 4-6 A concrete wall $L = 1$ ft thick, $k = 0.5$ Btu/h·ft·°F, and $\alpha = 0.02$ ft²/h is initially at a uniform temperature $T_i = 1100°$F. For times $t > 0$, the boundary surface at $x = 0$ is kept insulated while the boundary surface at $x = 1$ ft dissipates heat by convection into an environment at temperature $T_e = 100°$F with a heat-transfer coefficient of $h = 5$ Btu/h·ft²·°F. Determine the temperature of the insulated surface of the slab at times $t = 5$ h and 20 h after the start of cooling both by using the transient-temperature chart and by computing it from the analytical solution given previously.

Solution The mathematical formulation of this problem is exactly the same as that given by Eqs. (4-79), that is,

$$\frac{\partial^2 \theta}{\partial x^2} = \frac{1}{\alpha} \frac{\partial \theta}{\partial t} \qquad \text{in } 0 \leq x \leq L, t > 0 \tag{4-79a}$$

$$\frac{\partial \theta}{\partial x} = 0 \qquad \text{at } x = 0, t > 0 \tag{4-79b}$$

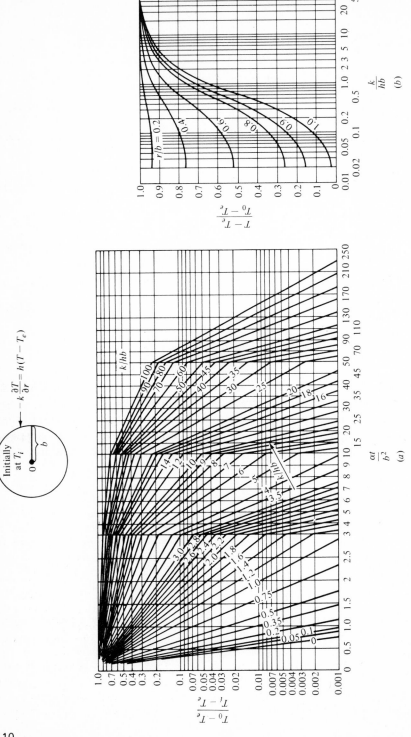

FIG. 4-7 Transient-temperature chart for a solid sphere of radius $r = b$ subjected to convection at the boundary surface $r = b$. (a) Temperature T_0 at the center of the sphere; (b) position correction for use with part a. (From Heisler [6].)

$$k \frac{\partial \theta}{\partial x} + h\theta = 0 \qquad \text{at } x = L,\ t > 0 \tag{4-79c}$$

$$\theta = 1 \qquad \text{for } t = 0 \text{ in } 0 \le x \le L \tag{4-79d}$$

where the dimensionless temperature $\theta(x, t)$ is defined as

$$\theta(x, t) = \frac{T(x, t) - T_e}{T_i - T_e} \tag{4-79e}$$

The solution of this problem was considered previously and can be obtained from Eq. (4-75) by setting $T_i = 1$ and replacing $T(x, t)$ by $\theta(x, t)$:

$$\theta(x, t) \equiv \frac{T(x, t) - T_e}{T_i - T_e} = \sum_{n=1}^{\infty} \frac{1}{N} e^{-\alpha\lambda_n^2 t} \frac{\sin \lambda_n L}{\lambda_n} \cos \lambda_n x \tag{4-82a}$$

where $1/N$ is given by Eq. (4-72b) as

$$\frac{1}{N} = \frac{2(\lambda_n^2 + H^2)}{L(\lambda_n^2 + H^2) + H} \qquad \text{with } H = \frac{h}{k} \tag{4-82b}$$

and λ_n's are the roots of the transcendental equation (4-66b), i.e.,

$$\lambda L \tan \lambda L = \frac{hL}{k} \tag{4-82c}$$

The numerical values of various quantities for the considered problem are

$$T_i = 1100°F \qquad T_e = 100°F \qquad L = 1 \text{ ft} \qquad \alpha = 0.02 \text{ ft}^2/\text{h}$$

$$k = 0.5 \text{ Btu/h·ft·°F} \qquad \text{and} \qquad h = 5 \text{ Btu/h·ft}^2 \cdot °F$$

Then, Eqs. (4-82) evaluated at $x = 0$ (i.e., at the insulated boundary) become

$$\theta(x, t)\Big|_{x=0} = \left[\frac{T(x, t) - 100}{1100 - 100} \right]_{x=0} = \sum_{n=1}^{\infty} \frac{1}{N} e^{-0.02\lambda_n^2 t} \frac{\sin \lambda_n}{\lambda_n} \tag{4-83a}$$

where

$$\frac{1}{N} = \frac{2(\lambda_n^2 + 100)}{\lambda_n^2 + 110} \tag{4-83b}$$

and λ_n's are the roots of the following transcendental equation:

$$\lambda \tan \lambda = 10 \tag{4-83c}$$

since $L = 1$ ft and $hL/k = 10$. The first six roots of Eq. (4-83c) are tabulated in Appendix B, or they can be computed by solving this equation numerically, using a digital computer.

In order to provide some insight into the relative order of magnitude of various terms in Eqs. (4-83), we present in Table 4-2 the first five terms of the series and the corresponding solution, together with the solutions determined from the transient-temperature chart given in Fig. 4-5a. It is apparent from Table 4-2 that the exponential term plays an important role in the convergence of the series, since the exponential term approaches zero very rapidly as the value of λ_n increases. For $t = 5$ h the calculation of temperature from Eqs. (4-83) by using only three terms in the series is sufficient to obtain convergence to four significant figures, whereas for $t = 20$ h two terms in the series can represent the temperature with the same degree of accuracy. Clearly, for larger times even the first term

TABLE 4-2   Computation of Temperature θ from Eqs. (4-83) and Fig. 4-5a

t, h	n	λ_n	$e^{-0.02\lambda_n^2 t}$	$\dfrac{\sin \lambda_n}{\lambda_n}$	$\dfrac{1}{N}$	$\dfrac{1}{N} e^{-0.02\lambda_n^2 t} \dfrac{\sin \lambda_n}{\lambda_n}$	$\theta = \sum\limits_{i=1}^{n} \dfrac{1}{N} e^{-0.02\lambda_n^2 t} \dfrac{\sin \lambda_n}{\lambda_n}$	θ Computed from Fig. 4-5a
5	1	1.4289	0.81535	0.69285	1.8215	1.02898	1.0289	Cannot be
	2	4.3058	0.15662	−0.21331	1.8444	-0.6162×10^{-1}	0.9674	computed
	3	7.2281	0.5383×10^{-2}	0.11212	1.8767	0.1133×10^{-2}	0.9685	accurately
	4	10.2003	0.3030×10^{-4}	−0.68630	1.9065	-0.3964×10^{-5}	0.9685	from
	5	13.2142	0.2610×10^{-7}	0.45663	1.9297	0.230×10^{-8}	0.9685	Fig. 4-5a
20	1	1.4289	0.444194	0.69285	1.8215	0.55774	0.5577	0.6
	2	4.3058	0.6016×10^{-3}	−0.21331	1.8444	-0.2367×10^{-3}	0.5575	
	3	7.2281	0.8398×10^{-9}	0.11212	1.8767	0.1767×10^{-9}	0.5575	
	4	10.2003	0.8425×10^{-18}	−0.68630	1.9065	-0.1102×10^{-18}	0.5575	
	5	13.2142	0.4541×10^{-30}	0.45663	1.9297	0.4089×10^{-31}	0.5575	

of the series is accurate enough to determine the temperature, but for smaller times many more terms are needed.   The transient-temperature chart in Fig. 4-5a gives the temperature within the accuracy of the scale for the time $t = 20$ h, but it is not accurate enough for the time $t = 5$ h.

Now, using the results from Table 4-2, the temperature of the insulated surface at various times as determined from the analytical computations is:

At $t = 5$ h:

$$\theta = \frac{T - 100}{1100 - 100} = 0.9685$$

or

$$T = 100 + 968.5 = 1068.5°\text{F} \ (575.8°\text{C})$$

At $t = 20$ h:

$$\theta = \frac{T - 100}{1100 - 100} = 0.5575$$

or

$$T = 100 + 557.5 = 657.5°\text{F} \ (347.5°\text{C})$$

The calculation of temperature at $t = 20$ h from the reading of Fig. 4-5a gives

$$\theta = 0.6 \qquad \text{or} \qquad T = 700°\text{F} \ (371.1°\text{C})$$

4-6   LUMPED-SYSTEM ANALYSIS FOR TRANSIENT HEAT FLOW

The results from the position-correction charts (i.e., Figs. 4-5b, 4-6b, and 4-7b) given in the preceding section reveal that for the values of the parameter k/hL (or k/hb) greater than about 10 or for Bi $\equiv hL/k$ less than about 0.1 the temperature in the medium is essentially uniform at any given time.   For such cases the spatial variation of temperature can be neglected, and the variation of temperature in the solid with time can be studied with a very simple method of analysis called the *lumped-system* analysis.   In this approach the geometry is immaterial since the temperature is considered to be a function of time only; hence the analysis becomes very simple.   The mathematical formulation of the problem, however, depends on the number of lumps considered for the system and on the type of the boundary conditions.   In the following analysis we consider one- and two-lump systems with different types of boundary conditions.

One-Lump System with Convection Boundaries

Consider that a solid of arbitrary shape, volume V, total surface area A, thermal conductivity k_s, density ρ, specific heat c_p, at a uniform temperature T_0 is suddenly immersed at the time $t = 0$ in a well-stirred fluid which is kept at a uniform temperature T_∞. Figure 4-8 illustrates the considered heat-transfer system.   Heat transfer between the solid and liquid takes place by convection with a heat-transfer coefficient h.   It is assumed that the temperature distribution

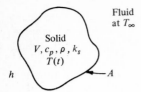

FIG. 4-8   Nomenclature for one-lump analysis of transient heat flow.

within the solid at any instant is sufficiently uniform so that the temperature of the solid can be considered to be a function of time only, that is, $T(t)$. To establish some criteria under which the temperature can be considered uniform within the body, we define the *characteristic length* L_s of the solid as the volume V divided by the surface area A, that is, $L_s = V/A$, and the *Biot number* as $Bi = hL_s/k_s$, where k_s is the thermal conductivity of the solid.   We have previously shown from the temperature-correction charts Figs. 4-5b, 4-6b, and 4-7b that for solids in the shape of a slab, long cylinder, and sphere, the temperature distribution within the solid at any instant is uniform, with an error less than about 5 percent, if $Bi < 0.1$.   Therefore, the lumped-system heat-transfer analysis may be considered applicable for situations in which $Bi = hL_s/k < 0.1$.

When the temperature distribution within the solid is assumed to be uniform, the variation of temperature takes place with time.   The energy equation for a solid of volume V may be stated as

$$\left(\begin{array}{l}\text{Rate of heat flow into}\\ \text{solid of volume } V \text{ through}\\ \text{boundary surfaces } A\end{array}\right) = \left(\begin{array}{l}\text{Rate of increase of}\\ \text{internal energy of solid}\\ \text{of volume } V\end{array}\right) \qquad (4\text{-}84)$$

By writing the appropriate mathematical expressions for each of these terms, Eq. (4-84) becomes

$$Ah[T_\infty - T(t)] = \rho c_p V \frac{dT(t)}{dt}$$

or

$$\frac{dT(t)}{dt} + \frac{Ah}{\rho c_p V}[T(t) - T_\infty] = 0 \qquad \text{for } t > 0 \qquad (4\text{-}85a)$$

subject to the initial condition

$$T(t) = T_0 \qquad \text{for } t = 0 \qquad (4\text{-}85b)$$

For convenience in the analysis a new temperature $\theta(t)$ is defined as

$$\theta(t) \equiv T(t) - T_\infty \qquad (4\text{-}86)$$

Then Eqs. (4-85) become

$$\frac{d\theta(t)}{dt} + m\theta(t) = 0 \qquad \text{for } t > 0 \tag{4-87a}$$

$$\theta(t) = T(t) - T_0 \equiv \theta_0 \qquad \text{for } t = 0 \tag{4-87b}$$

where we have defined

$$m \equiv \frac{Ah}{\rho c_p V} \tag{4-88}$$

Equation (4-87a) is an ordinary differential equation for the temperature $\theta(t)$, and its general solution is given as

$$\theta(t) = Ce^{-mt} \tag{4-89}$$

The application of the initial condition (4-87b) gives the integration constant as $C = \theta_0$. Then, the temperature of the solid as a function of time is given as

$$\frac{\theta(t)}{\theta_0} = \frac{T(t) - T_\infty}{T_0 - T_\infty} = e^{-mt} \tag{4-90}$$

The physical significance of the exponent mt is better envisioned if it is rearranged in the form

$$mt = \frac{Aht}{\rho c_p V} = \frac{hL_s}{k_s} \frac{k_s t}{\rho c_p L_s{}^2} = \frac{hL_s}{k_s} \frac{\alpha t}{L_s{}^2} \equiv \text{Bi Fo} \tag{4-91}$$

where the *Biot* and *Fourier numbers* are defined as

$$\text{Bi} = \frac{hL_s}{k} = \text{Biot number} \tag{4-92a}$$

$$\text{Fo} = \frac{\alpha t}{L_s{}^2} = \text{Fourier number} \tag{4-92b}$$

Here $\alpha = k_s/\rho c_p$ is the thermal diffusivity, and L_s is the characteristic dimension of the solid. Then, the solution is written in the form

$$\frac{\theta(t)}{\theta_0} = \frac{T(t) - T_\infty}{T_0 - T_\infty} = e^{-\text{Bi Fo}} \tag{4-93}$$

Figure 4-9 shows a plot of the dimensionless temperature given by Eq. (4-90) or (4-93) as a function of time. The temperature decays with time exponentially, and the shape of the curve is determined by the value of the exponent m.

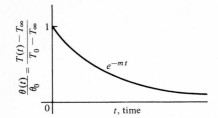

FIG. 4-9   Dimensionless temperature $\theta(t)/\theta_0$ as a function of time.

Example 4-7   A large aluminum plate ($k = 120$ Btu/h·ft·°F or 207.6 W/m·°C, $\rho = 170$ lb/ft³ or 2735 kg/m³, $c_p = 0.2$ Btu/lb·°F or 837.3 W·s/kg·°C) of thickness $L = 1.0$ in (2.54×10^{-2} m) at uniform temperature $T_0 = 200°F$ (93.3°C) is suddenly immersed at time $t = 0$ in a well-stirred fluid which is kept at a constant temperature $T_\infty = 40°F$ (4.4°C). The plate loses heat by convection from both of its surfaces into the fluid with a heat-transfer coefficient $h = 17$ Btu/h·ft²·°F (96.5 W/m²·°C).   Determine the time required for the center of the plate to reach a temperature of 80°F (26.7°C).

Solution   This problem can be solved with the lumped-system analysis with sufficient accuracy if Bi < 0.1.   To check the magnitude of the Biot number we determine the characteristic dimension L_s of the solid:

$$L_s = \frac{\text{volume}}{\text{area}} \cong \frac{LA}{2A} = \frac{L}{2} = 0.5 \text{ in } (1.27 \times 10^{-2} \text{ m})$$

Then the Biot number becomes

$$\text{Bi} = \frac{hL_s}{k_s} = \frac{17 \times 0.5}{120 \times 12} \simeq 5.9 \times 10^{-3}$$

Since the Biot number is a dimensionless quantity, the same result is obtained with the SI system of units, i.e.,

$$\text{Bi} = \frac{hL_s}{k_s} = \frac{96.5 \times (1.27 \times 10^{-2})}{207.6} \simeq 5.9 \times 10^{-3}$$

The Biot number being less than 0.1, the lumped-system analysis is applicable.   Equation (4-90) is now used to solve this problem.

$$\theta(t) \equiv \frac{T(t) - T_\infty}{T_0 - T_\infty} = e^{-mt}$$

where various quantities are given as

$$T(t) = 80°F \qquad T_\infty = 40°F \qquad T_0 = 200°F \qquad \text{hence } \theta(t) = 0.25$$

$$m = \frac{hA}{\rho c_p V} = \frac{h}{\rho c_p L_s} = \frac{17}{170 \times 0.2 \times 0.5/12} = 12 \qquad \text{hence } mt = 12t, \text{ where } t \text{ is in } hours$$

Substitution of these quantities into the above equation yields

$$0.25 = e^{-12t} \qquad \text{or} \qquad 12t = 1.385$$

Then the time required for the center of the plate to reach a temperature of 80°F becomes

$$t = \frac{1.385}{12} \, h = 6.92 \text{ min}$$

If the SI system of units were used in the above calculations, we should have

$$\theta(t) = \frac{26.7 - 4.4}{93.3 - 4.4} = 0.25$$

$$mt = \frac{h}{\rho c_p L_s} \, t = \frac{96.5}{2735 \times 837.3 \times (1.27 \times 10^{-2})} \, t = 0.00332 \, t, \text{ where } t \text{ is in } seconds$$

$$\cong 12 \, t, \text{ where } t \text{ is in } hours$$

Clearly, the result is the same in both systems of units.

One-Lump System with Convection and Prescribed Heat-Flux Boundaries

Consider a slab of thickness L, initially (that is, $t = 0$) at a uniform temperature T_0. For times $t > 0$, heat is supplied to the slab from one of its boundary surfaces at a constant rate of q Btu/h·ft², while heat is dissipated by convection from the other boundary surface into a medium at a uniform temperature T_∞ with a heat-transfer coefficient h. Figure 4-10 shows the geometry and the boundary conditions for the considered problem. When the value of the Biot number is less than about 0.1, this heat-transfer problem can be analyzed by lumped-system analysis. For the one-lump situation considered here, the statement of the energy-balance equation is the same as that given by Eq. (4-84). The substitution of the appropriate mathematical expressions into this equation, on the assumption that the surface areas are equal on both sides of the slab for the heat supply and convection, results in the energy equation for this problem:

$$Aq + Ah[T_\infty - T(t)] = \rho c_p \, AL \frac{dT(t)}{dt}$$

or

$$q + h[T_\infty - T(t)] = \rho c_p L \frac{dT(t)}{dt} \qquad \text{for } t > 0 \qquad (4\text{-}94a)$$

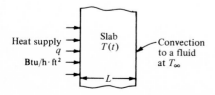

FIG. 4-10 Nomenclature for one-lump analysis of transient heat flow in a slab with a heat supply at one surface and convection from the other surface.

with the initial condition

$$T(t) = T_0 \qquad \text{for } t = 0 \tag{4-94b}$$

For convenience in the analysis a new temperature $\theta(t)$ is defined as

$$\theta(t) = T(t) - T_\infty \tag{4-95}$$

Then Eqs. (4-94) become

$$\frac{d\theta(t)}{dt} + m\theta(t) = Q \qquad \text{for } t > 0 \tag{4-96a}$$

$$\theta(t) = T_0 - T_\infty \equiv \theta_0 \qquad \text{for } t = 0 \tag{4-96b}$$

where we have defined

$$m \equiv \frac{h}{\rho c_p L} \qquad \text{and} \qquad Q \equiv \frac{q}{\rho c_p L} \tag{4-97}$$

The solution of Eq. (4-96a) is written as a sum of the solution of the homogeneous part of Eq. (4-96a) and a particular solution in the form

$$\theta(t) = Ce^{-mt} + \theta_p \tag{4-98}$$

where C is the integration constant. The particular solution θ_p is given by

$$\theta_p = \frac{Q}{m} \tag{4-99}$$

By combining Eqs. (4-98) and (4-99) we obtain

$$\theta(t) = Ce^{-mt} + \frac{Q}{m} \tag{4-100}$$

The integration constant C is determined by the application of the initial condition (4-96b) as

$$\theta_0 = C + \frac{Q}{m} \tag{4-101}$$

Substitution of Eq. (4-101) into (4-100) gives the solution of this heat-transfer problem:

$$\theta(t) = \theta_0 e^{-mt} + (1 - e^{-mt})\frac{Q}{m} \tag{4-102}$$

For $t \to \infty$ this solution simplifies to

$$\theta(\infty) = \frac{Q}{m} \tag{4-103}$$

which is the steady-state temperature in the slab.

Two-Lump System with Convection Boundaries

The lumped-system analysis described above is now extended to study temperature transients in a composite system consisting of two different bodies. To illustrate the basic approach for the analysis, we consider a well-stirred fluid at temperature $T_i(t)$ inside a closed container at temperature $T_c(t)$. The container is immersed in a well-stirred fluid at a constant temperature T_∞, as illustrated in Fig. 4-11. In this system the temperatures $T_i(t)$ and $T_c(t)$ of the inner fluid and the container, respectively, change with time as a result of convective heat transfer from the inner fluid to the outer fluid through the container walls. In this problem we are concerned with the determination of the temperatures $T_i(t)$ and $T_c(t)$ as a function of time if initially (that is, $t = 0$) the inner fluid and the container are both at the same temperature T_0.

Let the volume, density, and specific heat for the inside fluid and the container, be V_i, ρ_i, c_{pi}, and V_c, ρ_c, c_{pc}, respectively. The heat-transfer coefficient and the heat-transfer area for the inside and outside surfaces of the container are, respectively, h_i, A_i, and h_o, A_o. By treating the inside fluid and the container each as a single-lump system, the statement of the energy-balance equation for the inside fluid or the container is written as given by Eq. (4-84). Then the energy-balance equations for the inside fluid and the container become

Inside fluid:
$$A_i h_i [T_c(t) - T_i(t)] = \rho_i c_{pi} V_i \frac{dT_i(t)}{dt} \tag{4-104a}$$

Container:
$$A_i h_i [T_i(t) - T_c(t)] + A_o h_o [T_\infty - T_c(t)] = \rho_c c_{pc} V_c \frac{dT_c(t)}{dt} \tag{4-104b}$$

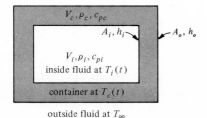

outside fluid at T_∞

FIG. 4-11   Nomenclature for two-lump analysis of transient heat flow.

with the initial conditions

$$T_i(t) = T_0 \quad \text{and} \quad T_c(t) = T_0 \quad \text{for } t = 0 \tag{4-105}$$

Now, new temperatures $\theta_i(t)$ and $\theta_c(t)$ are defined as

$$\theta_i(t) = T_i(t) - T_\infty \quad \text{and} \quad \theta_c(t) = T_c(t) - T_\infty \tag{4-106}$$

Then Eqs. (4-104) and (4-105) become, respectively,

$$\frac{d\theta_i(t)}{dt} + m_1[\theta_i(t) - \theta_c(t)] = 0 \qquad \text{for } t > 0 \tag{4-107a}$$

$$\frac{d\theta_c(t)}{dt} + m_2[\theta_c(t) - \theta_i(t)] + m_3\,\theta_c(t) = 0 \qquad \text{for } t > 0 \tag{4-107b}$$

with the initial conditions

$$\theta_i(t) = T_0 - T_\infty \equiv \theta_0 \quad \text{and} \quad \theta_c(t) = T_0 - T_\infty \equiv \theta_0 \qquad \text{for } t = 0 \tag{4-108}$$

where

$$m_1 \equiv \frac{A_i h_i}{\rho_i V_i c_{pi}} \qquad m_2 \equiv \frac{A_i h_i}{\rho_c V_c c_{pc}} \quad \text{and} \quad m_3 \equiv \frac{A_o h_o}{\rho_c V_c c_{pc}} \tag{4-109}$$

Equations (4-107) with the initial conditions (4-108) provide two coupled first-degree ordinary differential equations for the two unknown temperatures $\theta_i(t)$ and $\theta_c(t)$ of the inner fluid and the container. Analytical solution can be obtained for these equations in a straightforward manner, but as the results are rather lengthy they are not presented here. The purpose of the foregoing analysis for a two-lump system is to show that, for a composite system consisting of two or more bodies, the number of differential equations increases accordingly. For such cases the solution of the resulting coupled equations by analytical means is rather cumbersome, but their solution with a digital computer poses no problem.

4-7   LUMPED-SYSTEM ANALYSIS FOR PERIODIC HEAT FLOW

There are many engineering applications in which heat flow into a system varies periodically all the time. In such situations, after the transients have passed, the temperature in the medium follows the applied variation with some phase lag and continues to vary periodically. In this section one-lump-system analysis is used to determine the periodic variation of temperature in a medium resulting from a periodically varying boundary heat flux, after the transients have passed.

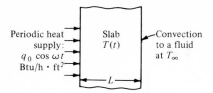

FIG. 4-12   Nomenclature for one-lump analysis of periodic heat flow in a slab.

Consider a slab of thickness L, with periodically varying applied heat flux $q_0 \cos \omega t$ at one face and convection into an environment at a constant temperature T_∞ with a heat-transfer coefficient h at the other face, as illustrated in Fig. 4-12.   The energy-balance equation for this case is immediately obtained from Eq. (4-94a) by replacing in that equation q by $q_0 \cos \omega t$.   We find

$$q_0 \cos \omega t + h[T_\infty - T(t)] = \rho c_p L \frac{dT(t)}{dt} \tag{4-110}$$

By defining a new variable $\theta(t)$ as

$$\theta(t) = T(t) - T_\infty \tag{4-111}$$

Eq. (4-110) becomes

$$\frac{d\theta(t)}{dt} + m\theta(t) = Q \cos \omega t \tag{4-112}$$

where

$$m \equiv \frac{h}{\rho c_p L} \quad \text{and} \quad Q \equiv \frac{q_0}{\rho c_p L} \tag{4-113}$$

The complete solution of Eq. (4-112) is taken as the sum of the homogeneous solution $\theta_h(t)$ and a particular solution $\theta_p(t)$ in the form

$$\theta(t) = \theta_h(t) + \theta_p(t) \tag{4-114}$$

where $\theta_h(t)$ is the solution of the homogeneous form of Eq. (4-112), that is, $d\theta_h/dt + m\theta_h = 0$, and its solution is given by

$$\theta_h(t) = C_1 e^{-mt} \tag{4-115}$$

A particular solution $\theta_p(t)$ of Eq. (4-112) is a solution which should satisfy only Eq. (4-112) but not its initial condition.   The method of finding a particular solution of ordinary differential equations with constant coefficients is described in the standard texts on mathematics [10].

For the first-order equation considered above a particular solution $\theta_p(t)$ is taken as a sum of the nonhomogeneous term $\cos \omega t$ and its first derivative in the form

$$\theta_p(t) = A \cos \omega t + B \sin \omega t \tag{4-116}$$

where the unknown coefficients A and B can be determined by constraining this particular solution to satisfy the differential equation (4-112). That is, the substitution of Eq. (4-116) into (4-112) yields

$$(-\omega A \sin \omega t + \omega B \cos \omega t) + m(A \cos \omega t + B \sin \omega t) = Q \cos \omega t$$

which is rearranged as

$$(\omega B + mA - Q) \cos \omega t + (mB - \omega A) \sin \omega t = 0 \tag{4-117}$$

If Eq. (4-117) should be valid for all t, the coefficients of $\cos \omega t$ and $\sin \omega t$ should be identically zero. This requirement leads to the following two algebraic equations for the determination of the coefficients A and B:

$$\omega B + mA - Q = 0 \tag{4-118a}$$

$$mB - \omega A = 0 \tag{4-118b}$$

A simultaneous solution of Eqs. (4-118) gives

$$A = \frac{m}{m^2 + \omega^2} Q \tag{4-119a}$$

$$B = \frac{\omega}{m^2 + \omega^2} Q \tag{4-119b}$$

Then a particular solution becomes

$$\theta_p(t) = \frac{Q}{m^2 + \omega^2} (m \cos \omega t + \omega \sin \omega t) \tag{4-120}$$

and the complete solution for $\theta(t)$ is obtained by substituting Eqs. (4-115) and (4-120) into Eq. (4-114):

$$\theta(t) = C_1 e^{-mt} + \frac{Q}{m^2 + \omega^2} (m \cos \omega t + \omega \sin \omega t) \tag{4-121}$$

If the complete transient solution were required, the unknown constant C_1 could be determined for a given initial condition. In this problem we are interested

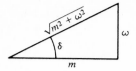

FIG. 4-13   Definition of the parameter δ.

in the solution for large times, that is, as $t \to \infty$.   Clearly, for very large values of t the exponential term in Eq. (4-121) vanishes and the solution for large times reduces to

$$\theta(t) = \frac{Q}{m^2 + \omega^2} (m \cos \omega t + \omega \sin \omega t) \qquad \text{for large } t \tag{4-122}$$

This solution can be expressed in a more compact form if a new parameter δ is defined as

$$\delta = \tan^{-1} \frac{\omega}{m} \qquad \text{or} \qquad \tan \delta = \frac{\omega}{m} \tag{4-123}$$

Then, by referring to the definition of $\tan \delta$ as illustrated in Fig. 4-13, we write

$$\sin \delta = \frac{\omega}{\sqrt{m^2 + \omega^2}} \qquad \text{or} \qquad \omega = \sqrt{m^2 + \omega^2} \sin \delta \tag{4-124a}$$

$$\cos \delta = \frac{m}{\sqrt{m^2 + \omega^2}} \qquad \text{or} \qquad m = \sqrt{m^2 + \omega^2} \cos \delta \tag{4-124b}$$

Substituting ω and m from Eqs. (4-124) into Eq. (4-122), we obtain

$$\theta(t) = \frac{Q}{\sqrt{m^2 + \omega^2}} (\cos \delta \cos \omega t + \sin \delta \sin \omega t)$$

or

$$\theta(t) = \frac{Q}{\sqrt{m^2 + \omega^2}} \cos (\omega t - \delta) \tag{4-125}$$

This result is the solution of the differential equation (4-112) after the temperature transients have passed.   Here the term $Q/\sqrt{m^2 + \omega^2}$ is the *amplitude* of the temperature oscillation in the slab, and δ is the *phase lag* of this oscillation in comparison with the oscillation of the applied heat flux.

   Example 4-8   A 1-in (2.54×10^{-2}-m) thick steel plate ($\rho = 490$ lb/ft³ or 7849 kg/m³, $c_p = 0.11$ Btu/lb·°F or 460.5 W·s/kg·°C, $k_s = 25$ Btu/h·ft·°F or 43.3 W/m·°C) is subjected to a periodically oscillating heat flux of magnitude $q = q_0 \cos \omega t$, where q_0 is in Btu/h·ft² (or W/m^2), at one of its surfaces and cooled by convection from the other surface into a medium at a constant temperature $T_\infty = 100°F$ (37.8°C) with a heat-transfer coefficient $h = 15$ Btu/h·ft^2·°F (85.2 W/m^2·°C). Determine the temperature of the slab after the transients have passed. Take $\omega = 2\pi$ h^{-1} (or $2\pi/3600$ s^{-1}) for the period of oscillation of the heat flux.

Solution This problem can be solved by using the foregoing result [i.e., Eq. (4-125)] for the periodic-heat-flow analysis if the Biot number is sufficiently small. We have

$$Bi = \frac{hL}{k_s} = \frac{15}{25 \times 12} = 0.05$$

If the SI system of units were used, the Biot number would be the same since Bi is a dimensionless quantity, (that is, Bi = $85.2 \times (2.54 \times 10^{-2})/43.3 \cong 0.05$). The Biot number being sufficiently small, the lump-system analysis is applicable. The temperature of the slab is given by Eq. (4-125):

$$T(t) - T_\infty = \frac{Q}{\sqrt{m^2 + \omega^2}} \cos(\omega t - \delta)$$

where

$$Q = \frac{q_0}{\rho c_p L} \qquad m = \frac{h}{\rho c_p L} \qquad \omega = 2\pi \qquad h^{-1}$$

$$\delta = \tan^{-1}\frac{\omega}{m} \qquad \text{and} \qquad T_\infty = 100°F \ (37.8°C)$$

Various quantities are evaluated as

$$m = \frac{15}{490 \times 0.11 \times \frac{1}{12}} = 3.34 \text{ h}^{-1} \qquad \left[\text{or } m = \frac{85.2}{7849 \times 460.5 \times (2.54 \times 10^{-2})} \right.$$

$$\left. = 9.280 \times 10^{-4} \text{ s}^{-1} \right]$$

$$\frac{\omega}{m} = \frac{2\pi}{3.37} = 1.88 \qquad \left[\text{or } \frac{\omega}{m} = \frac{2\pi}{3600 \times (9.280 \times 10^{-4})} = 1.88 \right]$$

Clearly, the dimensionless quantity ω/m is the same from alculations in both systems of units. Then,

$$\delta = \tan^{-1} 1.88 = 0.344\pi$$

$$\sqrt{m^2 + \omega^2} = \sqrt{3.37^2 + (2\pi)^2} = 7.1 \text{ h}^{-1}$$

If the English engineering system of units is used, we have

$$\frac{Q}{\sqrt{m^2 + \omega^2}} = \frac{q_0}{\rho c_p L \sqrt{m^2 + \omega^2}} = \frac{q_0}{4.5 \times 7.1} = \frac{q_0}{32} \qquad °F$$

Substituting, the temperature of the slab after the transients have passed becomes

$$T(t) = 100 + \frac{q_0}{32} \cos(2\pi t - 0.344\pi) \qquad °F$$

where t is measured in hours. We note that the amplitude of temperature oscillations in the slab is $q_0/32°F$, and the temperature oscillation in the medium follows the applied oscillations with a pulse lag of 0.344π rad.

REFERENCES

1 Churchill, Ruel V.: "Fourier Series and Boundary Value Problems," McGraw-Hill Book Company, New York, 1963.

2 Carslaw, H. S., and J. C. Jaeger: "Conduction of Heat in Solids," 2d ed., Oxford University Press, London, 1959.

3 Özışık, M. N.: "Boundary Value Problems of Heat Conduction," International Textbook Company, Scranton, Pa., 1968.

4 Arpaci, V. S.: "Conduction Heat Transfer," Addison-Wesley Publishing Company, Inc., Reading, Mass., 1966.

5 Grove, W. E.: "Brief Numerical Methods," Prentice-Hall, Inc., Englewood Cliffs, N.J., 1966.

6 Heisler, M. P.: Temperature Charts for Induction and Constant Temperature Heating, *Trans. ASME,* **69:**227–236 (1947).

7 Schneider, P. J.: "Temperature Response Charts," John Wiley & Sons, Inc., New York, 1963.

8 Holman, J. P.: "Heat Transfer," 3d ed., McGraw-Hill Book Company, New York, 1972.

9 Kreith, F.: "Principles of Heat Transfer," 3d ed., Intext Educational Publishers, New York, 1973.

10 Forsyth, A. R.: "A Treatise on Differential Equations," The Macmillan Company, New York, 1956.

Five

Conduction—
Numerical Methods of Solution

The numerical method of solution is used extensively in practical applications to determine the temperature distribution and heat flow in solids having complicated geometries and boundary conditions. A commonly used numerical scheme is the *finite-difference* method which is described in several references [1–10]. In this approach the partial differential equation of heat conduction is approximated by a set of algebraic equations for temperature at a number of nodal points over the region. Therefore, the first step in the analysis is the finite-difference representation or the transformation into a set of algebraic equations of the differential equation of heat conduction. This matter will be discussed in this chapter with emphasis on the heat-conduction equation in the rectangular coordinate system. The problems of two-dimensional steady-state heat conduction and one-dimensional time-dependent heat conduction will be considered. A brief discussion will be given of the methods of solution of the resulting set of algebraic equations. Before computing facilities were available, the *relaxation method* described by Southwell [8] was used to solve the resulting system of equations by hand calculations. This is a very time-consuming process and requires some experience before it can be applied effectively. On the other hand, the solution of simultaneous algebraic equations with the present-day computing facilities poses no problem. Therefore in this chapter attention will be focused on computer solutions.

In recent years another numerical scheme called the *finite-element* method, developed originally for the solution of structural problems, has also been applied to the solution of heat-conduction problems. For problems with complex geometries the finite-element method offers some advantage over the finite-difference method in the solution of heat-conduction problems. In the limited space available here, this method will not be discussed; however, the reader interested in this subject may consult Refs. [11–15].

5-1   FINITE-DIFFERENCE APPROXIMATION OF TWO-DIMENSIONAL, STEADY-STATE HEAT-CONDUCTION EQUATION

Consider the two-dimensional steady-state heat-conduction equation with the energy-generation term given in the form

$$\frac{\partial^2 T}{\partial x^2} + \frac{\partial^2 T}{\partial y^2} + \frac{g(x, y)}{k} = 0 \qquad \text{in region R} \tag{5-1}$$

subject to a set of boundary conditions. In the exact solution of this equation by an analytical method, the differential equation is satisfied everywhere in the region R. In the numerical method of solution, however, the region R is replaced by a set of discrete points, various derivatives are represented by a finite-difference approximation between the nodal points, and the partial differential

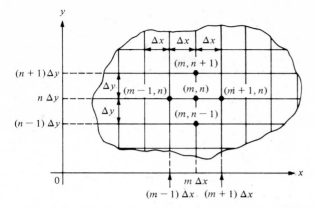

FIG. 5-1 Rectangular net of mesh Δx, Δy for the finite difference of derivatives.

equation is transformed into a set of algebraic equations for each nodal point; hence, the equation is satisfied only at the nodal points. To illustrate the finite-difference approximation of the above equation, we consider a rectangular net of mesh size Δx, Δy constructed over the region as shown in Fig. 5-1. The symbol m, n is used to denote the location of a nodal point whose coordinates are $x = m\,\Delta x$, $y = n\,\Delta y$. We focus our attention on a nodal point (m, n) and its four neighboring points as shown in Fig. 5-2. Let $T_{m, n}$ be the temperature at the node m, n, $T_{m+1, n}$ at the node $m + 1, n$, and so forth. The first derivative of temperature with respect to x at the node $m + \frac{1}{2}, n$ can be approximated as

$$\frac{\partial T}{\partial x}\bigg|_{m+1/2, n} \cong \frac{T_{m+1, n} - T_{m, n}}{\Delta x} \tag{5-2a}$$

and at the node $m - \frac{1}{2}, n$ as

$$\frac{\partial T}{\partial x}\bigg|_{m-1/2, n} \cong \frac{T_{m, n} - T_{m-1, n}}{\Delta x} \tag{5-2b}$$

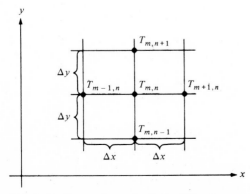

FIG. 5-2 Nomenclature for finite difference.

Then, the second derivative of temperature with respect to the x variable at m, n becomes

$$\frac{\partial^2 T}{\partial x^2}\bigg|_{m, n} \cong \frac{\partial T/\partial x|_{m+1/2, n} - \partial T/\partial x|_{m-1/2, n}}{\Delta x} = \frac{T_{m-1, n} + T_{m+1, n} - 2T_{m, n}}{(\Delta x)^2} \tag{5-3}$$

Similarly, the second derivative of temperature with respect to the y variable at the node m, n is given as

$$\frac{\partial^2 T}{\partial y^2}\bigg|_{m, n} \cong \frac{T_{m, n-1} + T_{m, n+1} - 2T_{m, n}}{(\Delta y)^2} \tag{5-4}$$

The substitution of the results in Eqs. (5-3) and (5-4) into Eq. (5-1) gives the finite-difference approximation of the differential equation of heat conduction at the node m, n as

$$\frac{T_{m-1, n} + T_{m+1, n} - 2T_{m, n}}{(\Delta x)^2} + \frac{T_{m, n-1} + T_{m, n+1} - 2T_{m, n}}{(\Delta y)^2} + \frac{g_{m, n}}{k} = 0 \tag{5-5}$$

For a square mesh

$$\Delta x = \Delta y = l \tag{5-6}$$

Equation (5-5) simplifies to

$$(T_{m-1, n} + T_{m+1, n} + T_{m, n-1} + T_{m, n+1} - 4T_{m, n}) + \frac{g_{m, n} l^2}{k} = 0 \tag{5-7}$$

Equation (5-7) is the *finite-difference representation* of the heat-conduction equation (5-1) for the node m, n. Note that the terms inside the parentheses in Eq. (5-7) are an addition of the temperatures at the four nodes surrounding the node m, n and subtracting from it four times the temperature $T_{m, n}$ of the node m, n. Here, l is the mesh size, $g_{m, n}$ is the heat-generation rate per unit volume at m, n, and k is the thermal conductivity.

The finite-difference equation (5-7) is *applicable for nodes in the interior of the region but not applicable if the node m, n is on a boundary.* We now illustrate the derivation of the finite-difference equation for a node m, n on an adiabatic boundary, on a convection boundary, and at the intersection of two such boundaries.

Node m, n on an Adiabatic (Insulated) Boundary

Consider a node m, n on an adiabatic (or insulated) boundary which is parallel to the y axis, as shown in Fig. 5-3. Let $T_{m, n}$ be the temperature at the node m, n, and $T_{m, n-1}$, $T_{m, n+1}$, and $T_{m+1, n}$ be the temperatures at the three neighboring

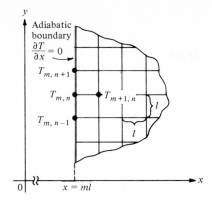

FIG. 5-3   Nodal point (m,n) on an adiabatic boundary parallel to the y axis.

nodes, as illustrated in Fig. 5-3. The adiabatic boundary for which $\partial T/\partial x = 0$ is equivalent to a symmetry condition about this boundary. Then, we consider the existence of a mirror-image point $m - 1$, n of the node $m + 1$, n with respect to this boundary. If the temperature at this image node is $T_{m-1, n}$, then by symmetry we have $T_{m-1, n} = T_{m+1, n}$. When the region and its mirror image with respect to the boundary are considered as a single region, the node m, n becomes an interior point and the finite-difference equation (5-7) is applicable with $T_{m-1, n} = T_{m+1, n}$. Then, *the finite-difference equation for the node m, n on an adiabatic boundary parallel to the y axis, as shown in Fig. 5-3, becomes*

$$(2T_{m+1, n} + T_{m, n-1} + T_{m, n+1} - 4T_{m, n}) + \frac{g_{m, n}l^2}{k} = 0 \tag{5-8}$$

If the node m, n is on an adiabatic boundary parallel to the x axis, as shown in Fig. 5-4, the finite-difference equation is determined with similar symmetry consideration by setting, in Eq. (5-7), $T_{m, n-1} = T_{m, n+1}$; one finds

$$(T_{m-1, n} + T_{m+1, n} + 2T_{m, n+1} - 4T_{m, n}) + \frac{g_{m, n}l^2}{k} = 0 \tag{5-9}$$

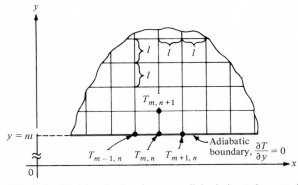

FIG. 5-4   Nodal point (m,n) on an adiabatic boundary parallel to the x axis.

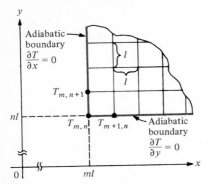

FIG. 5-5   Nodal point (m,n) at the intersection of two adiabatic boundaries.

Node m, n at the Intersection of Two Adiabatic Boundaries

Consider a node m, n at the intersection of two adiabatic boundaries as illustrated in Fig. 5-5.   Let $T_{m,\,n+1}$ and $T_{m+1,\,n}$ be the temperatures of the two neighboring nodes.   Considering the existence of the image points $m, n-1$ and $m-1, n$ and the applicability of Eq. (5-7) with the requirement $T_{m-1,\,n} = T_{m+1,\,n}$ and $T_{m,\,n-1} = T_{m,\,n+1}$, one obtains the following finite-difference equation for the node m, n at the intersection of the two adiabatic boundaries shown in Fig. 5-5.

$$(2T_{m+1,\,n} + 2T_{m,\,n+1} - 4T_{m,\,n}) + \frac{g_{m,\,n}\,l^2}{k} = 0 \qquad (5\text{-}10)$$

Node m, n on a Convection Boundary

We now consider a node m, n on a boundary subjected to heat exchange by convection with an environment at temperature T_∞ and with a heat-transfer coefficient h, as illustrated in Fig. 5-6.   In this case the finite-difference equation for the node m, n cannot be derived from Eq. (5-7); it should be determined

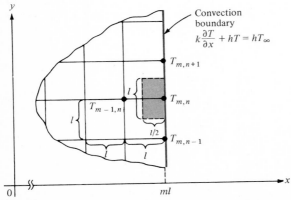

FIG. 5-6   Nodal point (m,n) at the convection boundary.

either by the finite-difference representation of the heat-conduction equation with proper cognizance of the convection-boundary condition or by writing an energy balance on the shaded volume element illustrated in Fig. 5-6. We prefer the latter approach and write the statement of the energy balance as

$$\begin{pmatrix}\text{Rate of heat entering volume} \\ \text{element through boundaries}\end{pmatrix} + \begin{pmatrix}\text{rate of heat generated} \\ \text{in volume element}\end{pmatrix} = 0 \qquad (5\text{-}11)$$

By considering a unit depth perpendicular to the plane of Fig. 5-6, we write the appropriate mathematical expressions for the two terms in this energy equation and obtain

$$\left[k\,\frac{T_{m-1,\,n} - T_{m,\,n}}{l}\,1l + k\,\frac{T_{m,\,n-1} - T_{m,\,n}}{l}\,1\frac{l}{2} + k\,\frac{T_{m,\,n+1} - T_{m,\,n}}{l}\,1\frac{l}{2} \right.$$
$$\left. + h(T_\infty - T_{m,\,n})1l \right] + \left(g_{m,\,n}\,l\frac{l}{2}\,1 \right) = 0$$

and, after simplification, the *finite-difference equation for the node m, n at the convection boundary* shown in Fig. 5-6 becomes

$$\left[2T_{m-1,\,n} + T_{m,\,n-1} + T_{m,\,n+1} - \left(4 + \frac{2hl}{k} \right)T_{m,\,n} \right] + \left(\frac{2hlT_\infty}{k} + \frac{g_{m,\,n}\,l^2}{k} \right) = 0 \qquad (5\text{-}12)$$

Node m, n at the Intersection of an Adiabatic and a Convection Boundary

We now consider a node m, n at the intersection of an adiabatic and a convection boundary, as shown in Fig. 5-7. The finite-difference equation for this point is derived by writing an energy-balance equation on the shaded area shown in this figure. The statement of this energy-balance equation is the same as given by Eq. (5-11); then, by writing the appropriate mathematical expressions, we obtain

$$\left[k\,\frac{T_{m-1,\,n} - T_{m,\,n}}{l}\,1\frac{l}{2} + k\,\frac{T_{m,\,n+1} - T_{m,\,n}}{l}\,1\frac{l}{2} + h(T_\infty - T_{m,\,n})1\frac{l}{2} \right]$$
$$+ \left(g_{m,\,n}\,\frac{l}{2}\frac{l}{2}\,1 \right) = 0$$

and, after simplification, the *finite-difference equation for the node m, n* in Fig. 5-7 becomes

$$\left[2T_{m-1,\,n} + 2T_{m,\,n+1} - \left(4 + \frac{2hl}{k} \right)T_{m,\,n} \right] + \left(\frac{2hlT_\infty}{k} + \frac{g_{m,\,n}\,l^2}{k} \right) = 0 \qquad (5\text{-}13)$$

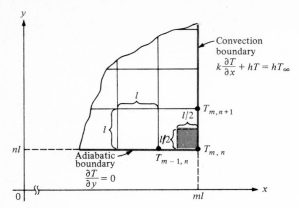

FIG. 5-7 Nodal point (m,n) at the intersection of a convection boundary and an adiabatic boundary.

Node m, n Next to a Curved Boundary

Consider a node m, n next to a curved boundary which intersects two of the strings from the point m, n, as shown in Fig. 5-8. It is assumed that temperature is prescribed at the curved boundary so that temperatures T_C and T_D are known at the points C and D, respectively, where the boundary intersects the strings. If the lengths of the strings from the point m, n to the points C and D are, respectively, ηl and ξl, where $\eta, \xi \leq 1$, then the finite-difference equation for the node m, n is given by [16, p. 425]

$$2\left[\frac{T_D}{\xi(1 + \xi)} + \frac{T_C}{\eta(1 + \eta)} + \frac{T_{m-1, n}}{1 + \xi} + \frac{T_{m, n-1}}{1 + \eta} - \left(\frac{1}{\xi} + \frac{1}{\eta}\right)T_{m, n}\right] + \frac{g_{m, n}l^2}{k} = 0 \qquad (5\text{-}14)$$

where

$$0 < \xi \leq 1 \qquad \text{and} \qquad 0 < \eta \leq 1$$

We note that for $\xi = \eta = 1$ this equation reduces to Eq. (5-7).

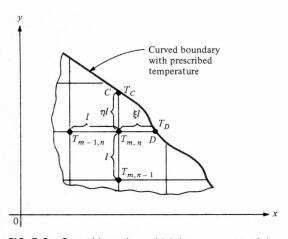

FIG. 5-8 Curved boundary which intersects two of the strings from the nodal point (m,n).

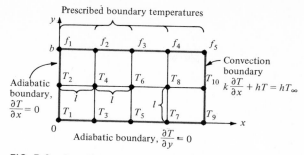

FIG. 5-9 Boundary conditions and the network for Example 5-1.

Finite-difference equations for many other special situations that are not considered here can be derived in a manner discussed above. We now illustrate the finite-difference representation of a simple heat-conduction problem with the following example.

Example 5-1 Consider a rectangular region in $0 \le x \le a$, $0 \le y \le b$, as shown in Fig. 5-9. The boundaries at $x = 0$ and $y = 0$ are adiabatic (i.e., insulated), the boundary at $y = b$ is kept at a prescribed temperature, the boundary at $x = a$ dissipates heat by convection into a medium at a temperature T_∞ with a heat-transfer coefficient h, and heat is generated in the medium at a constant rate of g Btu/h·ft³. The thermal conductivity k of the medium is constant. Write the finite-difference representation of this heat-conduction problem by using the relatively coarse network drawn over the region as shown in Fig. 5-9.

Solution In the considered heat-conduction problem the temperatures f_1, f_2, f_3, f_4, and f_5 at the five nodes along the boundary $y = b$ are prescribed and known, but the temperatures T_i, $i = 1$ to 10, at the nodes in the region and on the adiabatic and convective boundaries are unknown. Therefore, the problem involves 10 unknown temperatures T_i's $(i = 1$ to $10)$. The 10 equations needed for the determination of these temperatures are obtained by writing the appropriate finite-difference equation for each of these nodes. These equations are immediately obtainable from the finite-difference equations given above; we summarize below these 10 equations and indicate from which equations they are obtained.

Node Number	Finite-Difference Equation	Obtained from
1	$2T_2 + 2T_3 - 4T_1 + gl^2/k = 0$	Eq. (5-10)
2	$T_1 + f_1 + 2T_4 - 4T_2 + gl^2/k = 0$	Eq. (5-8)
3	$T_1 + T_5 + 2T_4 - 4T_3 + gl^2/k = 0$	Eq. (5-9)
4	$T_2 + T_3 + T_6 + f_2 - 4T_4 + gl^2/k = 0$	Eq. (5-7)
5	$T_3 + T_7 + 2T_6 - 4T_5 + gl^2/k = 0$	Eq. (5-9)
6	$T_4 + T_5 + T_8 + f_3 - 4T_6 + gl^2/k = 0$	Eq. (5-7)
7	$T_5 + T_9 + 2T_8 - 4T_7 + gl^2/k = 0$	Eq. (5-9)
8	$T_6 + T_{10} + T_7 + f_4 - 4T_8 + gl^2/k = 0$	Eq. (5-7)
9	$2T_7 + 2T_{10} - \left(4 + \dfrac{2hl}{k}\right)T_9 + \left(\dfrac{2hlT_\infty}{k} + \dfrac{gl^2}{k}\right) = 0$	Eq. (5-13)
10	$2T_8 + T_9 + f_5 - \left(4 + \dfrac{2hl}{k}\right)T_{10} + \left(\dfrac{2hlT_\infty}{k} + \dfrac{gl^2}{k}\right) = 0$	Eq. (5-12)

These 10 equations can be expressed in the matrix form:

$$
\begin{bmatrix}
-4 & 2 & 2 & 0 & 0 & 0 & 0 & 0 & 0 & 0 \\
1 & -4 & 0 & 2 & 0 & 0 & 0 & 0 & 0 & 0 \\
1 & 0 & -4 & 2 & 1 & 0 & 0 & 0 & 0 & 0 \\
0 & 1 & 1 & -4 & 0 & 1 & 0 & 0 & 0 & 0 \\
0 & 0 & 1 & 0 & -4 & 2 & 1 & 0 & 0 & 0 \\
0 & 0 & 0 & 1 & 1 & -4 & 0 & 1 & 0 & 0 \\
0 & 0 & 0 & 0 & 1 & 0 & -4 & 2 & 1 & 0 \\
0 & 0 & 0 & 0 & 0 & 1 & 1 & -4 & 0 & 1 \\
0 & 0 & 0 & 0 & 0 & 0 & 2 & 0 & -(4+H) & 2 \\
0 & 0 & 0 & 0 & 0 & 0 & 0 & 2 & 1 & -(4+H)
\end{bmatrix}
\begin{bmatrix}
T_1 \\ T_2 \\ T_3 \\ T_4 \\ T_5 \\ T_6 \\ T_7 \\ T_8 \\ T_9 \\ T_{10}
\end{bmatrix}
=
\begin{bmatrix}
-G \\
-G - f_1 \\
-G \\
-G - f_2 \\
-G \\
-G - f_3 \\
-G \\
-G - f_4 \\
-G - HT_\infty \\
-G - HT_\infty - f_5
\end{bmatrix}
\tag{5-15}
$$

where we have defined

$$
G \equiv \frac{gl^2}{k} \quad \text{and} \quad H \equiv \frac{2hl}{k} \tag{5-16}
$$

Once the numerical values of the parameters G and H are specified, Eqs. (5-15) can be solved for the 10 unknown nodal-point temperatures. A very coarse network used in this example resulted in only 10 simultaneous equations; if a finer network had been used for improved accuracy the number of equations to be solved would have increased accordingly.

It is now apparent that, in the finite-difference approach, the heat-conduction equation is replaced by a set of simultaneous algebraic equations for temperatures at the nodal points of a network constructed over the region. These equations can readily be solved with a high-speed digital computer. In order to give some idea of the method of solution of these equations we give below a brief discussion of the gaussian elimination and the iterative methods of solution.

5-2 GAUSSIAN ELIMINATION AND ITERATIVE METHODS OF SOLUTION

Gaussian Elimination Method

The finite-difference representation of a two-dimensional, steady-state heat-conduction equation results in a set of algebraic equations which form a *banded matrix*, as shown in Eqs. (5-15). That is, the nonzero elements of the matrix are in a band on either side of the diagonal.   A system of coupled algebraic equations which form a banded matrix is solved efficiently with a digital computer by using the gaussian elimination process.   To illustrate the basis of this method we consider a banded matrix as shown in Eqs. (5-17).   This matrix is transformed into an upper diagonal form shown in Eqs. (5-18) in the following manner.   The first equation in the system of Eqs. (5-17) is used to eliminate the nonzero elements a_{21} and a_{31} in the first column.   That is, the first equation is multiplied by a_{21}/a_{11}, and the resulting equation is subtracted from the second equation in order to eliminate a_{21}; the first equation is then utilized in a similar manner to eliminate a_{31}.   The second equation is then used to eliminate a_{32} and a_{42}. The third equation is used to eliminate a_{43}, and so forth.   When this process is carried out to the last equation, the resulting equation forms an upper diagonal matrix as shown in Eqs. (5-18).   Then, the last equation in Eqs. (5-18) immediately gives T_n.   Knowing T_n, the temperature T_{n-1} is determined from the $(n-1)$th equation, and the calculations are carried out until T_1 is determined from the first equation.

$$
\begin{bmatrix}
a_{11} & a_{12} & a_{13} & 0 & \cdots & 0 & 0 \\
a_{21} & a_{22} & a_{23} & a_{24} & \cdots & 0 & 0 \\
a_{31} & a_{32} & a_{33} & a_{34} & \cdots & \cdots & 0 \\
0 & a_{42} & a_{43} & a_{44} & \cdots & \cdots & 0 \\
\cdots & \cdots & \cdots & \cdots & \cdots & \cdots & \cdots \\
0 & 0 & 0 & \cdots & a_{n,n-2} & a_{n,n-1} & a_{m,n}
\end{bmatrix}
\begin{bmatrix}
T_1 \\ T_2 \\ T_3 \\ T_4 \\ \vdots \\ T_n
\end{bmatrix}
=
\begin{bmatrix}
C_1 \\ C_2 \\ C_3 \\ C_4 \\ \vdots \\ C_n
\end{bmatrix}
\tag{5-17}
$$

$$
\begin{bmatrix}
a_{11}^* & a_{12}^* & a_{13}^* & 0 & \cdots & \cdots & 0 \\
0 & a_{22}^* & a_{23}^* & a_{24}^* & \cdots & \cdots & 0 \\
0 & 0 & a_{33}^* & a_{34}^* & \cdots & \cdots & \cdots \\
0 & 0 & 0 & a_{44}^* & \cdots & \cdots & \cdots \\
\cdots & \cdots & \cdots & \cdots & \cdots & \cdots & \cdots \\
0 & \cdots & \cdots & \cdots & 0 & 0 & a_{n-1,n-1}^* & a_{n-1,n}^* \\
0 & 0 & 0 & \cdots & 0 & 0 & 0 & a_{n,n}^*
\end{bmatrix}
\begin{bmatrix}
T_1 \\ T_2 \\ T_3 \\ T_4 \\ \vdots \\ T_{n-1} \\ T_n
\end{bmatrix}
=
\begin{bmatrix}
C_1^* \\ C_2^* \\ C_3^* \\ C_4^* \\ \vdots \\ C_{n-1}^* \\ C_n^*
\end{bmatrix}
\tag{5-18}
$$

Computer subroutines are available for solving simultaneous algebraic equations by means of an elimination process as described above or by methods which are variations of the gaussian elimination process.

Iterative Methods

When the number of equations is very large, the matrix is not sparse, and the computer storage is critical, an iterative method is frequently used to solve the equations.   The iteration begins with an arbitrary initial approximation to the solution and then successively modifies the approximation according to some rule.   The Gauss-Seidel iterative process, for example, is one method frequently used for iterative solutions.   Rapid convergence of iteration is, of course, very important; a method is not considered effective unless the convergence is rapid.

Various techniques are available to accelerate the convergence of the iteration process. The reader should consult Refs. [3,4,10] for a discussion of iterative methods of solution of a system of coupled algebraic equations.

5-3 FINITE-DIFFERENCE APPROXIMATION OF TIME-DEPENDENT, ONE-DIMENSIONAL HEAT-CONDUCTION PROBLEMS

In this section we consider the finite-difference representation of time-dependent, one-dimensional heat-conduction problems given in the form

$$\frac{\partial T}{\partial t} = \alpha \frac{\partial^2 T}{\partial x^2} \qquad \text{in } 0 \le x \le L, \text{ for } t > 0 \tag{5-19}$$

subject to the boundary conditions

$$-k\frac{\partial T}{\partial x} + h_1 T = f_1 \qquad \text{at } x = 0, \text{ for } t > 0 \tag{5-20a}$$

$$k\frac{\partial T}{\partial x} + h_2 T = f_2 \qquad \text{at } x = L, \text{ for } t > 0 \tag{5-20b}$$

$$T = F(x) \text{ for } t = 0 \text{ in } 0 \le x \le L \tag{5-20c}$$

Various finite-difference schemes that are available in the literature for the finite-difference representation of the above time-dependent heat-conduction problem may be divided into two categories: (1) an *explicit scheme* and (2) an *implicit scheme*. Both of these schemes transform the differential equation and its boundary conditions into a set of algebraic equations, but each method has its own advantages and disadvantages. For example, equations arising from the explicit method are uncoupled and simple to solve, but their solution is stable under very restrictive conditions. On the other hand, the equations arising from the implicit method are coupled and more difficult to solve, but their solution is not restricted by the stability criteria. We now describe the finite-difference representation of the above heat-conduction problem with both the explicit and implicit methods, and discuss the advantages and disadvantages of each of these approaches.

An Explicit Method of Finite Differencing

The heat-conduction problem given by Eqs. (5-19) and (5-20) is now expressed in the finite-difference form by using an explicit method. The x and t domains are divided into small intervals of Δx and Δt, as shown in Fig. 5-10, so that

$$x = n\,\Delta x \qquad n = 0, 1, 2, \ldots, N \text{ with } L = N\,\Delta x$$
$$t = i\,\Delta t \qquad i = 0, 1, 2, 3, \ldots \tag{5-21}$$

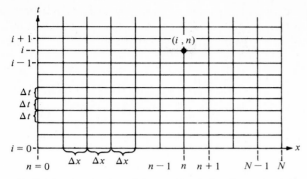

FIG. 5-10 Subdivision of the x-t domain into intervals of Δx and Δt for finite-difference representation of the one-dimensional, time-dependent heat-conduction equation.

Then, the temperature $T(x, t)$ at a location x and a time t is denoted by the symbol T_n^i, that is,

$$T(x, t) = T(n \, \Delta x, i \, \Delta t) \equiv T_n^i \qquad (5\text{-}22)$$

Various derivatives appearing in the above heat-conduction equation are approximated as now described.

The second derivative of temperature with respect to x, at a position $n \, \Delta x$ and at a time $i \, \Delta t$, is represented in the finite-difference form as described previously [see Eq. (5-3)]; we find

$$\left. \frac{\partial^2 T}{\partial x^2} \right|_{n, \, i} = \frac{T_{n-1}^i + T_{n+1}^i - 2T_n^i}{(\Delta x)^2} \qquad (5\text{-}23)$$

where T_{n-1}^i and T_{n+1}^i are the two neighboring points of the node T_n^i, and all of which are evaluated at the time $i \, \Delta t$.

The first derivative of temperature with respect to the time variable t at a position $n \, \Delta x$ and at a time $i \, \Delta t$ is represented by

$$\left. \frac{\partial T}{\partial t} \right|_{n, \, i} = \frac{T_n^{i+1} - T_n^i}{\Delta t} \qquad (5\text{-}24)$$

where T_n^{i+1} is the temperature at the location $n \, \Delta x$ at the time $(i + 1) \, \Delta t$.

Substituting Eqs. (5-23) and (5-24) into Eq. (5-19), the finite-difference form of the time-dependent heat-conduction equation becomes

$$\frac{T_n^{i+1} - T_n^i}{\Delta t} = \alpha \, \frac{T_{n-1}^i + T_{n+1}^i - 2T_n^i}{(\Delta x)^2} \qquad (5\text{-}25)$$

The finite-difference representation of the boundary and initial conditions (5-20a), (5-20b), and (5-20c) is given by

$$-k \frac{T_1^{i+1} - T_0^{i+1}}{\Delta x} + h_1 T_0^{i+1} = f_1 \tag{5-26a}$$

$$k \frac{T_N^{i+1} - T_{N-1}^{i+1}}{\Delta x} + h_2 T_N^{i+1} = f_2 \tag{5-26b}$$

$$T_n^0 = F(n \Delta x) \qquad n = 0, 1, \ldots, N \tag{5-26c}$$

Equations (5-25) and (5-26) are now rearranged, respectively, in the form

$$T_n^{i+1} = r T_{n-1}^i + (1 - 2r) T_n^i + r T_{n+1}^i \qquad n = 1, 2, 3, \ldots, N - 1 \tag{5-27}$$

$$T_0^{i+1} = \frac{1}{1 + h_1 \Delta x/k} \left(T_1^{i+1} + \frac{f_1 \Delta x}{k} \right) \qquad \text{at } n = 0 \tag{5-28a}$$

$$T_N^{i+1} = \frac{1}{1 + h_2 \Delta x/k} \left(T_{N-1}^{i+1} + \frac{f_2 \Delta x}{k} \right) \qquad \text{at } n = N \tag{5-28b}$$

$$T_n^0 = F(n \Delta x) \qquad n = 0, 1, 2, \ldots, N, \qquad \text{for } i = 0 \tag{5-28c}$$

where

$$r \equiv \frac{\alpha \Delta t}{(\Delta x)^2} \tag{5-29}$$

Equations (5-27) and (5-28) are the explicit forms of finite-difference representation of the time-dependent, one-dimensional heat-conduction problem given by Eqs. (5-19) and (5-20). We note that Eqs. (5-27) and (5-28) are a set of algebraic equations for the $N + 1$ unknown temperatures of the nodes $n = 0, 1, 2, \ldots, N$, for each time step i. It can be shown that the *solution of these equations is stable if the value of the parameter r is chosen as* [2, p. 93]

$$0 < r \leq \tfrac{1}{2} \tag{5-30}$$

The subject of the stability of the numerical methods of solution of partial differential equations is beyond the scope of this book; the reader interested in the derivation of various stability criteria should consult Refs. [1–3]. However, to provide some insight into the physical significance of the stability criteria given by Eq. (5-30), we present here a qualitative discussion of the stability of the solution of Eqs. (5-27). Suppose at any instant i the temperatures T_{n-1}^i and T_{n+1}^i at the nodes $n - 1$ and $n + 1$ are equal but less than T_n^i at the node n at time i. Then, if the value of r exceeds $\tfrac{1}{2}$, the coefficient $1 - 2r$ of T_n^i becomes negative and, according to Eq. (5-27), the temperature T_n^{i+1} at the node n after a time step Δt should be less than that at the neighboring nodes; but this is not possible thermodynamically. Therefore, to obtain meaningful solutions the value of the parameter r should be chosen as $0 < r \leq \tfrac{1}{2}$.

We now describe the method of solution of the algebraic equation given by the system of Eqs. (5-27) and (5-28). Once the boundary-condition functions f_1, f_2, the heat-transfer coefficients h_1, h_2, and the thermal conductivity k are given and a suitable value is chosen for the parameter r consistent with the stability criteria, the calculations are performed in the following order:

1   The calculations are started by setting $i = 0$. Then the right-hand side of Eqs. (5-27) becomes the initial conditions T_n^0 ($n = 0, 1, \ldots, N$), which are known by Eqs. (5-28c); then the temperatures T_n^1 ($n = 1, 2, \ldots, N - 1$) at the $N - 1$ internal nodes are calculated from Eqs. (5-27). The temperatures T_0^1 and T_N^1 at the nodes $n = 0$ and $n = N$ (i.e., at the two boundaries) are determined from Eqs. (5-28a) and (5-28b), respectively.

2   Knowing the temperatures T_n^1 ($n = 0, 1, \ldots, N$), at the end of the first time step, the temperatures T_n^2 ($n = 1, 2, \ldots, N - 1$) at the $N - 1$ internal nodes are determined from Eqs. (5-27) by setting $i = 1$, since the right-hand side of these equations is now known. The temperatures T_0^2 and T_N^2 at the two boundaries are calculated from Eqs. (5-28a) and (5-28b), respectively.

3   Knowing the temperatures T_n^2 ($n = 0, 1, \ldots, N$) at the end of the second time step, the temperature T_n^3 is determined in a similar manner, and the procedure is repeated to calculate the temperatures at the following time step.

To perform the above calculations with a digital computer is a trivial matter. We note that, Eqs. (5-27) being uncoupled, their solution with hand computation is also very straightforward. The implications of the restriction imposed by the stability consideration on the upper limit of r should also be recognized. The definition of r given by Eq. (5-29) implies that when r is fixed there is an upper limit on the permissible value of the time step Δt for given values of α and Δx. Then, if a small space step Δx is used to improve the accuracy of the computations, the corresponding Δt becomes much smaller and the calculations are to be performed for a large number of time steps if the solution is required for long times.

We now illustrate the application of the finite-difference equations (5-27) and (5-28) with the following example.

Example 5-2   A slab of thickness $L = 0.5$ ft, thermal conductivity $k = 0.5$ Btu/h·ft·°F, and thermal diffusivity $\alpha = 0.02$ ft^2/h is initially at a uniform temperature $T_0 = 400°F$. For times $t > 0$ the slab is cooled by convection with a heat-transfer coefficient of $h = 10$ Btu/h·ft^2·°F from both of its surfaces into a medium at a temperature $T_\infty = 100°F$. Assuming that the region is divided into 10 equal subintervals, write the explicit finite-difference equations for this heat-conduction problem for the cases (a) $r = \frac{1}{4}$ and (b) $r = \frac{1}{2}$, and (c) determine the temperature distribution at the nodes for a few consecutive time steps, using the equations for $r = \frac{1}{2}$.

Solution   The mathematical formulation of this heat-conduction problem is

$$\frac{\partial T}{\partial t} = \alpha \frac{\partial^2 T}{\partial x^2} \quad \text{in } 0 \leq x \leq L, t > 0 \tag{5-31a}$$

$$-k\frac{\partial T}{\partial x} + hT = hT_\infty \qquad \text{at } x = 0, \ t > 0 \tag{5-31b}$$

$$k\frac{\partial T}{\partial x} + hT = hT_\infty \qquad \text{at } x = L, \ t > 0 \tag{5-31c}$$

$$T = T_0 \qquad \text{for } t = 0, \text{ in } 0 \leq x \leq L \tag{5-31d}$$

This problem is similar to the one considered in Eqs. (5-19) and (5-20), with $h_1 = h_2 = h$, $f_1 = f_2 = hT_\infty$. Then the finite-difference equations for a subdivision of $N = 10$ are immediately obtained from Eqs. (5-27) and (5-28) as

$$T_n^{i+1} = rT_{n-1}^i + (1 - 2r)T_n^i + rT_{n+1}^i \qquad n = 1, 2, 3, \ldots, 9 \tag{5-32a}$$

$$T_0^{i+1} = \frac{1}{1 + h\,\Delta x/k}\left(T_1^{i+1} + \frac{hT_\infty\,\Delta x}{k}\right) \qquad \text{at } n = 0 \tag{5-32b}$$

$$T_{10}^{i+1} = \frac{1}{1 + h\,\Delta x/k}\left(T_9^{i+1} + \frac{hT_\infty\,\Delta x}{k}\right) \qquad \text{at } n = 10 \tag{5-32c}$$

$$T_n^0 = T_0 \qquad n = 0, 1, 2, \ldots, 10 \qquad \text{for } i = 0 \tag{5-32d}$$

where

$$r \equiv \frac{\alpha\,\Delta t}{(\Delta x)^2} \tag{5-32e}$$

For the specific case considered, various parameters are given:

$k = 0.5$ Btu/h·ft·°F (or 0.865 W/m·°C) $\alpha = 0.02$ ft²/h (or 0.516×10^{-6} m²/s)

$h = 10$ Btu/h·ft²·°F (or 56.8 W/m²·°C) $L = \frac{1}{2}$ ft (or 0.152 m)

$T_0 = 400$°F (or 204.4°C) $T_\infty = 100$°F (or 37.8°C)

When the region $0 \leq x \leq L$ is divided into 10 equal intervals we have

$$N = 10 \qquad \Delta x = \frac{L}{N} = \frac{1}{20} \text{ ft } (1.524 \times 10^{-2} \text{ m})$$

$$\frac{h\,\Delta x}{k} = \frac{10 \times 1}{0.5 \times 20} = 1 \qquad \frac{h\,\Delta x T_\infty}{k} = \frac{10 \times 100}{0.5 \times 20} = 100 \text{ (or 37.8)}$$

It is to be noted that the dimensionless quantity $h\,\Delta x/k$ is the same in both systems of units; but the magnitude of $(h\,\Delta x T_\infty)/k$ depends on whether the English or SI system of units is used. In the following analysis we perform the computations in °F.

The above numerical values are now substituted into Eqs. (5-32) to yield the following equations for the nodal-point temperatures.

(a)   For $r = \frac{1}{4}$, Eqs. (5-32) become

$$T_n^{i+1} = \tfrac{1}{4}T_{n-1}^i + \tfrac{1}{2}T_n^i + \tfrac{1}{4}T_{n+1}^i \qquad n = 1, 2, \ldots, 9 \tag{5-33a}$$

$$T_0^{i+1} = \tfrac{1}{2}(T_1^{i+1} + 100) \qquad \text{at } n = 0 \tag{5-33b}$$

$$T_{10}^{i+1} = \tfrac{1}{2}(T_9^{i+1} + 100) \qquad \text{at } n = 10 \tag{5-33c}$$

$$T_n^0 = 400 \qquad n = 0, 1, 2, \ldots, 10, \text{ for } i = 0 \tag{5-33d}$$

In this case the corresponding time step is

$$\Delta t = \frac{r(\Delta x)^2}{\alpha} = \frac{1}{4 \times 0.02 \times 20^2} = \frac{1}{32} \text{ h} = 1.875 \text{ min}$$

(b) For $r = \frac{1}{2}$, Eqs. (5-32) become

$$T_n^{i+1} = \tfrac{1}{2}(T_{n-1}^i + T_{n+1}^i) \qquad n = 1, 2, \ldots, 9 \tag{5-34a}$$

$$T_0^{i+1} = \tfrac{1}{2}(T_1^{i+1} + 100) \tag{5-34b}$$

$$T_{10}^{i+1} = \tfrac{1}{2}(T_9^{i+1} + 100) \tag{5-34c}$$

$$T_n^0 = 400 \qquad n = 0, 1, 2, \ldots, 10 \tag{5-34d}$$

and the corresponding time step becomes

$$\Delta t = \frac{r(\Delta x)^2}{\alpha} = \frac{1}{2 \times 0.02 \times 20^2} = \frac{1}{16} \text{ h} = 3.75 \text{ min}$$

(c) We now calculate the temperature distribution in the medium for a few consecutive time steps for the case of $r = \frac{1}{2}$ from Eqs. (5-34). Table 5-1 summarizes the results of these calculations for five consecutive time steps. The *first row* in the table for $i = 0$ (that is, $t = 0$) is the initial temperature distribution in the medium, which shows all the nodes at temperature 400°F as specified by the initial condition (5-34d).

The *second row* gives the temperature at the end of the first time step, that is, at $i = 1$ or $t = \Delta t = 3.75$ min. The temperatures at the nodes $n = 1, 2, \ldots, 9$ are calculated from Eqs. (5-34a), and the temperatures at the nodes $n = 0$ and $n = 10$ (i.e., at the boundaries $x = 0$ and $x = L$) are determined from Eqs. (5-34b) and (5-34c), respectively. We note that, to calculate the temperatures at the time step $i = 1$, the initial temperatures at the time $i = 0$ are utilized.

The *third row* gives the temperatures at the end of the second time step, that is, at $i = 2$ or $t = 2\Delta t = 7.5$ min. The temperatures at the nodes $1, 2, 3, \ldots, 9$ are calculated from Eqs. (5-34a), and the temperatures at the nodes $n = 0$ and $n = 10$ are determined from Eqs. (5-34b) and (5-34c), respectively. To perform the calculations for the second time step, the temperatures at the end of the first time step $i = 1$ are utilized.

The calculations for the following time steps are performed in a similar manner. In the foregoing calculations we used the finite-difference equations (5-34) corresponding to the choice of $r = \frac{1}{2}$, because the resulting system of equations is simpler for performing numerical calculations. However, the numerical computations have shown that the results for the temperature are more accurate with the values of r less than $\frac{1}{2}$.

We note that the above problem possesses symmetry about the midpoint $n = 5$; therefore calculations should be performed only for one-half the region.

An Implicit Method of Finite Differencing (the Crank-Nicolson Method)

The explicit method of finite differencing as described above is disadvantageous in that, if the space step Δx is chosen very small to improve the accuracy of the calculations, the computational problem becomes enormous because the time step Δt should also be chosen very small by the stability considerations. The

TABLE 5-1   The Calculation of Temperatures for Example 5-2

	$n=0$ $x=0$	1 Δx	2 $2\Delta x$	3 $3\Delta x$	4 $4\Delta x$	5 $5\Delta x$	6 $6\Delta x$	7 $7\Delta x$	8 $8\Delta x$	9 $9\Delta x$	10 $10\Delta x$
$i=0, t=0$	400	400	400	400	400	400	400	400	400	400	400
$i=1, t=\Delta t$	250	400	400	400	400	400	400	400	400	400	250
$i=2, t=2\Delta t$	212.5	325	400	400	400	400	400	400	400	325	212.5
$i=3, t=3\Delta t$	203	306	362	400	400	400	400	400	362	306	203
$i=4, t=4\Delta t$	191	282	353	381	400	400	400	381	353	282	191
$i=5, t=5\Delta t$	186	272	313.5	376.5	390.5	400	390.5	376.5	331.5	272	186

$(\Delta t = 3.75 \text{ min})$

implicit method of finite differencing now described imposes no restriction on the size of the time step Δt. Numerous implicit schemes have been described in the literature. In the so-called *fully implicit scheme*, the finite-difference representation of the heat-conduction equation (5-19) is given as

$$\frac{T_n^{i+1} - T_n^i}{\Delta t} = \alpha \frac{T_{n-1}^{i+1} + T_{n+1}^{i+1} - 2T_n^{i+1}}{(\Delta x)^2} \tag{5-35}$$

A comparison of this equation with the explicit form given by Eq. (5-25) reveals that the right-hand side of the equations is evaluated at the time step i for the explicit form and at $i + 1$ for the fully implicit form.

Here we consider a *modified implicit form* proposed by Crank and Nicolson [10]. In this formulation the left-hand side of the finite-difference equation is retained as given above, but for the right-hand side the arithmetic average of the right-hand sides of Eqs. (5-25) and (5-35) are taken. That is, in the *Crank-Nicolson implicit method* of finite differencing, the differential equation of heat conduction (5-19) becomes

$$\frac{T_n^{i+1} - T_n^i}{\Delta t} = \frac{\alpha}{2}\left[\frac{T_{n-1}^{i+1} + T_{n+1}^{i+1} - 2T_n^{i+1}}{(\Delta x)^2} + \frac{T_{n-1}^i + T_{n+1}^i - 2T_n^i}{(\Delta x)^2}\right] \tag{5-36}$$

which can be rearranged in the form

$$-rT_{n-1}^{i+1} + (2 + 2r)T_n^{i+1} - rT_{n-1}^{i+1} = rT_{n-1}^i + (2 - 2r)T_n^i + rT_{n+1}^i \tag{5-37a}$$

for $n = 0, 1, 2, \ldots, N$, where

$$r \equiv \frac{\alpha\,\Delta t}{(\Delta x)^2} \tag{5-37b}$$

The finite-difference representation of the boundary conditions (5-20a) and (5-20b) are given, respectively, as

$$-k\frac{T_1^i - T_{-1}^i}{2\Delta x} + h_1 T_0^i = f_1 \tag{5-38a}$$

$$k\frac{T_{N+1}^i - T_{N-1}^i}{2\Delta x} + h_2 T_N^i = f_2 \tag{5-38b}$$

and the initial condition (5-20c) becomes

$$T_n^0 = F(n\,\Delta x) \qquad n = 0, 1, 2, \ldots, N \tag{5-38c}$$

We note that Eqs. (5-37a) provide $N + 1$ coupled algebraic equations for the $N + 1$ unknown nodal-point temperatures T_n^{i+1} ($n = 0, 1, 2, \ldots, N$), since the fictitious temperatures T_{-1}^i, T_{-1}^{i+1}, T_{N+1}^i, and T_{N+1}^{i+1} appearing in these equations can be determined from the boundary-condition equations (5-38) by computing them for the times i and $i + 1$. Therefore, when these fictitious quantities are determined from the boundary conditions (5-38) and introduced into Eqs. (5-37), we obtain $N + 1$ simultaneous algebraic equations for the $N + 1$ unknown nodal-point temperatures T_n^{i+1} ($n = 0, 1, \ldots, N$). These equations are given in the matrix form

$$
\begin{bmatrix}
(2 + 2r\beta_1) & -2r & 0 & 0 & \cdots & 0 & 0 & 0 \\
-r & (2 + 2r) & -r & 0 & \cdots & 0 & 0 & 0 \\
0 & -r & (2 + 2r) & -r & \cdots & 0 & 0 & 0 \\
\multicolumn{8}{c}{\cdots\cdots\cdots\cdots} \\
0 & 0 & 0 & 0 & \cdots & -r & (2 + 2r) & -r \\
0 & 0 & 0 & 0 & \cdots & 0 & -2r & (2 + 2r\beta_2)
\end{bmatrix}
\begin{bmatrix}
T_0^{i+1} \\
T_1^{i+1} \\
T_2^{i+1} \\
\vdots \\
T_{N-1}^{i+1} \\
T_N^{i+1}
\end{bmatrix}
$$

$$
=
\begin{bmatrix}
(2 - 2r\beta_1) & 2r & 0 & 0 & \cdots & 0 & 0 & 0 \\
r & (2 - 2r) & r & 0 & \cdots & 0 & 0 & 0 \\
0 & r & (2 - 2r) & r & \cdots & 0 & 0 & 0 \\
\multicolumn{8}{c}{\cdots\cdots\cdots\cdots} \\
0 & 0 & 0 & 0 & \cdots & r & (2 - 2r) & r \\
0 & 0 & 0 & 0 & \cdots & 0 & 2r & (2 - 2r\beta_2)
\end{bmatrix}
$$

$$
\times
\begin{bmatrix}
T_0^i \\
T_1^i \\
T_2^i \\
\vdots \\
T_{N-1}^i \\
T_N^i
\end{bmatrix}
+
\begin{bmatrix}
\dfrac{4r\,\Delta x f_1}{k} \\
0 \\
0 \\
\vdots \\
0 \\
\dfrac{4r\,\Delta x f_2}{k}
\end{bmatrix}
\qquad (5\text{-}39a)
$$

where $i = 0, 1, 2, 3, \ldots$, and β_1 and β_2 are defined as

$$
\beta_1 = 1 + \frac{h_1\,\Delta x}{k} \qquad \text{and} \qquad \beta_2 = 1 + \frac{h_2\,\Delta x}{k} \qquad (5\text{-}39b)
$$

Once the values of the boundary-condition functions f_1, f_2, the heat-transfer coefficients h_1, h_2, the thermal conductivity k, the space step Δx, and the parameter r are chosen, the system of Eqs. (5-39a) is solved for each time step in the following manner.

1 The calculations are started by setting $i = 0$. Then the temperatures $T_0{}^0$, $T_1{}^0$, $T_2{}^0$, ..., $T_N{}^0$ on the right-hand side of Eqs. (5-39a) are the initial conditions and are known. Then, $N + 1$ algebraic equations (5-39a) are simultaneously solved, and the temperatures $T_0{}^1$, $T_1{}^1$, $T_2{}^1$, ..., $T_N{}^1$ for the nodes $n = 0, 1, ..., N$ at the end of the first time step are determined.
2 By setting $i = 1$, the temperatures $T_0{}^1$, $T_1{}^1$, $T_2{}^1$, ..., $T_N{}^1$ on the right-hand side of Eqs. (5-39a) are known from the calculations in step 1. Then, the equations are solved simultaneously and the temperatures $T_0{}^2$, $T_1{}^2$, $T_2{}^2$, ..., $T_N{}^2$ at the end of the second time step are determined.
3 By setting $i = 2$, the temperatures $T_0{}^2$, $T_1{}^2$, ..., $T_N{}^2$ on the right-hand side of Eqs. (5-39a) become known from step 2. Then, a simultaneous solution of the equations give the temperatures $T_0{}^3$, $T_1{}^3$, ..., $T_N{}^3$ at the end of the third time step.

The temperatures at the following time steps are computed in a similar manner.

We note that the implicit method results in a set of coupled equations to be solved for each time step whereas with the explicit method equations are uncoupled. Although it is more difficult to solve a system of coupled equations than uncoupled equations, the implicit method has no restriction imposed on the size of the time step Δt and has the advantage for situations where larger Δt may be needed to proceed with the solution more rapidly.

Other Methods of Finite Differencing

The implicit method of formulation discussed above removes the restriction on the size of the time step, but when this procedure is applied to N mesh points, a system of N simultaneous equations with N unknown variables must be solved. When two-dimensional transient heat conduction is to be solved with this method, the procedure is excessively time-consuming. To circumvent this difficulty, other methods of finite differencing have been developed. The *alternating-direction implicit method* was proposed by Peaceman and Rachford [17] and improved by D'Yakanov [18] and Fairweather and Mitchell [19] for the solution of two-dimensional time-dependent heat-conduction problems. An explicit finite-difference method which is claimed to be stable was proposed by Barakat and Clark [20] for solving multidimensional, nonhomogeneous, time-dependent heat-conduction problems. The reader should consult the original references for detailed discussions of these different finite-difference schemes.

REFERENCES

1 Fox, L.: "Numerical Solution of Ordinary and Partial Differential Equations," Addison-Wesley Publishing Company, Inc., Reading, Mass., 1962.

2 Smith, G. D.: "Numerical Solution of Partial Differential Equations with Exercises and Worked Solutions," Oxford University Press, London, 1965.

3 Richtmeyer, R. D.: "Difference Methods for Initial Value Problems," Interscience Publishers, Inc., New York, 1957.

4 Dusinberre, G. M.: "Heat Transfer Calculations by Finite Differences," International Textbook Company, Scranton, Pa., 1961.

5 Forsythe, G. E., and W. R. Wasow: "Finite Differences Method for Partial Differential Equations," John Wiley & Sons, Inc., New York, 1960.

6 Larkin, B. K.: Some Finite Difference Methods for Problems in Transient Heat Flow, *Chem. Eng. Prog., Symp. Ser.* 59, **61**: (1965).

7 Macon, N.: "Numerical Analysis," John Wiley & Sons, Inc., New York, 1963.

8 Southwell, R. V.: "Relaxation Methods in Engineering Science," Oxford University Press, New York, 1940.

9 Ames, W. F.: "Nonlinear Partial Differential Equations in Engineering," pp. 365–389, Academic Press, Inc., New York, 1965.

10 Crank, J., and P. Nicolson: A Practical Method for Numerical Evaluation of Solutions of P.D.E. of the Heat Conduction Type, *Proc. Cambridge Philos. Soc.*, **43**: 50–67 (1947).

11 Nickell, R. E., and E. Wilson: Application of the Finite Element Method to Heat Conduction Analysis, *Nucl. Eng. Des.*, **4**: 276–286 (1966).

12 Oktay, Ural: "Finite Element Method: Basic Concepts and Applications," International Textbook Company, Scranton, Pa., 1973.

13 Zienkiewicz, O. C., and I. K. Cheung: "The Finite Element Method in Engineering Science," McGraw-Hill Book Company, New York, 1971.

14 Martin, H. C., and G. F. Carey: "Introduction to Finite Element Analysis," McGraw-Hill Book Company, New York, 1973.

15 Huebner, K. H.: "Finite Element Method for Engineers," John Wiley & Sons, Inc., New York, 1975.

16 Özışık, M. N.: "Boundary Value Problems of Heat Conduction," International Textbook Company, Scranton, Pa., 1968.

17 Peaceman, D. W., and H. H. Rachford: The Numerical Solution of Parabolic and Elliptic Differential Equations, *J. Soc. Ind. Appl. Math.*, **3**: 28–41 (1955).

18 D'Yakonov, Y. G.: *Zh. Vychisl. Mat. Mat. Fiz.*, **2**: 549 (1962).

19 Fairweather, G., and A. R. Mitchell: A New Computational Procedure for A.D.I. Methods, *SIAM J. Numer. Anal.*, **4**: 1963–1970 (1967).

20 Barakat, H. Z., and J. A. Clark: On the Solution of Diffusion Equation by Numerical Methods, *J. Heat Transfer*, **88C**: 421–427 (1966).

Six

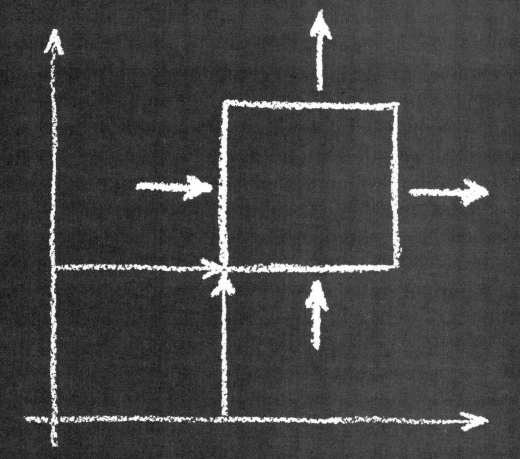

Convection—
Basic Equations

The analysis of heat transfer by convection is complicated by the fact that the motion of the fluid plays an important part in heat transfer. Therefore, in the determination of the temperature field in fluid flow, a knowledge of the velocity distribution is essential. Since heat transfer in fluid flow takes place in such a way that mass, momentum, and energy are all conserved, this requirement forms the basis for the derivation of the basic conservation equations which comprise the continuity, the momentum, and the energy equations. The purpose of this chapter is to provide a good understanding of the physical significance and the implication of these equations, so that in the following chapters on the applications of convective heat transfer, the reader will be able to obtain the equations needed for the formulation of simple problems, by proper simplification of the equations given in this chapter. To achieve this aim we consider the steady, two-dimensional flow of an incompressible, constant-property fluid and present the basic steps in the derivation of the governing equations, with emphasis on the physical significance of the individual terms in the equations. The dimensionless parameters affecting convective heat transfer are also determined by expressing these equations in the dimensionless form. The boundary-layer simplification is discussed, and a summary of the equations is presented for ready-reference purposes.

6-1   THE CONTINUITY EQUATION

The continuity equation is essentially the equation for the conservation of mass; it is derived by a mass balance on the fluid entering and leaving a volume element taken in the flow field. Consider a differential volume element $\Delta x \, \Delta y \, \Delta z$ about a point (x, y, z) in the flow field as illustrated in Fig. 6-1. For simplicity in the analysis we assume steady, two-dimensional flow with velocity components $u \equiv u(x, y)$ and $v \equiv v(x, y)$ in the x and y directions, respectively.

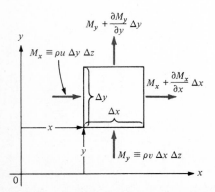

FIG. 6-1   Nomenclature for the derivation of the continuity equation.

The equation for the conservation of mass may be stated as

$$\begin{pmatrix} \text{Net rate of mass flow entering} \\ \text{volume element in} \\ x \text{ direction} \end{pmatrix} + \begin{pmatrix} \text{net rate of mass flow entering} \\ \text{volume element in} \\ y \text{ direction} \end{pmatrix} = 0 \qquad (6\text{-}1)$$

If $M_x \equiv \rho u\, \Delta y\, \Delta z$ is the mass-flow rate into the element in the x direction through the surface at x, then $M_x + (\partial M_x / \partial x)\, \Delta x$ is the mass-flow rate leaving the element in the x direction through the surface at $x + \Delta x$.   The net rate of mass flow into the element in the x direction is the difference between the entering and leaving flow rates given by

$$\begin{pmatrix} \text{Net rate of mass flow entering} \\ \text{element in } x \text{ direction} \end{pmatrix} = -\frac{\partial M_x}{\partial x}\, \Delta x = -\frac{\partial(\rho u)}{\partial x}\, \Delta x\, \Delta y\, \Delta z \qquad (6\text{-}2a)$$

Similarly, the net rate of mass flow entering the volume element in the y direction is given by

$$-\frac{\partial(\rho v)}{\partial y}\, \Delta x\, \Delta y\, \Delta z \qquad (6\text{-}2b)$$

Substituting Eqs. (6-2) into Eq. (6-1) and canceling the term $\Delta x\, \Delta y\, \Delta z$, one obtains the continuity equation in the rectangular coordinate system for a two-dimensional, steady flow as

$$\frac{\partial(\rho u)}{\partial u} + \frac{\partial(\rho v)}{\partial y} = 0 \qquad (6\text{-}3)$$

When density ρ is treated as constant, Eq. (6-3) simplifies to

$$\frac{\partial u}{\partial x} + \frac{\partial v}{\partial y} = 0 \qquad (6\text{-}4)$$

Equation (6-4) is called the *continuity equation for a two-dimensional, steady, incompressible flow* in the rectangular coordinate system.

6-2   THE MOMENTUM EQUATIONS

The momentum equations are derived from *Newton's second law* of motion which states that mass times the acceleration in a given direction is equal to the external forces acting on the body in the same direction.   The *external forces* acting on a volume element in a flow field are considered to consist of the *body forces* and the *surface forces*.   The body forces may result from such effects as the gravitational, electric, and magnetic fields acting on the body of the fluid, and the surface forces result from the stresses acting on the surface of the volume element. With this consideration Newton's second law may be stated as

$$(\text{Mass})\begin{pmatrix}\text{acceleration in}\\ \text{direction } i\end{pmatrix} = \begin{pmatrix}\text{body forces acting}\\ \text{in direction } i\end{pmatrix} + \begin{pmatrix}\text{surface forces}\\ \text{acting in}\\ \text{direction } i\end{pmatrix} \qquad (6\text{-}5)$$

For a three-dimensional flow, for example, in the rectangular coordinate system $i = x, y,$ and z; hence Eq. (6-5) provides three independent momentum equations. For simplicity in the analysis we consider here two-dimensional, steady, incompressible flow with constant properties having velocity components $u \equiv u(x, y)$ and $v \equiv v(x, y)$ in the x and y directions, respectively. Therefore, for this particular case with $i = x$ and y, Eq. (6-5) will provide two independent momentum equations, one for the x direction and the other for the y direction. The mathematical expressions for the various terms in Eq. (6-5) can be determined by considering a differential volume element $\Delta x \, \Delta y \, \Delta z$ about a point (x, y, z) in the flow field as described below.

Mass

If ρ is the density of the fluid, the mass of a differential volume element $\Delta x \, \Delta y \, \Delta z$ is given by

$$\text{Mass} = \Delta x \, \Delta y \, \Delta z \rho \qquad (6\text{-}6)$$

Acceleration

The acceleration term for an unsteady, one-dimensional flow is readily obtained simply by taking the derivative with respect to time of the velocity. For a flow field having velocity components in more than one direction, the rate of change of a velocity component, say u, is also associated with the convective motion of the fluid in other directions. In general, for *a three-dimensional, unsteady flow field*, having velocity components $u, v,$ and w in the $x, y,$ and z directions, respectively, the rate of change of a property ϕ in the flow field is given by the *total* (or *substantial*) *derivative* $D\phi/Dt$ defined as

$$\frac{D\phi}{Dt} \equiv \frac{\partial \phi}{\partial t} + u \frac{\partial \phi}{\partial x} + v \frac{\partial \phi}{\partial y} + w \frac{\partial \phi}{\partial z} \qquad (6\text{-}7)$$

A brief discussion of the total derivative is given in *note 1* at the end of this chapter.

For a *steady, two-dimensional flow* with velocity components u and v, we set $\partial/\partial t$ and w equal to zero in Eq. (6-7), and the total derivative becomes

$$\frac{D}{Dt} \equiv u\,\frac{\partial}{\partial x} + v\,\frac{\partial}{\partial y} \qquad \text{(for two-dimensional, steady flow)} \qquad (6\text{-}8a)$$

Then, for the steady, two-dimensional flow considered here the acceleration in the x direction is

$$\frac{Du}{Dt} \equiv u\,\frac{\partial u}{\partial x} + v\,\frac{\partial u}{\partial y} \qquad \text{(for two-dimensional, steady flow)} \qquad (6\text{-}8b)$$

and the acceleration in the y direction is

$$\frac{Dv}{Dt} \equiv u\,\frac{\partial v}{\partial x} + v\,\frac{\partial v}{\partial y} \qquad \text{(for two-dimensional, steady flow)} \qquad (6\text{-}8c)$$

Clearly, the total derivative Du/Dt expresses the rate of change in u associated with the fluid motion in the x and y directions.

Body Forces

There are many practical situations in which body forces act on the fluid element. In free convection when the density change due to temperature variation is considered, the body force is caused by the gravitational attraction because the magnitude of this force is proportional to the fluid density. If an electrically conducting fluid such as mercury moves through a magnetic field, there is the body force acting on the fluid due to the presence of a magnetic field. Here, without specifying the nature of the body forces, we simply denote by the symbols F_x and F_y the *body forces acting per unit volume* of the fluid in the x and y directions, respectively.   Then,

$$\begin{pmatrix} \text{Body forces acting on} \\ \Delta x\,\Delta y\,\Delta z \text{ in } x \text{ direction} \end{pmatrix} = F_x\,\Delta x\,\Delta y\,\Delta z \qquad (6\text{-}9a)$$

$$\begin{pmatrix} \text{Body forces acting on} \\ \Delta x\,\Delta y\,\Delta z \text{ in } y \text{ direction} \end{pmatrix} = F_y\,\Delta x\,\Delta y\,\Delta z \qquad (6\text{-}9b)$$

Surface Forces

The surface forces acting per unit area are called *stresses*.   When the stress acts normal to the surface it is called the *normal stress*, and when it acts along the surface it is called the *shear stress*.   Figure 6-2 shows various stresses acting on the surfaces of a differential volume element.   In this figure σ_x and σ_y denote the normal stresses in the x and y directions, respectively.

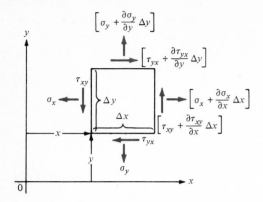

FIG. 6-2   Nomenclature for the various stresses acting on the surfaces of the volume element.

The shear stresses are denoted by τ_{xy} and τ_{yx}, where *the first subscript indicates the axis to which the surface is perpendicular and the second subscript indicates the direction of the shear stress.* Thus, τ_{xy} is the shear stress acting on the surface $\Delta y\,\Delta z$ (i.e., the surface perpendicular to the x axis) at x in the direction y. Then, *the net normal surface force acting on the element in the positive x direction* is $(\partial/\partial y)(\sigma_x\,\Delta y\,\Delta z)\,\Delta x$, and *the net shear force acting on the element in the positive x direction* is $(\partial/\partial y)(\tau_{yx}\,\Delta x\,\Delta z)\,\Delta y$. Hence, the net surface forces acting on the element in the positive x direction becomes

$$\begin{pmatrix}\text{Net surface forces acting}\\\text{in } x \text{ direction}\end{pmatrix} = \left(\frac{\partial\sigma_x}{\partial x} + \frac{\partial\tau_{yx}}{\partial y}\right)\Delta x\,\Delta y\,\Delta z \tag{6-10a}$$

Similarly the net surface force acting in the y direction is

$$\begin{pmatrix}\text{Net surface forces acting}\\\text{in } y \text{ direction}\end{pmatrix} = \left(\frac{\partial\sigma_y}{\partial y} + \frac{\partial\tau_{xy}}{\partial x}\right)\Delta x\,\Delta y\,\Delta z \tag{6-10b}$$

When Eqs. (6-6), (6-8b), (6-9a), and (6-10a) are substituted into Eq. (6-5), the x-momentum equation becomes

$$x\text{ Momentum:}\quad \rho\left(u\frac{\partial u}{\partial x} + v\frac{\partial u}{\partial y}\right) = F_x + \frac{\partial\sigma_x}{\partial x} + \frac{\partial\tau_{yx}}{\partial y} \tag{6-11a}$$

and the substitution of Eqs. (6-6), (6-8c), (6-9b), and (6-10b) into Eq. (6-5) gives the y-momentum equation as

$$y\text{ Momentum:}\quad \rho\left(u\frac{\partial v}{\partial x} + v\frac{\partial v}{\partial y}\right) = F_y + \frac{\partial\sigma_y}{\partial y} + \frac{\partial\tau_{xy}}{\partial x} \tag{6-11b}$$

If Eqs. (6-11) are to be used for the prediction of the velocity components in the flow field, the various stresses appearing in these equations are to be related to the velocity component; a discussion of this matter is given by Schlichting [1]. For the two-dimensional, incompressible, constant-property flow field considered here, various stresses in Eqs. (6-11) are related to the velocity components u and v by [1]

$$\tau_{xy} = \tau_{yx} = \mu\left(\frac{\partial u}{\partial y} + \frac{\partial v}{\partial x}\right) \tag{6-12}$$

$$\sigma_x = -p + 2\mu\frac{\partial u}{\partial x} \tag{6-13}$$

$$\sigma_y = -p + 2\mu\frac{\partial v}{\partial y} \tag{6-14}$$

where p is the pressure and μ is the viscosity of the fluid.

When the stresses given by Eqs. (6-12) to (6-14) are substituted into Eqs. (6-11) we obtain (*see note 2 for details of computation*)

$$x \text{ Momentum:} \quad \rho\left(u\frac{\partial u}{\partial x} + v\frac{\partial u}{\partial y}\right) = F_x - \frac{\partial p}{\partial x} + \mu\left(\frac{\partial^2 u}{\partial x^2} + \frac{\partial^2 u}{\partial y^2}\right) \tag{6-15a}$$

$$y \text{ Momentum:} \quad \rho\left(u\frac{\partial v}{\partial x} + v\frac{\partial v}{\partial y}\right) = F_y - \frac{\partial p}{\partial y} + \mu\left(\frac{\partial^2 v}{\partial x^2} + \frac{\partial^2 v}{\partial y^2}\right) \tag{6-15b}$$

where F_x and F_y are the body forces per unit volume acting in the x and y directions, respectively. Equations (6-15) are called *the x- and y-momentum equations for the steady, two-dimensional flow of an incompressible fluid with constant properties.*

The physical significance of the various terms in Eqs. (6-15) is as follows: The terms on the left-hand side represent the *inertia forces*, the first term on the right-hand side is the *body force*, the second term is the *pressure force*, and the last term in the parentheses is the *viscous forces* acting on the fluid element. If the body forces F_x and F_y are known, the continuity equation (6-4) and the two momentum equations (6-15) provide three independent equations for the determination of the three unknown quantities u, v, and p for the steady, two-dimensional flow of an incompressible fluid. The analytical solution of these equations is extremely difficult except for very simple situations. However, a good understanding of the physical significance of the various terms in these equations is very important because, in the following chapters on convective heat transfer, the equations governing the velocity distribution needed for the simple problems to be considered will be obtained directly from these equations by appropriate simplification.

6-3   THE ENERGY EQUATION

The temperature distribution in the flow field is governed by the energy equation which can be derived by writing an energy balance according to the first law of thermodynamics for a differential volume element in the flow field. If radiation is absent and there are no distributed heat sources in the fluid, the energy balance on a differential volume element $\Delta x\, \Delta y\, \Delta z$ about a point (x, y, z) may be stated as

$$
\begin{pmatrix} \text{Rate of heat addition} \\ \text{into element by} \\ \text{conduction} \\ \text{I} \end{pmatrix} + \begin{pmatrix} \text{rate of energy input into element due to} \\ \text{work done by surface stresses and body forces} \\ \text{II} \end{pmatrix}
$$

$$
= \begin{pmatrix} \text{rate of increase of energy} \\ \text{stored in element} \\ \text{III} \end{pmatrix} \qquad (6\text{-}16)
$$

The mathematical expressions for the various terms in this energy equation are now derived below for *a steady, two-dimensional, constant-property flow* in which the velocity components and the temperature variation are in the x and y directions (i.e., no flow and no temperature variation in the z direction).

Heat Addition by Conduction

If q_x and q_y are the heat fluxes in the x and y directions, the net rate of heat addition into the volume element $\Delta x\, \Delta y\, \Delta z$ by conduction is obtained, by referring to the nomenclature in Fig. 6-3, as

$$
I = -\left(\frac{\partial Q_x}{\partial x}\Delta x + \frac{\partial Q_y}{\partial y}\Delta y\right) = -\left(\frac{\partial q_x}{\partial x} + \frac{\partial q_y}{\partial y}\right)\Delta x\, \Delta y\, \Delta z \qquad (6\text{-}17)
$$

where the heat fluxes are given by the Fourier law as

$$
q_x = -k\frac{\partial T}{\partial x} \qquad \text{and} \qquad q_y = -k\frac{\partial T}{\partial y} \qquad (6\text{-}18)
$$

Substituting Eqs. (6-18) into (6-17) and assuming constant k, we obtain

$$
I = k\left(\frac{\partial^2 T}{\partial x^2} + \frac{\partial^2 T}{\partial y^2}\right)\Delta x\, \Delta y\, \Delta z \qquad (6\text{-}19)
$$

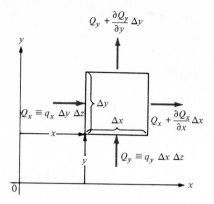

FIG. 6-3 Nomenclature for heat addition by conduction.

Energy Input Due to Work Done by Surface Stresses and Body Forces

If F_x and F_y are the body forces acting per unit volume of the fluid in the x and y directions, respectively, the rate of energy input into the volume element $\Delta x\,\Delta y\,\Delta z$ due to the increase of the potential energy is

$$(uF_x + vF_y)\,\Delta x\,\Delta y\,\Delta z \tag{6-20a}$$

where u and v are the x- and y-direction velocity components, respectively. The rate of energy input into the volume element due to work done by the surface stresses are computed as described in *note 3* at the end of this chapter and given by (see Fig. 6-4 for the nomenclature)

$$\left[\frac{\partial}{\partial x}(u\sigma_x) + \frac{\partial}{\partial y}(v\sigma_y) + \frac{\partial}{\partial y}(u\tau_{yx}) + \frac{\partial}{\partial x}(v\tau_{xy})\right]\Delta x\,\Delta y\,\Delta z \tag{6-20b}$$

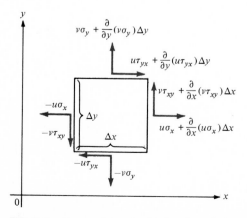

FIG. 6-4 Nomenclature for frictional work done by the surface forces.

Then, the total rate of energy input into the volume element due to the work done by the body forces and surface stresses is obtained by summing up the above two quantities:

$$\text{II} = \left[uF_x + vF_y + \frac{\partial}{\partial x}(u\sigma_x) + \frac{\partial}{\partial y}(v\sigma_y) + \frac{\partial}{\partial y}(u\tau_{yx}) + \frac{\partial}{\partial x}(v\tau_{xy}) \right] \Delta x\, \Delta y\, \Delta z \quad \text{(6-20c)}$$

Rate of Increase of Energy Stored

The energy contained in the fluid consists of the *specific internal energy e per unit mass* and the *kinetic energy* which is $\frac{1}{2}(u^2 + v^2)$ *per unit mass* of the fluid. Then, the internal energy of the volume element $\Delta x\, \Delta y\, \Delta z$ becomes

$$\rho[e + \tfrac{1}{2}(u^2 + v^2)]\, \Delta x\, \Delta y\, \Delta z$$

and the rate of increase of the stored energy is obtained by taking the *total derivative* of this quantity:

$$\text{III} = \rho\left[\frac{De}{Dt} + \frac{1}{2}\frac{D}{Dt}(u^2 + v^2) \right] \Delta x\, \Delta y\, \Delta z \quad \text{(6.21)}$$

where the total derivative D/Dt is defined as

$$\frac{D}{Dt} \equiv u\frac{\partial}{\partial x} + v\frac{\partial}{\partial y} \qquad \text{(for two-dimensional, steady flow)} \quad \text{(6-22)}$$

Equations (6-19), (6-20c), and (6-21) are substituted into Eq. (6-16), and the resulting expression is simplified as described in *note 4* at the end of this chapter by combining it with the momentum equations (6-11) and by introducing the various stress terms given by Eqs. (6-12) to (6-14). Finally, the energy equation for the two-dimensional, steady, incompressible, constant-property flow takes the form (*see note 4 for the derivation*)

$$\rho\frac{De}{Dt} = k\left(\frac{\partial^2 T}{\partial x^2} + \frac{\partial^2 T}{\partial y^2} \right) + \mu\Phi \quad \text{(6-23)}$$

where the *viscous-energy-dissipation function* Φ is defined as

$$\Phi \equiv 2\left[\left(\frac{\partial u}{\partial x}\right)^2 + \left(\frac{\partial v}{\partial y}\right)^2 \right] + \left(\frac{\partial v}{\partial x} + \frac{\partial u}{\partial y}\right)^2 \quad \text{(6-24)}$$

For constant density flow one may approximate the term De/Dt as

$$\frac{De}{Dt} \cong c_p\frac{DT}{Dt} \quad \text{(6-25)}$$

Then, the energy equation (6-23) becomes

$$\rho c_p\left(u\frac{\partial T}{\partial x} + v\frac{\partial T}{\partial y}\right) = k\left(\frac{\partial^2 T}{\partial x^2} + \frac{\partial^2 T}{\partial y^2}\right) + \frac{\mu}{g_c J}\Phi \tag{6-26}$$

where

$$\Phi = 2\left[\left(\frac{\partial u}{\partial x}\right)^2 + \left(\frac{\partial v}{\partial y}\right)^2\right] + \left(\frac{\partial v}{\partial x} + \frac{\partial u}{\partial y}\right)^2 \tag{6-27}$$

and the conversion factors J and g_c are introduced to allow viscosity μ to be taken in the dimension of (mass)/(length)(time).

The physical significance of various terms in the energy equation (6-26) is as follows: The left-hand side represents the *convective heat transfer*; on the right-hand side the first term in parentheses is for the *conductive heat transfer*, and the last term is for the *viscous energy dissipation* due to friction in the fluid.

For most engineering applications the flow velocities are moderate; hence the viscous-energy-dissipation term becomes small and can be neglected. Then Eq. (6-26) simplifies to

$$\rho c_p\left(u\frac{\partial T}{\partial x} + v\frac{\partial T}{\partial y}\right) = k\left(\frac{\partial^2 T}{\partial x^2} + \frac{\partial^2 T}{\partial y^2}\right) \tag{6-28}$$

For the case of no flow (that is, $u = v = 0$), the energy equation (6-28) reduces to the two-dimensional, steady-state heat-conduction equation with no heat generation.

The temperature distribution in the flow field is obtained from the solution of the energy equation subject to appropriate boundary conditions. However, before the energy equation can be solved, the continuity and momentum equations should be solved and the velocity components should be determined, since the energy equation contains the velocity components u and v.

6-4 SUMMARY OF EQUATIONS OF MOTION AND ENERGY

We now summarize the equations of motion and energy for the two-dimensional steady flow of an incompressible fluid, which were derived above in the rectangular coordinate system, and present the equivalent equations in the cylindrical coordinate system with cylindrical symmetry.

Equations in Rectangular Coordinates

The continuity, momentum, and energy equations are given as

Continuity: $\dfrac{\partial u}{\partial x} + \dfrac{\partial v}{\partial y} = 0$ \hfill (6-29)

$$x \text{ Momentum:} \quad \rho\left(u\frac{\partial u}{\partial x} + v\frac{\partial u}{\partial y}\right) = F_x - \frac{\partial p}{\partial x} + \mu\left(\frac{\partial^2 u}{\partial x^2} + \frac{\partial^2 u}{\partial y^2}\right) \tag{6-30}$$

$$y \text{ Momentum:} \quad \rho\left(u\frac{\partial v}{\partial x} + v\frac{\partial v}{\partial y}\right) = F_y - \frac{\partial p}{\partial y} + \mu\left(\frac{\partial^2 v}{\partial x^2} + \frac{\partial^2 v}{\partial y^2}\right) \tag{6-31}$$

$$\text{Energy:} \quad \rho c_p\left(u\frac{\partial T}{\partial x} + v\frac{\partial T}{\partial y}\right) = k\left(\frac{\partial^2 T}{\partial x^2} + \frac{\partial^2 T}{\partial y^2}\right) + \frac{\mu}{g_c J}\Phi \tag{6-32a}$$

where the viscous-dissipation function Φ is defined as

$$\Phi \equiv 2\left[\left(\frac{\partial u}{\partial x}\right)^2 + \left(\frac{\partial v}{\partial y}\right)^2\right] + \left(\frac{\partial v}{\partial x} + \frac{\partial u}{\partial y}\right)^2 \tag{6-32b}$$

The physical significance of various terms is as follows: In the momentum equations the terms on the left-hand side represent the *inertia forces*; on the right-hand side the first term is the *body force*, the second term is the *pressure force*, and the last term in the parentheses is for the *viscous forces* acting on the fluid. In the energy equation the left-hand side represents the *convective heat transfer*; on the right-hand side the first term in parentheses is for *conductive heat transfer*, and the last term is for *viscous energy dissipation* due to fluid friction.

Equations in Cylindrical Coordinates

Many problems in heat transfer involve flow inside circular tubes. The analysis of heat transfer for such situations requires the energy equation to be transformed into the cylindrical coordinate system. For problems with *cylindrical symmetry*, the velocity and temperature vary, in general, radially and along the tube axis; hence r and z are the two independent variables. Let $v_r \equiv v_r(r, z)$ and $v_z \equiv v_z(r, z)$ be the velocity components and F_r and F_z be the body forces acting on the fluid in the r and z directions, respectively. Then, the equivalent of Eqs. (6-29) to (6-32) in the cylindrical coordinate system (that is, r and z) is

$$\text{Continuity:} \quad \frac{1}{r}\frac{\partial}{\partial r}(rv_r) + \frac{\partial v_z}{\partial z} = 0 \tag{6-33}$$

$$r \text{ Momentum:} \quad \rho\left(v_r\frac{\partial v_r}{\partial r} + v_z\frac{\partial v_r}{\partial z}\right) = F_r - \frac{\partial p}{\partial r} + \mu\left(\frac{\partial^2 v_r}{\partial r^2} + \frac{1}{r}\frac{\partial v_r}{\partial r} - \frac{v_r}{r^2} + \frac{\partial^2 v_r}{\partial z^2}\right) \tag{6-34}$$

$$z \text{ Momentum:} \quad \rho\left(v_r\frac{\partial v_z}{\partial r} + v_z\frac{\partial v_z}{\partial z}\right) = F_z - \frac{\partial p}{\partial z} + \mu\left(\frac{\partial^2 v_z}{\partial r^2} + \frac{1}{r}\frac{\partial v_z}{\partial z} + \frac{\partial^2 v_z}{\partial z^2}\right) \tag{6-35}$$

$$\text{Energy:} \quad \rho c_p\left(v_r\frac{\partial T}{\partial r} + v_z\frac{\partial T}{\partial z}\right) = k\left(\frac{\partial^2 T}{\partial r^2} + \frac{1}{r}\frac{\partial T}{\partial r} + \frac{\partial^2 T}{\partial z^2}\right) + \frac{\mu}{g_c J}\Phi \tag{6-36a}$$

where the viscous-dissipation function is defined as

$$\Phi \equiv 2\left[\left(\frac{\partial v_r}{\partial r}\right)^2 + \frac{v_r{}^2}{r^2} + \left(\frac{\partial v_z}{\partial z}\right)^2\right] + \left(\frac{\partial v_z}{\partial r} + \frac{\partial v_r}{\partial z}\right)^2 \qquad (6\text{-}36b)$$

The physical significance of various terms in these equations is the same as given above for the rectangular coordinate system.

The reader should consult Refs. [1–6] for the more general, three-dimensional form of the above equations.

6-5   DIMENSIONLESS GROUPS

Because of the complexity of the equations of motion and energy, it is extremely difficult to solve convective heat-transfer problems except for idealized, simple situations. Therefore, for most cases of practical interest the convective heat transfer is studied experimentally and the results are presented in the form of empirical equations that involve dimensionless groups. The utility of using dimensionless groups in such correlations is that several variables are combined into a few dimensionless parameters; hence the number of variables to be studied is reduced. Therefore the establishment of dimensionless groups that are appropriate for a given heat-transfer problem is most important. Two different methods are generally used to determine the dimensionless groups. In one of these methods all the pertinent variables affecting the physical process are listed, and the number of independent dimensionless groups are determined by a rule such as the Buckingham π theorem discussed in Refs. [7,8]. The procedure in this approach is straightforward, but the analysis may lead to incorrect results if one or more of the pertinent variables are omitted in the listing. In the other approach, the dimensionless groups are determined directly from the dimensionless form of the differential equations governing the physical process; therefore it is less likely to omit the pertinent variables if proper equations are considered in the analysis. Here we determine the dimensionless groups for convective heat transfer from the nondimensional form of the governing equations.

We restrict the analysis to the two-dimensional, steady, incompressible, constant-property flow, assume that *the main flow is along the x direction,* and for simplicity neglect the body forces. However, the effects of body forces will be considered separately in Chap. 10 on free convection in which the buoyancy becomes the body force that sets the fluid into motion. Then the equations of motion and energy are obtained from Eqs. (6-29) to (6-32):

Continuity: $\dfrac{\partial u}{\partial x} + \dfrac{\partial v}{\partial y} = 0$ \qquad (6-37)

x Momentum: $\rho\left(u\dfrac{\partial u}{\partial x} + v\dfrac{\partial u}{\partial y}\right) = -\dfrac{\partial p}{\partial x} + \mu\left(\dfrac{\partial^2 u}{\partial x^2} + \dfrac{\partial^2 u}{\partial y^2}\right)$ (6-38)

y Momentum: $\rho\left(u\dfrac{\partial v}{\partial x} + v\dfrac{\partial v}{\partial y}\right) = -\dfrac{\partial p}{\partial y} + \mu\left(\dfrac{\partial^2 v}{\partial x^2} + \dfrac{\partial^2 v}{\partial y^2}\right)$ (6-39)

Energy: $\rho c_p\left(u\dfrac{\partial T}{\partial x} + v\dfrac{\partial T}{\partial y}\right) = k\left(\dfrac{\partial^2 T}{\partial x^2} + \dfrac{\partial^2 T}{\partial y^2}\right)$

$$+ \frac{\mu}{g_c J}\left[2\left(\frac{\partial u}{\partial x}\right)^2 + 2\left(\frac{\partial v}{\partial y}\right)^2 + \left(\frac{\partial v}{\partial x} + \frac{\partial u}{\partial y}\right)^2\right] \quad (6\text{-}40)$$

To nondimensionalize the above equations we select a characteristic length L, a reference velocity u_∞, a reference temperature T_∞, a reference temperature difference ΔT, and introduce the following new dimensionless variables:

$$X = \frac{x}{L} \qquad Y = \frac{y}{L} \qquad P = \frac{p}{\rho u_\infty^2}$$

$$U = \frac{u}{u_\infty} \qquad V = \frac{v}{u_\infty} \qquad \theta = \frac{T - T_\infty}{\Delta T}$$

(6-41)

Here the quantity ρu_∞^2 used to nondimensionalize the pressure term represents the double of the *dynamic head* (i.e., the quantity $\frac{1}{2}\rho u_\infty^2$ is called the dynamic head). By introducing these new variables into Eqs. (6-37) to (6-40), we obtain the continuity, momentum, and energy equations in the dimensionless form:

Continuity: $\dfrac{\partial U}{\partial X} + \dfrac{\partial V}{\partial Y} = 0$ (6-42)

X momentum: $U\dfrac{\partial U}{\partial X} + V\dfrac{\partial U}{\partial Y} = -\dfrac{\partial P}{\partial X} + \dfrac{1}{Re}\left(\dfrac{\partial^2 U}{\partial X^2} + \dfrac{\partial^2 U}{\partial Y^2}\right)$ (6-43)

Y momentum: $U\dfrac{\partial V}{\partial X} + V\dfrac{\partial V}{\partial Y} = -\dfrac{\partial P}{\partial Y} + \dfrac{1}{Re}\left(\dfrac{\partial^2 V}{\partial X^2} + \dfrac{\partial^2 V}{\partial Y^2}\right)$ (6-44)

Energy: $U\dfrac{\partial \theta}{\partial X} + V\dfrac{\partial \theta}{\partial Y} = \dfrac{1}{Re\ Pr}\left(\dfrac{\partial^2 \theta}{\partial X^2} + \dfrac{\partial^2 \theta}{\partial Y^2}\right)$

$$+ \frac{E}{Re}\left[2\left(\frac{\partial U}{\partial X}\right)^2 + 2\left(\frac{\partial V}{\partial Y}\right)^2 + \left(\frac{\partial V}{\partial X} + \frac{\partial U}{\partial Y}\right)^2\right]$$

(6-45)

where the new *dimensionless groups* are defined as

$$E \equiv \frac{u_\infty^2}{c_p \Delta T g_c J} = \text{Eckert number}$$
(6-46a)

$$\text{Pr} \equiv \frac{c_p \mu}{k} = \text{Prandtl number}$$
(6-46b)

$$\text{Re} \equiv \frac{\rho u_\infty L}{\mu} = \text{Reynolds number}$$
(6-46c)

and the conversion factors g_c, J are included in the Eckert number to allow the specific heat c_p to be taken in the units of Btu/lb·°F; generally the conversion factors g_c, J are omitted. For example, if c_p is in Btu/lb·°F, u_∞ in ft/s, ΔT in °F, the conversion factors are taken as $g_c = 32.17$ ft·lb/s²·lb$_f$ and $J = 778.2$ ft·lb$_f$/Btu. By direct substitution it is verified that the Eckert number has no dimensions, i.e.,

$$E = \frac{u_\infty^2}{c_p \Delta T g_c J} \sim \left(\frac{\text{ft}}{\text{s}}\right)^2 \frac{\text{lb}\cdot°\text{F}}{\text{Btu}} \frac{1}{°\text{F}} \frac{\text{s}^2\cdot\text{lb}_f}{32.2 \text{ ft}\cdot\text{lb}} \frac{\text{Btu}}{778.2 \text{ ft}\cdot\text{lb}_f}$$

If the SI system of units is used we have

$$E = \frac{u_\infty^2}{c_p \Delta T g_c} \sim \left(\frac{\text{m}}{\text{s}}\right)^2 \frac{\text{kg}\cdot°\text{C}}{\text{W}\cdot\text{s}} \frac{1}{°\text{C}} \left(1 \frac{\text{N}\cdot\text{s}^2}{\text{kg}\cdot\text{m}}\right) = \frac{\text{N}\cdot\text{m}}{\text{W}\cdot\text{s}}$$

which is dimensionless since 1 W·s = 1 N·m.

The above system of dimensionless equations contain *Eckert*, *Prandtl*, and *Reynolds* numbers as independent parameters; hence the temperature distribution or the heat transfer in forced convection depends on these three dimensionless groups. In practice, to compute the heat transfer between the fluid and the wall surface a heat-transfer coefficient h is defined as

$$q = h \Delta T$$
(6-47)

where q is the heat flux and ΔT is the difference between the wall-surface and the mean fluid temperatures. For flow over bodies, the main-stream temperature T_∞ is taken as the mean fluid temperature; for flow inside tubes, a *bulk-fluid temperature* that is defined in the next chapter is taken as the mean fluid temperature. If the main flow is in the x direction and the y axis is normal to the wall surface, the heat flux q is related to the temperature gradient by

$$q = -k \frac{\partial T}{\partial y}\bigg|_{\text{wall}}$$
(6-48)

From Eqs. (6-47) and (6-48) we obtain

$$h \, \Delta T = -k \frac{\partial T}{\partial y}\bigg|_{y=0} \tag{6-49a}$$

which is arranged in the dimensionless form

$$\mathrm{Nu} \equiv \frac{hL}{k} = -\frac{\partial \theta}{\partial Y}\bigg|_{\mathrm{wall}} \tag{6-49b}$$

where various dimensionless variables are defined as $Y = y/L$, $\theta \equiv (T - T_\infty)/\Delta T$, and the dimensionless group hL/k is called the *Nusselt number*. Clearly, the Nusselt number depends on the same dimensionless groups as the temperature; then the Nusselt number for forced convection depends on Eckert, Prandtl, and Reynolds numbers, and the functional relationship may be written

$$\mathrm{Nu} = f(\mathrm{Re}, \mathrm{Pr}, \mathrm{E}) \tag{6-50}$$

The Eckert number enters the problem because of the viscous-dissipation term in the energy equation (6-45). At moderate flow velocities the viscous-dissipation term is neglected; then the functional relationship for the Nusselt number in forced convection simplifies to

$$\mathrm{Nu} = f(\mathrm{Re}, \mathrm{Pr}) \tag{6-51}$$

According to this relation, for geometrically similar surfaces the heat-transfer coefficient (or the Nusselt number) for forced convection at moderate flow velocities can be correlated in terms of two dimensionless groups instead of several variables that enter the problem. This is very important in experimental investigations because the number of variables to be studied is significantly reduced.

We now discuss the physical significance of the above dimensionless groups. If the inertia forces are characterized by $u_\infty{}^2/L$ and the viscous forces by $v u_\infty/L^2$, then the Reynolds number can be arranged as

$$\mathrm{Re} = \frac{\rho u_\infty L}{\mu} = \frac{u_\infty L}{v} = \frac{u_\infty{}^2/L}{v u_\infty/L^2} = \frac{\text{inertia force}}{\text{viscous force}} \tag{6-52}$$

that is, the Reynolds number represents the ratio of the inertia to viscous force. The physical significance of the inertia and viscous forces as defined above is better envisioned by considering the x-momentum equation (6-30) divided by ρ. Then, the left-hand side of this equation which represents the inertial forces contains terms such as $u(\partial u/\partial x)$ which can be characterized by $u_\infty{}^2/L$; similarly the viscous-force term $v(\partial^2 u/\partial x^2)$ on the right-hand side of this equation can be characterized by vu_∞/L^2. The reader should consult Ref. [6, p.136] for a detailed discussion of this matter.

The Prandtl number can be arranged as

$$\Pr = \frac{c_p \mu}{k} = \frac{(\mu/\rho)}{(k/\rho c_p)} = \frac{v}{\alpha} = \frac{\text{molecular diffusivity of momentum}}{\text{molecular diffusivity of heat}} \tag{6-53}$$

Hence it represents the ratio of molecular diffusivity of momentum to molecular diffusivity of heat.

The Eckert number can be arranged as

$$E = \frac{u_\infty{}^2}{c_p \,\Delta T} = \frac{u_\infty{}^2/c_p}{\Delta T} = \frac{\text{dynamic temperature due to fluid motion}}{\text{temperature difference}} \tag{6-54}$$

since $\tfrac{1}{2}u_\infty{}^2/c_p$ is the dynamic temperature with c_p chosen in proper units.

6-6 BOUNDARY-LAYER CONCEPT

The mathematical difficulties associated with the solutions of the equations of motion and energy have prompted investigators to develop concepts that will lead to the simplification of these equations. The *boundary-layer* concept which was originally proposed by Prandtl [9] proved to be most successful in achieving simplification of the equations of motion and energy and has been applied to a large variety of practical situations.   In the boundary-layer concept, the flow over a body is divided into two regions: (1) a very thin layer in the neighborhood of the body, called the boundary layer, where the velocity and temperature gradients are steep, and (2) the region outside the boundary layer, called the *potential-flow* or the *external-flow* region, where the velocity and temperature gradients are small. In general, the boundary-layer concept provides a good description of the velocity and the temperature fields, provided that the velocity and temperature gradients in the streamwise direction are much smaller than those in the direction perpendicular to the wall.   Figure 6-5

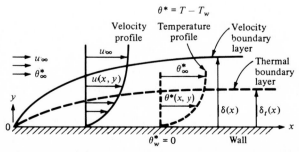

FIG. 6-5   Velocity and thermal boundary layers for laminar flow over a flat plate.

illustrates a *velocity* and a *thermal boundary layer* for laminar flow over a flat plate. In this figure the velocity and temperature are uniform in front of the leading edge of the plate (that is, $x = 0$) and in the external-flow region. The thickness of the velocity and thermal boundary layers increases with the distance from the leading edge. For practical purposes the edge of the velocity boundary layer is defined as the locus of points where the streamwise velocity component $u(x, y)$ in the velocity boundary layer reaches 99 percent of the velocity u_∞ of the external flow. In the case of the thermal boundary layer, let $\theta^*(x, y)$ and θ_∞^* denote, respectively, the temperatures in the thermal boundary layer and in the external flow in excess of the wall temperature. Then the edge of the thermal boundary layer is defined as the locus of the points where the temperature $\theta^*(x, y)$ in the boundary layer reaches 99 percent of temperature θ_∞^* of the external flow. The definition of the thickness of the boundary layer is rather arbitrary because the transition from the velocity and the temperature boundary layers to the conditions in the external flow takes place asymptotically.

In most applications the thickness of the velocity boundary layer $\delta(x)$ and the thermal boundary layer $\delta_t(x)$ are of the same order of magnitude but not necessarily equal; their relative thickness depends on the magnitude of the Prandtl number. That is, $\delta(x) > \delta_t(x)$ when $\mathrm{Pr} > 1$, $\delta(x) < \delta_t(x)$ when $\mathrm{Pr} < 1$, and $\delta(x) = \delta_t(x)$ when $\mathrm{Pr} = 1$. This matter will be discussed further in the following chapters.

The *laminar* and the *turbulent boundary layers* should be distinguished. Figure 6-6 illustrates these two types of velocity boundary layers for flow over a flat plate. Starting from the leading edge of the plate, the laminar boundary layer continues to develop until some critical distance x_c beyond which small disturbances start and grow inside the boundary layer and the transition from laminar to turbulent boundary layer takes place. This critical distance beyond which the flow can no longer keep its laminar character is specified in terms of the Reynolds number defined as $\mathrm{Re} = \rho x_c u_\infty / \mu$. In average situations, the flow over a flat plate begins to change from laminar to turbulent at a Reynolds number approximately equal to 5×10^5. When the surface is rough the transition to turbulent flow may start at a Reynolds number as small as 10^5; and when the flow is very calm and the surface is smooth the laminar boundary layer may exist up to Reynolds number 5×10^6. As illustrated in

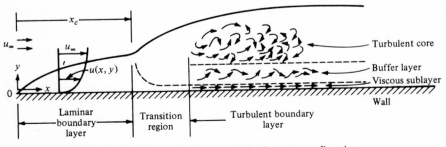

FIG. 6-6 Laminar and turbulent boundary layers for flow over a flat plate.

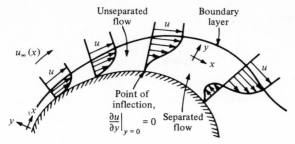

FIG. 6-7 Boundary-layer separation for flow over a curved body.

Fig. 6-6, the laminar boundary layer is followed by the *turbulent boundary layer* after a *transition region* in which flow changes from laminar to turbulent. In the turbulent boundary layer, next to the wall the flow is characterized by a very thin viscous-flow region called the *viscous sublayer*. Adjacent to the sublayer is a highly turbulent region called the *buffer layer* in which there is fine-grained turbulence, and the mean axial velocity increases rapidly with the distance from the wall. The buffer layer is followed by the *turbulent core* in which there is relatively lower intensity and larger-scale turbulence, and the velocity changes relatively little with the distance from the wall.

Figure 6-7 shows a boundary-layer flow over a curved body. In this case the boundary layer has the peculiar property that under certain conditions it separates from the wall, as illustrated in the figure. Beyond the point of separation the fluid particles near the wall move in a direction opposite to the external stream, the flow pattern becomes very complicated, and the boundary-layer equations are not applicable. The reader should consult Ref. [1] for detailed information on the subject of flow separation.

6-7 THE BOUNDARY-LAYER EQUATIONS

The *boundary-layer equations* are obtainable from the equations of motion and energy given previously, with simplification by an order-of-magnitude study of each term in these equations and then dropping the terms which are too small. To illustrate the approach, we present the simplification of the equations of motion and energy for the two-dimensional, steady flow of an incompressible fluid with constant properties given previously by Eqs. (6-42) to (6-45).

The basic assumptions made in the boundary-layer simplification include that the boundary-layer thicknesses δ and δ_t are small compared with a characteristic dimension L of the body, that is, $\delta \ll L$ and $\delta_t \ll L$, or

$$\Delta \equiv \frac{\delta}{L} \ll 1 \quad \text{and} \quad \Delta_t \equiv \frac{\delta_t}{L} \ll 1 \tag{6-55}$$

and the Reynolds number defined as

$$\text{Re} \equiv \frac{\rho u_\infty L}{\mu} \tag{6-56}$$

is assumed very large. Also the Reynolds number is assumed of the order $1/\Delta^2$ and the product of Reynolds and Prandtl numbers of the order of $1/\Delta_t^2$, that is,

$$\frac{1}{\text{Re}} \sim \Delta^2 \quad \text{and} \quad \frac{1}{\text{Re Pr}} \sim \Delta_t^2 \tag{6-57}$$

Since we are interested in the effects of small values of Δ and Δ_t on Eqs. (6-42) to (6-45), all other quantities in these equations can be measured in units of 1, Δ, and Δ_t. With this consideration, the variables U, X, and θ are assumed of the order unity,

$$U \sim 1 \quad X \sim 1 \quad \theta \sim 1 \tag{6-58}$$

and Y of the order Δ (or Δ_t),

$$Y \sim \Delta \quad (\text{or } \Delta_t) \tag{6-59}$$

The order of magnitude of the velocity component V is determined by inspecting the continuity equation (6-42). In this equation the two terms $\partial U/\partial X$ and $\partial V/\partial Y$ must be of the same order of magnitude. Since U and X are of the order unity, the derivative $\partial U/\partial X$ is of the order unity, and $\partial V/\partial Y$ must be of the order unity. Then, Y being assumed of the order Δ, the velocity component V must be of the order of Δ, that is,

$$V \sim \Delta \tag{6-60}$$

The foregoing analysis shows that the orders of magnitude of the quantities U, X, θ, Y, V, $1/\text{Re}$, and $1/\text{Re Pr}$ can be characterized in terms of 1 and Δ (or $\Delta_t \simeq \Delta$) such that $\Delta \ll 1$. These results are then applied to determine the orders of magnitude of the various terms in Eqs. (6-42) to (6-45) as described in *note 5* at the end of this chapter. These equations are simplified by dropping out the terms of the order of Δ in comparison with those of the order of 1. An interesting outcome of this order-of-magnitude study is that *the Y-momentum equation (6-44) is not needed in the boundary-layer analysis* because this equation merely implies that the pressure P across the boundary layer is practically constant. Also, the X-momentum and the energy equations are significantly simplified. The resulting *boundary-layer equations* for the two-dimensional, steady, incompressible, constant-property flow are given in the dimensionless term as

Continuity:
$$\frac{\partial U}{\partial X} + \frac{\partial V}{\partial Y} = 0 \tag{6-61}$$

X momentum:
$$U\frac{\partial U}{\partial X} + V\frac{\partial U}{\partial Y} = -\frac{dP}{dX} + \frac{1}{Re}\frac{\partial^2 U}{\partial Y^2} \tag{6-62}$$

Energy:
$$U\frac{\partial \theta}{\partial X} + V\frac{\partial \theta}{\partial Y} = -\frac{1}{Re\,Pr}\frac{\partial^2 \theta}{\partial Y^2} + \frac{E}{Re}\left(\frac{\partial U}{\partial Y}\right)^2 \tag{6-63}$$

where the dimensionless quantities are defined by Eqs. (6-41) and (6-46). In the framework of boundary-layer theory, the pressure term is regarded known; then Eqs. (6-61) to (6-63) provide three independent equations for the determination of three unknowns, U, V, and θ. If the Eckert number is small, that is, $E \ll 1$, the viscous-energy-dissipation term $(E/Re)(\partial U/\partial Y)^2$ in the energy equation can be neglected.

The boundary-layer equations (6-61) to (6-63) are much easier to solve than the original equations (6-42) to (6-45). The removal of the second derivative in the X direction (i.e., in the streamwise direction) from the X-momentum and the energy equations implies that in the boundary-layer equations the dependent variables U and θ are not bounded on the downstream side. That is, the velocity and temperature at a point in the boundary layer are not affected by the behavior of the fluid downstream. The Y-momentum equation is not needed in the boundary-layer analysis because under the boundary-layer assumptions this equation merely states that the pressure is constant across the boundary layer.

To compare the thicknesses of the velocity and the thermal boundary layers, we combine the two relations given in Eq. (6-57) and obtain

$$\left(\frac{\Delta_t}{\Delta}\right)^2 \sim \frac{1}{Pr}$$

or

$$\frac{\Delta_t}{\Delta} = \frac{\delta_t}{\delta} \sim \frac{1}{\sqrt{Pr}} \tag{6-64}$$

which relates the relative thickness of the thermal and velocity boundary layers to the Prandtl number of the fluid. For gases, Pr is of the order of unity; hence the two boundary layers are almost of the same thickness. For liquids, the Prandtl number ranges from about 10 to 1000; hence the thermal boundary-layer thickness is smaller than the velocity boundary-layer thickness. For liquid metal, Pr varies from about 0.003 to 0.03; then the thermal boundary layer is much thicker than the velocity boundary layer.

Finally, we present below the boundary-layer equations (6-61)—(6-63) in the dimensional form:

Continuity:
$$\frac{\partial u}{\partial x} + \frac{\partial v}{\partial y} = 0 \tag{6-65}$$

x Momentum:
$$\rho\left(u\frac{\partial u}{\partial x} + v\frac{\partial u}{\partial y}\right) = -\frac{dp}{dx} + \mu\frac{\partial^2 u}{\partial y^2} \tag{6-66}$$

Energy:
$$\rho c_p\left(u\frac{\partial T}{\partial x} + v\frac{\partial T}{\partial y}\right) = k\frac{\partial^2 T}{\partial y^2} + \mu\left(\frac{\partial u}{\partial y}\right)^2 \tag{6-67}$$

The pressure term in the x-momentum equation (6-66) [or (6-62)] can be related to the external stream velocity $u_\infty(x)$ by evaluating this equation at the edge of the velocity boundary layer, where $u \sim u_\infty(x)$. We find

$$-\frac{dp}{dx} = \rho u_\infty(x)\frac{du_\infty(x)}{dx} \tag{6-68}$$

since $u_\infty(x)$ is considered to be a function of x only. In the boundary-layer analysis the external-flow velocity $u_\infty(x)$ is assumed to be available from the solution of the velocity problem for flow outside the boundary layer; hence the term dp/dx is considered to be known. For flow along a flat plate, for example, the external-flow velocity u_∞ is constant, then

$$\frac{dp}{dx} = 0 \tag{6-69}$$

Thus the pressure gradient term $\partial p/\partial x$ does not appear in the x-momentum equation for flow along a flat plate.

REFERENCES

1 Schlichting, H.: "Boundary Layer Theory," 6th ed., McGraw-Hill Book Company, New York, 1968.
2 Bird, R. B., W. E. Stewart, and E. N. Lightfoot: "Transport Phenomena," John Wiley & Sons, Inc., New York, 1960.
3 Pai, S. I.: "Viscous Flow Theory," D. Van Nostrand Company, Inc., New York, 1956.
4 Eckert, E. R. G., and R. M. Drake: "Analysis of Heat and Mass Transfer," McGraw-Hill Book Company, New York, 1972.
5 Kays, W. M.: "Convective Heat and Mass Transfer," McGraw-Hill Book Company, New York, 1966.
6 Knudsen, J. G., and D. L. Katz: "Fluid Dynamics and Heat Transfer," McGraw-Hill Book Company, New York, 1958.

7 Langhaar, H. L.: "Dimensional Analysis and Theory of Models," John Wiley & Sons, Inc., New York, 1951.

8 Van Driest, E. R.: One Dimensional Analysis of the Presentation of Data in Fluid Flow Problems, *J. Appl. Mech.,* **13**: A-34 (1940).

9 Prandtl, L.: Über Flüssigkeitsbewegung bei sehr kleiner Reibung, *Proc. Third Int. Math. Congr. Heidelberg,* pp. 484–491, 1904; also *NACA Tech. Memo.* 452, 1928.

NOTES

1 The *total derivative* of a function $\phi(x, y, t)$, where ϕ may be a velocity component or temperature or pressure, etc., of the fluid, is obtained in the following manner. As the fluid element moves from a point (x, y) to another point $(x + dx, y + dy)$, the total differential change $d\phi$ of function ϕ is given by

$$d\phi(x, y, t) = \frac{\partial \phi}{\partial t}\, dt + \frac{\partial \phi}{\partial x}\, dx + \frac{\partial \phi}{\partial y}\, dy \tag{1}$$

or the total rate of change of ϕ is

$$\frac{d\phi}{dt} \equiv \frac{D\phi}{Dt} = \frac{\partial \phi}{\partial t} + \frac{\partial \phi}{\partial x} \frac{dx}{dt} + \frac{\partial \phi}{\partial y} \frac{dy}{dt} \tag{2}$$

The velocity components u and v are defined by

$$u = \frac{dx}{dt} \quad \text{and} \quad v = \frac{dy}{dt} \tag{3}$$

Then, Eq. (2) becomes

$$\frac{D\phi}{Dt} \equiv \frac{\partial \phi}{\partial t} + u \frac{\partial \phi}{\partial x} + v \frac{\partial \phi}{\partial y} \quad \text{(for two-dimensional, time-dependent)} \tag{4}$$

Here, the term $\partial \phi / \partial t$ represents the time rate of change of ϕ at a point in the fluid, and the remaining terms on the right-hand side account for changes associated with the convective motion of the fluid in the x and y directions. If the function ϕ *does not depend on time*, we have $\partial \phi / \partial t = 0$ but $D\phi / Dt \neq 0$. Then, for the *steady* case Eq. (4) simplifies to

$$\frac{D\phi}{Dt} = u \frac{\partial \phi}{\partial x} + v \frac{\partial \phi}{\partial y} \quad \text{(for two-dimensional, steady state)} \tag{5}$$

By replacing ϕ by u and v in Eq. (5) one recovers Eqs. (6-8b) and (6-8c), respectively.

2 When the stresses are given as in Eqs. (6-12) to (6-14), the term $\partial \sigma_x / \partial x + \partial \tau_{yx} / \partial y$ in the x-momentum equation (6-11a), for constant viscosity, becomes

$$\frac{\partial \sigma_x}{\partial x} + \frac{\partial \tau_{yx}}{\partial y} = \left(-\frac{\partial p}{\partial x} + 2\mu \frac{\partial^2 u}{\partial x^2} \right) + \mu \left(\frac{\partial^2 u}{\partial y^2} + \frac{\partial^2 v}{\partial y\, \partial x} \right)$$

$$= -\frac{\partial p}{\partial x} + \mu \left(\frac{\partial^2 u}{\partial x^2} + \frac{\partial^2 u}{\partial y^2} \right) + \mu \left(\frac{\partial^2 u}{\partial x^2} + \frac{\partial^2 v}{\partial y\, \partial x} \right)$$

$$\frac{\partial \sigma_x}{\partial x} + \frac{\partial \tau_{yx}}{\partial y} = -\frac{\partial p}{\partial x} + \mu\left(\frac{\partial^2 u}{\partial x^2} + \frac{\partial^2 u}{\partial y^2}\right) + \mu\frac{\partial}{\partial x}\left(\frac{\partial u}{\partial x} + \frac{\partial v}{\partial y}\right)$$

$$= -\frac{\partial p}{\partial x} + \mu\left(\frac{\partial^2 u}{\partial x^2} + \frac{\partial^2 u}{\partial y^2}\right)$$

since $\partial u/\partial x + \partial v/\partial y = 0$ by the continuity equation for incompressible flow.

The term $\partial \sigma_y/\partial y + \partial \tau_{xy}/\partial x$ for the y-momentum equation is evaluated in a similar manner.

3 The rate of energy input into the volume element $\Delta x\,\Delta y\,\Delta z$ due to work done by the surface stresses is determined by referring to Fig. 6-4. The rate of work done due to the normal stress σ_x is given by

$$\left\{-u\sigma_x + \left[u\sigma_x + \frac{\partial}{\partial x}(u\sigma_x)\,\Delta x\right]\right\}\Delta y\,\Delta z = \Delta x\,\Delta y\,\Delta z\,\frac{\partial}{\partial x}(u\sigma_x) \tag{1}$$

and, due to the normal stress σ_y, is given by

$$\left\{-v\sigma_y + \left[v\sigma_y + \frac{\partial}{\partial y}(v\sigma_y)\,\Delta y\right]\right\}\Delta x\,\Delta z = \Delta x\,\Delta y\,\Delta z\,\frac{\partial}{\partial y}(v\sigma_y) \tag{2}$$

Similarly, the rate of work done due to the shear stresses τ_{yx} and τ_{xy} are given, respectively, by

$$-u\tau_{yx} + \left[u\tau_{yx} + \frac{\partial}{\partial y}(u\tau_{yx})\,\Delta y\right]\Delta x\,\Delta z = \Delta x\,\Delta y\,\Delta z\,\frac{\partial}{\partial y}(u\tau_{yx}) \tag{3}$$

$$-v\tau_{xy} + \left[v\tau_{xy} + \frac{\partial}{\partial x}(v\tau_{xy})\,\Delta x\right]\Delta y\,\Delta z = \Delta x\,\Delta y\,\Delta z\,\frac{\partial}{\partial x}(v\tau_{xy}) \tag{4}$$

Then the rate of energy input due to the frictional work done by stresses on the volume element is obtained by summing up the above four quantities given by Eqs. (1) to (4).

$$\binom{\text{Rate of frictional}}{\text{work done on element}} = \left[\frac{\partial}{\partial x}(u\sigma_x) + \frac{\partial}{\partial y}(v\sigma_y) + \frac{\partial}{\partial y}(u\tau_{yx}) + \frac{\partial}{\partial x}(v\tau_{xy})\right]\Delta x\,\Delta y\,\Delta z \tag{5}$$

which is the relation given by Eq. (6-20b).

4 Equations (6-19) to (6-21) are substituted into Eq. (6-16) to yield

$$\rho\frac{De}{Dt} + \frac{\rho}{2}\frac{D}{Dt}(u^2 + v^2) = k\left(\frac{\partial^2 T}{\partial x^2} + \frac{\partial^2 T}{\partial y^2}\right) + \left[uF_x + vF_y + \frac{\partial}{\partial x}(u\sigma_x) + \frac{\partial}{\partial y}(v\sigma_y)\right.$$

$$\left. + \frac{\partial}{\partial y}(u\tau_{yx}) + \frac{\partial}{\partial x}(v\tau_{xy})\right] \tag{1}$$

This equation is now simplified by making use of the momentum equations (6-11a) and (6-11b). That is, Eq. (6-11a) is multiplied by u, Eq. (6-11b) is multiplied by v, and the results are added to give

$$\frac{\rho}{2}\frac{D}{Dt}(u^2 + v^2) + \left(uF_x + vF_y + u\frac{\partial \sigma_x}{\partial x} + v\frac{\partial \sigma_y}{\partial y} + u\frac{\partial \tau_{yx}}{\partial y} + v\frac{\partial \tau_{xy}}{\partial x}\right) \tag{2}$$

since

$$u\frac{Du}{Dt} = \frac{1}{2}\frac{Du^2}{Dt} \qquad \text{and} \qquad v\frac{Dv}{Dt} = \frac{1}{2}\frac{Dv^2}{Dt}$$

Subtracting Eq. (2) from Eq. (1) we obtain

$$\rho\frac{De}{Dt} = k\left(\frac{\partial^2 T}{\partial x^2} + \frac{\partial^2 T}{\partial y^2}\right) + \left(\sigma_x\frac{\partial u}{\partial x} + \sigma_y\frac{\partial v}{\partial y} + \tau_{yx}\frac{\partial u}{\partial y} + \tau_{xy}\frac{\partial v}{\partial x}\right) \tag{3}$$

since

$$\frac{\partial}{\partial x}(u\sigma_x) - u\frac{\partial\sigma_x}{\partial x} = \sigma_x\frac{\partial u}{\partial x}$$

$$\frac{\partial}{\partial y}(v\sigma_y) - v\frac{\partial\sigma_y}{\partial y} = \sigma_y\frac{\partial v}{\partial y}$$

$$\frac{\partial}{\partial y}(u\tau_{yx}) - u\frac{\partial\tau_{yx}}{\partial y} = \tau_{yx}\frac{\partial u}{\partial y}$$

$$\frac{\partial}{\partial x}(v\tau_{xy}) - v\frac{\partial\tau_{xy}}{\partial x} = \tau_{xy}\frac{\partial v}{\partial x}$$

Various stresses in Eq. (3) are given by

$$\tau_{xy} = \tau_{yx} = \mu\left(\frac{\partial u}{\partial y} + \frac{\partial v}{\partial x}\right) \tag{4a}$$

$$\sigma_x = -p + 2\mu\frac{\partial u}{\partial x} \tag{4b}$$

$$\sigma_y = -p + 2\mu\frac{\partial v}{\partial y} \tag{4c}$$

When the stresses in Eqs. (4) are substituted into Eq. (3) we obtain

$$\begin{aligned}
\rho\frac{De}{Dt} &= k\left(\frac{\partial^2 T}{\partial x^2} + \frac{\partial^2 T}{\partial y^2}\right) + \left[-p\frac{\partial u}{\partial x} + 2\mu\left(\frac{\partial u}{\partial x}\right)^2 - p\frac{\partial v}{\partial y} + 2\mu\left(\frac{\partial v}{\partial y}\right)^2 + \mu\left(\frac{\partial u}{\partial y} + \frac{\partial v}{\partial x}\right)^2\right] \\
&= k\left(\frac{\partial^2 T}{\partial x^2} + \frac{\partial^2 T}{\partial y^2}\right) + \left[-p\left(\frac{\partial u}{\partial x} + \frac{\partial v}{\partial y}\right) + 2\mu\left(\frac{\partial u}{\partial x}\right)^2 + 2\mu\left(\frac{\partial v}{\partial y}\right)^2 + \mu\left(\frac{\partial u}{\partial y} + \frac{\partial v}{\partial x}\right)^2\right] \\
&= k\left(\frac{\partial^2 T}{\partial x^2} + \frac{\partial^2 T}{\partial y^2}\right) + \mu\left[2\left(\frac{\partial u}{\partial x}\right)^2 + 2\left(\frac{\partial v}{\partial y}\right)^2 + \left(\frac{\partial u}{\partial y} + \frac{\partial v}{\partial x}\right)^2\right]
\end{aligned} \tag{5}$$

since $\partial u/\partial x + \partial v/\partial y = 0$ by the continuity equation.

Equation (5) is written more compactly as

$$\rho\frac{De}{Dt} = k\left(\frac{\partial^2 T}{\partial x^2} + \frac{\partial^2 T}{\partial y^2}\right) + \mu\Phi \tag{6}$$

where the *viscous-energy-dissipation term* Φ is defined as

$$\Phi \equiv 2\left[\left(\frac{\partial u}{\partial x}\right)^2 + \left(\frac{\partial v}{\partial y}\right)^2\right] + \left(\frac{\partial u}{\partial y} + \frac{\partial v}{\partial x}\right)^2 \tag{7}$$

which are the same as those given by Eqs. (6-23) and (6-24).

5 An order-of-magnitude study of the various terms in Eqs. (6-42) to (6-45) is now performed by utilizing the results in Eqs. (6-57) to (6-60) on the orders of magnitude of U, X, θ, Y, V, $1/\text{Re}$, and $1/\text{Re Pr}$. We write below Eqs. (6-42) to (6-45) and indicate under each term its order of magnitude in the units of 1, Δ, and Δ_t.

Continuity:
$$\frac{\partial U}{\partial X} + \frac{\partial V}{\partial Y} = 0 \tag{1}$$

$$\frac{1}{1} \qquad \frac{\Delta}{\Delta}$$

X momentum:
$$U\frac{\partial U}{\partial X} + V\frac{\partial U}{\partial Y} = -\frac{\partial P}{\partial X} + \frac{1}{\text{Re}}\left(\frac{\partial^2 U}{\partial X^2} + \frac{\partial^2 U}{\partial Y^2}\right) \tag{2}$$

$$1\,\frac{1}{1} \quad \Delta\,\frac{1}{\Delta} \qquad\qquad \Delta^2\!\left(\frac{1}{1}\quad \frac{1}{\Delta^2}\right)$$

Y momentum:
$$U\frac{\partial V}{\partial Y} + V\frac{\partial V}{\partial Y} = -\frac{\partial P}{\partial Y} + \frac{1}{\text{Re}}\left(\frac{\partial^2 V}{\partial X^2} + \frac{\partial^2 V}{\partial Y^2}\right) \tag{3}$$

$$1\,\frac{\Delta}{1} \quad \Delta\,\frac{\Delta}{\Delta} \qquad\qquad \Delta^2\!\left(\frac{\Delta}{1}\quad \frac{\Delta}{\Delta^2}\right)$$

Energy:
$$U\frac{\partial \theta}{\partial X} + V\frac{\partial \theta}{\partial Y} = \frac{1}{\text{Re Pr}}\left(\frac{\partial^2 \theta}{\partial X^2} + \frac{\partial^2 \theta}{\partial Y^2}\right)$$

$$1\,\frac{1}{1} \quad \Delta\,\frac{1}{\Delta} \quad \Delta_t{}^2\left(\frac{1}{1}\quad \frac{1}{\Delta_t{}^2}\right)$$

$$+ \, E\,\frac{1}{\text{Re}}\left[2\left(\frac{\partial U}{\partial X}\right)^2 + 2\left(\frac{\partial V}{\partial Y}\right)^2 + \left(\frac{\partial V}{\partial X} + \frac{\partial U}{\partial Y}\right)^2\right] \tag{4}$$

$$\Delta^2\left[\frac{1}{1} \qquad \frac{\Delta^2}{\Delta^2} \qquad \left(\frac{\Delta}{1},\ \frac{1}{\Delta}\right)^2\right]$$

An examination of these equations reveals the following simplifications:

a The continuity equation (1) remains unchanged.

b In the X-momentum equation (2), the term $\partial^2 U/\partial X^2$ can be neglected in comparison with the term $\partial^2 U/\partial Y^2$. With the choice of $1/\text{Re}$ of the order Δ^2, the viscous-force term $(1/\text{Re})(\partial^2 U/\partial Y^2)$ is of the order unity or of the same order of magnitude as the inertia forces on the left-hand side of this equation.

c In the Y-momentum equation (3), the pressure-gradient term $\partial P/\partial Y$ must be of the order of Δ since all other terms are of the order of Δ. This implies that $\partial P/\partial Y$ is very small, or the pressure P across the boundary layer is practically constant. Therefore, the Y-momentum equation is not needed in the boundary-layer analysis.

d In the energy equation (4) the term $\partial^2 \theta/\partial X^2$ is negligible in comparison with the term $\partial^2 \theta/\partial Y^2$. Then, the term $(1/\text{Re Pr})(\partial^2 \theta/\partial Y^2)$ becomes of the order of unity since we have chosen $1/\text{Re Pr}$ of the order of $\Delta_t{}^2$. In the viscous-dissipation function inside the brackets, an examination of the orders of magnitude of various terms reveals that all other terms in the brackets are to be neglected in comparison with $\partial U/\partial Y$. Then, the term $(E/Re)(\partial U/\partial Y)^2$ becomes of the order of unity if the Eckert number is chosen to be of the order of unity, because we have previously chosen $1/Re$ of the order of Δ^2.

We now summarize the results of the boundary-layer simplifications: The *boundary-layer equations* for two-dimensional, steady flow of an incompressible, constant-property flow include the continuity, the X-momentum, and the energy equations, which are given in the dimensionless form as

Continuity:
$$\frac{\partial U}{\partial X} + \frac{\partial V}{\partial Y} = 0 \tag{5}$$

X momentum:
$$U \frac{\partial U}{\partial X} + V \frac{\partial U}{\partial Y} = -\frac{dP}{dX} + \frac{1}{Re} \frac{\partial^2 U}{\partial Y^2} \tag{6}$$

Energy:
$$U \frac{\partial \theta}{\partial X} + V \frac{\partial \theta}{\partial Y} = -\frac{1}{Re\,Pr} \frac{\partial^2 \theta}{\partial Y^2} + \frac{E}{Re} \left(\frac{\partial U}{\partial Y}\right)^2 \tag{7}$$

Seven

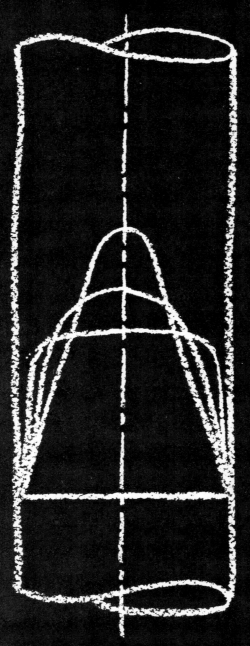

Convection—
Internal Laminar Forced Flow

In this chapter we illustrate with simple examples the determination of velocity and temperature distributions and heat transfer between the fluid and the walls for forced convection in laminar flow inside conduits. Although turbulent flow is more widely found than laminar flow in engineering applications, there are many situations in which laminar flow is important. For example, in liquid-metal-type nuclear reactors laminar flow may be desirable to reduce pumping power, since heat transfer with liquid metals is sufficiently high. In the design of heat exchangers for very viscous fluids such as oils, it may be economical to reduce the pumping-power requirement by lowering the flow velocity even though the heat transfer is less with laminar flow. In heat-transfer problems between a bearing and its journal the lubricant oil is sometimes in laminar flow. Finally, the laminar-flow examples considered in this chapter will provide a good insight into the physical significance of various parameters affecting heat transfer in forced flow. The governing equations for the specific problems to be solved here will be obtained directly from the general equations given in the preceding chapter by making appropriate simplifications in these equations.

7-1 COUETTE FLOW

Couette flow provides the simplest model for the analysis of heat transfer for flow between parallel plates. Figure 7-1 illustrates the geometry and the coordinates. The space between the two infinite parallel plates separated by a distance L is filled with a liquid having viscosity μ, density ρ, thermal conductivity k. The upper plate at $y = L$ moves with a constant velocity u_1 and sets the fluid particles moving in the direction parallel to the plates while the lower plate remains stationary. The lower and upper plates are kept at uniform temperatures T_0 and T_1, respectively. The heat-transfer problem characterized with this simple model is important for a journal and its bearing in which one of the surfaces is stationary while the other is rotating and the clearance between them is filled with a lubricant oil of high viscosity. When the clearance is small in comparison with the radius of the bearing, the geometry can be considered as two parallel plates. The oil being viscous, the temperature rise in

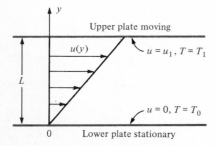

FIG. 7-1 Heat transfer in Couette flow.

the fluid due to friction (i.e., viscous energy dissipation) may become considerable even at moderate flow velocities. Therefore, the temperature rise in the fluid and the amount of heat transfer through the walls are of interest in engineering applications. In solving this heat-transfer problem we first determine the velocity distribution in the flow and then the temperature distribution, since the velocity profile is needed in the energy equation. The determination of the velocity and temperature distributions are now described.

Velocity Distribution

For an incompressible, constant-property flow the equations governing fluid motion are obtained from Eqs. (6-29) to (6-31) with the following considerations. Since all fluid particles are considered moving in the direction parallel to the plates, the velocity component v normal to the plates must be zero. By setting $v = 0$ in the continuity equation (6-29) we find

$$\frac{du}{dx} = 0 \tag{7-1}$$

Thus $u = u(y)$. The y-momentum equation (6-31) is not needed because $v = 0$. The x-momentum equation (6-30) is simplified by setting $v = 0$ and $F_x = 0$ for no body forces; we obtain

$$-\frac{dp}{dx} + \mu \frac{d^2u}{dy^2} = 0 \tag{7-2}$$

In Couette flow the fluid motion is set by simple *shear flow*, and no pressure gradient is involved in the direction of motion. With this consideration, the pressure-gradient term dp/dx is equal to zero, and Eq. (7-2) reduces to

$$\frac{d^2u}{dy^2} = 0 \quad \text{in } 0 \le y \le L \tag{7-3}$$

The boundary conditions for this equation are taken as velocity u equal to zero at the surface of the lower plate at $y = 0$ and equal to u_1 at the surface of the upper plate at $y = L$, that is,

$$u = 0 \quad \text{at } y = 0 \tag{7-4a}$$

$$u = u_1 \quad \text{at } y = L \tag{7-4b}$$

The solution of Eq. (7-3) subject to the boundary conditions (7-4) gives the velocity distribution in the Couette flow as

$$u(y) = \frac{y}{L} u_1 \tag{7-5}$$

Temperature Distribution

The equation governing the temperature distribution in the flow is now obtained from the energy equation (6-32) with the following simplifications.   It is shown above that $v = 0$ and $u = (y/L)u_1$; assuming that temperature varies in the y direction only, we have $T = T(y)$.   Introducing these simplifications into the energy equation (6-32), we obtain

$$k\frac{d^2T(y)}{dy^2} + \frac{\mu}{g_c J}\left(\frac{du}{dy}\right)^2 = 0 \quad \text{or} \quad k\frac{d^2T(y)}{dy^2} + \frac{\mu}{g_c J}\left(\frac{u_1}{L}\right)^2 = 0$$

or

$$\frac{d^2T(y)}{dy^2} = -\frac{\mu u_1^{\,2}}{g_c JkL^2} \quad \text{in } 0 \le y \le L \tag{7-6}$$

Here the conversion factors g_c and J are included in order to have the dimensions of the right-hand side consistent with those of the left-hand side if the viscosity μ is taken in the units of lb/ft·h.   To illustrate the physical significance of the conversion factors, we examine the dimensions of Eq. (7-6) in the English system.

$$\underbrace{\frac{°F}{ft^2}}_{\left(\frac{d^2T}{dy^2}\right)} = \underbrace{\frac{lb}{ft \cdot h}}_{(\mu)} \underbrace{\frac{ft^2}{h^2}}_{(u_1^{\,2})} \underbrace{\frac{h \cdot ft \cdot °F}{Btu}}_{\left(\frac{1}{k}\right)} \underbrace{\frac{1}{ft^2}}_{\left(\frac{1}{L^2}\right)} \underbrace{\frac{lb_f \cdot h^2}{4.17 \times 10^8 \; ft \cdot lb}}_{\left(\frac{1}{g_c}\right)} \underbrace{\frac{Btu}{778.2 \; ft \cdot lb_f}}_{\left(\frac{1}{J}\right)}$$

After simplification we find that the right-hand side of this equation has also the dimensions of $°F/ft^2$.

A similar analysis in the SI system gives

$$\underbrace{\frac{°C}{m^2}}_{\left(\frac{d^2T}{dy^2}\right)} = \underbrace{\frac{kg}{m \cdot s}}_{(\mu)} \underbrace{\frac{m^2}{s^2}}_{(u_1^{\,2})} \underbrace{\frac{m \cdot °C}{W}}_{\left(\frac{1}{k}\right)} \underbrace{\frac{1}{m^2}}_{\left(\frac{1}{L^2}\right)} \underbrace{\left(1\frac{N \cdot s^2}{kg \cdot m}\right)}_{\left(\frac{1}{g_c}\right)} \sim \frac{°C}{m^2}\frac{N \cdot m}{W \cdot s}$$

Clearly, both sides have the same dimensions since $1 \; N \cdot m = 1 \; W \cdot s$.

The boundary conditions for Eq. (7-6) are taken as temperature equals the lower-plate temperature T_0 at $y = 0$ and the upper-plate temperature T_1 at $y = L$, that is,

$$T(y) = T_0 \quad \text{at } y = 0 \tag{7-7a}$$

$$T(y) = T_1 \quad \text{at } y = L \tag{7-7b}$$

The solution of Eq. (7-6) subject to the boundary conditions (7-7) gives the temperature distribution in the fluid.   Integrating Eq. (7-6) twice, one obtains

$$T(y) = -\frac{1}{2}\frac{\mu u_1^{2}}{g_c J k L^2}\, y^2 + C_1 y + C_2 \tag{7-8}$$

when the integration contants C_1 and C_2 are evaluated by the application of the boundary condition (7-7); the temperature distribution in the fluid becomes

$$T(y) - T_0 = \frac{y}{L}\left[(T_1 - T_0) + \frac{\mu u_1^{2}}{2g_c Jk}\left(1 - \frac{y}{L}\right)\right] \tag{7-9}$$

We now examine the physical significance of this solution and heat flow at the wall for the two cases $T_0 \neq T_1$ and $T_0 = T_1$.

The Case $T_0 \neq T_1$ It is convenient to rearrange Eq. (7-9) in the dimensionless form.   Both sides of Eq. (7-9) are divided by $T_1 - T_0$ to yield

$$\frac{T(y) - T_0}{T_1 - T_0} = \frac{y}{L}\left[1 + \frac{1}{2}\frac{\mu u_1^{2}}{g_c Jk(T_1 - T_0)}\left(1 - \frac{y}{L}\right)\right] \tag{7-10a}$$

which is written more compactly in the form

$$\theta(\eta) = \eta[1 + \tfrac{1}{2}\mathrm{Pr}\, E(1 - \eta)] \tag{7-10b}$$

where various dimensionless quantities are defined as

$$\theta(\eta) = \frac{T(y) - T_0}{T_1 - T_0}$$

$$\eta = \frac{y}{L}$$

$$\mathrm{Pr} = \frac{c_p \mu}{k} = \text{Prandtl number} \tag{7-11}$$

$$E = \frac{u_1^{2}}{c_p(T_1 - T_0)g_c J} = \text{Eckert number}$$

Figure 7-2 shows a plot of the dimensionless temperature $\theta(\eta)$ as a function of η for $T_1 > T_0$ for several different values of the parameter Pr E.   The case Pr E = 0 corresponds to no flow condition; hence there is no viscous dissipation in the medium, and the temperature distribution is a straight line which characterizes pure conduction across the fluid layer.   The physical significance of other curves is envisioned better if heat transfer at the wall is considered.

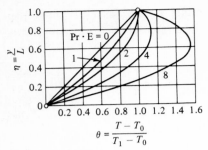

$$\theta = \frac{T - T_0}{T_1 - T_0}$$

FIG. 7-2   Temperature distribution in Couette flow $(T_1 > T_0)$.

The heat flux at the wall is determined from the definition

$$q_{wall} = -k \frac{dT(y)}{dy}\bigg|_{wall} \qquad (7\text{-}12)$$

and in terms of $\theta(\eta)$ it is written

$$q_{wall} = -\frac{k(T_1 - T_0)}{L} \frac{d\theta(\eta)}{d\eta}\bigg|_{wall} \qquad (7\text{-}13)$$

where the derivative of temperature is obtained from Eq. (7-10b) as

$$\frac{d\theta(\eta)}{d\eta} = 1 + Pr\, E\left(\frac{1}{2} - \eta\right) \qquad (7\text{-}14)$$

Then the heat flux, for example, at the upper wall is obtained from Eqs. (7-13) and (7-14) by setting $\eta = 1$. We find

$$q_{upper\ wall} = -\frac{k(T_1 - T_0)}{L}\left(1 - \frac{1}{2} Pr\, E\right) \qquad (7\text{-}15)$$

We now examine the direction of heat flow at the upper wall for the case $T_1 > T_0$ shown in Fig. 7-2 for different values of the parameter Pr E by considering the heat-transfer relation given by Eq. (7-15). The following cases are of interest:

1 Pr E > 2. In this case the term $1 - \frac{1}{2} Pr\, E$ in Eq. (7-15) is negative and $T_1 - T_0$ is positive for $T_1 > T_0$; then $q_{upper\ wall} > 0$. This implies that the heat flow is in the positive y direction, or heat flows from the liquid into the upper wall even though the upper wall is at a higher temperature than the lower wall. The reason for this is that the energy generation by viscous dissipation is so large that it cannot be all removed from the lower plate.

2 $Pr\ E < 2$.   In this case both terms $1 - \frac{1}{2} Pr\ E$ and $T_1 - T_0$ are positive in Eq. (7-15); then $q_{\text{upper wall}} < 0$ and the heat flow is in the negative y direction, or from the upper wall into the fluid.

3 $Pr\ E = 2$.   In this case the term $1 - \frac{1}{2} Pr\ E$ vanishes and there is no heat transfer at the upper wall.   Now referring to the temperature profile shown in Fig. 7-2, we infer that the derivative of temperature at the upper wall should be zero for $Pr\ E = 2$ because the heat flow is zero.

The Case $T_0 = T_1$ When both plates are at the same temperature Eq. (7-9) simplifies to

$$T(y) - T_0 = \frac{\mu u_1^2}{2g_c Jk} \frac{y}{L} \left(1 - \frac{y}{L}\right)$$

(7-16)

The maximum temperature in the fluid occurs at the midpoint between the plates; by setting $y = L/2$, Eq. (7-16) becomes

$$T_{\text{max}} - T_0 = \frac{\mu u_1^2}{8g_c Jk}$$

(7-17)

By combining Eqs. (7-16) and (7-17), the temperature distribution in the fluid is expressed as

$$\frac{T(\eta) - T_0}{T_{\text{max}} - T_0} = 4\eta(1 - \eta)$$

(7-18a)

where

$$\eta = \frac{y}{L}$$

(7-18b)

The heat flux at the walls is obtained from the definition of heat flux given by Eq. (7-12).

Example 7-1   A heavy lubricating oil ($\mu = 0.2$ lb/ft·s or 0.298 kg/m·s, $k = 0.072$ Btu/h·ft·°F or 0.125 W/m·°C) at room temperature flows in the clearance between a journal and its bearing.   Assuming both the bearing and the journal are kept at the same temperature, determine the maximum temperature rise in the lubricant for a velocity of $u_1 = 20$ ft/s (6.1 m/s).

Solution   The maximum temperature rise in the fluid for the case $T_0 = T_1$ is obtained from Eq. (7-17) as

$$\Delta T_{\text{max}} \equiv T_{\text{max}} - T_0 = \frac{\mu u_1^2}{8g_c Jk}$$

We now perform the computation by using both the English and the SI systems of units.

If the English system of units is used, we take $g_c = 32.2 \text{ ft} \cdot \text{lb/s}^2 \cdot \text{lb}_f$ and $J = 778 \text{ ft} \cdot \text{lb}_f/\text{Btu}$ and obtain

$$\Delta T_{\text{max}} = \frac{1}{8}\left(0.2 \frac{\text{lb}}{\text{ft} \cdot \text{s}}\right)\left(400 \frac{\text{ft}^2}{\text{s}^2}\right)\left(\frac{1}{32.2} \frac{\text{s}^2 \cdot \text{lb}_f}{\text{ft} \cdot \text{lb}}\right)\left(\frac{1}{778} \frac{\text{Btu}}{\text{ft} \cdot \text{lb}_f}\right)\left(\frac{1}{2 \times 10^{-5}} \frac{\text{s} \cdot \text{ft} \cdot {}^\circ\text{F}}{\text{Btu}}\right)$$

$$\Delta T_{\text{max}} = 20{}^\circ\text{F} \ (11.1{}^\circ\text{C})$$

If the SI system of units is used, by taking $g_c = 1 \text{ kg} \cdot \text{m/N} \cdot \text{s}^2$ and noting that $1 \text{ N} \cdot \text{m} = 1 \text{ W} \cdot \text{s}$, we find

$$\Delta T_{\text{max}} = \frac{1}{8}\left(0.298 \frac{\text{kg}}{\text{m} \cdot \text{s}}\right)\left(6.1^2 \frac{\text{m}^2}{\text{s}^2}\right)\left(\frac{1}{0.125} \frac{\text{m} \cdot {}^\circ\text{C}}{\text{W}}\right)\left(1 \frac{\text{N} \cdot \text{s}^2}{\text{kg} \cdot \text{m}}\right) = 11.1{}^\circ\text{C}$$

7-2 FULLY DEVELOPED VELOCITY- AND TEMPERATURE-PROFILE CONCEPTS FOR FLOW INSIDE CONDUITS

Figure 7-3 illustrates schematically the development of a velocity profile for the steady laminar flow of an incompressible fluid inside a circular tube. The changes in the velocity profile along the tube take place as now described. The fluid has a uniform velocity u_0 at the tube inlet; as it enters the conduit the velocity immediately adjacent to the walls becomes zero while it increases above u_0 in the central portions by the requirement of the continuity of flow. Then, one envisions that a *velocity boundary layer* starts to develop along the wall surface, and the thickness of this boundary layer continues to grow until the thickness reaches the tube center. The region from the tube inlet to a little beyond the hypothetical location where the boundary layer reaches the tube center is called the *hydrodynamic entry length* in which the velocity profile in flow changes axially and radially. The region beyond the hydrodynamic entry length in which the *velocity profile is considered invariant with the distance* along the tube is called the *region of fully developed velocity*. The subdivision of the tube into a *hydrodynamic entry region* and a *region of fully developed velocity* is illustrated in Fig. 7-3. A distinction should also be made in the concept of a fully developed velocity profile between the cases of laminar and turbulent flows. If the boundary layer remains laminar until its thickness reaches the tube center, *fully developed laminar flow* prevails beyond that point. If the boundary layer changes

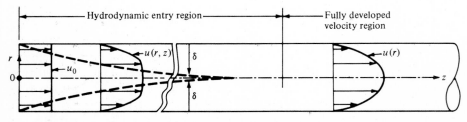

FIG. 7-3   Development of a velocity profile in laminar flow inside a tube.

into turbulent before its thickness reaches the tube center, *fully developed turbulent flow* exists beyond the hydrodynamic entry region.

In the case of temperature distribution it is more difficult to visualize the existence of a fully developed temperature profile in regions away from the entrance than the fully developed velocity profile discussed above. However, under certain heating or cooling conditions such as *constant heat flux* or *uniform temperature* at the tube walls, it is possible to consider the existence of a dimensionless temperature profile that remains invariant with the distance along the tube in regions away from the entrance. A qualitative discussion of this matter leading to the concept of a fully developed dimensionless profile is as follows:

Consider a laminar flow inside a circular tube subjected to uniform heat flux at the walls. Let r and z be the radial and axial coordinate axis, respectively. A dimensionless temperature $\theta(r, z)$ is defined as

$$\theta(r, z) = \frac{T(r, z) - T_w(z)}{T_m(z) - T_w(z)}$$

(7-19a)

where $T(r, z) =$ local temperature in fluid
$T_w(z) =$ tube wall temperature
$T_m(z) =$ mean temperature of fluid over cross-sectional area of tube

Clearly, $\theta(r, z)$ is zero at the wall and attains some value at the tube center. Then, one envisions the development of a thermal boundary layer along the wall surface such that the dimensionless temperature profile $\theta(r, z)$ in the flow continues to vary both radially and axially until the thickness of the thermal boundary layer reaches the tube center. The region from the tube inlet to the hypothetical location where the thermal boundary layer reaches the tube center is called the *thermal entry length* or the *thermal entry region*. In the region beyond the thermal entry length, in analogy with the development of the velocity profile, the above dimensionless temperature profile is considered invariant with the distance z. In the region where the dimensionless temperature profile θ does not vary with the distance z along the tube, the temperature profile is called the *fully developed temperature profile* which is a function of the r variable only, i.e.,

$$\theta(r) = \frac{T(r, z) - T_w(z)}{T_m(z) - T_w(z)}$$

(7-19b)

In the foregoing discussion we argued qualitatively, by analogy with the velocity development, the existence of a *thermally developed region* in a conduit for certain heating or cooling conditions such as *constant heat flux* or *uniform temperature* at the walls. However, the existence of a fully developed temperature profile $\theta(r)$ as defined by Eq. (7-19b) can also be shown mathematically by considering the asymptotic solution for $\theta(r, z)$ in regions away from the inlet. *The reader interested in the details of this matter is referred to note 1 at the end of this chapter.*

The fully-developed-temperature concept as defined by Eq. (7-19b) will be utilized later in this chapter in heat-transfer analysis of flow inside tubes in the thermally developed regions.

7-3 HEAT TRANSFER AND PRESSURE DROP IN HYDRODYNAMICALLY AND THERMALLY DEVELOPED LAMINAR FLOW IN CONDUITS

The problem of steady-state heat transfer in laminar forced convection inside conduits in the regions away from the inlet is of interest in numerous practical applications. The analysis of such problems for simple geometries such as for flow inside a circular tube or in a channel between two parallel plates is relatively easy because both the velocity and the temperature profiles are considered fully developed. In this section we determine the velocity and temperature distribution for hydrodynamically and thermally developed laminar forced flow inside a circular tube for the case of uniformly applied heat flux at the walls. The velocity distribution determined in this manner is then utilized to establish the *friction factor f* which is useful in the calculation of pressure drop for flow inside a tube. The *heat-transfer coefficient h* in the thermally developed region under uniform wall heat-flux conditions is determined from the knowledge of temperature distribution. The details of the analysis are as follows:

Velocity Distribution

Consider an incompressible, constant-property fluid in laminar flow inside a circular tube in regions away from the inlet where the velocity profile is considered fully developed. The radial velocity component v_r is zero for the fully developed region. This fact is now utilized to simplify the continuity and momentum equations (6-33) and (6-35) for laminar flow inside circular tubes.

The continuity equation (6-33) for $v_r = 0$ simplifies to

$$\frac{\partial u}{\partial z} = 0 \tag{7-20}$$

where v_z of Eq. (6-33) is replaced by u for convenience. Equation (7-20) implies that the axial velocity component does not depend on z, or we have $u = u(r)$.

The r-momentum equation (6-34) is not needed since $v_r = 0$.

The z-momentum equation (6-35) is now simplified by utilizing the above facts and by considering that no body forces are acting on the fluid. Then by setting $v_r = 0$, $F_z = 0$, noting that $\partial v_z/\partial z = 0$, $\partial^2 v_z/\partial z^2 = 0$, and replacing v_z by u Eq. (6-35) simplifies to

$$\frac{d^2u}{dr^2} + \frac{1}{r}\frac{du}{dr} = \frac{1}{\mu}\frac{dp}{dz} \qquad (7\text{-}21a)$$

which is written in the form

$$\frac{1}{r}\frac{d}{dr}\left(r\frac{du}{dr}\right) = \frac{1}{\mu}\frac{dp}{dz} \qquad \text{in } 0 \le r \le R \qquad (7\text{-}21b)$$

where R is the tube radius. The boundary conditions for this equation are taken as

$$u = 0 \qquad \text{at } r = R \qquad (7\text{-}22)$$

and u should remain finite at the tube center $r = 0$. (Because of symmetry about the tube axis, the boundary condition $du/dr = 0$ at $r = 0$ is also permissible; it leads to the same result, as u remains finite at $r = 0$.)

The solution of Eq. (7-21b) for constant $(1/\mu)(dp/dz)$, subject to the above boundary conditions gives the velocity profile $u(r)$ in the fully developed flow region:

$$u(r) = -\left(\frac{1}{4\mu}\frac{dp}{dz}\right)R^2\left[1 - \left(\frac{r}{R}\right)^2\right] \qquad (7\text{-}23)$$

Here the velocity $u(r)$ is always a positive quantity because for flow in the positive z direction the pressure gradient dp/dz is a negative quantity.

The *mean flow velocity* u_m over the tube cross section is determined from the definition as

$$u_m = \frac{1}{\pi R^2}\int_0^R 2\pi r u(r)\,dr = -\frac{R^2}{8\mu}\frac{dp}{dz} \qquad (7\text{-}24)$$

since $u(r)$ is given by Eq. (7-23).

From Eqs. (7-23) and (7-24) one finds

$$\frac{u(r)}{u_m} = 2\left[1 - \left(\frac{r}{R}\right)^2\right] \qquad (7\text{-}25)$$

This relation shows that for fully developed laminar flow inside a circular tube the velocity distribution over the tube cross section is parabolic.

The velocity u_0 at the tube axis is obtained from Eq. (7-23) by setting $r = 0$,

$$u_0 = -\frac{1}{4\mu}\frac{dp}{dz}R^2 \tag{7-26}$$

and combining Eqs. (7-23) and (7-26) one finds

$$\frac{u(r)}{u_0} = 1 - \left(\frac{r}{R}\right)^2 \tag{7-27}$$

A comparison of the results in Eqs. (7-24) and (7-26) shows that the velocity at the tube axis is equal to twice the mean flow velocity, i.e.,

$$u_0 = 2u_m \tag{7-28}$$

Friction Factor

For flow in tubes a *friction factor f* is defined as

$$f = -\frac{dp/dz}{(\frac{1}{2}\rho u_m^2)/D} \tag{7-29}$$

where D is the tube diameter. Substituting dp/dz from Eq. (7-24) into Eq. (7-29) we find the friction factor is given by

$$f = 64\frac{\mu}{\rho u_m D} = \frac{64}{\text{Re}} \tag{7-30a}$$

where

$$\text{Re} = \frac{\rho u_m D}{\mu} = \text{Reynolds number} \tag{7-30b}$$

The friction factor in Eq. (7-29) is *defined on the basis of the tube inside diameter D.* In the literature the friction factor has also been defined on the basis of the *hydraulic radius.* If f_r is the friction factor based on the hydraulic radius, it is related to the friction factor f by $f = 4f_r$. That is, Eqs. (7-30) on the basis of f_r would be $f_r = 16/\text{Re}$, where $\text{Re} = \rho u_m D/\mu$. This result is sometimes referred to as the Hagen-Poiseuille relation for friction factor in tubes, because Hagen's [1] experimental data were later verified theoretically by Poiseuille [2].

The utility of the friction factor in the calculation of pressure drop for fully developed flow inside a tube is now described. Let p_1 and p_2 be the pressures at the locations z_1 and z_2, respectively, along the tube. Equation (7-29) is integrated from z_1 to z_2 as

$$\int_{p_1}^{p_2} dp = -f\frac{\rho u_m^2}{2D}\int_{z_1}^{z_2} dz$$

or the pressure drop Δp becomes

$$\Delta p = f \frac{L}{D} \frac{\rho u_m^2}{2}$$

(7-31a)

where we define $\Delta p = p_1 - p_2$ and $L = z_2 - z_1$.

In the English system of units the pressure drop is generally given in lb_f/ft^2 or lb_f/in^2 (that is, psi) and in the SI system it is given in newtons/m². Therefore, in order to have the right-hand side of Eq. (7-31a) have the same dimension a conversion factor g_c is included, and Eq. (7-31a) is written

$$\Delta p = f \frac{L}{D} \frac{\rho u_m^2}{2g_c}$$

(7-31b)

To illustrate the physical significance of the conversion factor g_c we examine the dimensions of Eq. (7-31b) in both the English system and the SI system of units.

In the English system, we have

$$\frac{lb_f}{ft^2} = \frac{ft}{ft} \frac{lb}{ft^3} \frac{ft^2}{s^2} \frac{lb_f \cdot s^2}{32.2 \, lb \cdot ft} \sim \frac{lb_f}{ft^2}$$

$$(\Delta p) \qquad \left(\frac{L}{D}\right) (\rho)(u_m^2) \qquad \left(\frac{1}{g_c}\right)$$

Clearly both sides have the same dimensions.

In the SI system, we have

$$\frac{N}{m^2} = \frac{m}{m} \frac{kg}{m^3} \frac{m^2}{s^2} \frac{N \cdot s^2}{1 \, kg \cdot m} \sim \frac{N}{m^2}$$

$$(\Delta p) \qquad \left(\frac{L}{D}\right) (\rho)(u_m^2) \qquad \left(\frac{1}{g_c}\right)$$

Again both sides have the same dimensions, but we note that in the SI system the conversion factor g_c is equal to 1.

The pressure drop Δp inside a circular tube of length L, diameter D, for laminar flow with a mean velocity u_m can be determined from Eq. (7-31) if the friction factor f is known; the friction factor f is given by Eqs. (7-30). It is, however, to be pointed out that the friction factor f given by Eqs. (7-30) is applicable only for fully developed laminar flow inside circular tubes; i.e., for the Reynolds number, $Re = \rho u_m D/\mu < 2100$. Theoretical relations for the friction factor for fully developed laminar flow inside rectangular conduits have been derived by Cornish [3] and Lea and Tadros [4]; the results are also given in the text by Knudsen and Katz [5].

Example 7-2 A light oil at room temperature ($\rho = 56$ lb/ft^3 or 897 kg/m^3, $v = 5 \times 10^{-4}$ ft^2/s or 0.465×10^{-4} m^2/s) flows with a mean velocity of $u_m = 1.2$ ft/s (0.366 m/s) inside a circular tube. Determine the pressure drop for the length $L = 100$ ft (30.48 m) of the tube in regions away from the inlet for tube inside diameters (*a*) $D = \frac{1}{2}$ in (1.27×10^{-2} m) ID and (*b*) $D = 1$ in (2.54×10^{-2} m) ID.

Solution The pressure drop Δp for flow inside a tube is determined by Eq. (7-31*b*):

$$\Delta p = f\,\frac{L}{D}\,\frac{\rho u_m^{\,2}}{2g_c}$$

where the friction factor f for laminar flow is given by Eqs. (7-30) as

$$f = \frac{64}{\text{Re}} \qquad \text{and} \qquad \text{Re} = \frac{u_m D}{v}$$

(*a*) For the tube $D = \frac{1}{2}$ in ID:

$$\text{Re} = \frac{1.2}{(5 \times 10^{-4})(2 \times 12)} = 10^2$$

and the flow is laminar. Then, $f = \frac{64}{100}$. When the English system of units is used, we find

$$\Delta p = \frac{64}{100}(100 \times 2 \times 12)\,\frac{56 \times 1.2^2}{2 \times 32.17} = 1925\,\frac{\text{lb}_f}{\text{ft}^2}\,(92.2 \times 10^3\ \text{N/m}^2)$$

When the SI system of units is used, we obtain

$$\Delta p = \frac{64}{100}\,\frac{30.48}{1.27 \times 10^{-2}}\,(897)(0.366^2)\left(\frac{1}{2}\right) = 92.2 \times 10^3\ \text{N/m}^2$$

(*b*) For the tube $D = 1$ in ID,

$$\text{Re} = \frac{1.2}{(5 \times 10^{-4})(12)} = 2 \times 10^2$$

and the flow is laminar. Then, $f = \frac{64}{200}$.

$$\Delta p = \frac{64}{200}(100 \times 12)\,\frac{56 \times 1.2^2}{2 \times 32.17} = 481.3\,\frac{\text{lb}_f}{\text{ft}^2}\,(23 \times 10^3\ \text{N/m}^2)$$

We note that increasing the tube diameter by a factor of 2 decreases the pressure drop by a factor of 4. This is shown better by the relation for the pressure drop

$$\Delta p = \frac{64}{\text{Re}}\,\frac{L}{D}\,\frac{\rho u_m^{\,2}}{2g_c} = 64v\,\frac{L}{D^2}\,\frac{\rho u_m}{2g_c} \sim \frac{1}{D^2}$$

that is, the pressure drop is inversely proportional to D^2.

Temperature Distribution

The temperature distribution in the flow is determined from the solution of the energy equation subject to appropriate boundary conditions. The energy equation for flow inside circular tubes was given previously by Eqs. (6-36), but for the specific problem considered here, this equation is simplified with the following considerations: $v_r = 0$ for fully developed velocity and $\Phi = 0$ for moderate flow velocities.   Then the energy equations (6-36) reduce to

$$\frac{1}{\alpha} u(r) \frac{\partial T}{\partial z} = \frac{\partial^2 T}{\partial r^2} + \frac{1}{r} \frac{\partial T}{\partial r} + \frac{\partial^2 T}{\partial z^2} \tag{7-32}$$

where we replaced v_z by $u(r)$ for convenience.   The velocity profile $u(r)$ is already available by Eq. (7-25) from the solution of the velocity problem.   Equation (7-32) is a partial differential equation for the temperature $T(r, z)$ and is valid over the entire heated region $z > 0$ inside the tube.   In the regions away from the inlet where a fully developed dimensionless temperature profile is considered to exist, this equation can be reduced to an ordinary differential equation in the following manner.

We consider the dimensionless temperature $\theta(r)$ for the thermally developed region given by Eq. (7-19b)

$$\theta(r) = \frac{T(r, z) - T_w(z)}{T_m(z) - T_w(z)} \tag{7-33}$$

and differentiate it with respect to z to obtain

$$\frac{d\theta(r)}{dz} = \frac{\partial}{\partial z} \left[\frac{T(r, z) - T_w(z)}{T_m(z) - T_w(z)} \right] = 0 \tag{7-34}$$

since $\theta(r)$ is independent of z.   The terms in the brackets are differentiated with respect to z, and the constant heat-flux condition at the wall is utilized; we find (*see note 2 for details*)

$$\frac{dT(r, z)}{\partial z} = \frac{dT_m(z)}{dz} = \text{const} \tag{7-35}$$

Equation (7-35) implies that in the thermally developed region the average fluid temperature $T_m(z)$ increases linearly with z.   The substitution of Eq. (7-35) into Eq. (7-32) and noting that the term $\partial^2 T/\partial z^2$ vanishes for constant $\partial T/\partial z$ by Eq. (7-35) yield the following ordinary differential equation for the fluid temperature T:

$$\frac{1}{\alpha} u(r) \frac{dT_m(z)}{dz} = \frac{1}{r} \frac{d}{dr} \left(r \frac{dT}{dr} \right) \qquad \text{in } 0 \le r \le R \tag{7-36a}$$

The velocity $u(r)$ appearing in this equation is obtained from Eq. (7-25) as

$$u(r) = 2u_m \left[1 - \left(\frac{r}{R} \right)^2 \right] \tag{7-36b}$$

Equations (7-36) are written more compactly in the form

$$\frac{d}{dr}\left(r\frac{dT}{dr}\right) = Ar\left[1 - \left(\frac{r}{R}\right)^2\right] \qquad \text{in } 0 \le r \le R \tag{7-37a}$$

where the constant A is defined as

$$A \equiv \frac{2u_m}{\alpha}\frac{dT_m(z)}{dz} = \text{const} \tag{7-37b}$$

The boundary conditions for Eq. (7-37a) are taken as

$$\frac{dT}{dr} = 0 \qquad \text{at } r = 0 \tag{7-38a}$$

$$T = T_w(z) \qquad \text{at } r = R \tag{7-38b}$$

The first of these boundary conditions is the symmetry condition about the tube axis at $r = 0$. The second boundary condition merely states that the fluid temperature equals the wall temperature $T_w(z)$ at the wall surface. Here $T_w(z)$ is an unknown quantity; but this is immaterial because in the subsequent analysis we need the temperature of the fluid with reference to the wall-surface temperature.

The integration of Eq. (7-37a) once and the application of the boundary condition (7-38a) give

$$\frac{dT}{dr} = A\left(\frac{1}{2}r - \frac{r^3}{4R^2}\right) \tag{7-39}$$

When this result is integrated and the boundary condition (7-38b) is applied we find the fluid temperature $T(r, z)$:

$$T(r, z) - T_w(z) = -AR^2\left[\frac{3}{16} + \frac{1}{16}\left(\frac{r}{R}\right)^4 - \frac{1}{4}\left(\frac{r}{R}\right)^2\right] \tag{7-40}$$

The *mean fluid temperature* (or the *bulk-fluid temperature*) $T_m(z)$ across the tube cross section is obtained from the definition (*see note 3*)

$$T_m(z) - T_w(z) = \frac{\int_0^R 2\pi r u(r)[T(r, z) - T_w(z)]\, dr}{\int_0^R 2\pi r u(r)\, dr}$$

$$= \frac{1}{\pi R^2 u_m}\int_0^R 2\pi r u(r)[T(r, z) - T_w(z)]\, dr \tag{7-41}$$

where $T(r, z) - T_w(z)$ is given by Eq. (7-40) and $u(r)$ is given by Eq. (7-25).

Now substituting Eqs. (7-25) and (7-40) into Eq. (7-41) we obtain

$$T_m(z) - T_w(z) = -4A \int_0^R r \left(1 - \frac{r^2}{R^2} \right) \left(\frac{3}{16} + \frac{1}{16} \frac{r^4}{R^4} - \frac{1}{4} \frac{r^2}{R^2} \right) dr \tag{7-42}$$

When the integration with respect to r is performed, we find

$$T_m(z) - T_w(z) = -\frac{11}{48} \frac{AR^2}{2} \tag{7-43}$$

The constant A can be related to the wall heat flux q_w by utilizing the relation

$$k \frac{dT}{dr} \bigg|_{r=R} = q_w \tag{7-44a}$$

and from Eq. (7-39) we have

$$\frac{dT}{dr} \bigg|_{r=R} = \frac{1}{4} AR \tag{7-44b}$$

By combining Eqs. (7-44a) and (7-44b), the value of A becomes

$$A = \frac{4q_w}{kR} \tag{7-45}$$

In practical applications the heat-transfer coefficient or the Nusselt number for the flow is of interest. The above relations for the temperature are now used to determine the Nusselt number.

Nusselt Number

The heat-transfer coefficient h between the fluid flow and the wall is defined as

$$h[T_m(z) - T_w(z)] = -k \frac{dT}{dr} \bigg|_{r=R} \tag{7-46a}$$

or

$$h = -\frac{k}{T_m(z) - T_w(z)} \frac{dT}{dr} \bigg|_{r=R} \tag{7-46b}$$

The substitution from Eqs. (7-43) and (7-44b) into Eq. (7-46b) gives the heat-transfer coefficient as

$$h = \frac{48}{11} \frac{k}{D} \tag{7-47}$$

where $D = 2R$ is the tube inside diameter.

This result is rearranged in the form of the Nusselt number as

$$\text{Nu} \equiv \frac{hD}{k} = \frac{48}{11} = 4.364 \tag{7-48}$$

Thus, *for laminar flow inside a circular tube in the thermally developed region and under a constant heat-flux boundary condition at the wall the Nusselt number is constant and equal to* $\frac{48}{11}$.

Nusselt Number for Other Geometries and Boundary Conditions

In the foregoing example we considered heat transfer in the thermally developed region for laminar flow inside a circular tube under uniformly applied wall heat-flux boundary conditions.   The analysis can be extended in a similar manner to determine the Nusselt number under uniform wall-temperature boundary conditions.   The solutions are available in the literature for rectangular and triangular tubes under both uniform temperature and uniform wall heat-flux boundary conditions.   The reader should consult Ref. [6] for a discussion of such solutions. In Table 7-1 we present the Nusselt number for laminar flow in the thermally developed region inside tubes having circular, triangular, and rectangular cross sections for both constant wall temperature and constant wall heat-flux boundary conditions.   In this table the Nusselt number is defined as

$$\text{Nu} = \frac{hD_e}{k} \tag{7-49a}$$

where the *equivalent diameter* D_e is given by

$$D_e = \frac{4 \times (\text{flow area})}{\text{wetted perimeter}} \tag{7-49b}$$

Clearly, for a circular tube D_e is equal to the tube inside diameter D, that is,

$$D_e = \frac{4 \times (\pi/4)D^2}{\pi D} = D \tag{7-50}$$

Example 7-3 Calculate the heat-transfer coefficient for laminar flow of a fluid ($k = 0.1$ Btu/hr·ft·°F or 0.173 W/m·°C) inside a $\frac{1}{4}$ in (0.635×10^{-2} m) ID tube in the hydrodynamically and thermally developed region under uniform wall-temperature boundary conditions.   Also determine the heat-transfer rate between the tube walls and the fluid for the $L = 24$-ft (7.31-m) length of the tube if the mean temperature difference between the wall and the fluid is $\Delta T = 100°F$ (55.6°C).

Solution The Nusselt number for laminar flow inside a circular tube in the hydrodynamically and thermally developed regions under uniform wall-temperature boundary conditions is obtained from Table 7-1 as

$$\text{Nu} = \frac{hD}{k} = 3.66 \quad \text{or} \quad h = \frac{3.66k}{D}$$

TABLE 7-1   Nusselt Number for Laminar Forced Convection in Conduits of Various Cross Sections for Fully Developed Velocity and Temperature Profiles[†]

Shape of Channel Cross Section	$\dfrac{b}{a}$	Nu[‡] for Constant Wall Heat Flux	Nu[‡] for Constant Wall Temperature
circle		4.364	3.66
square	1.0	3.63	2.98
rectangle	1.4	3.78	
rectangle	2.0	4.11	3.39
rectangle	4.0	5.35	4.44
rectangle	8.0	6.60	5.95
parallel plates	∞	8.235	7.54
triangle 60° 60°		3.00	2.35

[†] Obtained from Kays [6].

[‡] Nusselt number is defined as $Nu = hD_e/k$.

In the English system of units, the heat-transfer coefficient h becomes

$$h = \frac{3.66 \times 0.1}{1/(4 \times 12)} = 17.57 \ \text{Btu/h} \cdot \text{ft}^2 \cdot {}^\circ\text{F}$$

or, using the SI system of units, we find

$$h = \frac{3.66 \times 0.173}{0.635 \times 10^{-2}} = 99.7 \ \text{W/m}^2 \cdot {}^\circ\text{C}$$

The heat-transfer rate Q between the tube walls and the fluid is determined from

$$Q = \text{area} \times h \times \Delta T = (\pi DL)h \, \Delta T$$

Using the English system of units, we find

$$Q = \pi \frac{1}{4 \times 12} \times 24 \times 17.57 \times 100 = 2760 \text{ Btu/h (809 W)}$$

or, using the SI system of units, we obtain

$$Q = \pi \times (0.635 \times 10^{-2}) \times 7.31 \times 99.7 \times 55.6 = 809 \text{ W}$$

7-4   HEAT-TRANSFER RESULTS FOR THERMAL ENTRY REGION IN LAMINAR FLOW THROUGH CONDUITS

In the preceding section we considered heat transfer in laminar flow through conduits in the region of fully developed velocity and temperature profiles; it was shown that the Nusselt number was constant.   In the entry region, however, the analysis is more involved, because the velocity and temperature profiles, in general, vary both radially and in the direction of flow.   When a hydrodynamic entry length is provided before the heat transfer starts, the velocity profile is assumed to be fully developed (i.e., velocity varies radially but not axially) while the temperature profile varies radially and axially; then some simplification is achieved in the heat-transfer analysis.   A classic solution of heat transfer for laminar flow inside a circular tube in the thermal entry region with the assumption of fully established parabolic velocity profile and subject to uniform tube wall temperature was given by Graetz [7].   A discussion of the Graetz solution can be found, in addition to the original work, in several references [5,6,8,9].   The Graetz problem was extended to flow inside circular tubes [10] for other boundary conditions, including uniform wall heat flux and linearly varying wall temperature, and to flow between parallel plates [11].   In the case of other geometries, however, the extension of the original Graetz problem is more involved.   When the Prandtl number is high relative to unity, the velocity profile is fully established much before the temperature profile.   For such situations the assumption of the fully developed velocity profile, as has been made in the Graetz solution and its extensions, is reasonably well justified.   Figure 7-4 shows the range of the Prandtl number tor liquid metals, gases, water, organic liquid, and oils.   Therefore, the Graetz solution and its extensions are applicable to predict heat transfer in the entry regions for the laminar flow of fluids such as oils for which the Prandtl number is high.

The heat transfer in the entry regions while both velocity and temperature profiles are developing simultaneously is of interest in many practical applications.   Sparrow [12] analyzed laminar heat transfer in the entrance region of

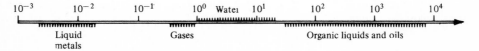

FIG. 7-4   Range of Prandtl number for liquid metals, gases, water, organic liquids, and oil.

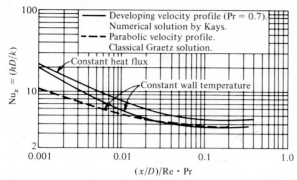

FIG. 7-5 Local Nusselt number determined by Kays [13] for simultaneous velocity and temperature development for laminar flow in a circular tube. (Pr = 0.7.)

two parallel plates for developing velocity and temperature profiles. Kays [13] considered the flow inside a circular tube and solved the heat-transfer problem numerically for Pr = 0.7 by employing Langhaar's velocity profile for the axial-velocity component but omitted the radial-velocity term which is important only very near the inlet. Figure 7-5 shows the local Nusselt number for laminar flow in a circular tube computed by Kays for developing velocity and temperature profiles with Pr = 0.7. This result is strictly applicable to the flow of air or the common gases because a Prandtl number 0.7 is used. Included in this figure is the classic Graetz solution for a parabolic velocity profile and constant wall temperature. A comparison of these two results reveals that the Graetz solution underestimates the heat-transfer coefficient at the entry region where the velocity and temperature profiles are simultaneously developing. In practice the mean value of the Nusselt number Nu_m over the length $x = 0$ to $x = L$ is of more interest than the local value of the Nusselt number Nu_x. The mean value of the Nusselt number is defined as

$$\mathrm{Nu}_m = \frac{1}{L} \int_{x=0}^{L} \mathrm{Nu}_x \, dx \qquad (7\text{-}51)$$

Figure 7-6 shows the mean Nusselt number Nu_m as obtained by Kays plotted as a function of the parameter $(x/D)/\mathrm{Re}\ \mathrm{Pr}$ for the case of developing velocity and temperature profiles for Pr = 0.7. Experimental data reported by Kays [14] were in good agreement with these numerical solutions. In Figs. 7-5 and 7-6 the asymptotic value of the Nusselt number is 3.66 for the case of constant wall temperature; this result is the same as that shown in Table 7-1 for flow in a circular tube with fully developed velocity and temperature profiles and the constant tube-wall temperature. The asymptotic Nusselt number for the case of constant wall heat flux is 4.364. The thermal entry length for laminar flow in a circular tube may be taken as approximately

$$\frac{x/D}{\mathrm{Re}\ \mathrm{Pr}} \simeq 0.05 \qquad (7\text{-}52)$$

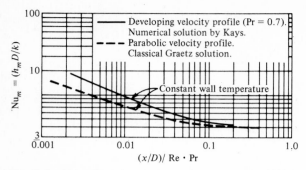

FIG. 7-6   Mean Nusselt number determined by Kays [13] for simultaneous velocity and temperature development for laminar flow in a circular tube. (Pr = 0.7)

Thus for $x/D > 0.05$ Re Pr the asymptotic value of the Nusselt number may be used to determine the local heat-transfer coefficient in a tube. For example, for gases with Pr $\simeq$ 1 flowing with Re = 500 a distance $x/D = 25$ is required to attain a fully developed temperature profile, whereas for an oil with Pr = 100 flowing with Re = 500 a distance $x/D = 2500$ is required. Therefore in oil heat exchangers a fully developed temperature profile is hardly attained.

Hausen [15] reported the following formula for the mean Nusselt number in the thermal entry region as representing the Graetz solution for a parabolic velocity profile and constant wall temperature:

$$Nu_m = 3.66 + \frac{0.0668(D/x)Re\ Pr}{1 + 0.04[(D/x)Re\ Pr]^{2/3}} \tag{7-53}$$

where $Nu_m = h_m D/k$, $Re = u_m D/v$, and D is the tube inside diameter. This equation is useful for viscous fluids such as oils, which require a long thermal entry length, and approximates the heat transfer at the entry region. In this expression the dimensionless group Re Pr(D/x) is called the *Graetz number*, i.e.,

$$Gz \equiv Re\ Pr\ \frac{D}{x}$$

All the heat-transfer results given above are based on an analysis which assumes that thermal properties of the fluid do not vary with temperature. If the viscosity varies substantially from the wall to the center of the fluid stream because of large temperature differences, the velocity profile is altered as indicated in Fig. 7-7. That is, when the temperature distribution in the flow is such that the viscosity near the wall is higher than that in the tube center, the fully developed velocity profile for uniform viscosity is distorted in a manner that increases velocity in the tube center and decreases it near the wall. The reason for this is that the increased viscosity increases the drag which decreases the velocity. The viscosity of liquids decreases with increasing temperature, but the viscosity of gases

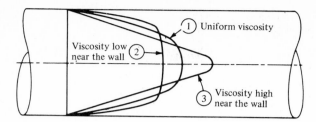

FIG. 7-7 Distortion of a fully developed velocity profile by viscosity variation. Curve 1: fully developed velocity profile for uniform viscosity; curve 2: liquid is heated or gas is cooled by the wall; curve 3: liquid is cooled or gas is heated by the wall.

increases with increasing temperature. Therefore, for a cold liquid heated by a hot tube wall, the viscosity is less near the wall than at the tube center, but the situation is reversed for gases.

Empirical correlations are generally used to correlate the heat-transfer data based on a constant-property assumption for the effects of viscosity variation with temperature. For liquids, the Nusselt number in the presence of temperature gradients may be obtained by multiplying the average Nusselt number for the isothermal case by the factor $(\mu_b/\mu_w)^{0.14}$, that is,

$$(\text{Nu}_m)_{\text{nonisothermal}} = (\text{Nu}_m)_{\text{isothermal}}\left(\frac{\mu_b}{\mu_w}\right)^{0.14} \tag{7-54}$$

where μ_b is the fluid viscosity at bulk temperature and μ_w the fluid viscosity at tube wall temperature.

A study of heat transfer in internal flow with gas-property variation was made by Swearingen and McEligot [16].

Example 7-4 Ethylene glycol at a mean temperature 60°F (15.6°C) flows with a mean velocity of 2 ft/s (0.61 m/s) through a $\frac{1}{4}$-in (0.635×10^{-2}-m) ID tube which is maintained at a uniform temperature 100°F (37.8°C). Determine the mean heat-transfer coefficient over the first 5-ft (1.52-m) length of the tube.

Solution The heat-transfer coefficient can be determined by using Eq. (7-53) with a modification for the effects of viscosity variation according to Eq. (7-54).

$$\text{Nu}_m = \left\{3.66 + \frac{0.0668(D/x)\text{Re Pr}}{1 + 0.04[(D/x)\text{Re Pr}]^{2/3}}\right\}\left(\frac{\mu_b}{\mu_w}\right)^{0.14}$$

The physical properties of ethylene glycol are given as:

μ_b (at 60°F) = 62.1 lb/ft·h   (25.67×10^{-3} kg/m·s)

μ_w (at 100°F) = 25.1 lb/ft·h   (10.38×10^{-3} kg/m·s)

$k = 0.169$ Btu/h·ft·°F   (0.292 W/m·°C)

$\rho = 68.7$ lb/ft^3 (1100.5 kg/m^3)

Pr (at 60°F) = 204

Then,

$$\text{Re} = \frac{\rho u D}{\mu} = \frac{68.7 \times 2}{62.1/3600} \frac{1}{4 \times 12} \simeq 166$$

or

$$\text{Re} = \frac{1100.5 \times 0.61 \times (0.635 \times 10^{-2})}{25.67 \times 10^{-3}} \simeq 166$$

$$\frac{D}{x} = \frac{1}{4 \times 12 \times 5} = \frac{1}{240}$$

$$\text{Nu}_m = \left[3.66 + \frac{0.0668 \times \frac{1}{240} \times 166 \times 204}{1 + 0.04(\frac{1}{240} \times 166 \times 204)^{2/3}}\right]\left(\frac{62.1}{25.1}\right)^{0.14} = 8.79$$

The mean heat-transfer coefficient h_m is determined as follows:

Using the English system of units, we find

$$h_m = \frac{k\text{Nu}_m}{D} = \frac{0.169 \times 8.79}{1/(4 \times 12)} = 71.3 \text{ Btu/h·ft}^2\text{·°F (404 W/m}^2\text{·°C)}$$

or, using the SI system of units, we obtain

$$h_m = \frac{0.292 \times 8.79}{0.635 \times 10^{-2}} = 404 \text{ W/m}^2\text{·°C}$$

7-5   HEAT TRANSFER IN LIQUID METALS FOR LAMINAR FLOW IN CONDUITS

Heat transfer in liquid metals requires special consideration because of their high thermal conductivity and correspondingly very low Prandtl number which varies approximately from 0.003 to 0.03. Lithium, sodium, potassium, mercury, bismuth, lead, sodium-potassium, and lead-bismuth are among common low-melting metals which are suitable for heat-transfer purposes as liquid metals. Their high thermal conductivity is advantageous in heat-transfer applications, in that large amounts of heat can be transferred at high temperatures with relatively low temperature difference between the fluid and the wall surface. Table 7-2 shows physical properties of some common liquid metals. Since the difference between the melting and boiling points is quite large (that is, 1000°F or greater except for mercury), they can be used as a heat-transfer medium over a wide range of temperatures at about atmospheric pressure. For this reason there has been considerable interest in the use of liquid metals as heat-transfer media in nuclear reactors and many other high-temperature, high-heat-flux applications. The major disadvantage of their use is the difficulty in handling them.

The Prandtl number for liquid metals being much smaller than unity, the temperature profile is established much faster than the velocity profile. The thermal conductivity being very high, the axial heat conduction becomes important

TABLE 7-2 Physical Properties of Some Common Liquid Metals

Metal	Melting Point	Normal Boiling Point	Temperature °F	Temperature °C	c_p Btu/lb·°F	$c_p \times 10^{-3}$ W·s/kg·°C	$\mu \times 10^3$ lb/ft·s	$\mu \times 10^3$ kg/m·s	k Btu/h·ft·°F	k W/m·°C	ρ lb/ft³	ρ kg/m³	$Pr = \dfrac{c_p \mu}{k}$
Bismuth	520°F (271.1°C)	2691°F (1477°C)	600	315.6	0.0345	0.144	1.09	1.62	9.5	16.4	625	10,120	0.014
			1000	537.8	0.0369	0.155	0.74	1.10	9.0	15.6	608	9729	0.011
			1400	760.0	0.0393	0.165	0.53	0.79	9.0	15.6	591	9467	0.008
Lead	621°F (327.2°C)	3159°F (1737°C)	700	371.1	0.038	0.159	1.61	2.40	9.3	16.1	658	10,540	0.024
			1300	704.4	0.037	0.155	0.92	1.37	8.6	14.9	633	10,140	0.014
Lithium	354°F (178.9°C)	2403°F (1317°C)	400	204.4	1.04	4.354	0.36	0.54	26.8	46.4	31.6	506	0.051
			1000	537.8	1.00	4.187	0.23	0.34	17.6	30.5	29.6	474	0.048
Mercury	−38°F (−38.9°C)	675°F (357°C)	50	10.0	0.033	0.138	1.07	1.59	4.7	8.1	847	13,568	0.027
			600	315.6	0.032	0.140	0.58	0.86	8.1	14.0	802	12,847	0.008
Potassium	147°C (63.9°C)	1400°F (760°C)	800	426.7	0.183	0.766	0.14	0.21	22.8	39.5	46.1	738	0.0041
			1400	760.0	0.187	0.783	0.09	0.13	18.0	31.1	41.5	665	0.0033
Sodium	208°F (97.8°C)	1621°F (883°C)	400	204.4	0.320	1.340	0.30	0.45	46.7	80.8	56.7	908	0.0075
			1400	760.0	0.303	1.269	0.12	0.18	32.7	56.6	47.7	764	0.0039
NaK (22%Na, 78%K)	66°F (18.9°C)	1518°F (826°C)	200	93.3	0.226	0.946	0.33	0.49	14.1	24.4	53.0	849	0.019
			1400	760.0	0.211	0.883	0.098	0.146			43.1	690	
NaK (56%Na, 44%K)	12°F (−11.1°C)	1443°F (784°C)	200	93.3	0.270	1.130	0.390	0.58	14.8	25.6	55.4	887	0.026
			1400	760.0	0.249	1.042	0.108	0.16	16.7	28.9	46.2	740	0.058

and cannot be neglected in the analysis. Johnson, Hartnett, and Clabaugh [17] investigated experimentally Nusselt numbers for mercury and lead-bismuth eutectic for laminar flow inside circular tubes, and the measured value of Nusselt number was considerably lower than the limiting value of Nu = 4.36 for ordinary liquids under constant wall heat-flux conditions. Measurements by various investigators in the laminar-flow region also deviate from each other. The nonwetting of some liquid metals on solid surfaces has been considered a reason for the heat-transfer coefficient for liquid metals being lower than the theoretical predictions. Opinions are divided on this matter, and no satisfactory explanation is yet available in the literature.

REFERENCES

1   Hagen, G.: Über die Bewegung des Wassers in engen zylindrischen Röhren, *Pogg. Ann.*, **46**:423 (1839).
2   Poiseuille, J.: Recherches experimentelles sur le mouvement des Liquides dans les tubes de tres petits diamètres, *C. R.*, **11**:961 (1840).
3   Cornish, R. J.: Flow in Pipe of Rectangular Cross-Section, *Proc. R. Soc. London, Ser. A*, **120**:691–700 (1928).
4   Lea, F. C., and A. G. Tadros: C. VI. Flow of Water Through A Circular Tube with a Central Core and Through Rectangular Tubes, *Philos. Mag.*, **11**:1235–1247 (1931).
5   Knudsen, J. G., and D. L. Katz: "Fluid Dynamics and Heat Transfer," McGraw-Hill Book Company, New York, 1958.
6   Kays, W. M.: "Convective Heat and Mass Transfer," McGraw-Hill Book Company, New York, 1966.
7   Graetz, L.: Über die Wärmeleitfähigkeit von Flüssigkeiten, *Ann. Phys.*, **25**:337 (1885).
8   Jakob, M.: "Heat Transfer," vol. 1, John Wiley & Sons, Inc., New York, 1949.
9   Eckert, E. R. G., and R. M. Drake, Jr.: "Analysis of Heat and Mass Transfer," McGraw-Hill Book Company, New York, 1972.
10 Sellars, S. R., M. Tribus, and J. S. Klein: Heat Transfer to Laminar Flow in a Round Tube or Flat Plate—The Graetz Problem Extended, *Trans. ASME*, **78**:441–448 (1956).
11 Norris, R. H., and D. D. Streid: Laminar-Flow Heat-Transfer Coefficient for Ducts, *Trans. ASME*, **62**:525 (1940).
12 Sparrow, E. M.: *NACA Tech. Note* 3331, 1955.
13 Kays, W. M.: Numerical Solutions for Laminar Flow Heat Transfer in Circular Tubes, *Trans. ASME*, **77**:1265–1274 (1955).
14 Kays, W. M.: *Stanford Univ. Dep. Mech. Eng. Tech. Rep.* 17, Navy Contract N6-Onr-251, Aug. 15, 1953.
15 Hausen, H.: *Verfahrenstechnik Beih. Z. Ver. Heut. Ing.*, **4**:91–98 (1943).
16 Swearingen, T. W., and D. M. McEligot: Internal Laminar Heat Transfer with Gas-Property Variation, *J. Heat Transfer*, **93C**:432–440 (1971).
17 Johnson, H. A., J. P. Hartnett, and W. J. Clabaugh: *ASME Paper* 53-A-188, 1953.

NOTES

1 Consider that a fluid at a uniform temperature T_0 enters at the origin of the axial coordinate $z = 0$ into a circular tube of radius R whose walls are kept at a uniform temperature T_w. If this temperature problem is solved by assuming a *slug flow* (i.e., velocity u is uniform over the cross section of the tube), neglecting the axial conduction and the viscous dissipation, the solution for the temperature distribution $T(r, z)$ in the fluid is given by

$$\frac{T(r, z) - T_w}{T_0 - T_w} = 2 \sum_{m=1}^{\infty} e^{-(\alpha/u)\beta_m^2 z} \frac{J_0(\beta_m r)}{R\beta_m J_1(\beta_m R)} \tag{1}$$

where α is the thermal diffusivity of the fluid, J_0 and J_1 are the Bessel functions of the zero and first order, respectively, and the eigenvalues β_m are the roots of

$$J_0(\beta R) = 0 \tag{2}$$

We can now show that θ is independent of z in the regions away from the inlet. At regions away from the inlet, that is, for large values of z, only the first term of the series in Eq. (1) is sufficient to represent the temperature distributions; then neglecting all the terms in the series except the first one, Eq. (1) simplifies to

$$\Psi(r, z) \equiv \frac{T(r, z) - T_w}{T_0 - T_w} = 2e^{-(\alpha/u)\beta_1^2 z} \frac{J_0(\beta_1 r)}{R\beta_1 J_1(\beta_1 r)} \tag{3}$$

where β_1 is the first eigenvalue as obtained from Eq. (2). The mean value of $\Psi(r, z)$ over the tube cross section is determined as

$$\Psi_m(z) = \frac{\int_0^R 2\pi r u \Psi(r, z)\, dr}{\int_0^R 2\pi r u\, dr} = \frac{2}{\pi R^2} \int_0^R r\Psi(r, z)\, dr = 4e^{-(\alpha/u)\beta_1^2 z} \frac{1}{(R\beta_1)^2} \tag{4}$$

Then, from Eqs. (3) and (4), $\theta(r)$ becomes

$$\theta(r) = \frac{T(r, z) - T_w}{T_m(z) - T_w} = \frac{\Psi(r, z)}{\Psi_m(z)} = \frac{1}{2} \frac{R\beta_1 J_0(\beta_1 r)}{J_1(\beta_1 r)} \tag{5}$$

which is independent of z.

2 The result given in Eq. (7-35) is obtained from Eq. (7-34) as follows:

$$\frac{\partial}{\partial z} \frac{T(r, z) - T_w(z)}{T_m(z) - T_w(z)} = \frac{(T_m - T_w)(\partial/\partial z)(T - T_w) - (T - T_w)(\partial/\partial z)(T_m - T_w)}{(T_m - T_w)^2} = 0 \tag{1}$$

or

$$\frac{\partial}{\partial z}(T - T_w) - \frac{T - T_w}{T_m - T_w} \frac{\partial}{\partial z}(T_m - T_w) = 0 \qquad \text{for } T_m \neq T_w \tag{2}$$

The constant heat flux q_w at the wall is related to the heat-transfer coefficient h between the fluid and the wall surface as

$$q_w = h(T_w - T_m) = \text{const} \tag{3}$$

For constant h we conclude that

$$T_w - T_m = \text{const} \tag{4}$$

Differentiating this relation with respect to z we find

$$\frac{d}{dz}(T_w - T_m) = 0 \tag{5}$$

or

$$\frac{dT_w}{dz} = \frac{dT_m}{dz} = \text{constant (since } q \text{ is constant)} \tag{6}$$

The substitution of Eq. (5) into Eq. (2) yields

$$\frac{\partial}{\partial z}(T - T_w) = 0 \tag{7}$$

or

$$\frac{\partial T}{\partial z} = \frac{dT_w}{dz} \tag{8}$$

From Eqs. (6) and (8) we obtain

$$\frac{\partial T}{\partial z} = \frac{dT_m}{dz} = \text{const} \tag{9}$$

which is the result given by Eq. (7-35).

3 The bulk temperature T_m for flow inside a circular tube of radius R is defined, in general, as

$$T_m = \frac{\int_0^R 2\pi\rho u c_p \, Tr \, dr}{\int_0^R 2\pi\rho u c_p r \, dr} \tag{1}$$

The numerator represents the total energy-flow rate through the tube at a given cross section, and the denominator represents the product of the mass-flow rate and the specific heat integrated over the cross section. For constant ρ and c_p, Eq. (1) simplifies to

$$T_m = \frac{\int_0^R uTr \, dr}{\int_0^R ur \, dr} \tag{2}$$

Eight

Convection—
External Laminar Forced Flow

In the preceding chapter we discussed the analysis of velocity and temperature distribution and the determination of friction factor and heat-transfer coefficient for laminar flow inside conduits.   This chapter is devoted to the analysis of similar problems for external laminar forced flow (i.e., flow over bodies) which has numerous applications in several areas including flow over aircraft wings, turbine and compressor blades, etc.   To illustrate the basic approach in the analysis, the flow over a flat plate is considered because a flat plate provides the simplest geometry that is capable of revealing important aspects of the problem. The quantities of practical interest such as the drag and heat-transfer coefficients are determined from the knowledge of the velocity and the temperature distributions.   In the course of the analysis given in this chapter the pertinent equations of motion and energy are directly obtained from those given in Chap. 6.

A brief discussion is also given of the determination of heat transfer in high-speed laminar flow along a flat plate, and pertinent heat-transfer results are presented.

The analysis of the heat transfer for flow over bodies such as cylinders, spheres, bank of tubes, etc., is difficult because of the phenomenon known as *flow separation*, but empirical heat-transfer relations are available for many geometries of practical interest.   The discussion of this matter and the pertinent relations for the drag and heat-transfer coefficients will be given in the next chapter on turbulent flow.

8-1   APPROXIMATE ANALYSIS OF VELOCITY AND DRAG FOR LAMINAR FLOW ALONG A FLAT PLATE

Consider two-dimensional, steady flow of an incompressible constant-property fluid along a flat plate as illustrated in Fig. 8-1.   The x axis is chosen along the plate with the origin $x = 0$ at the leading edge and the y axis perpendicular to the plate surface.   Let $u(x, y)$ and $v(x, y)$ be the velocity components in the x and y directions, respectively, u_∞ the free-stream velocity, and $\delta(x)$ the thickness of the velocity boundary layer.   The velocity components $u(x, y)$ and $v(x, y)$ satisfy the continuity and momentum equations for a boundary layer given in Chap. 6 [see Eqs. (6-65) and (6-66)]:

$$\text{Continuity:} \qquad \frac{\partial u}{\partial x} + \frac{\partial v}{\partial y} = 0 \tag{8-1}$$

$$x \text{ Momentum:} \qquad u \frac{\partial u}{\partial x} + v \frac{\partial u}{\partial y} = v \frac{\partial^2 u}{\partial y^2} \tag{8-2}$$

The pressure term dp/dx in the momentum equation vanishes for flow along a flat plate, as discussed in Chap. 6.

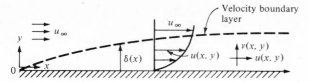

FIG. 8-1   Coordinates for the velocity problem for forced laminar flow along a flat plate.

The boundary conditions for these equations are

$$u = 0 \qquad v = 0 \qquad \text{at } y = 0 \tag{8-3a}$$

$$u \to u_\infty \qquad\qquad \text{at } y = \delta(x) \tag{8-3b}$$

The boundary conditions (8-3a) state that the velocity components are zero at the wall surface (i.e., wall surface is impermeable to flow), and the boundary condition (8-3b) implies that the axial velocity component is almost equal to the external flow velocity u_∞ at the edge of the velocity boundary layer at $y = \delta(x)$.

The velocity problem described by Eqs. (8-1) to (8-3) will now be solved by the approximate *integral method* originally developed by von Kármán [1]. This velocity problem will also be solved exactly in Sec. 8-4. The purpose of presenting here this approximate method of analysis is to illustrate the application of this powerful mathematical technique to obtain an analytical solution to the velocity problem, hence to provide some insight into the significance of various parameters. Approximate methods are useful for solving analytically more complicated problems which cannot readily be solved exactly; but the accuracy of an approximate method cannot be assessed until the approximate solution is compared with the exact solution.

The basic steps in the analysis with this method are as follows:

1   The x-momentum equation (8-2) is integrated with respect to y over the boundary-layer thickness $\delta(x)$, and the velocity component $v(x, y)$ appearing in this equation is eliminated by means of the continuity equation (8-1). The resulting equation is called the *momentum integral equation*.

2   A suitable profile is chosen for the velocity component $u(x, y)$ over the boundary layer $0 \le y \le \delta(x)$. A polynomial profile is generally chosen for this purpose, and experience has shown that there is no significant improvement in the accuracy of the solution by choosing a polynomial greater than the fourth degree. Suppose a cubic profile is chosen in the form

$$u(x, y) = a_0 + a_1 y + a_2 y^2 + a_3 y^3 \tag{8-4}$$

where the coefficients a_i are a function of x. These coefficients are determined in terms of the boundary-layer thickness $\delta(x)$ by making use of the conditions at $y = 0$ and $y = \delta(x)$; then the velocity profile given by Eq. (8-4) becomes a function of y and $\delta(x)$ in the form

$$u(x, y) = f[y, \delta(x)] \qquad \text{in } 0 \le y \le \delta(x) \tag{8-5}$$

3   The velocity profile $u(x, y)$ given by Eq. (8-5) is substituted into the *momentum integral equation* derived in step 1, and the integration with respect to the y variable is carried out.   The resulting expression is an ordinary differential equation for $\delta(x)$; when this ordinary differential equation is solved subject to the boundary condition

$$\delta(x) = 0 \qquad \text{for } x = 0 \tag{8-6}$$

the boundary-layer thickness $\delta(x)$ is determined.

4   Once $\delta(x)$ is known, the velocity distribution $u(x, y)$ is determined from Eq. (8-5).

When the velocity distribution is available from step 4, the drag coefficient is readily determined from its definition.

The integral method of analysis described here provides a very straightforward method of solution to the boundary-layer equations.   Although the analysis is an approximate one, it will be apparent later in this chapter that the drag coefficient determined with this method is sufficiently close to the exact results for most practical purposes.   We now illustrate the application of this method to the solution of the foregoing velocity problem.

The Solution of Velocity Problem by the Integral Method

The boundary-layer equations (8-1) and (8-2) subject to the boundary conditions (8-3) are now solved by the integral method following the steps described above.

Step 1   The momentum equation (8-2) is integrated with respect to the y variable over the boundary-layer thickness $\delta(x)$; we obtain

$$\int_0^{\delta(x)} u \frac{\partial u}{\partial x} \, dy + \int_0^{\delta(x)} v \frac{\partial u}{\partial y} \, dy = v \left(\frac{\partial u}{\partial y} \bigg|_{y=\delta} - \frac{\partial u}{\partial y} \bigg|_{y=0} \right) = -v \frac{\partial u}{\partial y} \bigg|_{y=0} \tag{8-7}$$

since $\partial u / \partial y|_{y=\delta} = 0$ by the boundary-layer concept.   The velocity component v appearing in Eq. (8-7) is eliminated by making use of the continuity equation (8-1).   After some manipulations as described in *note 1 at the end of this chapter*, Eq. (8-7) reduces to

$$\frac{d}{dx} \left[\int_0^{\delta} u(u_\infty - u) \, dy \right] = v \frac{\partial u}{\partial y} \bigg|_{y=0} \qquad \text{in } 0 \le y \le \delta \tag{8-8}$$

where $u \equiv u(x, y)$ and $\delta = \delta(x)$. Equation (8-8) is called the *momentum integral equation*.

Step 2 In the present analysis we choose a cubic-polynomial representation for the velocity $u(x, y)$ in the form

$$u(x, y) = a_0 + a_1 y + a_2 y^2 + a_3 y^3 \qquad (8-9)$$

The four conditions needed to determine these four coefficients a_0, a_1, a_2, and a_3 are taken as

$$u\Big|_{y=0} = 0 \qquad u\Big|_{y=\delta} = u_\infty$$

$$\frac{\partial u}{\partial y}\Big|_{y=\delta} = 0 \qquad \text{and} \qquad \frac{\partial^2 u}{\partial y^2}\Big|_{y=0} = 0 \qquad (8-10)$$

Clearly, the first two of these relations are the boundary conditions for the problem, the third one results from the boundary-layer concept, and the last one is the derived condition which is obtained by evaluating the x-momentum equation (8-2) at $y = 0$, where $u = v = 0$. The application of these four conditions given by Eqs. (8-10) to Eq. (8-9) results in a velocity profile in the form

$$\frac{u(x, y)}{u_\infty} = \frac{3}{2}\left(\frac{y}{\delta}\right) - \frac{1}{2}\left(\frac{y}{\delta}\right)^3 \qquad (8-11)$$

The reader is referred to note 2 for a discussion of other polynomial representations of the velocity profile.

Step 3 The velocity profile given by Eq. (8-11) is substituted into the momentum integral equation (8-8):

$$u_\infty{}^2 \frac{d}{dx}\left\{\int_0^\delta \left[\frac{3}{2}\frac{y}{\delta} - \frac{1}{2}\left(\frac{y}{\delta}\right)^3\right]\left[1 - \frac{3}{2}\frac{y}{\delta} + \frac{1}{2}\left(\frac{y}{\delta}\right)^3\right] dy\right\} = vu_\infty \frac{3}{2\delta} \qquad (8-12)$$

where the right-hand side of this equation results from the relation $\partial u/\partial y\big|_{y=0} = 3u_\infty/2\delta$. When the integration with respect to y is performed, Eq. (8-12) becomes

$$\frac{d}{dx}\left[\frac{39}{280}\delta(x)\right] = \frac{3v}{2u_\infty \delta(x)}$$

or

$$\delta \, d\delta = \frac{140}{13}\frac{v}{u_\infty} dx \qquad (8-13)$$

This is an ordinary differential equation for the boundary-layer thickness $\delta(x)$ which should be solved subject to the boundary condition

$$\delta(x) = 0 \quad \text{at } x = 0 \tag{8-14}$$

The integration of Eq. (8-13) with the boundary condition (8-14) yields

$$\delta^2(x) = \frac{280}{13} \frac{vx}{u_\infty} \tag{8-15}$$

Then, the boundary-layer thickness becomes

$$\delta(x) = \sqrt{\frac{280}{13} \frac{vx}{u_\infty}} = 4.64 \sqrt{\frac{vx}{u_\infty}} \tag{8-16}$$

which can be rearranged in the dimensionless form as

$$\frac{\delta(x)}{x} = \frac{4.64}{\text{Re}_x^{1/2}} \tag{8-17}$$

where the local Reynolds number Re_x is defined by

$$\text{Re}_x = \frac{u_\infty x}{v} \tag{8-18}$$

Step 4 Equation (8-11) establishes the velocity profile in the boundary layer since the boundary-layer thickness $\delta(x)$ is available by Eq. (8-16). However, in practice, it is not the velocity profile $u(x, y)$ but the drag coefficient that is of interest. The velocity profile established in this step is utilized in the determination of the drag coefficient as described below.

Drag Coefficient

The local-drag force τ_x per unit area exerted by the fluid flowing over a flat plate is related to a *local-drag coefficient* c_x as

$$\tau_x = c_x \frac{\rho u_\infty^2}{2g_c} \quad \text{(lb}_f/\text{ft}^2 \text{ or N/m}^2) \tag{8-19}$$

where ρ is the density and u_∞ is the external-flow velocity. We note that τ_x is, in fact, the local shear stress which is related to the velocity gradient at the wall by

$$\tau_x = \frac{\mu}{g} \frac{\partial u}{\partial y}\bigg|_{y=0} \tag{8-20}$$

From Eqs. (8-19) and (8-20) we obtain

$$c_x = \frac{2v}{u_\infty^2} \frac{\partial u}{\partial y}\bigg|_{y=0} \tag{8-21}$$

The velocity gradient appearing in Eq. (8-21) is determined from the velocity profile given by Eq. (8-11) as

$$\frac{\partial u}{\partial y}\bigg|_{y=0} = \frac{3u_\infty}{2\delta(x)} \tag{8-22}$$

The substitution of Eq. (8-22) into Eq. (8-21) with $\delta(x)$ as given by Eq. (8-16) yields the local-drag coefficient for laminar flow along a flat plate:

$$c_x = \frac{3v}{u_\infty} \sqrt{\frac{13}{280} \frac{u_\infty}{vx}} = \sqrt{\frac{117}{280} \frac{v}{u_\infty x}}$$

$$= \sqrt{\frac{117}{280}} \frac{1}{Re_x^{1/2}} = \frac{0.648}{Re_x^{1/2}} \tag{8-23}$$

where

$$Re_x = \frac{u_\infty x}{v}$$

The mean value of the drag coefficient $c_{m,\,L}$ over the length $x = 0$ to L is defined as

$$c_{m,\,L} = \frac{1}{L} \int_{x=0}^{L} c_x \, dx \tag{8-24}$$

Substituting Eq. (8-23) into (8-24) and performing the integration yield

$$c_{m,\,L} = \frac{1}{L} \sqrt{\frac{117}{280} \frac{v}{u_\infty}} \int_0^L x^{-1/2} \, dx = 2 \sqrt{\frac{117}{280} \frac{v}{u_\infty L}} = 2c_x \bigg|_{x=L} \tag{8-25}$$

Thus, the mean value of the drag coefficient over the length $x = 0$ to L is equal to twice the value of the local-drag coefficient at $x = L$.

Knowing the mean drag coefficient $c_{m,\,L}$, the drag force F acting on the plate over the length from $x = 0$ to $x = L$ and for the width w is determined from

$$F = wLc_{m,\,L} \frac{\rho u_\infty^2}{2g_c} \quad \text{(lb}_f \text{ or N)} \tag{8-26a}$$

where

$$c_{m,L} = 2\sqrt{\frac{117}{280}\frac{v}{u_\infty L}} = \frac{1.296}{Re_L^{1/2}} \qquad (8\text{-}26b)$$

In the foregoing analysis an approximate velocity profile obtained by the integral method with a cubic profile was used to determine the drag coefficient given by Eq. (8-23). It will be shown later in this chapter that if the exact velocity profile obtained from the exact solution of the velocity problem has been used, the resulting local-drag coefficient becomes

$$c_x = \frac{0.664}{Re_x^{1/2}} \qquad \text{(exact)} \qquad (8\text{-}27a)$$

and its average value over the length $x = 0$ to $x = L$ is given by

$$c_{m,L} = \frac{1.328}{Re_L^{1/2}} \qquad \text{(exact)} \qquad (8\text{-}27b)$$

Thus the approximate drag coefficient based on a cubic-polynomial representation of velocity is within 2.2 percent of the exact value. In Table 8-1 we summarize the boundary-layer thickness $\delta(x)/x$ and the local-drag coefficient c_x obtained by the exact solution of the velocity problem as well as by approximate solutions using a second-degree, cubic-, and fourth-degree-polynomial representation of the velocity profile.

Finally, it is to be noted that the results given above are applicable for laminar flow along a flat plate; under average conditions the laminar flow is maintained up to a critical Reynolds number $Re_c = ux_c/v = 5 \times 10^5$, where x_c is the distance from the leading edge of the plate to the point of transition from

TABLE 8-1   A Comparison of Exact and Approximate Solutions for Boundary Layer Thickness and the Local-Drag Coefficient for Laminar Flow Along a Flat Plate

Velocity Profile	$\dfrac{\delta(x)}{x}$	c_x
Exact	$\dfrac{4.96}{Re_x^{1/2}}$	$\dfrac{0.664}{Re_x^{1/2}}$
Approximate: Second-degree polynomial	$\dfrac{5.5}{Re_x^{1/2}}$	$\dfrac{0.727}{Re_x^{1/2}}$
Approximate: Cubic polynomial	$\dfrac{4.64}{Re_x^{1/2}}$	$\dfrac{0.648}{Re_x^{1/2}}$
Approximate: Fourth-degree polynomial	$\dfrac{5.83}{Re_x^{1/2}}$	$\dfrac{0.686}{Re_x^{1/2}}$

laminar to turbulent flow. However, factors such as the surface roughness of the plate, the level of disturbance in the flow, and the heat-transfer conditions affect the value of this critical Reynolds number.

Example 8-1 Air at atmospheric pressure and at 100°F (37.8°C) temperature flows with a velocity of $u_\infty = 3$ ft/s (0.915 m/s) along a flat plate. Determine the boundary-layer thickness $\delta(x)$ and the local-drag coefficient c_x at a distance $x = 2$ ft (0.61 m) from the leading edge of the plate. What is the mean drag coefficient over the length $x = 0$ to 2 ft, and the drag force acting on the plate over the length $x = 0$ to 2 ft per foot width of the plate?

Solution The physical properties of air at atmospheric pressure and 100°F (37.8°C) are

$\rho = 0.07$ lb/ft³ (1.126 kg/m³)

$v = 0.18 \times 10^{-3}$ ft²/s (0.167 × 10⁻⁴ m²/s)

The Reynolds number at a distance $x = 2$ ft from the leading edge of the plate is

$$\text{Re}_x = \frac{u_\infty x}{v} = \frac{3 \times 2}{0.18 \times 10^{-3}} = 3.3 \times 10^4 \quad \left(\text{or } \text{Re}_x = \frac{0.915 \times 0.61}{0.167 \times 10^{-4}} = 3.3 \times 10^4 \right)$$

Therefore the flow is laminar since the Reynolds number is less than 5×10^5. The boundary-layer thickness $\delta(x)$ and the local-drag coefficient at $x = 2$ ft are determined by using the exact relations given in Table 8-1.

$$\delta(x) = \frac{4.96x}{\sqrt{\text{Re}_x}} = 0.054 \text{ ft}$$

$$c_x = \frac{0.664}{\sqrt{\text{Re}_x}} = 3.6 \times 10^{-3}$$

The mean value of the drag coefficient over the length $x = 0$ to 2 ft is given by

$$c_{m, L} = 2 \times c_x \Big|_{x=2} = 7.2 \times 10^{-3}$$

The drag force F acting on the plate over the length $x = 0$ to 2 ft and per foot (0.305 m) width of the plate is now determined as follows:

Using the English system of units in Eq. (8-26a) and setting $w = 1$ ft, $L = 2$ ft, we find

$$F = wLc_{m, L} \frac{\rho u_\infty^2}{2g_c} = 1 \times 2 \times (7.2 \times 10^{-3}) \times \frac{0.07 \times 3^2}{2 \times 32.2} = 1.41 \times 10^4 \text{ lb}_f \text{ (or 6.3 × 10⁻⁴ N)}$$

Using the SI system of units and setting $w = 0.305$ m, $L = 0.61$ m, we obtain

$$F = 0.305 \times 0.61 \times (7.2 \times 10^{-3}) \times \frac{1.126 \times 0.915^2}{2} = 6.3 \times 10^{-4} \text{ N}$$

8-2 APPROXIMATE ANALYSIS OF HEAT TRANSFER FOR LAMINAR FLOW ALONG A FLAT PLATE

The approximate integral method will now be used to determine the temperature distribution in laminar flow along a flat plate which is kept at a uniform temperature. Once the temperature distribution in the flow is known, the heat-transfer coefficient between the fluid and the plate surface can readily be determined from its definition.

Consider a fluid at temperature T_∞ flowing with a velocity u_∞ along a flat plate as shown in Fig. 8-2. The x axis is chosen along the plate in the direction of flow with the origin $x = 0$ at the leading edge, and the y axis is perpendicular to the plate. It is assumed that the heat transfer between the plate and the fluid does not start until the location $x = x_0$; that is, the plate is kept at a temperature T_∞ in the region $0 \le x \le x_0$, and at a uniform temperature T_w in the region $x > x_0$. Referring to Fig. 8-2, we note that the *velocity boundary layer* of thickness $\delta(x)$ starts to develop at $x = 0$ and the *thermal boundary layer* of thickness $\delta_t(x)$ starts to develop at $x = x_0$ where the heat transfer between the plate and the fluid begins to take place. Let $T(x, y)$ be the temperature of the fluid inside the thermal boundary layer. Then the energy equation for the steady, two-dimensional boundary-layer flow of an incompressible fluid with constant properties, neglecting the viscous-dissipation term, is obtained from Eq. (6-67) as

$$u \frac{\partial T}{\partial x} + v \frac{\partial T}{\partial y} = \alpha \frac{\partial^2 T}{\partial y^2} \qquad (8\text{-}28)$$

For convenience in the analysis we define a dimensionless temperature $\theta(x, y)$ as

$$\theta(x, y) = \frac{T(x, y) - T_w}{T_\infty - T_w} \qquad (8\text{-}29)$$

where $\theta(x, y)$ varies from a value of zero at the wall surface to unity at the edge of the thermal boundary layer. Then the energy equation is written in terms of $\theta(x, y)$ as

$$u \frac{\partial \theta}{\partial x} + v \frac{\partial \theta}{\partial y} = \alpha \frac{\partial^2 \theta}{\partial y^2} \qquad \text{for } x > x_0 \qquad (8\text{-}30)$$

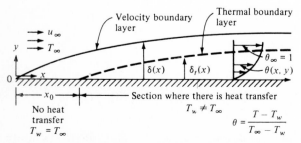

FIG. 8-2   Coordinates for the temperature problem for forced laminar flow along a flat plate.

and the boundary conditions are taken as

$\theta = 0 \qquad$ at $y = 0$ $\hspace{6cm}$ (8-31a)

$\theta = 1 \qquad$ at $y = \delta_t(x)$ $\hspace{5.5cm}$ (8-31b)

The integral method discussed previously in the solution of the velocity problem is now used to solve approximately the above energy equation (8-30) subject to the boundary conditions (8-31). The basic steps in this method of solution are as follows:

1 The energy equation (8-30) is integrated with respect to y over a distance H that exceeds the thickness of both the velocity and the thermal boundary layers, and the velocity component $v(x, y)$ is eliminated by means of the continuity equation (8-1). The resulting equation is called the *energy integral equation.*

2 A suitable profile is chosen for both the temperature distribution $\theta(x, y)$ and the velocity component $u(x, y)$. A polynomial approximation is generally used to represent these profiles within the boundary layers.

3 The velocity and temperature profiles determined in step 2 are substituted into the *energy integral equation* obtained in step 1 and the integration with respect to the y variable is performed. If the thickness of the thermal boundary layer is less than that of the velocity boundary layer (that is, $\delta_t < \delta$), which is the case to be considered here, one obtains an ordinary differential equation for a function $\Delta(x)$ defined as $\Delta(x) = \delta_t/\delta$. When this differential equation is solved subject to the boundary condition $\Delta = 0$ for $x = x_0$, the function $\Delta(x)$ is determined. Then the thermal boundary-layer thickness is computed from $\delta_t = \delta\Delta$, since the velocity boundary-layer thickness δ is available from the previous analysis.

4 Knowing δ_t, the temperature distribution in the boundary layer is determined from step 2.

Solution of Temperature Problem by the Integral Method

The energy equation (8-30) is solved by the integral method and the temperature distribution in the thermal boundary layer determined as now described.

Step 1 The energy equation (8-30) is integrated with respect to y over a distance H that exceeds the thicknesses of both boundary layers

$$\int_0^H u \frac{\partial \theta}{\partial x} dy + \int_0^H v \frac{\partial \theta}{\partial y} dy = \alpha \left(\frac{\partial \theta}{\partial y} \bigg|_{y=H} - \frac{\partial \theta}{\partial y} \bigg|_{y=0} \right) = -\alpha \frac{\partial \theta}{\partial y} \bigg|_{y=0} \qquad (8\text{-}32)$$

since $\partial\theta/\partial y|_{y=H} = 0$ by the definition of the boundary layer. The velocity component v appearing in Eq. (8-32) is eliminated by making use of the continuity equation (8-1). After some manipulations as described *in note 3 at the end of this chapter*, Eq. (8-32) reduces to

$$\frac{d}{dx}\left[\int_0^{H=\delta_t} u(1-\theta)\,dy\right] = \alpha \frac{\partial\theta}{\partial y}\bigg|_{y=0} \tag{8-33}$$

where the upper limit of the integration is restricted to $H = \delta_t$, because $\theta = 1$ for $H > \delta_t$ and the integrand vanishes for $H > \delta_t$. Equation (8-33) is called the *energy integral equation*.

Step 2 The velocity component $u(x, y)$ is represented in the velocity boundary layer by a cubic polynomial in the form [see Eq. (8-11)]

$$\frac{u(x, y)}{u_\infty} = \frac{3}{2}\left(\frac{y}{\delta}\right) - \frac{1}{2}\left(\frac{y}{\delta}\right)^3 \tag{8-34}$$

The temperature profile $\theta(x, y)$ in the thermal boundary layer δ_t can be represented by a cubic polynomial as

$$\theta(x, y) = c_0 + c_1(x)y + c_2(x)y^2 + c_3(x)y^3 \tag{8-35}$$

and the four conditions that are needed to determine the four coefficients are taken as

$$\theta = 0 \qquad \text{at } y = 0 \tag{8-36a}$$

$$\theta = 1 \qquad \text{at } y = \delta_t \tag{8-36b}$$

$$\frac{\partial\theta}{\partial y} = 0 \qquad \text{at } y = \delta_t \tag{8-36c}$$

$$\frac{\partial^2\theta}{\partial y^2} = 0 \qquad \text{at } y = 0 \tag{8-36d}$$

We note that the first two of these conditions are the boundary conditions for the problem given by Eqs. (8-31), the third condition is based on the definition of the thermal boundary layer, and the last condition is obtained by evaluating the energy equation (8-30) at $y = 0$ and noting that $u = v = 0$ at the wall surface. The application of conditions (8-36) to Eq. (8-35) gives the temperature profile in the form

$$\theta(x, y) = \frac{3}{2}\left(\frac{y}{\delta_t}\right) - \frac{1}{2}\left(\frac{y}{\delta_t}\right)^3 \tag{8-37}$$

Substitution of the velocity and temperature profiles given by Eq. (8-34) and (8-37) into the energy integral equation (8-33) gives

$$\frac{d}{dx}\left\{u_\infty \int_0^{\delta_t} \left[\frac{3}{2}\frac{y}{\delta} - \frac{1}{2}\left(\frac{y}{\delta}\right)^3\right]\left[1 - \frac{3}{2}\frac{y}{\delta_t} + \frac{1}{2}\left(\frac{y}{\delta_t}\right)^3\right] dy\right\} = \frac{3\alpha}{2\delta_t} \qquad (8\text{-}38a)$$

or

$$\frac{d}{dx}\left\{\int_0^{\delta_t}\left[\left(\frac{3}{2\delta}\right)y - \left(\frac{9}{4\delta\delta_t}\right)y^2 + \left(\frac{3}{4\delta\delta_t{}^3}\right)y^4 - \left(\frac{1}{2\delta^3}\right)y^3\right.\right.$$

$$\left.\left. + \left(\frac{3}{4\delta^3\delta_t}\right)y^4 - \left(\frac{1}{4\delta^3\delta_t{}^3}\right)y^6\right]dy\right\} = \frac{3\alpha}{2\delta_t u_\infty} \qquad (8\text{-}38b)$$

The integration with respect to y is performed:

$$\frac{d}{dx}\left(\frac{3}{4}\frac{\delta_t{}^2}{\delta} - \frac{3}{4}\frac{\delta_t{}^2}{\delta} + \frac{3}{20}\frac{\delta_t{}^2}{\delta} - \frac{1}{8}\frac{\delta_t{}^4}{\delta^3} + \frac{3}{20}\frac{\delta_t{}^4}{\delta^3} - \frac{1}{28}\frac{\delta_t{}^4}{\delta^3}\right) = \frac{3\alpha}{2\delta_t u_\infty} \qquad (8\text{-}39)$$

A new variable $\Delta(x)$ is now defined as the ratio of the thermal boundary-layer thickness to the velocity boundary-layer thickness as

$$\Delta(x) = \frac{\delta_t(x)}{\delta(x)} \qquad (8\text{-}40)$$

Then Eq. (8-39) becomes

$$\frac{d}{dx}\left[\delta\left(\frac{3}{20}\Delta^2 - \frac{3}{280}\Delta^4\right)\right] = \frac{3\alpha}{2\delta\Delta u_\infty} \qquad (8\text{-}41)$$

To simplify this equation we consider situations in which the *thermal boundary-layer thickness δ_t is smaller than the velocity boundary-layer thickness δ.* This is the case with fluids having a Prandtl number larger than unity. (The case $Pr \ll 1$ will be considered separately in the next section.) Then for $\Delta < 1$, the term $\frac{3}{280}\Delta^4$ can be neglected in comparison with the term $\frac{3}{20}\Delta^2$, and Eq. (8-41) simplifies to

$$\delta\Delta\frac{d}{dx}(\delta\Delta^2) = \frac{10\alpha}{u_\infty} \qquad (8\text{-}42)$$

Differentiation with respect to x is performed as

$$2\delta^2\Delta^2\frac{d\Delta}{dx} + \Delta^3\delta\frac{d\delta}{dx} = \frac{10\alpha}{u_\infty}$$

or

$$\frac{2}{3}\delta^2\frac{d\Delta^3}{dx} + \Delta^3\delta\frac{d\delta}{dx} = \frac{10\alpha}{u_\infty} \qquad (8\text{-}43)$$

since

$$\Delta^2 \frac{d\Delta}{dx} = \frac{1}{3} \frac{d\Delta^3}{dx}$$

The velocity boundary-layer thickness δ was previously determined as [see Eq. (8-15)]

$$\delta^2 = \frac{280}{13} \frac{vx}{u_\infty} \tag{8-44a}$$

and by differentiating we obtain

$$\delta \frac{d\delta}{dx} = \frac{140}{13} \frac{v}{u} \tag{8-44b}$$

Substitution of Eqs. (8-44) into Eq. (8-43) yields

$$x \frac{d\Delta^3}{dx} + \frac{3}{4} \Delta^3 = \frac{39}{56} \frac{\alpha}{v} \tag{8-45}$$

This is an ordinary differential equation of the first order in Δ^3, and its general solution is written as [see note 4 for a discussion of the solution of an equation of the form given by Eq. (8-45)]

$$\Delta^3(x) = Cx^{-3/4} + \frac{13}{14} \frac{\alpha}{v} \tag{8-46}$$

The integration constant C is determined by the application of the boundary condition $\delta_t = 0$ for $x = x_0$, which is equivalent to

$$\Delta(x) = 0 \qquad \text{for } x = x_0 \tag{8-47}$$

We find

$$\Delta^3(x) = \frac{13}{14} \text{Pr}^{-1} \left[1 - \left(\frac{x_0}{x} \right)^{3/4} \right] \tag{8-48}$$

where

$$\text{Pr} = \frac{v}{\alpha} = \text{Prandtl number}$$

If it is assumed that the heat transfer to the fluid starts at the leading edge of the plate, we set $x_0 \to 0$ and Eq. (8-48) simplifies to

$$\Delta(x) = \frac{\delta_t(x)}{\delta(x)} = \left(\frac{13}{14} \right)^{1/3} \text{Pr}^{-1/3} = 0.975 \, \text{Pr}^{-1/3} \tag{8-49}$$

This relation shows that the ratio of the thermal to velocity boundary-layer thickness for laminar flow along a flat plate is inversely proportional to the cube root of the Prandtl number. Substituting $\delta(x)$ from Eq. (8-44a) into Eq. (8-49), we obtain

$$\delta_t(x) = 4.51 \frac{x}{Re_x^{1/2} \, Pr^{1/3}}$$

(8-50)

where

$$Re_x = \frac{u_\infty x}{\nu}$$

Once $\delta_t(x)$ is known, the temperature distribution in the thermal boundary layer is determined from Eq. (8-37). However, in practice the heat-transfer coefficient between the fluid and the wall surface is of interest. The temperature profile established here is utilized to determine the heat-transfer coefficient as now described.

Heat-Transfer Coefficient

The local heat-transfer coefficient $h(x)$ between the flow and the wall surface is defined as

$$q(x) = h(x)(T_\infty - T_w)$$

(8-51)

The heat flux $q(x)$ in the fluid at the immediate vicinity of the wall surface is determined from

$$q(x) = -k \left.\frac{\partial T(x, y)}{\partial y}\right|_{y=0} = k(T_\infty - T_w) \left.\frac{\partial \theta}{\partial y}\right|_{y=0}$$

(8-52)

Since the flow velocity is zero at the wall, heat transfer is by conduction. From Eqs. (8-51) and (8-52) we obtain

$$h(x) = k \left.\frac{\partial \theta}{\partial y}\right|_{y=0}$$

(8-53)

The derivative of temperature at the wall is determined from Eq. (8-37) as

$$\left.\frac{\partial \theta}{\partial y}\right|_{y=0} = \frac{3}{2\delta_t}$$

(8-54)

which is substituted into Eq. (8-53) to give

$$h(x) = \frac{3}{2} \frac{k}{\delta_t}$$

(8-55)

Introducing the value of δ_t from Eq. (8-50) into Eq. (8-55), the local Nusselt number Nu_x becomes

$$Nu_x = \frac{h(x)x}{k} = 0.331 \, Pr^{1/3} \, Re_x^{1/2} \tag{8-56}$$

This approximate solution is remarkably close to the exact solution of this problem given by Pohlhausen [2] as

$$Nu_x = 0.332 \, Pr^{1/3} \, Re_x^{1/2} \qquad \text{(exact)} \tag{8-57}$$

It is of interest to note that the heat-transfer relation given by Eq. (8-56) was derived by an approximate analysis on the assumption $\delta_t < \delta$ or $Pr > 1$. However, its comparison with the exact results shows that it is valid in the range of Prandtl number $0.6 < Pr < 10$, which covers most gases and liquids.

For very large values of the Prandtl number Pohlhausen's exact calculations show that the local Nusselt number Nu_x is given by

$$Nu_x = 0.339 \, Pr^{1/3} \, Re_x^{1/2} \qquad \text{(exact for } Pr \to \infty \text{)} \tag{8-58}$$

In engineering applications an average heat-transfer coefficient h_m over the length of the plate from $x = 0$ to $x = L$ is defined as

$$h_m = \frac{1}{L} \int_0^L h(x) \, dx \tag{8-59}$$

Substituting $h(x)$ from Eq. (8-56) into Eq. (8-59) and performing the integration, one finds

$$h_m = 2h(x) \bigg|_{x=L} \tag{8-60a}$$

and similarly

$$Nu_m = 2Nu_x \bigg|_{x=L} \tag{8-60b}$$

That is, the average heat-transfer coefficient (or the Nusselt number) over the length $x = 0$ to L is equal to twice the value of the local heat-transfer coefficient (or the Nusselt number) at $x = L$.

To calculate the heat-transfer coefficient from the above relations it is recommended that the fluid properties be evaluated at the arithmetic mean of the wall temperature T_w and the external-flow temperature T_∞.

The above results are not applicable to liquid metals which have a very low Prandtl number; this subject is discussed in the next section.

Example 8-2 Air at atmospheric pressure and at a temperature 150°F (65.6°C) flows with a velocity of 3 ft/s (0.915 m/s) along a flat plate which is kept at a uniform temperature 250°F (121.1°C). Determine the local heat-transfer coefficient $h(x)$ at a distance $x = 2$ ft (0.61 m) from the leading edge of the plate and the average heat-transfer coefficient h_m over the length $x = 0$ to 2 ft (0.61 m). Calculate the total heat-transfer rate from the plate to the air over the region $x = 0$ to 2 ft per foot width of the plate.

Solution The physical properties of air at 200°F (i.e., arithmetic mean of T_w and T_∞) and at atmospheric pressure are

$v = 0.24 \times 10^{-3}$ ft²/s $(0.223 \times 10^{-4}$ m²/s$)$

$k = 0.0181$ Btu/h·ft·°F $(0.0313$ W/m·°C$)$

$Pr = 0.692$

The local Reynolds number at $x = 2$ ft (0.61 m) is

$$Re_x = \frac{u_\infty x}{v} = \frac{3 \times 2}{0.24 \times 10^{-3}} = 2.5 \times 10^4 \quad \left(or\ Re_x = \frac{0.915 \times 0.61}{0.223 \times 10^{-4}} = 2.5 \times 10^4 \right)$$

Then the flow is laminar, and the foregoing analysis is applicable.

The local Nusselt at $x = 2$ ft is determined by using the exact relation given by Eq. (8-57) as

$$Nu_x = \frac{h(x)x}{k} = 0.332\ Re_x^{1/2}\ Pr^{1/3}$$

and at $x = 2$ ft we have

$$\frac{h \times 2}{0.0181} = 0.332\ (2.5 \times 10^4)^{1/2}\ (0.692)^{1/3}$$

$$h\bigg|_{x=2} = 0.393\ \text{Btu/h·ft}^2\text{·°F}\ (2.23\ \text{W/m}^2\text{·s})$$

The average heat-transfer coefficient h_m over the length $x = 0$ to 2 ft is

$$h_m = 2h_x\bigg|_{x=2} = 0.786\ \text{Btu/h·ft}^2\text{·°F}\ (4.46\ \text{W/m}^2\text{·s})$$

The heat-transfer rate from the plate to the fluid becomes

$$Q = \text{area} \times h_m(T_w - T_\infty)$$
$$= 1 \times 2 \times 0.786\ (250 - 150) = 157.2\ \text{Btu/h}\ (46.2\ \text{W})$$

8-3 APPROXIMATE ANALYSIS OF HEAT TRANSFER TO LIQUID METALS (Pr ≪ 1) FOR LAMINAR FLOW ALONG A FLAT PLATE

For liquid metals the Prandtl number is very low; and as a result the thickness of the thermal boundary layer is much larger than that of the velocity boundary layer (that is, $\delta_t \gg \delta$). Therefore the heat-transfer analysis of liquid metals requires different considerations.

Consider the flow of a liquid metal at a temperature T_∞ with a velocity u_∞ along a flat plate which is kept at a uniform temperature T_w. As illustrated in Fig. 8-3, the thermal boundary layer is much thicker than the velocity boundary layer. The *energy integral equation* in this case is the same as that given previously by Eq. (8-33), that is,

$$\frac{d}{dx}\left[\int_0^{\delta_t} u(1-\theta)\,dy\right] = \alpha\left.\frac{\partial\theta}{\partial y}\right|_{y=0} \tag{8-61}$$

where

$$\theta(x, y) = \frac{T(x, y) - T_w}{T_\infty - T_w} \tag{8-62}$$

The velocity boundary layer being very thin, the flow velocity over the large portion of the thermal boundary layer is uniform and equal to u_∞. Therefore, as a first approximation, velocity is taken as

$$u = u_\infty \tag{8-63}$$

A cubic-polynomial approximation is assumed for the temperature profile $\theta(x, y)$ in the thermal boundary layer; the resulting expression for $\theta(x, y)$ is the same as that given by Eq. (8-37), that is,

$$\theta(x, y) = \frac{3}{2}\left(\frac{y}{\delta_t}\right) - \frac{1}{2}\left(\frac{y}{\delta_t}\right)^3 \tag{8-64}$$

Substitution of the profiles Eqs. (8-63) and (8-64) into the energy integral equation (8-61) yields

$$\frac{d}{dx}\left\{\int_0^{\delta_t} u_\infty\left[1 - \frac{3}{2}\frac{y}{\delta_t} + \frac{1}{2}\left(\frac{y}{\delta_t}\right)^3\right]dy\right\} = \alpha\frac{3}{2\delta_t} \tag{8-65}$$

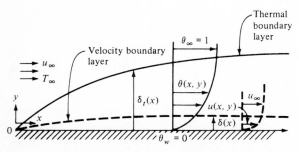

FIG. 8-3 Velocity and thermal boundary layers for liquid-metal heat transfer.

After performing the integration, we obtain

$$u_\infty \frac{3}{8} \frac{d\delta_t}{dx} = \frac{3\alpha}{2\delta_t}$$

or

$$\delta_t \, d\delta_t = \frac{4\alpha}{u_\infty} \, dx \tag{8-66}$$

The integration of Eq. (8-66) with the condition $\delta_t = 0$ for $x = 0$ gives the thermal boundary-layer thickness as

$$\delta_t^2 = \frac{8\alpha}{u_\infty} x \tag{8-67a}$$

or

$$\delta_t = \sqrt{\frac{8\alpha x}{u_\infty}} \tag{8-67b}$$

The local heat-transfer coefficient $h(x)$ for the cubic temperature profile is related to δ_t by Eq. (8-55) as

$$h(x) = \frac{3k}{2\delta_t} \tag{8-68}$$

Substitution of δ_t from Eq. (8-67b) into Eq. (8-68) gives the local heat-transfer coefficient as

$$h(x) = \frac{3k}{2\sqrt{8}} \sqrt{\frac{u_\infty}{\alpha x}} = \frac{3}{2\sqrt{8}} \frac{k}{x} \sqrt{\frac{u_\infty x}{v} \frac{v}{\alpha}} = \frac{3}{2\sqrt{8}} \frac{k}{x} \sqrt{\mathrm{Re}_x \, \mathrm{Pr}} \tag{8-69a}$$

and the local Nusselt number Nu_x becomes

$$\mathrm{Nu}_x = \frac{h(x)x}{k} = \frac{3}{2\sqrt{8}} \sqrt{\mathrm{Re}_x \, \mathrm{Pr}} = 0.530 \, \mathrm{Pe}_x^{1/2} \tag{8-69b}$$

where $\mathrm{Re}_x = \dfrac{u_\infty x}{v} = $ local Reynolds number

$\mathrm{Pr} = \dfrac{v}{\alpha} = $ Prandtl number

$\mathrm{Pe}_x = \mathrm{Re}_x \, \mathrm{Pr} = \dfrac{u_\infty x}{\alpha} = $ local Peclet number

The result in Eq. (8-69b) shows that the local Nusselt number for the laminar flow of liquid metals along a flat plate at uniform temperature is proportional to the square root of the Peclet number.

Pohlhausen's [2] exact solution of this heat-transfer problem for the limiting case of $Pr \to 0$ gives

$$Nu_x = 0.564 \, Pe_x^{1/2} \qquad (\text{exact for } Pr \to 0) \tag{8-70}$$

The approximate solution given by Eq. (8-69b), considering various approximations made in the analysis, is reasonably close to the above exact result.

8-4   EXACT ANALYSIS OF VELOCITY AND DRAG FOR LAMINAR FLOW ALONG A FLAT PLATE

The approximate boundary-layer analysis presented in the previous sections is relatively easy and straightforward, but the accuracy of the results cannot be assessed until they are compared with the exact results. Therefore, the exact solutions are essential to determine the accuracy of the approximate analysis. The exact analyses of the velocity and temperature problems considered in the previous sections are available in the literature. In order to give some idea of the exact methods of analysis, we discuss here the exact method of solution of the velocity and drag problem considered in Sec. 8-1.

The governing equations are taken as

Continuity: $\qquad \dfrac{\partial u}{\partial x} + \dfrac{\partial v}{\partial y} = 0$ $\qquad\qquad\qquad\qquad$ (8-71)

x Momentum: $\qquad u \dfrac{\partial u}{\partial x} + v \dfrac{\partial u}{\partial y} = v \dfrac{\partial^2 u}{\partial y^2}$ $\qquad\qquad\qquad$ (8-72)

subject to the boundary conditions

$u = v = 0 \qquad$ at $y = 0$ $\qquad\qquad\qquad\qquad\qquad\qquad$ (8-73a)

$u = u_\infty \qquad$ at $y \to \infty$ $\qquad\qquad\qquad\qquad\qquad\qquad$ (8-73b)

A stream function $\Psi(x, y)$ is defined as

$u = \dfrac{\partial \Psi}{\partial y} \qquad$ and $\qquad v = -\dfrac{\partial \Psi}{\partial x}$ $\qquad\qquad\qquad\qquad$ (8-74)

To introduce the stream function as defined above is useful in that the continuity equation (8-71) is identically satisfied and is *no longer needed*. The substitution of Eqs. (8-74) into the momentum equation (8-72) yields

$$\frac{\partial \Psi}{\partial y} \frac{\partial^2 \Psi}{\partial x \, \partial y} - \frac{\partial \Psi}{\partial x} \frac{\partial^2 \Psi}{\partial y \, \partial x} = v \frac{\partial^3 \Psi}{\partial y^3} \qquad (8\text{-}75a)$$

and the boundary conditions (8-73) become

$$\frac{\partial \Psi}{\partial y} = \frac{\partial \Psi}{\partial x} = 0 \qquad \text{at } y = 0 \qquad (8\text{-}75b)$$

$$\frac{\partial \Psi}{\partial y} = u_\infty \qquad \text{at } y \to \infty \qquad (8\text{-}75c)$$

The partial differential equation (8-75a) for the function $\Psi(x, y)$ and its boundary conditions (8-75b) and (8-75c) can be transformed into an ordinary differential equation for a function $f(\eta)$ in the independent variable $\eta \equiv \eta(x, y)$, if these new variables are defined as

$$\eta = y \sqrt{\frac{u_\infty}{vx}} \qquad (8\text{-}76a)$$

$$\Psi = f(\eta) \sqrt{xvu_\infty} \qquad (8\text{-}76b)$$

These variables are sometimes called the *similarity variables*. There are several methods for developing the similarity variables for the transformation of a partial differential equation into an ordinary differential equation, but this subject is beyond the scope of this book. The reader interested in this matter is referred to the texts by Hansen [3], Ames [4], Özışık [5], and to the paper by Morgan [6].

When the similarity variables given by Eqs. (8-76) are introduced into Eqs. (8-75) and (8-76), and the indicated operations are performed, the partial differential equation for Ψ is transformed into an ordinary differential equation for $f(\eta)$ subject to appropriate boundary conditions. Omitting the details of these operations, we give below the resulting differential equation for $f(\eta)$ and the boundary conditions as

$$2 \frac{d^3 f}{d\eta^3} + f \frac{d^2 f}{d\eta^2} = 0 \qquad (8\text{-}77)$$

$$\frac{df}{d\eta} = 0 \qquad f = 0 \text{ at } \eta = 0 \qquad (8\text{-}78a)$$

$$\frac{df}{d\eta} = 1 \qquad \text{at } \eta \to \infty \qquad (8\text{-}78b)$$

Thus, the velocity problem defined by the partial differential equations (8-71) and (8-72) is transformed into an ordinary differential equation (8-77) for the function $f(\eta)$. It is much easier to solve this ordinary differential equation than the original two coupled partial differential equations.

It can readily be shown that the velocity component $u(x, y)$ and the local-drag coefficient c_x are related to the function $f(\eta)$ by the relations (*see note 5 for the derivation of these relations*)

$$u(x, y) = u_\infty \frac{df(\eta)}{d\eta} \tag{8-79}$$

$$c_x = \frac{2}{\sqrt{Re_x}} \frac{d^2 f(\eta)}{d\eta^2}\bigg|_{\eta=0} \tag{8-80}$$

where

$$Re_x = \frac{x u_\infty}{\nu}$$

Therefore, knowing the derivatives $df/d\eta$ and $d^2f/d\eta^2|_{\eta=0}$, the velocity distribution $u(x, y)$ in the boundary layer and the local-drag coefficient c_x are determined from Eqs. (8-79) and (8-80), respectively. The third-order ordinary differential equation (8-77) was originally solved by Blasius [7] using series expansions, and later by Howarth [8] numerically. We present in Table 8-2 the values of $f(\eta)$, $df/d\eta$, and $d^2f/d\eta^2$ as a function of the independent variable η. We note that the value of $d^2f/d\eta^2|_{\eta=0}$ is 0.332. Then, the exact value of the local-drag coefficient c_x is obtained from Eq. (8-80) as

$$c_x = \frac{0.664}{\sqrt{Re_x}} \tag{8-81}$$

If the thickness of the velocity boundary layer $y = \delta(x)$ is defined as the location where $u(x, y)/u_\infty \cong 0.99$, then according to Table 8-2 the function $u(x, y)/u_\infty = df/d\eta$ has a value of about 0.99 when $\eta = 4.96$. Substituting this result into Eq. (8-76a), the boundary-layer thickness becomes

$$\delta \sqrt{\frac{u_\infty}{x\nu}} = \eta = 4.96$$

or

$$\frac{\delta(x)}{x} = \frac{4.96}{\sqrt{Re_x}} \tag{8-82}$$

The results given in Eqs. (8-81) and (8-82) have already been given in Table 8-1.

Figure 8-4 shows a plot of the velocity distribution $u(x, y)/u_\infty = df/d\eta$ in the boundary layer as a function of the parameter η.

TABLE 8-2   The Functions $f(\eta)$, $df/d\eta$, and $d^2f/d\eta^2$ for Laminar Flow Along a Flat Plate[†]

$\eta = y\sqrt{\dfrac{u_\infty}{vx}}$	f	$\dfrac{df}{d\eta} = \dfrac{u}{u_\infty}$	$d^2f/d\eta^2$
0	0	0	0.33206
0.2	0.00664	0.06641	0.33199
0.4	0.02656	0.13277	0.33147
0.6	0.05974	0.19894	0.33008
0.8	0.10611	0.26471	0.32739
1.0	0.16557	0.32979	0.32301
1.2	0.23795	0.39378	0.31659
1.4	0.32298	0.45627	0.30787
1.6	0.42032	0.51676	0.29667
1.8	0.52952	0.57477	0.28293
2.0	0.65003	0.62977	0.26675
2.2	0.78120	0.68132	0.24835
2.6	1.07252	0.77246	0.20646
3.0	1.39682	0.84605	0.16136
3.4	1.74696	0.90177	0.11788
3.8	2.11605	0.94112	0.08013
4.2	2.49806	0.96696	0.05052
4.6	2.88826	0.98269	0.02948
5.0	3.28329	0.99155	0.01591
5.4	3.68094	0.99616	0.00793
5.8	4.07990	0.99838	0.00365
6.2	4.47948	0.99937	0.00155
6.6	4.87931	0.99977	0.00061
7.0	5.27926	0.99992	0.00022
7.4	5.67924	0.99998	0.00007
7.8	6.07923	1.00000	0.00002
8.2	6.47923	1.00000	0.00001
8.6	6.87923	1.00000	0.00000

[†] From Howarth [8].

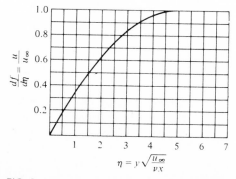

FIG. 8-4   Velocity distribution in the boundary-layer flow along a flat plate. (*After Blasius* [7].)

8-5   HEAT TRANSFER IN HIGH-SPEED LAMINAR BOUNDARY-LAYER FLOW ALONG A FLAT PLATE

In the previous sections we considered heat transfer in laminar flow along a flat plate by neglecting the effects of viscous energy dissipation in the boundary layer. When the free-stream velocity is high, the viscous-dissipation effects cannot be neglected; also the temperature gradients in the boundary layer become so large that the properties of the fluid vary significantly with temperature.   A number of special techniques have been developed for studying the heat-transfer problems of this nature, but such analysis is rather involved and beyond the scope of this book.   Fortunately, for most practical purposes, the heat-transfer rate in high-speed flow along a flat plate at uniform temperature can be determined by using the low-speed incompressible-flow heat-transfer coefficient h with a temperature difference $T_w - T_{aw}$, where T_w is the actual wall temperature and T_{aw} is the *adiabatic wall temperature*.   We now describe the basic idea behind this approach and the determination of T_{aw}.

Consider the high-speed flow of an incompressible, constant-property fluid at a temperature T_∞ with a velocity u_∞ along a flat plate.   The heat transfer can be determined from the solution of the boundary-layer equations given as [see Eqs. (6-65) to (6-67)]

Continuity: $$\frac{\partial u}{\partial x} + \frac{\partial v}{\partial y} = 0 \tag{8-83}$$

x Momentum: $$u \frac{\partial u}{\partial x} + v \frac{\partial u}{\partial y} = v \frac{\partial^2 u}{\partial y^2} \tag{8-84}$$

Energy: $$u \frac{\partial T}{\partial x} + v \frac{\partial T}{\partial y} = \alpha \frac{\partial^2 T}{\partial y^2} + \frac{\mu}{\rho c_p} \left(\frac{\partial u}{\partial y} \right)^2 \tag{8-85}$$

where the last term on the right-hand side of the energy equation is for the viscous-dissipation effects.   We first consider the solution of these equations for the special case of an *adiabatic plate*, that is, $\partial T/\partial y = 0$ at $y = 0$.   With this consideration the appropriate boundary conditions for Eqs. (8-83) to (8-85) are taken as

$$u = 0 \qquad v = 0 \qquad \frac{\partial T}{\partial y} = 0 \qquad \text{at } y = 0 \tag{8-86a}$$

$$u \to u_\infty \qquad T \to T_\infty \qquad \text{as } y \to \infty \tag{8-86b}$$

Equations (8-83) to (8-85) subject to the above boundary conditions were solved by Pohlhausen [2]; it was shown that the difference between the *adiabatic wall temperature* T_{aw} and the external-flow temperature T_∞ can be expressed in the form

$$T_{aw} - T_\infty = r \frac{u_\infty^2}{2 c_p g_c J} \tag{8-87a}$$

where the coefficient r is called the *recovery factor*, which is a function of the Prandtl number, and g_c and J are the conversion factors. Figure 8-5 shows a plot of the computed numerical values of the recovery factor against the Prandtl number for laminar flow along an adiabatic flat plate. For moderate Prandtl numbers (i.e., gas, water, etc.) the recovery factor r is related to the Prandtl number by the relation

$$r \cong Pr^{1/2} \qquad \text{for } 0.6 < Pr < 15 \tag{8-87b}$$

and for the limiting case $Pr \to \infty$ it can be represented by the relation [10]

$$r \cong 1.9\, Pr^{1/3} \qquad \text{(very large Pr)} \tag{8-87c}$$

Equation (8-87c) is applicable for fluids such as oils which have a very large Prandtl number.

The physical significance of the recovery factor is better envisioned by considering an ideal gas at a temperature T_∞ with a velocity u_∞ that is slowed down adiabatically to zero velocity. The conversion of kinetic energy in the gas into internal energy will result in a gas temperature T_0 given by the solution

$$T_0 - T_\infty = \frac{u_\infty{}^2}{2 c_p g_c J} \tag{8-88}$$

where T_0 is called the *stagnation temperature*. A comparison of Eqs. (8-87a) and (8-88) shows that the adiabatic wall temperature T_{aw} for $r = 1$ is equivalent to the stagnation temperature T_0. An examination of Fig. 8-5 reveals that $r = 1$ for $Pr = 1$; thus for a gas with $Pr = 1$, velocity u_∞ flowing along an adiabatic flat plate, the adiabatic wall temperature is the same as the stagnation temperature.

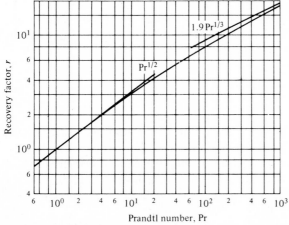

FIG. 8-5   Recovery factor for laminar flow along an adiabatic flat plate. (*From Eckert and Drewitz* [9]; *Schlichting* [12].)

For fluids with $\Pr > 1$, the recovery factor r is greater than unity and the adiabatic wall temperature exceeds the stagnation temperature. For a fluid $\Pr < 1$, the recovery factor r is less than unity, and the adiabatic wall temperature is less than the stagnation temperature.

Eckert and Weise [11] measured the recovery factor for air flow along an adiabatic flat plate as a function of the Reynolds number, and the agreement between the experimental and the analytical results was very good in the laminar-flow region (that is, $\mathrm{Re} < 5 \times 10^5$).

We now return to our original problem for the high-speed flow of an incompressible, constant-property fluid at a temperature T_∞ with a velocity u_∞ along a flat plate maintained at a uniform temperature T_w. It is shown that the local heat flux q_x for this problem is given by [12, p. 286; 13, p. 422]

$$q_x = 0.332 \, \Pr^{1/3} \, \mathrm{Re}_x^{1/2} \, \frac{k}{x} \, (T_w - T_{aw}) \tag{8-89}$$

where T_{aw} is the adiabatic wall temperature discussed previously and $\mathrm{Re}_x = u_\infty x / v$.

If a local heat-transfer coefficient $h(x)$ is now defined as

$$q_x = h(x)(T_w - T_{aw}) \tag{8-90}$$

then from Eqs. (8-89) and (8-90) we obtain

$$\mathrm{Nu}_x \equiv \frac{h(x)x}{k} = 0.332 \, \Pr^{1/3} \, \mathrm{Re}_x^{1/2} \tag{8-91}$$

A comparison of Eq. (8-91) with Eq. (8.57) reveals that the heat-transfer coefficient $h(x)$ based on the temperature difference $(T_w - T_{aw})$ for the high-speed flow considered above is exactly the same as the heat-transfer coefficient for the low-speed flow. The average value of the heat-transfer coefficient h_m over the length $0 \le x \le L$ of the plate is determined from

$$h_m = 2h(x) \Big|_{x=L} \tag{8-92}$$

The reader should consult Refs. [12–14] for details of the analysis leading to the heat-transfer relation given by Eq. (8-89).

We now summarize the foregoing analysis for high-speed heat transfer for flow over a flat plate at uniform temperature T_w: *The local heat flux q_x is given by Eq. (8-90) where the heat-transfer coefficient $h(x)$ is the same as for low-speed flow and determined from Eq. (8-91), the adiabatic wall temperature T_{aw} is defined by Eq. (8-87a), and the recovery factor r is obtained from Eq. (8-87b) or (8-87c) depending on the value of the Prandtl number.*

Effects of Variable Properties

In high-speed flow, temperature gradients in the boundary layer are generally large; hence the properties of the fluid vary significantly with temperature. Although the exact analysis of such problems is very involved, the effects of variation of properties may be approximately included in the heat-transfer relation given by Eq. (8-89) if the properties are evaluated at a properly chosen reference temperature. Eckert [15] recommended that the property variation may be approximately included in the heat-transfer coefficients given by Eq. (8-90) if the properties of the fluid are evaluated at the following reference temperature:

$$T_r = T_\infty + 0.5(T_w - T_\infty) + 0.22(T_{aw} - T_\infty) \tag{8-93}$$

Example 8-3 Air at a temperature $T_\infty = 460°R$ (255.6 K) and pressure $P = \frac{1}{30}$ atm flows with a velocity $u_\infty = 2000$ ft/s (609.6 m/s) along a 1-ft (0.305-m) long flat plate maintained at a uniform temperature $T_w = 560°R$ (311.1 K). Determine the heat-transfer rate to the plate over the 1-ft length per unit width.

Solution This is a high-speed flow because the free-stream velocity is higher than the acoustic velocity in the gas; i.e., the acoustic velocity u_s is obtainable from the relation

$$u_s = \sqrt{\frac{\gamma \mathscr{R} T g_c}{M}}$$

where u_s = acoustic velocity, ft/s
  γ = specific heat ratio (about 1.4 for air)
  M = molecular weight
  $\mathscr{R}$ = gas constant (1544 ft·lb$_f$/lb mol·°R)
  T = absolute temperature, °R
  g_c = gravitational conversion factor (32.17 ft·lb/lb$_f$·s^2)

Clearly, when the appropriate values are substituted in the above relation we find that $u_s = 1050$ ft/s; thus $u_\infty = 2000$ ft/s is higher than the acoustic velocity in the gas.

This high-speed-flow problem will now be solved by using the approach described above. The Prandtl number for air at 460°R is Pr = 0.711. Assuming laminar flow, the recovery factor r for moderate Prandtl numbers is determined by Eq. (8-87b) as

$$r = \text{Pr}^{1/2} = 0.711^{1/2} = 0.843$$

The adiabatic wall temperature T_{aw} is determined from Eq. (8-87a) by taking $c_p = 0.24$ as

$$T_{aw} = T_\infty + r\frac{u_\infty{}^2}{2c_p g_c J} = 460 + 0.843\frac{2000^2}{2 \times 0.24 \times 32.2 \times 778} = 460 + 280.4$$

$$= 740.4°R \ (411.3 \text{ K})$$

If the SI system of units is used, T_{aw} is obtained from Eq. (8-87a) be setting $c_p = 1005$ W·s/kg·°C, and we obtain

$$T_{aw} = 255.6 + 0.843\frac{609.6^2}{2 \times 1005 \times 1} = 255.6 + 155.7 = 411.3 \text{ K}$$

which is the same as that determined above by converting degrees Rankine °R into kelvins. The reference temperature T_r at which the properties are to be evaluated is determined by Eq. (8-93):

$$T_r = T_\infty + 0.5(T_w - T_\infty) + 0.22(T_{aw} - T_\infty)$$

$$= 460 + 0.5 \times 100 + 0.22 \times 280.4 \cong 572°R = 112°F$$

The physical properties of air at $T_r = 112°F$ and at $\frac{1}{30}$ atm pressure are taken as

$$\rho = \frac{PM}{RT} = \frac{144 \times 14.7}{30} \frac{28.7}{1544 \times 572} = 2.3 \times 10^{-3} \text{ lb/ft}^3 \ (0.037 \text{ kg/m}^3)$$

$$\mu = 1.29 \times 10^{-5} \text{ lb/ft·s} \ (1.92 \times 10^{-5} \text{ kg/m·s})$$

$$c_p = 0.24 \text{ Btu/lb·°F} \ (1005 \text{ W·s/kg·K})$$

$$k = 0.0169 \text{ Btu/h·ft·°F} \ (0.0292 \text{ W/m·s})$$

$$\text{Pr} = 0.70$$

We note that the Prandtl number at the reference temperature is sufficiently close to the one used above in computing the recovery factor; it it were significantly different the recovery-factor calculations should be repeated until sufficient agreement is obtained.

The Reynolds number at $x = 1$ ft becomes

$$\text{Re} = \frac{\rho u_\infty x}{\mu} = \frac{(2.3 \times 10^{-3}) \times 2000 \times 1}{1.29 \times 10^{-5}} = 3.58 \times 10^5$$

Then the laminar-flow analysis is considered applicable. The local heat-transfer coefficient at $x = 1$ ft is determined by Eq. (8-91) as

$$h(x) = \frac{k}{x} 0.332 \text{ Re}_x^{1/2} \text{ Pr}^{1/3} = \frac{0.0169}{1} \times 0.332 \times (3.58 \times 10^5)^{1/2} \times 0.70^{1/3}$$

$$= 2.98 \text{ Btu/h·ft}^2·°F \ (16.92 \text{ W/m}^2·°C)$$

The average heat-transfer coefficient over the length $x = 0$ to $x = 1$ ft becomes

$$h_m = 2h(x)\Big|_{x=1} = 2 \times 2.98 = 5.96 \text{ Btu/h·ft}^2·°F \ (33.85 \text{ W/m}^2·°C)$$

The total heat-transfer rate Q to the plate over the 1-ft length and per unit width becomes

$$Q = Ah_m(T_w - T_{aw}) = 1 \times 5.96 \times (560 - 740.4) \cong -1075 \text{ Btu/h} \ (-315 \text{ W})$$

that is, 1075 Btu/h (or 315 W) of cooling is needed at the plate surface in order to keep the surface at a temperature $T_w = 560°R$ (311.1 K).

REFERENCES

1 Von Kármán, T.: Über laminare und turbulente Reibung, *Z. Angew. Math. Mech.*, **1**:233 (1912); also (translation) *NACA Tech. Memo.* 1092, 1946.
2 Pohlhausen, E.: *Z. Angew. Math. Mech.*, **1**:115 (1921).

3   Hansen, Arthur G.: "Similarity Analysis of Boundary Value Problems in Engineering," Prentice-Hall, Inc., Englewood Cliffs, N.J., 1964.

4   Ames, W. F.: "Nonlinear Partial Differential Equations in Engineering," Academic Press, Inc., New York, 1965.

5   Özışık, M. N.: "Boundary Value Problems of Heat Conduction," chap. 8, International Textbook Company, Scranton, Pa., 1968.

6   Morgan, A. J. A.: The Reduction by One of the Number of Independent Variables in Some System of Partial Differential Equations, *Q. J. Math. (Oxford)*, **3**:250–259 (1951).

7   Blasius, H.: Grenzschleten in Flussigkeiten mit kleiner Reibung, *Z. Angew. Math. Phys.*, **56**:1 (1908).

8   Howarth, L.: On the Solution of the Laminar Boundary Layer Equations, *Proc. R. Soc. London, Ser. A*, **164**:547 (1938).

9   Eckert, E., and O. Drewitz: *Forsch. Geb. Ingenieurwes.*, **7**:116 (1940); (translation) *NACA Tech. Memo.* 1045, 1943.

10 Meksyn, D.: Plate Thermometer, *Z. Angew. Math. Phys.*, **11**:63–68 (1960).

11 Eckert, E., and H. Weise: *Forsch. Geb. Ingenieurwes.*, **13**:246 (1942).

12 Schlichting, H.: "Boundary Layer Theory," 6th ed., chap. 12, McGraw-Hill Book Company, New York, 1968.

13 Eckert, E. R. G., and R. M. Drake, Jr: "Analysis of Heat and Mass Transfer," chap. 10, McGraw-Hill Book Company, New York, 1972.

14 Gebhart, B.: "Heat Transfer," 2d ed., chap. 7, McGraw-Hill Book Company, New York, 1971.

15 Eckert, E. R. G.: Engineering Relations for Heat Transfer and Friction in High-Velocity Laminar and Turbulent Boundary Layer Flow over Surface with Constant Pressure and Temperature, *Trans. ASME*, **78**:1273–1284 (1956).

NOTES

1   Consider Eq. (8-7):

$$\int_0^\delta u \frac{\partial u}{\partial x} \, dy + \int_0^\delta v \frac{\partial u}{\partial y} \, dy = -v \left. \frac{\partial u}{\partial y} \right|_{y=0} \tag{1}$$

The second integral on the left-hand side is evaluated by parts as

$$\int_0^\delta v \frac{\partial u}{\partial y} \, dy = uv \Big|_0^\delta - \int_0^\delta u \frac{\partial v}{\partial y} \, dy = u_\infty v \Big|_\delta - \int_0^\delta u \frac{\partial v}{\partial y} \, dy \tag{2}$$

since $u = u_\infty$ at $y = \delta$ and $u = 0$ at $y = 0$. The terms on the right-hand side of this relation are now determined in the following manner: $\partial v/\partial y$ is immediately obtained from the continuity equation (8-1) as

$$\frac{\partial v}{\partial y} = -\frac{\partial u}{\partial x} \tag{3}$$

and $v|_\delta$ is determined by integrating Eq. (3) from $y = 0$ to δ as

$$v\Big|_0^\delta = -\int_0^\delta \frac{\partial u}{\partial x}\,dy$$

or

$$v\Big|_\delta = -\int_0^\delta \frac{\partial u}{\partial x}\,dy \tag{4}$$

since $v|_{y=0} = 0$. The substitution of Eqs. (3) and (4) into Eq. (2) yields

$$\int_0^\delta v\frac{\partial u}{\partial y}\,dy = -u_\infty \int_0^\delta \frac{\partial u}{\partial x}\,dy + \int_0^\delta u\frac{\partial u}{\partial x} \tag{5}$$

When this result is substituted into Eq. (1), we find

$$\int_0^\delta 2u\frac{\partial u}{\partial x}\,dy - u_\infty \int_0^\delta \frac{\partial u}{\partial x}\,dy = -v\frac{\partial u}{\partial y}\Big|_{y=0}$$

or

$$\int_0^\delta \frac{\partial u^2}{\partial x}\,dy - \int_0^\delta \frac{\partial(uu_\infty)}{\partial x}\,dy = -v\frac{\partial u}{\partial y}\Big|_{y=0}$$

or

$$\frac{d}{dx}\left[\int_0^\delta u(u_\infty - u)\,dy\right] = v\frac{\partial u}{\partial y}\Big|_{y=0} \tag{6}$$

which is the momentum integral equation given by Eq. (8-8).

2　Other polynomial representations of the velocity profile $u(x, y)$ are as follows:

Second-degree polynomial. If $u(x, y)$ is represented with a second-degree polynomial in the form

$$u(x, y) = a_0 + a_1 y + a_2 y^2 \tag{1}$$

the three conditions needed to determine the coefficients a_0, a_1, and a_2 are taken as

$$u\Big|_{y=0} = 0 \qquad u\Big|_{y=\delta} = u_\infty \qquad \text{and} \qquad \frac{\partial u}{\partial y}\Big|_{y=\delta} = 0 \tag{2}$$

Then the second-degree profile becomes

$$\frac{u(x, y)}{u_\infty} = 2\left(\frac{y}{\delta}\right) - \left(\frac{y}{\delta}\right)^2 \tag{3}$$

Fourth-degree polynomial. If $u(x, y)$ is represented with a fourth-degree polynomial in the form

$$u(x, y) = a_0 + a_1 y + a_2 y^2 + a_3 y^3 + a_4 y^4 \tag{4}$$

the five conditions needed to determine the five coefficients are taken as

$$u\Big|_{y=0} = 0 \qquad u\Big|_{y=\delta} = u_\infty$$

$$\frac{\partial u}{\partial y}\Big|_{y=\delta} = 0 \qquad \frac{\partial^2 u}{\partial y^2}\Big|_{y=0} = 0 \qquad \frac{\partial^2 u}{\partial y^2}\Big|_{y=\delta} = 0 \tag{5}$$

where the last two conditions are determined by evaluating the momentum equation (8-2) at $y = 0$ and $y = \delta$. By using these five conditions, the fourth-degree polynomial profile becomes

$$\frac{u(x, y)}{u_\infty} = 2\left(\frac{y}{\delta}\right) - 2\left(\frac{y}{\delta}\right)^3 + \left(\frac{y}{\delta}\right)^4 \tag{6}$$

3 Consider Eq. (8-32):

$$\int_0^H u \frac{\partial \theta}{\partial x} dy + \int_0^H v \frac{\partial \theta}{\partial y} dy = -\alpha \frac{\partial \theta}{\partial y}\bigg|_{y=0} \tag{1}$$

The second integral on the left-hand side of this equation is evaluated by parts as

$$\int_0^H v \frac{\partial \theta}{\partial y} dy = v\theta \bigg|_0^H - \int_0^H \theta \frac{\partial v}{\partial y} dy = v \bigg|_{y=H} - \int_0^H \theta \frac{\partial v}{\partial y} dy \tag{2}$$

since $v|_{y=0} = 0$ and $\theta|_{y=H} = 1$. The terms $v|_{y=H}$ and $\partial v/\partial y$ appearing in Eq. (2) are obtained from the continuity equation (8-1) as

$$\frac{\partial v}{\partial y} = -\frac{\partial u}{\partial x} \quad \text{and} \quad v\bigg|_{y=H} = -\int_0^H \frac{\partial u}{\partial x} dy \tag{3}$$

The substitution of Eqs. (3) into Eq. (2) yields

$$\int_0^H v \frac{\partial \theta}{\partial y} dy = -\int_0^H \frac{\partial u}{\partial x} dy + \int_0^H \theta \frac{\partial u}{\partial x} dy \tag{4}$$

Substituting Eq. (4) into Eq. (1) we obtain

$$\int_0^H \left(u \frac{\partial \theta}{\partial x} + \theta \frac{\partial u}{\partial x} - \frac{\partial u}{\partial x}\right) dy = -\alpha \frac{\partial \theta}{\partial y}\bigg|_{y=0}$$

or

$$\int_0^H \left[\frac{\partial(u\theta)}{\partial x} - \frac{\partial u}{\partial x}\right] dy = -\alpha \frac{\partial \theta}{\partial y}\bigg|_{y=0}$$

or

$$\frac{d}{dx}\left[\int_0^{H=\delta_t} u(1 - \theta) dy\right] = \alpha \frac{\partial \theta}{\partial y}\bigg|_{y=0} \tag{5}$$

which is the energy integral equation given by Eq. (8-33).

4 The differential equation (8-45) is of the following form:

$$x \frac{dy}{dx} + Ay = B \tag{1}$$

where A and B are constants. A particular solution Y_p of this equation is given as

$$Y_p = \frac{B}{A} \tag{2}$$

and the homogeneous solution Y_H that satisfies the homogeneous part of this equation is given as

$$Y_H = x^{-A} \tag{3}$$

Then the complete solution becomes

$$Y = Cx^{-A} + \frac{B}{A} \tag{4}$$

where C is the integration constant. This solution is of the same form as that given by Eq. (8-46).

5 The velocity component $u(x, y)$ is related to the function $f(\eta)$ as described below:

$$u(x, y) = \frac{\partial \Psi}{\partial y} = \frac{\partial \Psi}{\partial \eta} \frac{d\eta}{dy} \tag{1}$$

From Eqs. (8-76a) and (8-76b) we have, respectively,

$$\frac{d\eta}{dy} = \sqrt{\frac{u_\infty}{xv}} \tag{2}$$

$$\frac{d\Psi}{d\eta} = \sqrt{xvu_\infty} \frac{df}{d\eta} \tag{3}$$

Substituting Eqs. (2) and (3) into Eq. (1) we obtain

$$u(x, y) = \sqrt{xvu_\infty} \frac{df}{d\eta} \sqrt{\frac{u_\infty}{vx}} = u_\infty \frac{df}{d\eta} \tag{4}$$

The local-drag coefficient defined by Eq. (8-21) is related to $f(\eta)$ as follows:

$$c_x = \frac{2v}{u_\infty^2} \frac{\partial u}{\partial y}\bigg|_{y=0} = \frac{2v}{u_\infty^2} \frac{\partial u}{\partial \eta}\bigg|_{\eta=0} \frac{d\eta}{dy} \tag{5}$$

From Eqs. (2) and (4) we have, respectively,

$$\frac{d\eta}{dy} = \sqrt{\frac{u_\infty}{xv}} \quad \text{and} \quad \frac{\partial u}{\partial \eta}\bigg|_{\eta=0} = u_\infty \frac{d^2f}{d\eta^2}\bigg|_{\eta=0} \tag{6}$$

Substituting Eqs. (6) into Eq. (5) we obtain

$$c_x = \frac{2v}{u_\infty^2} u_\infty \frac{d^2f}{d\eta^2}\bigg|_{\eta=0} \sqrt{\frac{u_\infty}{xv}} = 2\sqrt{\frac{v}{xu_\infty}} \frac{d^2f}{d\eta^2}\bigg|_{\eta=0}$$

$$= \frac{2}{Re_x^{1/2}} \frac{d^2f}{d\eta^2}\bigg|_{\eta=0} \tag{7}$$

Nine

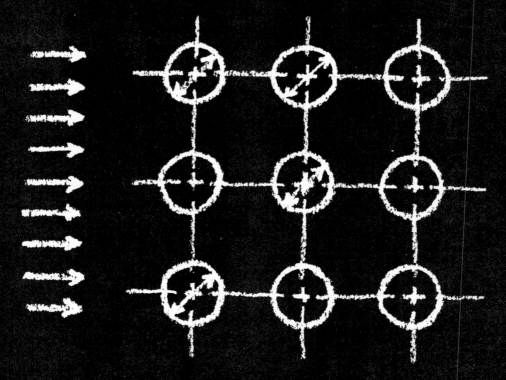

Convection—
Turbulent Internal and
External Flow

Turbulent flow is important in engineering applications because it is involved in the vast majority of fluid-flow and heat-transfer problems encountered in practice. In this chapter we present a qualitative discussion of turbulence, give some of the pertinent theories of the turbulent exchange process, and illustrate the application of the theory in the prediction of the friction factor for turbulent flow inside a circular tube. The determination of heat transfer in turbulent flow is a more complicated problem. However, the principal mechanism of momentum and heat transfer in turbulent flow being the crosswise mixing of fluid particles, the heat-transfer coefficient can be determined, for certain situations, from the knowledge of the friction factor by making use of the analogy between the momentum and the heat transfer. To give some idea of the physical significance of such an approach, a brief discussion is given of some of the simple analogies between the momentum and the heat transfer. For most practical problems encountered in engineering applications, however, the flow patterns are so complex that the engineer must turn to experimental and semi-empirical results for the determination of the friction factor (or the drag coefficient) and the heat-transfer coefficient. Various useful empirical relations for the determination of the friction factor and the heat-transfer coefficient are presented. Finally, the implications of heat transfer in high-speed turbulent flow are discussed, and a simple, approximate method of prediction of heat transfer for such situations is given.

9-1 FUNDAMENTALS OF TURBULENT FLOW

Osborne Reynolds [1] in his classic experiment of injecting dye into the water flowing through a transparent pipe showed that at low flow rates the flow was streamlined but as the flow rate was increased the streamlines became unstable and the laminar motion changed into turbulent flow. The term *turbulent* is used to denote that the motion of the fluid is chaotic in nature and involves crosswise mixing or eddying superimposed on the motion of the main stream. For flow inside a circular tube, for example, the transition from laminar to turbulent flow takes place at Reynolds number approximately

$$\text{Re} = \frac{u_m D}{v} \cong 2300$$

where u_m = mean flow velocity across tube cross section
        D = tube inside diameter
        v = kinematic viscosity

However, for rough inlet conditions the transition may take place at a Reynolds number as small as 2000, and under extremely calm inlet conditions it can be delayed up to a value of Reynolds number about 40,000.

For flow along a smooth flat plate the transition from laminar to turbulent flow takes place at Reynolds number approximately

$$\mathrm{Re} = \frac{u_\infty x}{\nu} \cong 5 \times 10^5$$

where u_∞ is the external stream velocity and x is the distance from the leading edge of the plate. Experiments [2] have shown that, depending on the inlet conditions, the transition may start as early as at Reynolds number 3×10^5 or may be delayed up to a value of about 10^6. For flow over a body, external pressure gradients, body forces, free-stream turbulence, geometry, etc., affect the onset of turbulence.

The eddying or crosswise mixing in turbulent flow is advantageous in that it assists greatly in improving the heat transfer in flow, but it has the disadvantage that it causes large resistance to flow. The flow patterns in turbulent flow are so complex that they cannot be completely predicted mathematically even for simple passage geometries. In a steady laminar flow, fluid particles follow definite streamlines, and properties such as velocity, pressure, and temperature at any point do not vary with time. In turbulent flow, however, because of chaotic crosswise eddying of fluid particles between the streamlines, the properties such as velocity, pressure, temperature, etc., at any point vary continuously with respect to time. Therefore in the analysis of turbulent flow the *instantaneous values of the properties are represented as a sum of a time-averaged mean part and a fluctuating part* in the form

$$u_i = u + u'$$

$$v_i = v + v'$$

$$T_i = T + T' \tag{9-1}$$

$$P_i = P + P'$$

where u_i, v_i, T_i, P_i are the *instantaneous values;* u, v, T, P are the *time-averaged* values; and u', v', T', P' are the fluctuations. For example, if a thermocouple with a sufficiently small time constant is placed at a given location in turbulent flow and the instantaneous value of temperature T_i is recorded as a function of time, temperature may show fluctuations as illustrated in Fig. 9-1. In this figure the instantaneous value of temperature T_i is considered to be composed of a fluctuation T' and a time-averaged value T defined as

$$T = \frac{1}{\Delta t} \int_t^{t + \Delta t} T_i \, dt \tag{9-2}$$

where Δt is a very small time interval which is large enough for recording the turbulent fluctuations but sufficiently small for the temperature to be unaffected

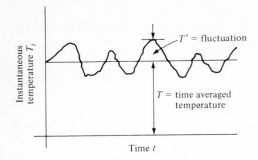

FIG. 9-1   Temperature fluctuation in turbulent flow.

by the external disturbances on the system.   It is apparent from the definition of the mean value given by Eq. (9-2) that the time average of the *fluctuation*, T', *is zero*, i.e.,

$$\frac{1}{\Delta t}\int_{t}^{t+\Delta t} T'\, dt \equiv \overline{T'} = 0 \tag{9-3}$$

On the other hand, the time average of the product of two fluctuations is *not zero*.   For example, the time average of the products $u'v'$ and $v'T'$ are written

$$\frac{1}{\Delta t}\int_{t}^{t+\Delta t} u'v'\, dt \equiv \overline{u'v'} \tag{9-4a}$$

$$\frac{1}{\Delta t}\int_{t}^{t+\Delta t} v'T'\, dt \equiv \overline{v'T'} \tag{9-4b}$$

where the bar denotes the time average.

To study momentum and heat transfer in turbulent flow, the equations of motion and energy are to be solved.   Although the instantaneous values of the properties such as velocity, temperature, and pressure satisfy these equations, the resulting equations cannot be solved because the fluctuations are random in nature.   To circumvent this difficulty, the properties are expressed in terms of a *mean* and a *fluctuating* part, and the equations are modified accordingly. To illustrate the implications of this approach and the utility of the results in the study of momentum and heat transfer in turbulent flow, we consider below the continuity, momentum, and energy equations for the boundary-layer flow of an incompressible, constant-property fluid.   (For simplicity in the analysis, we consider the boundary-layer equations for this purpose; actually, the complete momentum equations are to be considered.)

Continuity Equation

The continuity equation (6-65) for the instantaneous values of the velocity components u_i and v_i becomes

$$\frac{\partial u_i}{\partial x} + \frac{\partial v_i}{\partial y} = 0 \tag{9-5}$$

We let the instantaneous values of velocity components be represented as a sum of a time-averaged mean part and a fluctuating part in the form $u_i = u + u'$ and $v_i = v + v'$; then Eq. (9-5) becomes

$$\frac{\partial u}{\partial x} + \frac{\partial v}{\partial y} + \frac{\partial u'}{\partial y} + \frac{\partial v'}{\partial y} = 0 \tag{9-6}$$

When this equation is averaged over a small time interval Δt [that is, $(1/\Delta t) \int_0^{\Delta t}$], the time average of the fluctuating quantities u' and v' becomes zero [see Eq. (9-3)], and the average of the average quantities u and v remains the same. Then Eq. (9-6) reduces to

$$\frac{\partial u}{\partial x} + \frac{\partial v}{\partial y} = 0 \tag{9-7}$$

Thus, for turbulent flow the time-averaged velocity components u and v satisfy the same continuity equation as for the laminar flow given by Eq. (6-65).

x-Momentum Equation

The x-momentum equation (6-66) is written in terms of the instantaneous values of the properties u_i, v_i, and P_i as

$$u_i \frac{\partial u_i}{\partial x} + v_i \frac{\partial u_i}{\partial y} = -\frac{1}{\rho}\frac{dP_i}{dx} + \frac{\mu}{\rho}\frac{\partial^2 u_i}{\partial y^2} \tag{9-8a}$$

and, when the left-hand side of this equation is combined with the continuity equation (9-5), it is written in an alternative form as (*see note 1 for the derivation*)

$$\frac{\partial u_i^2}{\partial x} + \frac{\partial}{\partial y}(u_i v_i) = -\frac{1}{\rho}\frac{dP_i}{dx} + \frac{\mu}{\rho}\frac{\partial^2 u_i}{\partial y^2} \tag{9-8b}$$

In order to avoid the mathematical details and to concentrate only on the physical significance of the basic results in the analysis, we describe verbally the mathematical manipulations to be performed on Eq. (9-8b). We substitute in Eq. (9-8b) $u_i = u + u'$, $v_i = v + v'$, and $P_i = P + P'$, average the resulting equation over a small time interval Δt, utilize the rules of averaging given by Eqs. (9-3) and (9-4) to simplify the result, and separate out from the left-hand side the continuity equation. Then, we obtain the momentum equation for the boundary-layer flow of turbulent flow in the form

$$\rho\left(u\frac{\partial u}{\partial x} + v\frac{\partial u}{\partial y}\right) = -\frac{dP}{dx} + \frac{\partial}{\partial y}\left(\mu\frac{\partial u}{\partial y} - \rho\overline{u'v'}\right) \tag{9-9}$$

A comparison of this equation with the corresponding equation (6-66) for the laminar flow reveals that the momentum equation for turbulent flow includes an additional term $-\rho\overline{u'v'}$ which is called the *turbulent stress* or the *Reynolds stress*. The result in Eq. (9-9) implies that in turbulent flow *the total shear stress τ can be considered to be composed of a viscous component τ^l and a turbulent component τ^t in the form*

$$\tau = \tau^l + \tau^t \tag{9-10a}$$

where

$$\tau^l \equiv \mu\frac{\partial u}{\partial y} = \text{viscous shear stress} \tag{9-10b}$$

$$\tau^t \equiv -\rho\overline{u'v'} = \text{turbulent shear stress} \tag{9-10c}$$

If the turbulent-stress term in Eq. (9-9) were zero, the resulting equation would be the same as Eq. (6-66) for laminar flow. The turbulent stress is several orders of magnitude greater than the laminar stress, thus increasing the total shear stress, hence resistance to fluid flow in turbulent flow. The difficulty with the analysis of Eq. (9-9) is that the turbulent-stress term is not known and an additional relation is needed for its definition. This matter will be discussed later in this chapter.

Energy Equation

Neglecting the viscous-dissipation term, the energy equation (6-67) is now written in terms of the instantaneous values of the properties u_i, v_i, and T_i:

$$u_i\frac{\partial T_i}{\partial x} + v_i\frac{\partial T_i}{\partial y} = \frac{k}{\rho c_p}\frac{\partial^2 T_i}{\partial y^2} \tag{9-11a}$$

When the continuity equation (9-5) is multiplied by T_i and combined with the left-hand side of Eq. (9-11a), it is written in an alternative form as (*see note 2 for the derivation*)

$$\frac{\partial}{\partial x}(u_i T_i) + \frac{\partial}{\partial y}(v_i T_i) = \frac{k}{\rho c_p}\frac{\partial^2 T_i}{\partial y^2} \tag{9-11b}$$

To avoid the mathematical details, we describe verbally the mathematical manipulations to be performed on Eq. (9-11b). We substitute in Eq. (9-11b) $u_i = u + u'$, $v_i = v + v'$, and $T_i = T + T'$; average the resulting equation over a small time interval Δt; utilize the rules of averaging given by Eqs. (9-3) and (9-4) to simplify the result; and separate out from the left-hand side the continuity equation. Then, the energy equation for turbulent flow becomes

$$\rho c_p \left(u \frac{\partial T}{\partial x} + v \frac{\partial T}{\partial y} \right) = \frac{\partial}{\partial y} \left(k \frac{\partial T}{\partial y} - \rho c_p \overline{v'T'} \right) \tag{9-12}$$

A comparison of this equation with the corresponding energy equation (6-67) for laminar flow without the viscous-dissipation term reveals that the energy equation for turbulent flow includes an additional term $\rho c_p \overline{v'T'}$ which is called the *turbulent heat flux*. Thus, the result in Eq. (9-12) implies that in turbulent flow *the total heat flux q can be considered to be composed of a diffusive component q^l and a turbulent component q^t in the form*

$$q = q^l + q^t \tag{9-13a}$$

where

$$q^l = -k \frac{\partial T}{\partial y} = \text{diffusive (conductive) heat flux} \tag{9-13b}$$

$$q^t = \rho c_p \overline{v'T'} = \text{turbulent heat flux} \tag{9-13c}$$

If there were no turbulent-heat-flux term, Eq. (9-12) would be the same as that for laminar flow given by Eq. (6-67) for the case of no viscous dissipation. The turbulent-heat-flux term is much greater than the laminar-heat-flux term for ordinary fluids, thus increasing the heat transfer in turbulent flow. The turbulent-heat-flux term $\rho c_p \overline{v'T'}$ is an unknown quantity, and an additional relation is needed for its definition. This matter will be discussed later in this chapter.

9-2   EDDY-VISCOSITY AND EDDY-CONDUCTIVITY CONCEPTS

The momentum and energy equations for turbulent flow as given above cannot be used for computational purposes unless the *turbulent stress* $-\rho \overline{u'v'}$ and the *turbulent heat flux* $\rho c_p \overline{v'T'}$ are related to mean quantities. The complexity of the mechanism of turbulence, however, makes it almost impossible to obtain such relationships by purely mathematical means. Therefore, attempts have been made to study turbulence and turbulent exchange by semiempirical hypothesis. The *mixing-length* concept, for example, has been used extensively to relate the Reynolds stresses to the gradient of the local mean velocity. The basic idea in this concept is analogous to the *mean-free-path* concept for molecules in the kinetic theory of gases, but the main difference is that in turbulent motion macroscopic-size lumps are envisioned. That is, for turbulent flow in the x direction along a surface, it is postulated that fluid particles at a distance y from the wall surface coalesce into macroscopic lumps and then travel, on the average, a distance l in the direction normal to the main flow while retaining their x-direction momentum before they are dispersed. Thus, if the slow-moving lumps enter the fast-moving layer they act as a drag on it, and the momentum is transferred between layers as a result of transverse mixing. In kinetic theory the average distance between collisions is called the mean free path, and in turbulent flow the average distance l traveled by fluid lumps is called the *mixing length*. Of course, l is an unknown quantity, and in reality there is no such clearly defined distance. Although the concept lacks generality, it has been found useful in the study of turbulent exchange in most engineering applications. Prandtl [3] and von Kármán [4] proposed mixing-length theories to relate Reynolds stresses to the gradient of the local mean velocity and to establish relations defining the mixing length. Here we present a brief description of Prandtl's mixing-length theory and the result from von Kármán's theory. The reader is referred to the texts by Schlichting [5], Hinze [6], and Cebeci and Smith [7] for detailed discussions of these and various other concepts proposed to find models for Reynolds stresses.

Consider turbulent flow with a mean local velocity u in the x direction along a flat plate. Let y denote the coordinate axis perpendicular to the plate surface, and u' and v' denote the velocity fluctuations in the x and y directions, respectively. Prandtl suggested that the velocity fluctuation u' is related to du/dy by the relation

$$u' = l\frac{du}{dy} \tag{9-14a}$$

Then, by continuity considerations, he argued that the velocity fluctuation v' is of the same order of magnitude as u', that is,

$$v' \sim l\frac{du}{dy} \tag{9-14b}$$

and that u' and v' are of opposite sign. With this consideration, the time-averaged value of $\overline{u'v'}$ is different from zero and negative when du/dy is greater than zero. Thus $\overline{u'v'}$ is related to du/dy by

$$\overline{u'v'} = -l^2\left|\frac{du}{dy}\right|\frac{du}{dy} \equiv -\varepsilon_m\frac{du}{dy} \tag{9-15a}$$

where ε_m is called the *eddy viscosity of momentum* (or *turbulent diffusivity of momentum*) and defined as

$$\varepsilon_m = l^2\left|\frac{du}{dy}\right| \tag{9-15b}$$

A similar approach is now applied to relate the term $\overline{v'T'}$ to the gradient of the local mean temperature. By using the mixing-length concept the temperature fluctuation T' is related to dT/dy by

$$T' = l \frac{dT}{dy} \tag{9-16}$$

where l is the *mixing length for energy transport*, but it need not be equal to the mixing length l for momentum transport. Then by Eqs. (9-14b) and (9-16) we write

$$\overline{v'T'} = -l^2 \left| \frac{du}{dy} \right| \frac{dT}{dy} = -\varepsilon_h \frac{dT}{dy} \tag{9-17a}$$

where ε_h is called the *eddy conductivity* (or *turbulent diffusivity of heat*) and is defined as

$$\varepsilon_h = l^2 \left| \frac{dT}{dy} \right| \tag{9-17b}$$

Having established the relations defining $\overline{u'v'}$ and $\overline{v'T'}$ in terms of the averaged quantities, we now determine the relations for the turbulent stress τ^t and turbulent heat flux q^t.

The turbulent shear stress τ^t is determined from Eqs. (9-10c) and (9-15a) as

$$\tau^t = \rho \varepsilon_m \frac{du}{dy} \tag{9-18a}$$

where

$$\varepsilon_m = l^2 \left| \frac{du}{dy} \right| \tag{9-18b}$$

and the *turbulent heat flux* q^t is obtained from Eqs. (9-13c) and (9-17a) as

$$q^t = -\rho c_p \varepsilon_h \frac{dT}{dy} \tag{9-19a}$$

where

$$\varepsilon_h \equiv l^2 \left| \frac{dT}{dy} \right| \tag{9-19b}$$

It is to be noted that the mixing length l for the transport of momentum and heat appearing in Eqs. (9-18b) and (9-19b) need not be the same. The eddy diffusivity of momentum ε_m and the eddy conductivity ε_h as defined above are analogous, respectively, to the kinematic viscosity v and the thermal diffusivity α

used to define the viscous shear stress τ^l and the diffusive heat flux q^l given in the form

$$\tau^l = \mu \frac{du}{dy} = \rho v \frac{du}{dy} \tag{9-20}$$

$$q^l = -k \frac{dT}{dy} = -\rho c_p \alpha \frac{dT}{dy} \tag{9-21}$$

However, it is to be noted that v and α are properties of the fluid, whereas ε_m and ε_h are not properties of the fluid but are purely local functions in the fluid.

It is apparent from Eqs. (9-18) and (9-19) that the mixing length l is needed to complete the relations defining the turbulent stress and the heat flux. Various models have been proposed to determine the mixing length l, and a discussion of such models is given in several references [5–7]. The simplest of these models was proposed by Prandtl for turbulent flow along a flat plate. In this model the mixing length l is assumed to vanish at the wall surface, since transverse motion is not possible at the wall, and to increase linearly with the distance in the neighborhood of the wall. With this consideration, the relation defining l is taken in the form

$$l = \kappa y \tag{9-22}$$

where κ is a *universal constant* which must be determined from experiments. Indeed, experimental results confirmed Prandtl's hypothesis as given above for small distances from the wall. The value of the universal constant κ is found to be

$$\kappa \cong 0.4 \tag{9-23}$$

Improved relations defining the mixing length have been derived subsequently by other investigators; a discussion of this matter is given in Refs. [5,7].

The mixing lengths for momentum and heat transfer are not necessarily the same. To include this effect in the analysis, a *turbulent Prandtl number* Pr_t is generally defined as

$$Pr_t = \frac{\varepsilon_m}{\varepsilon_h} \tag{9-24}$$

Then, ε_h is determined from the knowledge of ε_m if the value of Pr_t is available. Various assumptions have been made and models derived for the determination of Pr_t. The simplest of these assumptions is the one due to Reynolds who assumed $Pr_t = 1$, which implies that heat and momentum transfer in turbulent flow takes place exactly by the same process. The assumption of $Pr_t = 1$ has been found successful in many heat-transfer analyses, and the results have been improved in some cases by choosing a value for Pr_t different from unity. However, recent measurements show that Pr_t varies with both Prandtl number and position.

Although the mixing-length and eddy-viscosity concepts have been used extensively and successfully in providing answers to many problems in engineering applications, the very basis of the method and the empirical constants associated with it are not universal. Thus, their scope is very limited with respect to the geometries and flow conditions experimentally investigated, and the results cannot be readily extrapolated to other flow situations. In order to overcome these deficiencies a new approach has been made in recent years to determine the transport properties for turbulent flow. In this approach the transport properties are described via a system of partial differential equations similar in form to the conservation equations, and a number of constants associated with such equations are determined experimentally. Although such equations for the turbulent-flow transport properties have existed since the mid-1940s and early 1950s, renewed interest in this area has been generated with the availability of high-speed large-memory digital computers. That is, the equations for the transport properties together with the conservation equations constitute a system of nonlinear partial differential equations which can now be solved numerically with the existing computing facilities. Launder and Spalding [8, 9] have summarized the governing equations for the calculation of transport properties for turbulent flow.

9-3 VELOCITY DISTRIBUTION IN TURBULENT FLOW FOR SMOOTH PIPES

Velocity distribution in turbulent flow has been investigated extensively because of its importance in practice, but no fundamental theory is yet available to determine this velocity distribution rigorously by purely theoretical approaches. Therefore empirical and semiempirical relations are used to correlate the velocity field in turbulent flow.

Nikuradse [10a,b] was an early investigator who presented careful measurement of velocity distribution in turbulent flow through a smooth pipe. Later on, experiments were performed by other investigators for turbulent flow along a flat plate [11,12] and also inside a pipe [13]. Attempts were then made to develop empirical relations that would fit the velocity distribution in turbulent flow [5,7,14,15]. Here we discuss the velocity-distribution law based on the concept of separating the flow field into three distinct layers as illustrated in Fig. 6-6. That is, (1) a very thin layer immediately adjacent to the wall in which laminar or viscous shear stress is dominant is called the *viscous sublayer;* (2) adjacent to this layer is the *buffer layer* in which viscous and turbulent shear stresses are equally important; and (3) the third layer that follows the buffer layer is called the *turbulent core* in which turbulent shear stress is dominant. We now examine the velocity-distribution laws for each of these layers for the steady, turbulent flow of an incompressible, constant-property fluid over a *smooth surface.*

In the analysis of velocity distribution in turbulent flow, Prandtl introduced a far-reaching assumption on the constancy of the total shear stress τ; that is, he assumed that the total shear stress τ in the flow is equal to the wall shear stress τ_0. With this consideration, we write the relation for the total shear stress in turbulent flow given by Eqs. (9-10) as

$$\tau_0 = \mu \frac{du}{dy} - \rho \overline{u'v'} \tag{9-25}$$

where τ_0 is the wall shear stress which is considered to be constant. This equation is now utilized to derive the velocity-distribution laws for the three layers.

Turbulent Core

Assuming that the laminar shear stress is negligible in comparison with the turbulent shear stress, Eq. (9-25) for the turbulent core simplifies to

$$\frac{\tau_0}{\rho} = -\overline{u'v'} \tag{9-26}$$

From Eqs. (9-15a) and (9-22) we have, respectively,

$$\overline{u'v'} = -l^2 \left(\frac{du}{dy}\right)^2 \qquad \text{and} \qquad l = \kappa y \tag{9-27}$$

By combining Eqs. (9-26) and (9-27) we obtain

$$\frac{\tau_0}{\rho} = \kappa^2 y^2 \left(\frac{du}{dy}\right)^2$$

or

$$\frac{du}{dy} = \frac{1}{\kappa} \sqrt{\frac{\tau_0}{\rho}} \frac{1}{y} \tag{9-28}$$

The integration of Eq. (9-28) gives

$$u = \frac{1}{\kappa} \sqrt{\frac{\tau_0}{\rho}} \ln y + C' \tag{9-29}$$

where C' is the integration constant. Equation (9-29) is written in the dimensionless form as

$$u^+ = \frac{1}{\kappa} \ln y^+ + C \tag{9-30}$$

where

$$u^+ = \frac{u}{\sqrt{\tau_0/\rho}} = \text{dimensionless velocity} \qquad (9\text{-}31a)$$

$$y^+ = \frac{y}{v}\sqrt{\frac{\tau_0}{\rho}} = \text{dimensionless distance} \qquad (9\text{-}31b)$$

Equation (9-30) is called the *logarithmic velocity-distribution law* for turbulent flow. Here the term $\sqrt{\tau_0/\rho}$ has the dimensions of velocity and is called the *friction velocity*, and κ is the universal constant which has a value $\kappa = 0.4$ as given by Eq. (9-23). The value of the constant C has been determined by the correlation of Eq. (9-30) with the measured velocity profile, and for turbulent flow inside a smooth pipe its value is found to be

$$C = 5.5 \qquad \text{for } y^+ > 30 \qquad (9\text{-}32)$$

Then, the *logarithmic velocity distribution* in the turbulent core region for turbulent flow over a smooth surface is given as

$$u^+ = 2.5 \ln y^+ + 5.5 \qquad \text{for } y^+ > 30 \qquad (9\text{-}33)$$

where ln denotes the natural logarithm.

Viscous Sublayer

In this region the laminar shear stress is dominant and the turbulent shear stress is virtually zero. With this consideration, Eq. (9-25) simplifies to

$$\frac{\tau_0}{\rho} = v\frac{du}{dy} \qquad (9\text{-}34)$$

The integration of this equation with $u = 0$ for $y = 0$ gives

$$u = \frac{\tau_0}{\rho}\frac{y}{v}$$

which can be rearranged as

$$\frac{u}{\sqrt{\tau_0/\rho}} = \frac{y}{v}\sqrt{\frac{\tau_0}{\rho}} \qquad (9\text{-}35)$$

It is found from experiments that this relation is valid for $y^+ < 5$. Then, by introducing the dimensionless variables u^+, y^+ as defined previously, Eq. (9-35) is written

$$u^+ = y^+ \qquad \text{for } y^+ < 5 \tag{9-36}$$

which is *the velocity-distribution law for the viscous sublayer.*

Buffer Layer

The region between the viscous sublayer and the turbulent core, extending from $y^+ = 5$ to 30, is the buffer layer. If a logarithmic velocity-distribution law is assumed for this region with the requirement of the continuity of velocity with those in the viscous sublayer and the turbulent core approximately at the locations $y^+ = 5$ and $y^+ = 30$, respectively, the resulting *velocity profile for the buffer layer becomes*

$$u^+ = 5.0 \ln y^+ - 3.05 \qquad \text{for } 5 \le y^+ \le 30 \tag{9-37}$$

Summary of Velocity Law for Turbulent Flow

We now summarize the foregoing velocity-distribution laws for turbulent flow along a smooth surface.

$$u^+ = y^+ \qquad\qquad\qquad \text{for viscous sublayer, } y^+ < 5 \tag{9-38a}$$

$$u^+ = 5.0 \ln y^+ - 3.05 \qquad \text{for buffer layer, } 5 \le y^+ \le 30 \tag{9-38b}$$

$$u^+ = 2.5 \ln y^+ + 5.5 \qquad \text{for turbulent core, } y^+ > 30 \tag{9-38c}$$

where

$$u^+ \equiv \frac{u}{\sqrt{\tau_0/\rho}} \qquad \text{and} \qquad y^+ \equiv \frac{y}{\nu}\sqrt{\frac{\tau_0}{\rho}} \tag{9-38d}$$

and τ_0 is the total wall shear stress. Figure 9-2 shows a correlation of the velocity-distribution law given by Eqs. (9-38) with Nikuradse's [10a] measured velocity distribution for turbulent flow inside smooth pipes.

Although the velocity-distribution law obtained by separating the flow field into three distinct layers as discussed above appears to be in reasonably good agreement with the experimental data, it should be recognized that the transition from a viscous to a turbulent flow regime in reality takes place gradually. Therefore, the representation of velocity distribution by three different curves having discontinuous slopes at locations where they join is not realistic. A more serious inconsistency of the logarithmic velocity-distribution law Eq. (9-38c) is that it does not give zero velocity gradient at the tube center. For this reason, the average velocity for flow inside a pipe as determined by using the above equations overestimates the velocity. Despite these shortcomings, the velocity-

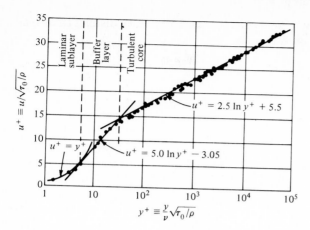

FIG. 9-2 Logarithmic velocity-distribution law and Nikuradse's [10a] experimental data for turbulent flow inside smooth pipes.

distribution laws given by Eqs. (9-38) have been extensively used in the literature to study the relation between momentum and heat transfer.

Effects of Surface Roughness

The velocity-distribution relations discussed above are applicable to turbulent flow over surfaces which are *hydrodynamically smooth*. A surface is considered hydrodynamically smooth if the heights λ of the protrusions are much smaller than the thickness of the viscous sublayer. Surfaces encountered in engineering applications are not generally perfectly smooth, and since for most cases the viscous sublayer is very thin the protrusions may penetrate it. Varied geometrical forms of roughness and the variety of ways that the protrusions may be distributed over the surface make it difficult to analyze the effects of roughness on velocity distribution. Nikuradse [10b] made extensive experiments with turbulent flow inside artificially roughened pipes over a wide range of *relative roughness* λ/D (i.e., protrusion height-to-diameter ratio) from about $\frac{1}{1000}$ to $\frac{1}{30}$. The *sand-grain* roughness used in these experiments has been adopted as a standard for the effects of roughness. These experiments showed that to study the effects of roughness it is desirable to introduce a *roughness Reynolds number* λ^+ (i.e., a dimensionless protrusion height) defined as

$$\lambda^+ \equiv \frac{\lambda}{v} \sqrt{\frac{\tau_0}{\rho}} \tag{9-39}$$

and it has been found that when λ^+ is less than about 5 the roughness has no effect on the friction due to flow. With this consideration three distinct situations are envisioned for the effects of roughness:

Hydrodynamically smooth: $\quad 0 \le \lambda^+ \le 5$

Transitional: $\qquad\qquad\quad 0 \le \lambda^+ \le 70$ $\hspace{4cm}$ (9-40)

Fully rough: $\qquad\qquad\quad \lambda^+ > 70$

For the hydrodynamically smooth case the heights of roughness are so small that all protrusions are covered by the viscous sublayer; hence roughness has no effect. For the transitional case the protrusions are partly outside the viscous sublayer and cause some additional resistance to flow. For the fully rough case the heights of protrusions are so large that all protrusions penetrate the viscous sublayer; hence the *viscous sublayer no longer exists*, and protrusions influence the turbulent mixing. For the fully rough regime the logarithmic velocity-distribution law of Eq. (9-30) is also applicable if y^+ is replaced by y/λ; then the equation becomes

$$u^+ = \frac{1}{\kappa} \ln \left(\frac{y}{\lambda}\right) + C \qquad\qquad (9\text{-}41)$$

where $\kappa = 0.4$ as in the case of smooth wall but the constant C is different. A correlation of this relation with Nikuradse's [10b] experiments for the sand-roughened wall-surface conditions has shown that $C = 8.5$. Then Eq. (9-41) takes the form

$$u^+ = 2.5 \ln \left(\frac{y}{\lambda}\right) + 8.5 \qquad\qquad (9\text{-}42)$$

where

$$u^+ \equiv \frac{u}{\sqrt{\tau_0/\rho}}$$

which is called *the logarithmic velocity-distribution law for turbulent flow in rough pipes in the fully rough region.*

9-4 FRICTION FACTOR FOR TURBULENT FLOW INSIDE PIPES

In engineering applications, to facilitate the pressure-drop calculations for turbulent flow inside a pipe, a *friction factor f* is defined as [see Eq. (7-31a)]

$$\Delta p = f \frac{L}{D} \frac{\rho u_m^{\,2}}{2} \qquad\qquad (9\text{-}43)$$

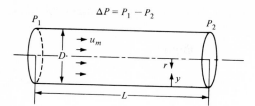

FIG. 9-3   Geometry for a force balance over length L of a pipe.

where Δp = pressure drop over length L
D = pipe inside diameter
u_m = mean velocity of turbulent flow over pipe cross section

We recall that for laminar flow inside a circular pipe we considered a similar relation for the pressure drop; the friction factor f was determined by a purely theoretical approach and was found to be $f = 64/\text{Re}$.   In the case of turbulent flow, however, some empiricism is introduced in the derivation of the friction factor because the semiempirical relation for the velocity distribution is used in the analysis.

We now describe the basic steps in the derivation of the friction factor for turbulent flow inside a smooth pipe by using the logarithmic velocity-distribution law given previously.

A force balance over the length L of a circular pipe shown in Fig. 9-3 is written

$$\pi D L \tau_0 = \Delta p \frac{\pi D^2}{4}$$

or

$$\Delta p = \frac{4L}{D} \tau_0 \tag{9-44}$$

where τ_0 is the total shear stress at the wall.   It is apparent from the derivation that this equation is strictly applicable to situations in which the only forces acting on the fluid are the wall friction and the pressure, and that the net momentum flux is zero.

By combining Eqs. (9-43) and (9-44) we obtain

$$f = \frac{8\tau_0}{\rho u_m{}^2} = \frac{8}{(u_m/\sqrt{\tau_0/\rho})^2} = \frac{8}{(u_m{}^+)^2} \tag{9-45}$$

where $u_m{}^+$ is the mean value of the dimensionless velocity u^+ over the cross-sectional area of flow through the tube.   We assume that the velocity distribution for turbulent flow is given by Eq. (9-38c), that is,

$$u^+ = 2.5 \ln y^+ + 5.5 \tag{9-46}$$

where y^+ is the dimensionless distance [that is, $y^+ = (y/v)\sqrt{\tau_0/\rho}$] measured from the tube wall. The average value of u^+ over the cross section of the flow through the tube is given by (*see note 3 for the derivation*)

$$u_m{}^+ = 2.5 \ln \left(\frac{R}{v} \sqrt{\frac{\tau_0}{\rho}} \right) + 1.75 \tag{9-47}$$

where R is the tube radius. The term $\sqrt{\tau_0/\rho}$ appearing in this expression is obtained from Eq. (9-45) as

$$\sqrt{\frac{\tau_0}{\rho}} = u_m \sqrt{\frac{f}{8}} \tag{9-48}$$

Substituting Eq. (9-48) into (9-47) to eliminate $\sqrt{\tau_0/\rho}$ and then introducing $u_m{}^+$ from Eq. (9-47) into Eq. (9-45), we obtain

$$f = \frac{8}{\{2.5 \ln [(Du_m/v)(1/4\sqrt{2})\sqrt{f}] + 1.75\}^2}$$

or, in terms of logarithm to base 10, this relation becomes

$$\frac{1}{\sqrt{f}} = 2.035 \log (\mathrm{Re} \sqrt{f}) - 0.91 \tag{9-49}$$

where

$$\mathrm{Re} \equiv \frac{u_m D}{v} = \text{Reynolds number}$$

Equation (9-49) is derived by utilizing a semiempirical relation for the velocity distribution; therefore it should be checked with experiments. A comparison with experiments suggests that if Eq. (9-49) is modified as follows,

$$\frac{1}{\sqrt{f}} = 2.0 \log (\mathrm{Re} \sqrt{f}) - 0.8 \tag{9-50}$$

it agrees with experiments extremely well. Figure 9-4 shows a comparison of Eq. (9-50) with the experiments of various investigators by plotting $1/\sqrt{f}$ against $\log (\mathrm{Re} \sqrt{f})$; here, Nikuradse's [10] experiments cover a range of Reynolds numbers up to 3.4×10^6.

Effects of Surface Roughness on Friction Factor

Moody [16] prepared a friction-factor chart for turbulent flow inside pipes by utilizing the velocity-distribution law for fully rough surfaces based on Nikuradse's sand-roughness conditions. He also calculated the friction factor for turbulent

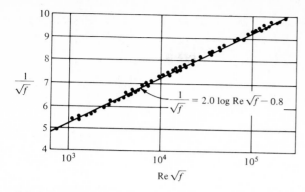

FIG. 9-4 Friction law for turbulent flow inside smooth pipes and experimental data of various investigators. (*From Schlichting* [5].)

flow inside smooth pipes by using the relation given previously. Figure 9-5 shows a plot of the friction factor f against the Reynolds number for several different values of the relative roughness λ/D (i.e., protrusion height-to-diameter ratio). To give some idea of the roughness of various commercial surfaces, we have included in this figure approximate values of roughness heights for typical surface preparations. In many surface preparations the roughness is much different from that of sand roughness considered by Nikuradse. To correlate other types of roughnesses with its *equivalent sand roughness*, Schlichting [5] gives the experimentally determined values of equivalent sand roughness for a large number of regular roughness patterns.

Once the friction factor f is established from the chart in Fig. 9-5, the pressure drop Δp over the length L of a pipe of diameter D is determined from [see Eq. (9-43)]

$$\Delta p = f \frac{L}{D} \frac{\rho u_m^2}{2g_c} \qquad (\text{lb}_f/\text{ft}^2 \text{ or } N/m^2) \tag{9-51}$$

where u_m is the mean velocity of flow over the pipe cross section.

Example 9-1 Air at 80 lb_f/in^2 abs. (5.44 atm) and 100°F (37.8°C) flows through a 1-in (2.54×10^{-2}-m) ID tube with a mean velocity of 50 ft/s (15.24 m/s). Determine the pressure drop per 100-ft (30.5-m) length of the tube for (a) a smooth tube and (b) a rough tube having a relative roughness $\lambda/D = 0.0002$.

Solution The density of air at 100°F and atmospheric pressure (that is, 14.7 lb_f/in^2 abs.) is 0.071 lb/ft^3; the density at 80 lb_f/in^2 abs. becomes

$$\rho = 0.07 \frac{80}{14.7} = 0.380 \text{ lb/ft}^3 \ (6.089 \text{ kg/m}^3)$$

The viscosity of air at 100°F is

$$\mu = 1.285 \times 10^{-5} \text{ lb/ft·s} \ (1.912 \times 10^{-5} \text{ kg/m·s})$$

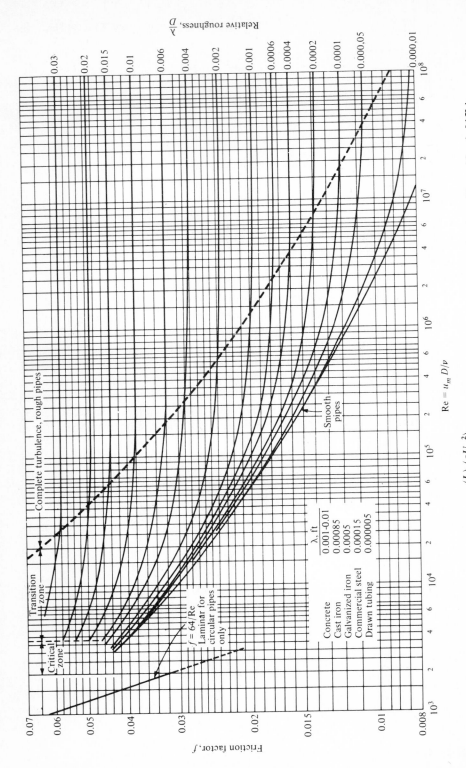

FIG. 9-5 Friction factor for use in the relation $\Delta P = f\left(\dfrac{L}{D}\right)\left(\dfrac{\rho U_m^{\,2}}{2g_c}\right)$ for pressure drop for flow inside circular pipes. (*From Moody* [16].)

The Reynolds number for flow inside the tube is given by

$$\text{Re} = \frac{\rho u_m D}{\mu} = \frac{0.380 \times 50 \times 1}{(1.285 \times 10^{-5}) \times 12} = 1.23 \times 10^5$$

$$\left(\text{or Re} = \frac{6.089 \times 15.24 \times (2.54 \times 10^{-2})}{1.912 \times 10^{-5}} = 1.23 \times 10^5 \right)$$

Thus the flow is turbulent. The friction factor f for $\text{Re} = 1.23 \times 10^5$ is obtained from Fig. 9-5 as

$f = 0.017$ for smooth tube

$f = 0.0185$ for rough tube with $\dfrac{\lambda}{D} = 0.0002$

The pressure drop per 100-ft (30.5-m) length of the tube is calculated by Eq. (9-51).
(a)   For the smooth tube:
      By using the English system of units, we find

$$\Delta p = f \frac{L}{D} \frac{\rho u_m^2}{2g_c} = 0.017 \frac{100}{\frac{1}{12}} \frac{0.380 \times 50^2}{2 \times 32.2} = 300.9 \text{ lb}_f/\text{ft}^2 \ (14.4 \times 10^3 \text{ N/m}^2)$$

or, by using the SI system of units, we obtain

$$\Delta p = 0.017 \frac{30.5}{2.54 \times 10^{-2}} \frac{6.089 \times 15.24^2}{2 \times 1} = 14.4 \times 10^3 \text{ N/m}^2 = 14.4 \text{ kN/m}^2$$

(b)   For the rough tube:

$$\Delta p = 0.0185 \frac{100}{\frac{1}{12}} \frac{0.380 \times 50^2}{2 \times 32.2} = 327.4 \text{ lb}_f/\text{ft}^2 \ (15.68 \times 10^3 \text{ N/m}^2)$$

9-5   ANALOGIES BETWEEN HEAT AND MOMENTUM TRANSFER IN TURBULENT FLOW

The principal mechanism of momentum and heat transfer in turbulent flow in regions away from the wall is the crosswise mixing of fluid particles. Therefore, the analogy between heat and momentum transfer can be utilized to determine the heat-transfer coefficient from the knowledge of the friction factor or the drag coefficient. The first and the simplest of such relations was derived by Reynolds and is now known as the *Reynolds analogy* for momentum and heat transfer. More refined analogies were later developed by Prandtl [17], von Kármán [18], Martinelli [19], Deissler [20a,b], and many others. In this section we give a derivation of Reynolds and Prandtl analogies for turbulent flow inside pipes and along flat plates in order to provide some insight into the physical significance of such relations.

Turbulent Flow Inside Pipes

In the previous sections we have shown that the total shear stress τ and the total heat flux q in turbulent flow are composed of a laminar and a turbulent component. If the variation with radius is neglected, the relations defining τ and q are obtained from Eqs. (9-10) and (9-13), respectively, as

$$\tau = \mu \frac{du}{dy} - \rho \overline{u'v'} \tag{9-52a}$$

$$q = -k \frac{dT}{dy} + \rho c_p \overline{v'\,T'} \tag{9-52b}$$

Here the terms $\overline{u'v'}$ and $\overline{v'T'}$ are obtained from Eqs. (9-15a) and (9-17a), respectively, as

$$\overline{u'v'} = -\varepsilon_m \frac{du}{dy} \tag{9-53a}$$

$$\overline{v'T'} = -\varepsilon_h \frac{dT}{dy} \tag{9-53b}$$

The substitution of Eqs. (9-53) into (9-52) gives the relations for the total shear stress τ and the total heat flux q as

$$\frac{\tau}{\rho} = (v + \varepsilon_m) \frac{du}{dy} \tag{9-54}$$

$$\frac{q}{\rho c_p} = -(\alpha + \varepsilon_h) \frac{dT}{dy} \tag{9-55}$$

These equations provide the basic relations from which the Reynolds and Prandtl analogies between momentum and heat transfer are derived as now described.

Reynolds Analogy Reynolds assumed that the entire flow field consisted of *a single zone of highly turbulent region* (i.e., the presence of a viscous sublayer and the buffer layer is neglected) such that the molecular diffusivities of momentum and heat are negligible in comparison with the turbulent diffusivities, i.e.,

$$v \ll \varepsilon_m \quad \text{and} \quad \alpha \ll \varepsilon_h \tag{9-56a}$$

and in addition the turbulent diffusivities of momentum and heat are equal, i.e.,

$$\varepsilon_m = \varepsilon_h \equiv \varepsilon \tag{9-56b}$$

With these assumptions Eqs. (9-54) and (9-55) simplify, respectively, to

$$\frac{\tau}{\rho} = \varepsilon \frac{du}{dy} \tag{9-57}$$

$$\frac{q}{\rho c_p} = -\varepsilon \frac{dT}{dy} \tag{9-58}$$

By combining these two equations we find

$$\frac{q}{\tau c_p} = -\frac{dT}{du}$$

or

$$dT = -\frac{q}{\tau c_p} du \tag{9-59}$$

The integration of Eq. (9-59) from the wall conditions $T = T_w$, $u = 0$ to the mean bulk-stream conditions $T = T_m$, $u = u_m$ with the assumption that q/τ remains constant (that is, q and τ are taken at the surface) results in

$$\int_{T_w}^{T_m} dT = -\frac{q}{\tau c_p} \int_0^{u_m} du$$

or

$$T_w - T_m = \frac{q u_m}{\tau c_p} \tag{9-60}$$

Now, the heat-transfer coefficient h and the friction factor f for flow inside a tube are defined as

$$q = h(T_w - T_m) \tag{9-61a}$$

$$\tau = f \frac{\rho u_m^2}{8} \qquad \text{[see Eq. (9-45)]} \tag{9-61b}$$

The substitution of Eqs. (9-61) into Eq. (9-60) yields

$$\text{St} \equiv \frac{h}{\rho c_p u_m} = \frac{f}{8} \tag{9-62a}$$

where the dimensionless group $\text{St} \equiv h/\rho c_p u_m$ is called the *Stanton number*. This result may be written in the alternative form

$$\text{St} = \frac{\text{Nu}}{\text{Re Pr}} = \frac{f}{8} \tag{9-62b}$$

where

$$\text{Nu} = \frac{hD}{k} \qquad \text{Re} = \frac{u_m D}{v} \qquad \text{and} \qquad \text{Pr} = \frac{v}{\alpha}$$

The result given by Eqs. (9-62) is called the *Reynolds analogy for momentum and heat transfer;* it is valid for $\text{Pr} \cong 1$ because in its derivation it is assumed that the heat and momentum are transported at the same rate, that is, $\varepsilon_m = \varepsilon_h$, and $v = \alpha$. It relates the heat-transfer coefficient h to the friction factor f; thus h can be determined from the knowledge of f. For fluids having a Prandtl number close to unity, such as gases, it is in reasonably good agreement with experiments.

Prandtl Analogy Prandtl considered the flow field to consist of *two layers*, a viscous sublayer where molecular diffusivities are dominant and a turbulent core where turbulent diffusivities are dominant, and performed the analysis for the two layers in the following manner.

Viscous Sublayer In this region, taking $\varepsilon_m \ll v$ and $\varepsilon_h \ll \alpha$, Eqs. (9-57) and (9-58) simplify to

$$\frac{\tau}{\rho} = v \frac{du}{dy} \qquad\qquad (9\text{-}63a)$$

$$\frac{q}{\rho c_p} = -\alpha \frac{dT}{dy} \qquad\qquad (9\text{-}63b)$$

These equations are now combined to give

$$dT = -\frac{v}{\alpha} \frac{q}{\tau c_p} du \qquad\qquad (9\text{-}64)$$

Equation (9-64) is integrated from the wall conditions $T = T_w$, $u = 0$ to the edge of the viscous sublayer where $T = T_1$, $u = u_1$, with the assumptions $q/\tau = \text{const}$ and $v/\alpha = \text{Pr} = \text{const}$, to obtain

$$T_w - T_1 = \text{Pr} \frac{q}{\tau c_p} u_1 \qquad\qquad (9\text{-}65)$$

Turbulent Core In this region it is assumed that $v \ll \varepsilon_m$, $\alpha \ll \varepsilon_h$, and $\varepsilon_m = \varepsilon_h = \varepsilon$. Equations (9-54) and (9-55) simplify to

$$\frac{\tau}{\rho} = \varepsilon \frac{du}{dy} \qquad\qquad (9\text{-}66a)$$

$$\frac{q}{\rho c_p} = -\varepsilon \frac{dT}{dy} \qquad\qquad (9\text{-}66b)$$

These equations are combined as

$$dT = -\frac{q}{\tau c_p} du \tag{9-67}$$

The integration of Eq. (9-67) from the edge of the viscous sublayer where $u = u_1$, $T = T_1$ to the mean bulk-stream conditions where $u = u_m$, $T = T_m$ for constant q/τ yields

$$T_1 - T_m = \frac{q}{\tau c_p} (u_m - u_1) \tag{9-68}$$

Equations (9-65) and (9-68) are combined to eliminate T_1:

$$T_w - T_m = \frac{q u_m}{\tau c_p} \left[1 + \frac{u_1}{u_m} (\mathrm{Pr} - 1) \right] \tag{9-69}$$

The heat-transfer coefficient h and the friction factor f for flow inside a tube are defined as

$$q = h(T_w - T_m) \tag{9-70a}$$

$$\tau = f \frac{\rho u_m^2}{8} \qquad [\text{see Eq. (9-45)}] \tag{9-70b}$$

The substitution of Eqs. (9-70) into (9-69) to eliminate q and τ gives

$$\mathrm{St} \equiv \frac{h}{\rho c_p u_m} = \frac{f}{8} \frac{1}{1 + (u_1/u_m)(\mathrm{Pr} - 1)} \tag{9-71}$$

Here the unknown velocity u_1 at the edge of the viscous sublayer is determined from the velocity-distribution law $u^+ = y^+$ [see Eq. (9-38a)] by setting in this relation $y^+ = 5$. We obtain

$$u^+ \equiv \frac{u_1}{\sqrt{\tau/\rho}} = 5 \tag{9-72}$$

The substitution of τ/ρ from Eq. (9-70b) into (9-72) yields

$$\frac{u_1}{u_m\sqrt{f/8}} = 5 \qquad \text{or} \qquad \frac{u_1}{u_m} = 5\sqrt{\frac{f}{8}} \tag{9-73}$$

Introducing Eq. (9-73) into Eq. (9-71) we obtain

$$\mathrm{St} \equiv \frac{h}{\rho c_p u_m} = \frac{f}{8} \frac{1}{1 + 5\sqrt{f/8}\,(\mathrm{Pr} - 1)} \tag{9-74}$$

Equation (9-74) is called the *Prandtl analogy for momentum and heat transfer for turbulent flow in a pipe*. We note that, for Pr = 1, the Prandtl analogy Eq. (9-74) reduces to the Reynolds analogy given by Eqs. (9-62).

The Prandtl analogy can be used to determine the heat-transfer coefficient h for turbulent flow inside a pipe from the knowledge of the friction factor f for fluids having a Prandtl number different from unity.

Von Kármán Analogy   Von Kármán extended Prandtl's analogy by separating the flow field into *three distinct layers:* a viscous sublayer, a buffer layer, and a turbulent core.   He made assumptions on the relative magnitudes of the molecular and turbulent diffusivities of heat and momentum in the viscous sublayer and the turbulent core similar to those made by Prandtl, but in addition he included the effects of the buffer layer by assuming that the molecular and eddy diffusivities in this layer are of the same order of magnitude.   *The von Kármán analogy between the momentum and heat transfer for turbulent flow inside a circular pipe* is given as

$$\text{St} \equiv \frac{h}{\rho c_p u_m} = \frac{f}{8} \frac{1}{1 + 5\sqrt{f/8}\{(\text{Pr} - 1) + \ln[(5\,\text{Pr} + 1)/6]\}} \tag{9-75}$$

We note that for Pr = 1 this result reduces to the Reynolds analogy.   This relation appears to be good for a Prandtl number up to about 30.

Other Analogies   Martinelli [19] extended von Kármán's analogy by considering that the molecular diffusivity of heat is not negligible and that the shear stress and the heat flux vary linearly with the distance from the tube axis in the turbulent core.   Therefore, this analysis is also applicable to fluids having a very low Prandtl number, such as liquid metals.

Deissler [20a,b] made extensions by taking into consideration the variation of viscosity of fluid with temperature according to a power-law relationship and also by making it applicable to low and very high Prandtl numbers.

We do not present here the results of these more refined analogies because the final expressions are rather complicated and require additional computations before the results can be of practical use.   The reader should consult the original references for the details of the analysis and the results.

Turbulent Flow Along a Flat Plate

In the foregoing analysis, we presented the results of various analogies for momentum and heat transfer for turbulent flow inside a pipe for which the friction factor f was related to the shear stress τ by

$$\tau = f \frac{\rho u_m^2}{8} \quad \text{[see Eq. (9-45)]} \tag{9-76}$$

In the case of turbulent flow along a flat plate, the local-drag coefficient c_x is related to the local shear tress τ_x by

$$\tau_x = c_x \frac{\rho u_\infty^2}{2} \qquad \text{[see Eq. (8-19)]} \qquad\qquad (9\text{-}77)$$

where u_∞ is the external stream velocity. A comparison of Eqs. (9-76) and (9-77) implies that the foregoing results for analogies are applicable for flow along a flat plate if f is replaced by $4c_x$, and u_m by u_∞. That is, the *Reynolds analogy* [Eqs. (9-62)] for turbulent flow along a flat plate becomes

$$\text{St}_x \equiv \frac{h_x}{\rho c_p u_\infty} = \frac{c_x}{2} \qquad\qquad (9\text{-}78a)$$

or the *von Kármán analogy* of Eq. (9-75) becomes

$$\text{St}_x \equiv \frac{h_x}{\rho c_p u_\infty} = \frac{c_x}{2} \frac{1}{1 + 5\sqrt{c_x/2}\{(\text{Pr} - 1) + \ln[(5\,\text{Pr} + 1)/6]\}} \qquad\qquad (9\text{-}78b)$$

The analogies are useful to determine the heat-transfer coefficient from the knowledge of the friction factor or the drag coefficient in flow for which no empirical or theoretical expressions are available for the heat-transfer coefficient. Here, c_x can be determined from the empirical relations [5,21]

$$c_x = 0.0592\,\text{Re}_x^{-1/5} \qquad\quad \text{for} \quad 5 \times 10^5 < \text{Re}_x < 10^7 \qquad\qquad (9\text{-}78c)$$

$$c_x = 0.370(\log \text{Re}_x)^{-2.584} \qquad \text{for} \quad 10^7 < \text{Re}_x < 10^9 \qquad\qquad (9\text{-}78d)$$

9-6 EMPIRICAL FORMULAS FOR HEAT TRANSFER AND FLUID FRICTION IN TURBULENT FLOW

For most practical problems encountered in engineering applications the flow patterns are so complicated that the determination of the friction factor and the heat-transfer coefficient by analysis is extremely difficult or impractical with the presently available methods. Therefore, in such cases the engineer must turn to the experimental and the semiempirical correlations for the determination of the friction factor and the heat-transfer coefficient in turbulent flow. A vast amount of empirical data are available in the literature on this subject. In this section we present representative empirical and semiempirical relations for fluid friction and the heat-transfer coefficient for both internal and external flow.

Turbulent Flow in Smooth Circular Tubes

The friction factor f for turbulent flow inside smooth circular tubes in regions away from the inlet is determined from the semiempirical relation given by Eq. (9-50) or from the friction-factor chart in Fig. 9-5. However, the analysis of heat transfer being more involved, a large number of experimental correlations have been developed for the determination of the heat-transfer coefficient h in the regions away from the thermal entry. These relations include, among others, the following:

1 The Dittus-Boelter [22] equation:

$$\text{Nu} = 0.023 \, \text{Re}^{0.8} \, \text{Pr}^n \tag{9-79}$$

where $n = 0.4$ for heating and 0.3 for cooling,

$$\text{Nu} = \frac{hD}{k} \qquad \text{Re} = \frac{u_m D}{v} \qquad \text{and} \qquad \text{Pr} = \frac{v}{\alpha}$$

It is applicable for small temperature differences and for $0.7 < \text{Pr} < 100$, $\text{Re} > 10,000$, $L/D > 60$. Fluid properties are evaluated at the average fluid temperature at the region considered.

2 The Colburn [23] equation: For fluids of high viscosity, instead of using a different exponent for the Prandtl number for heating and cooling as given above, Colburn proposed the following correlation:

$$\text{St} \, \text{Pr}^{2/3} = 0.023 \, \text{Re}^{-0.2} \tag{9-80}$$

where $\quad \text{St} \equiv \dfrac{\text{Nu}}{\text{Re} \, \text{Pr}} = \dfrac{h}{\rho u_m c_p} = \text{Stanton number}$

and Re, Nu, and Pr are as defined above. This relation is recommended for heat transfer to fluids with temperature-dependent properties and is applicable for $0.7 < \text{Pr} < 160$, $\text{Re} < 10,000$, and $L/D > 60$. The Stanton number is to be evaluated at the average fluid temperature, and the Reynolds and the Prandtl numbers at the arithmetic mean of the wall and fluid temperatures.

3 The Sieder and Tate [24] equation:

$$\text{Nu} = 0.027 \, \text{Re}^{0.8} \, \text{Pr}^{1/3} \left(\frac{\mu_m}{\mu_w} \right)^{0.14} \tag{9-81}$$

This relation is recommended for heat transfer to fluids whose viscosity changes sharply with temperature and is applicable for $0.7 < \text{Pr} < 16,700$,

Re > 10,000, and $L/D > 60$. Here the fluid properties for Nu, Re, and Pr numbers are to be evaluated at the mean fluid temperature, and μ_m and μ_w refer to the viscosity evaluated at the mean fluid and wall temperatures, respectively.

4 Notter and Sleicher [34] equation:

$$\text{Nu} = 5 + 0.016\,\text{Re}^a\,\text{Pr}^b \tag{9-82}$$

where

$$a = 0.88 - \frac{0.24}{4 + \text{Pr}} \qquad \text{and} \qquad b = 0.33 + 0.5e^{-0.6\,\text{Pr}}$$

valid for $0.1 < \text{Pr} < 10^4$ and $10^4 < \text{Re} < 10^6$. It appears to be in excellent agreement with the best data for air within 10 percent of the best data at Prandtl numbers as high as 10^5. The thermal-entry length varies with the Reynolds and Prandtl numbers, that is, for the Nusselt-number ratio Nu_x/Nu to reach a value of 1.05, the thermal-entry length L/D varies from 5 to 25, depending on the combination of the Reynolds and Prandtl numbers. Hence, Eq. (9-82) is valid for $L/D > 25$.

Thermal Entry Region The foregoing equations are applicable in the regions where both velocity and temperature profiles are considered fully developed. Nusselt [25] studied the experimental data in the range of $L/D = 10$ to 100 and concluded that h is proportional to $(D/L)^{1/8}$. Hence *he replaced Eq. (9-81) by the following relation* to take into account the entry effects:

$$\text{Nu} = 0.036\,\text{Re}^{0.8}\,\text{Pr}^{1/3}\left(\frac{D}{L}\right)^{0.055} \qquad \text{for } 10 < \frac{L}{D} < 400 \tag{9-83}$$

where L is the tube length measured from the inlet and the fluid properties are evaluated at the mean fluid temperature. The reader is referred to the analysis of Notter and Sleicher [34] for the variation of the Nusselt number in the thermal-entry region and variation of thermal-entry length with the Reynolds and Prandtl numbers.

Some experimental data on heat transfer at the entrance regions of a tube for water and oil are given by Hartnett [26].

Turbulent Flow in Smooth Noncircular Ducts

For moderate temperature differences, Eqs. (9-79) and (9-80) can be used to predict heat transfer for turbulent flow in noncircular smooth ducts in regions away from the inlet, provided that the tube diameter D is replaced by the *equivalent diameter* D_e defined as

$$D_e = \frac{4 \times (\text{cross-sectional area for flow})}{\text{wetted perimeter}} \tag{9-84}$$

The equivalent-diameter concept is, in general, satisfactory for the prediction of fluid friction and heat-transfer coefficient for turbulent flow in noncircular ducts for many practical situations, but there are also many situations where the method is not satisfactory. Irvine [27] discusses some of the problems on heat transfer in noncircular ducts.

Turbulent Flow in Rough Circular Tubes

The heat-transfer coefficient for turbulent flow in rough-walled tubes is higher than that for smooth-walled tubes because roughness disturbs the viscous sublayer. The increased heat transfer due to roughness is achieved at the expense of increased friction to fluid flow. The correlation of heat transfer for turbulent flow in rough-walled tubes is very sparse in the literature. The friction factor f can be obtained from the friction-factor chart given in Fig. 9-5 if the relative roughness of the surface is known. In the case of the heat-transfer coefficient, it is recommended that the analogy between the momentum and heat transfer should be used to determine the heat-transfer coefficient from the knowledge of the friction factor.

Liquid Metals in Turbulent Flow Inside Smooth Tubes

The empirical relations given above for predicting the heat-transfer coefficient in turbulent flow are not applicable to liquid metals because they are developed for Prandtl numbers greater than about 0.7, whereas the Prandtl number for liquid metals ranges from 0.003 to 0.03. We summarize below some of the empirical and semiempirical relations for heat transfer to liquid metals in fully developed turbulent flow inside smooth circular tubes under uniform wall heat-flux and uniform wall-temperature conditions.

Uniform Wall Heat Flux   The Martinelli analogy [19] can be used to determine the heat-transfer coefficient from the knowledge of the friction factor. Lubarsky and Kaufman [28] proposed the following empirical relation:

$$\mathrm{Nu} = 0.625\,\mathrm{Pe}^{0.4} \tag{9-85a}$$

where

$$\text{Peclet number} \equiv \mathrm{Pe} = \mathrm{Re}\,\mathrm{Pr}$$

for $10^2 < \mathrm{Pe} < 10^4$, $L/D > 60$, and properties are evaluated at bulk fluid temperature. Skupinski et al. [29] recommend the following relation, based on heat-transfer experiments with sodium-potassium mixtures:

$$\mathrm{Nu} = 4.82 + 0.0185\,\mathrm{Pe}^{0.827} \tag{9-85b}$$

for $3.6 \times 10^3 < \mathrm{Re} < 9.05 \times 10^5$ and $10^2 < \mathrm{Pe} < 10^4$. Equation (9-85b) may be preferred to Eq. (9-85a).

Uniform Wall Temperature Seban and Shimazaki [30] utilized the analogy between momentum and heat transfer and proposed the following expression for the heat-transfer coefficient

$$Nu = 5.0 + 0.025\, Pe^{0.8} \tag{9-86}$$

for $Pe > 100$, $L/D > 60$, and physical properties are evaluated at bulk fluid temperature. Knudsen and Katz [31] compared Eq. (9-86) with that obtainable from the Martinelli analogy for uniform wall heat flux and found that for high values of Peclet number there is little difference between the two results, but for low values of Peclet number, the Nusselt number with uniform wall temperature was lower than that with uniform wall heat flux.

Other expressions developed by various investigators include the following: Sleicher and Tribus [32],

$$Nu = 4.8 + 0.015\, Pe^{0.91}\, Pr^{0.30} \qquad \text{for } Pr < 0.05; \tag{9-87}$$

Azer and Chao [33],

$$Nu = 5.0 + 0.05\, Pe^{0.77}\, Pr^{0.25} \qquad \text{for } Pr < 0.1 \text{ and } Pe < 15{,}000; \tag{9-88}$$

and Notter and Sleicher [34],

$$Nu = 4.8 + 0.0156\, Pe^{0.85}\, Pr^{0.08} \qquad \text{for } 0.004 < Pr < 0.1 \text{ and } Re < 500{,}000. \tag{9-89}$$

It is to be noted that these equations have been obtained by empirical fits to the calculation and the results differ from author to author primarily because of eddy-diffusivity distributions assumed in the analysis. Figure 9-6 shows a comparison of the Nusselt number calculated from the equations of Seban and Shimazaki [Eq. (9-86)], Sleicher and Tribus [Eq. (9-87)], Azer and Chao [Eq. (9-88)], and Notter and Sleicher [Eq. (9-89)] with the experimental data of Sleicher, Awad, and Notter [35] for heat transfer to NaK in turbulent flow inside a circular tube with uniform wall temperature.

Thermal-Entry Region Sleicher et al. [35] examined the heat-transfer calculations of Notter and Sleicher [34] in the thermal-entry region of both uniform wall-temperature and uniform wall heat-flux conditions and noted that the thermal-entry-length effects can be correlated within 20 percent with the following relations:

$$Nu_x = Nu\left(1 + \frac{2}{x/D}\right) \qquad \text{for } \frac{x}{D} > 4 \tag{9-90a}$$

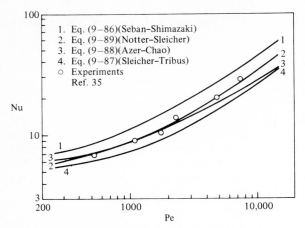

FIG. 9-6   Comparison of Nusselt number from various correlations and the experimental results for heat transfer to NaK in turbulent flow inside a circular tube with uniform wall temperature. (*From Sleicher et al.* [*35*].)

where

$$Nu = 6.3 + 0.0167 \, Pe^{0.85} \, Pr^{0.08} \qquad \text{for uniform wall heat flux} \qquad (9\text{-}90b)$$

$$Nu = 4.8 + 0.0156 \, Pe^{0.85} \, Pr^{0.08} \qquad \text{for uniform wall temperature} \qquad (9\text{-}90c)$$

and applicable in the range $0.004 < Pr < 0.1$.

The reader should consult Refs. [36,37] for extensive data on liquid-metal heat transfer and the heat-transfer characteristics of liquid metals.

Example 9-2   Water at a mean temperature of 100°F (37.8°C) flows through a 1-in (2×10^{-2}-m) ID smooth-walled tube with a mean velocity of 5 ft/s (1.52 m/s). The tube wall is 30°F (16.67°C) above the water temperature. Determine the heat-transfer coefficient and the total heat-transfer rate to the water for a 10-ft length of the tube, assuming the velocity and temperature profiles are fully developed.

Solution   The physical properties of water at 100°F are $\nu = 0.74 \times 10^{-5}$ ft²/s (0.687×10^{-6} m²/s), $k = 0.364$ Btu/h·ft·°F (0.63 W/m·°C), and $Pr = 4.52$. Then the Reynolds number becomes

$$Re = \frac{uD}{\nu} = \frac{5}{(0.74 \times 10^{-5}) \times 12} = 56,300$$

and the flow is turbulent. We can use Eq. (9-79) to determine the heat-transfer coefficient because the temperature difference is sufficiently small and the effects of viscosity variation with temperature can be neglected. Taking $n = 0.4$ for heating of fluid, we have

$$Nu \equiv \frac{hD}{k} = 0.023 \, Re^{0.8} \, Pr^{0.4}$$

$$\frac{h}{0.364 \times 12} = 0.023 \times 56,300^{0.8} \times 4.52^{0.4}$$

$$h = 1160 \, \text{Btu/h·ft}^2\text{·°F} \, (6585 \, \text{W/m}^2\text{·s})$$

The heat-transfer rate from the wall to the fluid for $L = 10$-ft length of the tube is

$$Q = (\pi DL)h(T_w - T_m)$$

$$= \left(\pi \frac{1}{12} 10\right) \times 1160 \times 30 = 91,100 \text{ Btu/h } (26,690 \text{ W})$$

Flow Across a Single Circular Cylinder

Flow across a single circular cylinder frequently occurs in practice, but the determination of the drag and heat-transfer coefficients by analytical means is very complicated because of flow separation. Figure 9-7a and b illustrates flow patterns around a circular cylinder with and without flow separation. Defining the Reynolds number for flow across the cylinder as

$$\text{Re} = \frac{u_\infty D}{\nu} \tag{9-91}$$

where D is the cylinder diameter and u_∞ is the undisturbed free-stream velocity, the flow is laminar and remains attached to the surface of the cylinder as illustrated in Fig. 9-7a only at very low Reynolds number (that is, Re < 1). At higher Reynolds number the flow separates along the cylinder and vortex motion starts in the wake as illustrated in Fig. 9-7b. Therefore the drag coefficient is affected by different behavior of flow at different Reynolds numbers. Figure 9-8 shows the average value of the drag coefficient c_D for flow across a long cylinder. If F is the total drag force due to flow acting on a single cylinder of length L and diameter D, then the drag coefficient c_D is defined by the relation

$$\frac{F}{LD} = c_D \frac{\rho u_\infty^2}{2g_c} \tag{9-92}$$

where u_∞ is the free-stream velocity. We note that F/LD is the drag force per unit *frontal area* of the cylinder. A scrutiny of Fig. 9-8 reveals that at very low

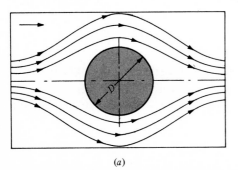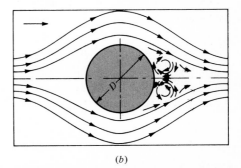

(a) (b)

FIG. 9-7 Flow patterns around a circular cylinder. (a) Fluid adheres to the surface everywhere at a low Reynolds number (Re < 1); (b) flow separation and vortex motion in the wake at a high Reynolds number.

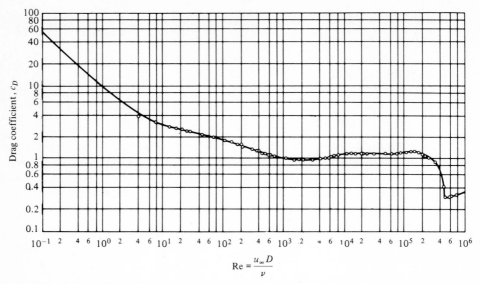

$$\mathrm{Re} = \frac{u_\infty D}{\nu}$$

FIG. 9-8 Drag coefficient for flow across a single circular cylinder. (*From Schlichting* [5].)

Reynolds number, of the order of unity or less, the drag is caused by viscous forces since the boundary layer remains attached to the cylinder surface everywhere. At Reynolds number of the order of about 10, the drag curve starts to deviate from a straight line, indicating the formation of eddies in the rear of the cylinder. In this region the contribution of viscous friction and the drag caused by the formation of weak eddies on the total drag are of the same order of magnitude. At Reynolds number in the range of 10^3 to 10^5 the drag is caused predominantly by turbulent eddies in the wake of the cylinder; hence the drag coefficient remains essentially constant in this region. The flow separation lies at about 80 to 85° measured from the direction of flow. At Reynolds number above 10^5 we note a sudden reduction in the total drag; this is caused because the point of flow separation moves toward the rear of the cylinder, thus reducing the size of the wake and the drag.

Heat transfer for flow across a cylinder is important in many applications. Figure 9-9 shows McAdams' [38] correlation of the average heat-transfer coefficient for the heating or cooling of *air* flowing across a single cylinder. In the case of *liquids* flowing across a single cylinder, Knudsen and Katz [31] recommend the following relation for the average heat-transfer coefficient:

$$\mathrm{Nu}_m \equiv \frac{h_m D}{k} = c \, \mathrm{Re}^n \, \mathrm{Pr}^{1/3} \tag{9-93}$$

where the values of the constant c and the exponent n are given in Table 9-1, and the properties are to be evaluated at the arithmetic mean of the free-stream and wall temperatures.

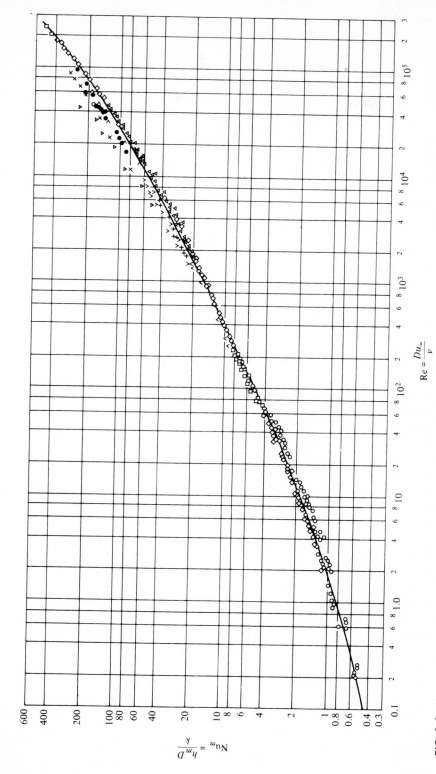

FIG. 9-9   Average Nusselt number for heating or cooling of air flowing across a single circular cylinder. *(From McAdams [38].)*

TABLE 9-1   Constants c and n of
Eq. (9-93)†

$\mathrm{Re} = u_\infty D/v$	c	n
1–4	0.989	0.330
4–40	0.911	0.385
40–4,000	0.683	0.466
4,000–40,000	0.193	0.618
40,000–250,000	0.0266	0.805

† Based on Ref. 31, p. 505.

Experimental work by Fand [39] for water supports McAdams' [38] correlation for the average heat-transfer coefficient for liquids flowing across a single cylinder given by the relation

$$\mathrm{Nu}_m \equiv \frac{h_m D}{k} = (0.35 + 0.56\ \mathrm{Re}^{0.52})\ \mathrm{Pr}^{0.30} \qquad \text{for } 10^{-1} < \mathrm{Re} < 10^5 \qquad (9\text{-}94)$$

where all properties are evaluated at the arithmetic mean of the free-stream and wall temperatures.

More recently Whitaker [40] presented a more general correlation that takes into account contributions to the average heat-transfer coefficient from the undetached boundary-layer region, the wake region around the cylinder, and the temperature effects.   This relation, which is applicable to heat transfer for flow across a single cylindrical tube or wire, is given as

$$\mathrm{Nu}_m \equiv \frac{h_m D}{k} = \left(0.4\ \mathrm{Re}^{0.5} + 0.06\ \mathrm{Re}^{0.67}\right)\ \mathrm{Pr}^{0.4} \left(\frac{\mu}{\mu_w}\right)^{0.25} \qquad (9\text{-}95)$$

where the physical properties are to be evaluated at the free-stream temperature except μ_w which is the viscosity at the wall-surface temperature.   This relation correlated the experimental data within ± 25 percent in the range $40 < \mathrm{Re} < 10^5$ and $0.67 < \mathrm{Pr} < 300$.   In Eq. (9-95) the term $\mathrm{Re}^{0.5}$ characterizes the contribution to the heat-transfer coefficient from the undetached boundary-layer region and the term $\mathrm{Re}^{0.67}$ from the wake region around the cylinder.

In the above discussion we focused our attention on the determination of the average value of the heat-transfer coefficient for the cylinder.   Actually the local value of the heat-transfer coefficient $h(\theta)$ varies with the angle θ around the cylinder. It has a fairly high value at the stagnation point $\theta = 0$ and decreases around the cylinder as the boundary layer thickens.   The decrease of the heat-transfer coefficient is continuous until the boundary layer separates from the wall surface or the laminar boundary layer changes into turbulent; then an increase occurs with the distance around the cylinder.   Figures 9-10 and 9-11 show the variation of

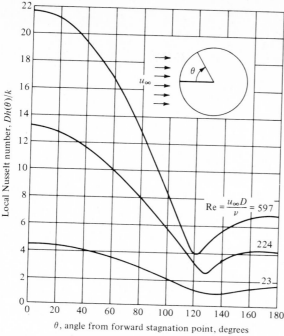

FIG. 9-10   Variation of the local heat-transfer coefficient $h(\theta)$ around a circular cylinder at low Reynolds numbers. (*From Eckert and Soehngen* [41].)

the local heat-transfer coefficient $h(\theta)$ with angle θ around a circular cylinder at low and high Reynolds numbers, respectively.   Eckert and Soehngen's [41] data in Fig. 9-10 for low Reynolds numbers show a variation of the local heat-transfer coefficient around the cylinder as described above.   Giedt's [42] data in Fig. 9-11 for high Reynolds numbers, on the other hand, show that the heat-transfer coefficient exhibits two minimums around the cylinder.   The reason for this is that at very large Reynolds number the boundary-layer transition from laminar to turbulent takes place before the flow separation.   The first minimum occurs because during the transition from laminar to turbulent flow the local heat-transfer coefficient begins to increase and then starts to decrease as the fully developed turbulent boundary-layer thickness increases.   The second minimum occurs at the point of separation beyond which the local heat-transfer coefficient begins to increase.   In Fig. 9-11 the boundary-layer separation takes place at $\theta \cong 80$ for the curve Re = 70,800, and the increase of $h(\theta)$ for $\theta > 80$ is due to the vortex motion in the wake of the cylinder.   The curve for Re = 101,300 shows a similar behavior.   In the curve Re = 140,000 the first minimum occurs at the transition from laminar to turbulent boundary layer at $\theta \cong 80$, but the separation does not take place until $\theta \cong 130$ where the second minimum occurs.   The curve for Re = 186,000 shows these effects more clearly.

It is now apparent that the variation of the local heat-transfer coefficient $h(\theta)$ around the cylinder depends on a number of parameters, and its analysis is a very

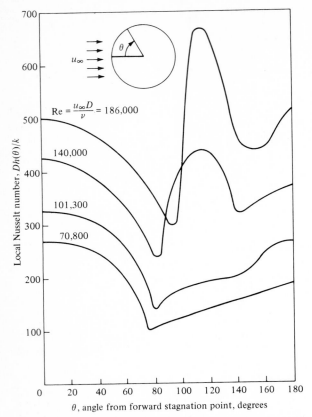

FIG. 9-11 Variation of the local heat-transfer coefficient $h(\theta)$ around a circular cylinder at high Reynolds numbers. (*From Giedt* [42].)

complex matter. For this reason most correlations of the heat-transfer coefficient for flow normal to a cylinder available in the literature are for the average value around the entire cylinder; furthermore for most practical applications only the average value of the heat-transfer coefficient is needed.

Example 9-3 Water at 50°F (10°C) with a free-stream velocity $u_\infty = 5$ ft/s (1.524 m/s) flows across a single cylinder of 1-in (2.54×10^{-2}-m) OD whose surface is kept at 150°F (65.6°C). Determine the average heat-transfer coefficient h_m and heat-transfer rate to the water per linear foot of the tube.

Solution The properties of water at the arithmetic mean of the free-stream and wall temperatures, $(50 + 150)/2 = 100$°F, are

$c_p = 0.998$ Btu/lb·°F (4178 W·s/kg·°C)

$v = 0.74 \times 10^{-5}$ ft²/s (0.687×10^{-6} m²/s)

$k = 0.364$ Btu/h·ft·°F (0.63 W/m·°C)

Pr = 4.52

Then the Reynolds number for the flow is

$$\text{Re} = \frac{u_\infty D}{v} = \frac{5}{(0.74 \times 10^{-5}) \times 12} \cong 56,300$$

For simplicity, Eq. (9-93) is used to calculate the heat-transfer coefficient h_m. That is,

$$\frac{h_m D}{k} = c \, \text{Re}^n \, \text{Pr}^{1/3}$$

where the constants c and n, for the Reynolds number considered here, are obtained from Table 9-1 as $c = 0.0266$ and $n = 0.805$. Then

$$\frac{h_m}{0.364 \times 12} = 0.0266 \times 56,300^{0.805} \times 4.52^{1/3}$$

or

$$h_m = 1281.5 \text{ Btu/h} \cdot \text{ft}^2 \cdot {}^\circ\text{F} \ (7275 \text{ W/m}^2 \cdot \text{s})$$

The heat-transfer rate per *linear foot* of the cylinder is determined as follows: By using the English system of units, we find

$$Q = h_m(1\pi D)(T_w - T_\infty) = (1281.5)\left(\pi \, \frac{1}{12}\right)(150 - 50) \cong 33.550 \text{ Btu/h}$$

or, by using the SI system of units, we obtain

$$Q = (7275)[0.348 \times \pi \times (2.54 \times 10^{-2})](55.56) = 9833 \text{ W}$$

Flow Across a Single, Noncircular Cylinder

The results of experiments for the average heat-transfer coefficient h_m for the flow of gases across a single, noncircular, long cylinder of various geometries have been correlated by Jakob [43] with the following simple relationship:

$$\text{Nu}_m \equiv \frac{h_m D_e}{k} = c\left(\frac{u_\infty D_e}{v}\right)^n \tag{9-96}$$

where the constant c, the exponent n, and the characteristic dimension D_e for various geometries are presented in Table 9-2. The physical properties of the fluid are evaluated at the arithmetic mean of the free-stream and the wall temperatures.

Flow Across a Single Sphere

If F is the total drag force due to flow across a single sphere, the average drag coefficient c_D is defined by the relation

$$\frac{F}{A} = c_D \frac{\rho u_\infty^2}{2g_c} \tag{9-97}$$

where A is the *frontal area* (that is, $A = \pi D^2/4$) and u_∞ is the free-stream velocity.

TABLE 9-2   Constants c and n of Eq. (9-96)[†]

Flow Direction and Geometry	$\text{Re} = \dfrac{u_\infty D_e}{\nu}$	n	c
$u_\infty \longrightarrow$ ◇ D_e	5000–100,000	0.588	0.222
$u_\infty \longrightarrow$ ◯ D_e	2500–15,000	0.612	0.224
$u_\infty \longrightarrow$ ◇ D_e	2500–7500	0.624	0.261
$u_\infty \longrightarrow$ ⬡ D_e	5000–100,000	0.638	0.138
$u_\infty \longrightarrow$ ⬡ D_e	5000–19,500	0.638	0.144
$u_\infty \longrightarrow$ ☐ D_e	5000–100,000	0.675	0.092
$u_\infty \longrightarrow$ ☐ D_e	2500–8000	0.699	0.160
$u_\infty \longrightarrow$ │ D_e	4000–15,000	0.731	0.205
$u_\infty \longrightarrow$ ⬡ D_e	19,500–100,000	0.782	0.035
$u_\infty \longrightarrow$ ◖◗ D_e	3000–15,000	0.804	0.085

† From Jakob [43].

We note that F/A is the drag force per unit frontal area of the sphere. Figure 9-12 shows the average drag coefficient c_D for flow across a single sphere. A comparison of the drag-coefficient curves in Figs. 9-8 and 9-12 for a single cylinder and sphere, respectively, reveals that the two curves have similar general characteristics.

The average value of the heat-transfer coefficient h_m for flow across a single sphere is also of practical interest. For the flow of *gases* across a single sphere McAdams [38] recommends the following relation:

$$\frac{h_m D}{k} = 0.37 \, \text{Re}^{0.6} \qquad \text{for } 17 < \text{Re} < 70,000 \tag{9-98}$$

and for the flow of *liquids* Kramers [44] proposes the following relation:

$$\frac{h_m D}{k} = (0.97 + 0.68 \, \text{Re}^{0.5}) \, \text{Pr}^{0.3} \qquad 1 < \text{Re} < 2000 \tag{9-99}$$

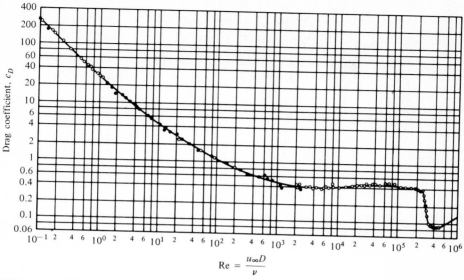

FIG. 9-12   Drag coefficient for flow over a single sphere. (*From Schlichting* [5].)

where $Re = u_\infty D/v$
D = diameter of sphere
u_∞ = free-stream velocity

In both these relations the fluid properties are evaluated at the arithmetic mean of free-stream and wall-surface temperatures.

On the basis of experiments with the flow of water and oils across a single sphere in the range of Reynolds 1 to 50,000, Vliet and Leppert [45] proposed the following more general relation:

$$\frac{h_m D}{k} = (1.2 + 0.53\, Re^{0.54})\, Pr^{0.3} \left(\frac{\mu}{\mu_w}\right)^{0.25} \tag{9-100}$$

where $Re = u_\infty D/v$.   The authors suggest that this correlation is applicable to Reynolds numbers in the range $1 < Re < 300,000$ and represents all Kramers' [44] oil and water data.

The foregoing relations are not applicable to liquid metals which have very low Prandtl number.   In the case of *liquid metals*, Witte [46] carried out experiments with liquid sodium flowing across a single sphere and proposed the following correlation:

$$\frac{h_m D}{k} = 2 + 0.386(Re\, Pr)^{0.5} \tag{9-101}$$

for

$$3.56 \times 10^4 \le Re \le 1.53 \times 10^5$$

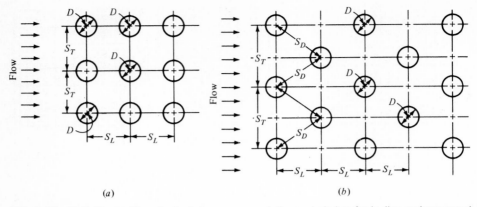

FIG. 9-13   Definitions of longitudinal, transverse, and diagonal pitches for in-line and staggered tube-bundle arrangements. (*a*) In-line arrangement; (*b*) staggered arrangement.

Flow Across Tube Banks

Heat transfer in flow across tube banks has numerous applications in the design and analysis of heat exchangers.   For example, a common type of heat exchanger consists of tube banks with one fluid passing through the tubes and the other across the tubes.   Frequently used tube-bank arrangements include the *in-line* and the *staggered* arrangements as illustrated in Fig. 9-13*a* and *b*.   The tube-bundle geometry is characterized by the *transverse pitch* S_T and the *longitudinal pitch* S_L between the tube centers; the *diagonal pitch* S_D between the centers of the tubes in the diagonal row is sometimes used for the staggered arrangement. To define the Reynolds number for flow through the tube bank, the flow velocity is based on the *minimum free-flow area* available for flow, whether the minimum area occurs between the tubes in a transverse row or in a diagonal row.   Then the Reynolds number for flow across a tube bank is defined as

$$\text{Re} = \frac{DG_{\text{max}}}{\mu} \tag{9-102}$$

where

$$G_{\text{max}} = \rho u_{\text{max}} = \text{maximum mass-flow velocity} \tag{9-103}$$

D is the outside diameter of the tube, ρ is the density, and u_{max} the maximum velocity based on the minimum free-flow area available for fluid flow.   If u_∞ is the flow velocity measured at a point in the heat exchanger before the fluid enters the tube bank (or the flow velocity based on flow inside the heat-exchanger shell without the tubes), then the maximum flow velocity u_{max} for the *in-line arrangement* is determined from

$$u_{\text{max}} = u_\infty \frac{S_T}{S_T - D} \tag{9-104}$$

where S_T is the transverse pitch and D is the outer diameter of the tube. Clearly, for the in-line arrangement $S_T - D$ is the minimum free-flow area between the adjacent tubes in a transverse row per unit length of the tube.

For the *staggered arrangement* shown in Fig. 9-13b the minimum free-flow area may occur either between adjacent tubes in a transverse row or between adjacent tubes in a diagonal row. In the former case u_{max} is determined as given above; in the latter case it is determined from

$$u_{max} = u_\infty \frac{S_T}{2(S_D - D)} \tag{9-105}$$

In general, if M (in lb/h) is the total mass-flow rate of fluid through the tube bundle in a heat exchanger and A_{min} (in ft²) is the minimum total free-flow area available for flow regardless where the minimum occurs, then the *maximum mass-flow velocity* G_{max} is readily calculated from

$$G_{max} = \frac{M}{A_{min}} \quad \text{lb/h·ft}^2 \quad \text{(or kg/m}^2\text{·s)} \tag{9-106}$$

The flow patterns through a tube bundle are so complicated that it is virtually impossible to predict heat transfer and pressure drop for flow across tube banks by pure analysis. Therefore, an experimental approach is the only alternative, and a wealth of experimental data are available in the literature. Bergelin et al. [47] studied the friction factor and the heat-transfer coefficient for the flow of oil across tube banks of $\frac{3}{8}$-in-diameter tubes for five different tube-bundle arrangements in both the *laminar- and the transition-flow* regimes. Figure 9-14 shows their results for flow across 10 rows of tubes. In this figure the average friction factor f and the average heat-transfer coefficient h_m represented in a dimensionless group in the form

$$j \equiv \frac{h_m}{c_p G_{max}} \text{Pr}^{2/3} \left(\frac{\mu_w}{\mu}\right)^{0.14} \tag{9-107}$$

are plotted against the Reynolds number $\text{Re} = DG_{max}/\mu$. Here μ_w is the viscosity at the tube wall temperature, and properties are evaluated at the bulk-fluid temperature. The authors specify the *transition-flow region* in the range $200 < \text{Re} < 5000$. A straight-line behavior of the curves up to a Reynolds number of about 200 indicates that viscous flow is predominant in this region. The curves for the in-line and staggered arrangements in the transition region not only deviate from a straight line but they behave differently, indicating that the transition process for these two arrangements is not the same. At Reynolds number of about 5000 and higher the curves for both the in-line and staggered arrangements appear to converge into a straight line, indicating that heat-transfer data for the turbulent-flow regime can be correlated with a simpler relation.

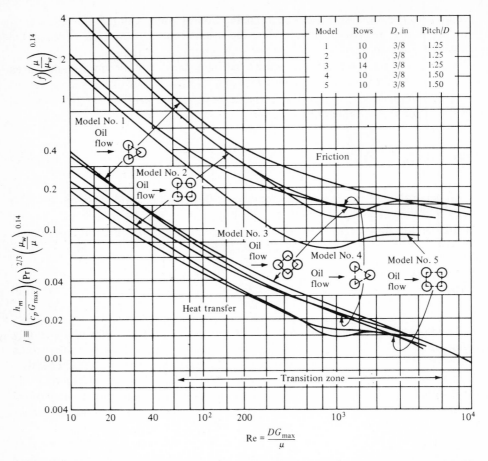

FIG. 9-14 . Average heat-transfer coefficient h_m and friction factor f for flow across tube banks of five different arrangements. (*From Bergelin, Brown, and Doberstein* [47].)

The results in Fig. 9-14 are for a tube bundle 10 rows deep in the direction of flow; they may also be applicable for more than 10 rows, but for bundles less than 10 rows deep some correction is needed to the results given in this figure. Meece [48] investigated the effects of the number of rows on the average heat-transfer coefficient in the laminar-flow regime with square in-line tube arrangements for tube bundles one, two, four, six, eight, and ten rows deep. For example, he found that, for a given Reynolds number, the average heat-transfer coefficient for a single row of tubes was about 50 percent larger than that given in Fig. 9-14 for 10 rows.

Grimison [49] correlated the data of various investigators and proposed the following equation for the average heat-transfer coefficient for the staggered and in-line arrangements *10 rows or more deep* in the direction of flow:

$$\frac{h_m D}{k} = c\left(\frac{DG_{max}}{\mu}\right)^n \qquad \text{for air} \qquad 2000 < \text{Re} < 40{,}000 \qquad (9\text{-}108a)$$

This relation may be generalized for fluids other than air in the following manner:

$$\frac{h_m D}{k} = 1.13c\left(\frac{DG_{max}}{\mu}\right)^n \text{Pr}^{1/3} \qquad \text{for } 2000 < \text{Re} < 40{,}000 \qquad (9\text{-}108b)$$

where the coefficient c and the exponent n are listed in Table 9-3 for both staggered and in-line arrangements for various values of the pitch-to-diameter ratio. The data given in this table are applicable to tube bundles having 10·or more rows of tubes in the direction of flow. They may be applicable to tube bundles as small as 4 rows deep within about 10 percent. The reader should consult references by Kays and Lo [50] and Kays and London [51] where extensive correlations are presented for heat transfer and friction for the flow of air over banks of $\frac{1}{4}$- and $\frac{3}{8}$-in tubes in various geometrical arrangements. The pressure drop Δp for flow over tube banks may be calculated with the relation

$$\Delta p = 2f\,\frac{G_{max}^2 N}{\rho g_c}\left(\frac{\mu_w}{\mu}\right)^{0.14} \qquad \text{lb}_f/\text{ft}^2 \text{ (or N/m}^2) \qquad (9\text{-}109)$$

TABLE 9-3   Constants c and n of Eq. (9-108)[†]

Arrange-ment	$\dfrac{S_L}{D}$	S_T/D							
		1.25		1.50		2.0		3.0	
		c	n	c	n	c	n	c	n
Stag-gered	0.6							0.213	0.636
	0.9					0.446	0.571	0.401	0.581
	1.0			0.497	0.558				
	1.125					0.478	0.565	0.518	0.560
	1.250	0.518	0.556	0.505	0.554	0.519	0.556	0.522	0.562
	1.50	0.451	0.568	0.460	0.562	0.452	0.568	0.488	0.568
	2.0	0.404	0.572	0.416	0.568	0.482	0.556	0.449	0.570
	3.0	0.310	0.592	0.356	0.580	0.440	0.562	0.421	0.574
In-line	1.25	0.348	0.592	0.275	0.608	0.100	0.704	0.0633	0.752
	1.50	0.367	0.586	0.250	0.620	0.101	0.702	0.0678	0.744
	2.0	0.418	0.570	0.299	0.602	0.229	0.632	0.198	0.648
	3.0	0.290	0.601	0.357	0.584	0.374	0.581	0.286	0.608

† From Grimison [49].

where G_{max} = maximum mass-flow velocity, $lb/ft^2 \cdot s$ $(kg/m^2 \cdot s)$
N = number of rows in direction of flow
g_c = 32.17 $ft \cdot lb/lb_f s^2$ $(1 \ kg \cdot m/N \cdot s^2)$
ρ = density, lb/ft^3 (kg/m^3)
μ, μ_w = viscosity evaluated at bulk-stream and wall-surface conditions, respectively

and the friction factor f is given by Jakob [43] as

$$f = \left[0.25 + \frac{0.118}{(S_T/D - 1)^{1.08}} \right] Re^{-0.16} \qquad \text{for staggered arrangement} \qquad (9\text{-}110a)$$

$$f = \left[0.044 + \frac{0.08(S_L/D)}{(S_T/D - 1)^{0.43 + 1.13(D/S_L)}} \right] Re^{-0.15} \qquad \text{for in-line arrangement}$$

$$(9\text{-}110b)$$

Both are for $5000 < Re < 40{,}000$.

Liquid Metals   The above relations for heat transfer are not applicable to fluids having very low Prandtl number, such as liquid metals.   Hoe et al. [52] and Richards et al. [53] reported experimental data on heat-transfer rates for mercury flowing across staggered-tube banks.   In these experiments the mercury flow was across a tube bank 60 to 70 rows deep and consisting of $\frac{1}{2}$-in tubes arranged in equilateral-triangular array having a pitch-to-diameter ratio of 1.375. The average heat-transfer coefficient for these experiments were correlated by the following relation:

$$\frac{h_m D}{k} = 4.03 + 0.228 \, (Re \, Pr)^{0.67} \qquad (9\text{-}111)$$

where

$$Re = \frac{D G_{max}}{\mu}$$

for $20{,}000 < Re < 80{,}000$.   The physical properties are evaluated at the arithmetic average of the bulk-fluid and wall-surface temperatures.

Kalish and Dwyer [54] give heat-transfer data for NaK flowing through tube bundles.

Example 9-4   Air at atmospheric pressure and at temperature $T_\infty = 100°F$ (37.8°C) flows across a tube bank consisting of $D = \frac{3}{4}$-in (1.9×10^{-2}-m) OD tubes at temperature $T_w = 200°F$ (93.3°C) and in-line arrangement with transverse and longitudinal pitches $S_T = S_L = 2D$.   The bank consists of $L = 2$-ft (0.61-m) long tubes arranged in $N_L = 10$ rows deep in the direction of flow and $N_T = 20$ rows high perpendicular to the flow. The velocity of air just before the air enters the tube bank (i.e., if there were no tubes in the shell) is $u_\infty = 25$ ft/s. (*a*) Determine the pressure drop across the tube bank. (*b*) Determine the average heat-transfer coefficient h_m and the total heat-transfer rate Q from the tubes to the gas.

Solution The physical properties of air at atmospheric pressure and at a mean temperature $(100 + 200)/2 = 150°F$ are:

$c_p = 0.240$ Btu/lb·°F $(1005$ W·s/kg·°C$)$

$\rho = 0.0652$ lb/ft³ $(1.044$ kg/m³$)$

$k = 0.0182$ Btu/h·ft·°F $(0.0315$ W/m·°C$)$

$\mu = 1.37 \times 10^{-5}$ lb/ft·s $(2.04 \times 10^{-5}$ kg/m·s$)$

$Pr = 0.696$

The maximum flow velocity u_{max} for the in-line arrangement considered here is determined by Eq. (9-104):

$$u_{max} = u_\infty \frac{S_T}{S_T - D} = 25 \frac{2}{2 - 1} = 50 \text{ ft/s } (15.24 \text{ m/s})$$

The maximum mass-flow velocity G_{max} becomes

$$G_{max} = \rho u_{max} = 0.0652 \times 50 = 3.26 \text{ lb/ft}^2\cdot\text{s } (15.91 \text{ kg/m}^2\cdot\text{s})$$

and the Reynolds number is determined as

$$Re = \frac{DG_{max}}{\mu} = \frac{\frac{3}{4} \times 3.26}{12 \times (1.3 \times 10^{-5})} \cong 15{,}673$$

(a) The friction factor f is evaluated by Eq. (9-110b)

$$f = \left(0.044 + \frac{0.08 \times 2}{1}\right)(15{,}673^{-0.15}) = 0.0479$$

and the pressure drop Δp is evaluated by Eq. (9-109). Neglecting the variation of viscosity by temperature, Eq. (9-109) is computed as follows:
By using the English system of units, we find

$$\Delta p = 2f \frac{G_{max}^2 N_L}{\rho g_c} = 2 \times 0.0479 \times \frac{3.26^2 \times 10}{0.065 \times 32.2} = 4.86 \text{ lb}_f/\text{ft}^2$$

or, by using the SI system of units, we obtain

$$\Delta p = 2 \times 0.0479 \times \frac{15.91^2 \times 10}{1.044 \times 1} = 232 \text{ N/m}^2$$

(b) The average heat-transfer coefficient h_m can be determined by Eq. (9-108a) with constants c and n obtained from Table 9-3. For the in-line arrangement considered here with $S_T/D = S_L/D = 2$, from Table 9-3 we have $c = 0.229$ and $n = 0.632$. Then Eq. (9-108a) becomes

$$\frac{h_m D}{k} = 0.229 Re^{0.632}$$

$$\frac{h_m \frac{3}{4}}{0.0182 \times 12} = 0.229 \times 15{,}673^{0.632}$$

$$h_m = 29.5 \text{ Btu/h·ft}^2\cdot°F \ (169.7 \text{ W/m}^2\cdot°C)$$

and the total surface area A for heat transfer is $A = \pi DL$ (number of tubes) $= \pi[3/(4 \times 12)] \times 2 \times (20 \times 10) = 78.5$ ft².

To calculate the total heat-transfer rate Q, the temperature difference between the tube-surface and the mean gas temperatures is needed; but this is not known because the temperature rise ΔT of air through the bank is not known. Therefore, a *first approximation* to $Q^{(1)}$ is obtained by assuming that free-stream temperature remains constant throughout the bank and is taken as T_∞. Then

$$Q^{(1)} = Ah_m(T_w - T_\infty)$$

$$Q^{(1)} = 78.5 \times 29.4 \times (200 - 100) \cong 234{,}750 \text{ Btu/h } (68{,}795 \text{ W})$$

To improve the approximation, the above value of $Q^{(1)}$ is used to determine the temperature rise ΔT of air through the bank by writing the energy balance

$$Q^{(1)} = \rho u_\infty c_p A_\infty \Delta T$$

where $\quad A_\infty = $ flow area without tubes $= LN_T S_T = 2 \times 20 \times \dfrac{1.5}{12} = 5 \text{ ft}^2$

$\quad u_\infty = $ air velocity without tubes $= 25 \times 3600 \text{ ft/h}$

Then,

$$199{,}390 = 0.0652 \times (25 \times 3600) \times 0.240 \times 5 \times \Delta T$$

$$\Delta T = 28.3°\text{F}$$

This value of ΔT is now included in the analysis to obtain a better approximation to the total heat-transfer rate Q as

$$Q = Ah_m\left[T_w - \left(T_\infty + \frac{\Delta T}{2}\right)\right]$$

$$Q = 78.5 \times 29.9 \times \left[200 - \left(100 + \frac{28.3}{2}\right)\right] = 210{,}503 \text{ Btu/h } (56{,}060 \text{ W})$$

For better accuracy the logarithmic-mean-temperature-difference concept that will be discussed in Chap. 15 on heat exchangers should be used to determine the temperature difference between the wall-surface and gas temperatures, and the effects of ΔT on the heat-transfer coefficient h_m should also be considered.

9-7 HEAT TRANSFER IN HIGH-SPEED TURBULENT BOUNDARY-LAYER FLOW ALONG A FLAT PLATE

At flow velocities approaching or exceeding the velocity of sound encountered in applications such as high-speed aircraft, missiles, and reentry vehicles, the effects of compressibility, viscous dissipation, and property variation with temperature become important. The general analysis of such problems is very involved. On the other hand, the heat-transfer rate in high-speed turbulent flow along a flat plate at a uniform temperature T_w can be predicted, for most practical purposes, by using the low-speed heat-transfer coefficient h_x in conjunction with the temperature difference $(T_w - T_{aw})$ as discussed previously in Sec. 8-5. That is, the local heat flux q_x at the wall surface is computed from the relation

$$q_x = h_x(T_w - T_{aw}) \qquad (9\text{-}112)$$

where T_w is the wall temperature, h_x is the local heat-transfer coefficient for turbulent flow of an incompressible, constant-property fluid along a plate, and T_{aw} is the adiabatic wall temperature. The local heat-transfer coefficient h_x for moderate Prandtl numbers and for Reynolds numbers in the range $5 \times 10^5 < \mathrm{Re}_x < 10^7$ can be determined from the relation [55]

$$\frac{h_x}{c_p \rho u_\infty} = 0.0296 \mathrm{Re}_x^{-1/5} \, \mathrm{Pr}^{-2/3} \qquad (9\text{-}113)$$

and for Reynolds numbers in the range $10^7 < \mathrm{Re}_x < 10^9$ from the relation

$$\frac{h_x}{c_p \rho u_\infty} = 0.185(\log \mathrm{Re}_x)^{-2.584} \, \mathrm{Pr}^{-2/3} \qquad (9\text{-}114)$$

The adiabatic wall temperature T_{aw}, as discussed in Sec. 8-5, is determined from the relation

$$T_{aw} = T_\infty + r \frac{u^2}{2c_p g_c J} \qquad (9\text{-}115)$$

and T_∞ and u_∞ are the external-flow temperature and velocity, respectively, and the recovery factor r for turbulent boundary layer along a flat plate is determined from [56]

$$r \cong \mathrm{Pr}^{1/3} \qquad (9\text{-}116)$$

The effects of variation of properties of fluid with temperature can be included in the analysis in an approximate way if the properties of the fluid are evaluated at a reference temperature T_r given as [57,58]

$$T_r = T_\infty + 0.5(T_w - T_\infty) + 0.22(T_{aw} - T_\infty) \qquad (9\text{-}117)$$

The reader should consult Refs. [59,60] for a discussion of high-speed heat transfer problems.

REFERENCES

1　Reynolds, O.: On the Experimental Investigation of the Circumstances Which Determine Whether the Motion of Water Shall be Direct or Sinuous, and the Law of Resistance in Parallel Channels, *Philos. Trans. R. Soc. London, Ser. A*, **174**:935 (1883).

2　Burgess, J. M.: The Motion of Fluid in the Boundary Layer Along a Plane Smooth Surface, *Proc. First Int. Congr. Appl. Mech., Delf*, 1924.

3　Prandtl, L.: *Z. Angew. Math. Mech.*, **5**:136 (1925); *NACA Tech. Memo.* 1231, 1949; *Proc. Second Int. Congr. Appl. Mech., Zurich*, pp. 62–75, 1926.

4　Von Kármán, Th.: *Nach. Ges. Wiss. Göttingen, Math.-Phys. Klasse*, p. 58, 1930; *NACA Tech. Memo.* 611, 1931.

5　Schlichting, H.: "Boundary Layer Theory," 6th ed., McGraw-Hill Book Company, New York, 1968.

6　Hinze, J. O.: "Turbulence," McGraw-Hill Book Company, New York, 1959.

7　Cebeci, T., and A. M. O. Smith: "Analysis of Turbulent Boundary Layers," Academic Press, Inc., New York, 1974.

8　Launder, B. E., and D. B. Spalding: "Mathematical Models of Turbulence," Academic Press, Inc., New York, 1972.

9　Launder, B. E., and D. B. Spalding: *Heat and Fluid Flow*, **2**:43–54 (1972).

10a　Nikuradse, J.: *Forsch. Arb. Ing. Wes.*, no. 346, 1932.

10b　Nikuradse, J.: *Forsch. Arb. Ing. Wes.*, no. 361, 1933; also, Laws of Flow in Rough Pipes (translation), *NACA Tech. Memo.* 1292, 1950.

11　Wieghardt, K.: Zum Reibungswiderstand rauher Platten, *Kaiser-Wilhelm-Institut fur Stromunsforschung*, Göttingen, UM-6612, 1944.

12　Klebanoff, P. S.: Characteristics of Turbulence in a Boundary Layer with Zero Pressure Gradient, *NACA Tech. Note* 3178, 1954.

13　Laufer, J.: The Structure of Turbulence in Fully Developed Pipe Flow, *NACA Tech. Note* 1174, 1954.

14　Coles, D.: The Law of Wake in the Turbulent Boundary Layers, *J. Fluid Mech.*, **1**:191–226 (1956).

15　Spalding, D. B.: Heat Transfer to a Turbulent Stream from a Surface with a Step-wise Discontinuity in Wall Temperature, *Conf. Int. Dev. Heat Transfer, ASME*, Boulder, Colo., pt. II, pp. 439–446, 1961.

16　Moody, L. F.: Friction Factor for Pipe Flow, *Trans. ASME*, **66**:671–684 (1944).

17　Prandtl, L.: *Z. Phys.*, **11**:1072 (1910).

18　Von Kármán, Th.: The Analogy Between Fluid Friction and Heat Transfer, *Trans. ASME*, **61**:705–711 (1939).

19　Martinelli, R. C.: Heat Transfer in Molten Metals, *Trans. ASME*, **69**:947–959 (1947).

20a　Deissler, R. G.: Investigation of Turbulent Flow and Heat Transfer in Smooth Tubes Including the Effects of Variable Properties, *Trans. ASME*, **73**:101 (1951).

20b　Deissler, R. G.: Analysis of Turbulent Heat Transfer, Mass Transfer and Friction in Smooth Tubes at High Prandtl and Schmidt Numbers, *NACA Tech. Note* 3145, 1954.

21　Schultz-Grunow, F.: *NACA Tech. Memo.* 986, 1941.

22　Dittus, F. W., and L. M. K. Boelter: *Univ. Calif., Berkeley, Publ. Eng.*, **2**:443 (1930).

23　Colburn, A. P.: A Method of Correlating Forced Convection Heat Transfer Data and a Comparison with Liquid Frictions, *Trans. AIChE*, **29**:174–210 (1933).

24　Sieder, E. N., and G. E. Tate: Heat Transfer and Pressure Drop of Liquids in Tubes, *Ind. Eng. Chem.*, **28**:1429 (1936).

25 Nusselt, W.: Der Warmeaustausch Zwischen Wand und Wasser im Rohr, *Forsch. Geb. Ingenieurwes.*, **2**:309 (1931).

26 Hartnett, J. P.: Experimental Determination of the Thermal Entrance Length for the Flow of Water and Oil in Circular Pipes, *Trans. ASME*, **77**:1211–1220 (1955).

27 Irvine, T. R.: Noncircular Convective Heat Transfer, in W. Ible (ed.), "Modern Developments in Heat Transfer," Academic Press, Inc., New York, 1963.

28 Lubarsky, B., and S. J. Kaufman: Review of Experimental Investigation of Liquid Metal Heat Transfer, *NACA Tech. Note* 3336, 1955.

29 Skupinski, E. S., J. Tortel, and L. Vautrey: Determination des coefficients de convection d'un alliage sodium-potassium dans un tube circulaire, *Int. J. Heat Mass Transfer*, **8**:937 (1965).

30 Seban, R. A., and T. T. Shimazaki: Heat Transfer to Fluid Flowing Turbulently in a Smooth Pipe with Walls of Constant Temperature, *Trans. ASME*, **73**:803–808 (1951).

31 Knudsen, J. G., and D. L. Katz: "Fluid Dynamics and Heat Transfer," McGraw-Hill Book Company, New York, 1958.

32 Sleicher, C. A., Jr., and M. Tribus: Heat Transfer in a Pipe with Turbulent Flow and Arbitrary Wall-Temperature Distribution, *Trans. ASME*, **79**:789–797 (1957).

33 Azer, N. Z., and B. T. Chao: Turbulent Heat Transfer in Liquid Metals—Fully Developed Pipe Flow with Constant Wall Temperature, *Int. J. Heat Mass Transfer*, **3**:77–83 (1961).

34 Notter, R. H., and C. A. Sleicher: A Solution to the Turbulent Graetz Problem. III. Fully Developed and Entry Region Heat Transfer Rates, *Chem. Eng. Sci.*, **27**:2073–2093 (1972).

35 Sleicher, C. A., A. S. Awad, and R. H. Notter: Temperature and Eddy Diffusivity Profiles in NaK, *Int. J. Heat Mass Transfer*, **16**:1565–1575 (1973).

36 Lyon, R. D. (ed.): "Liquid Metals Handbook," 3d ed., U.S. Atomic Energy Commission and Department of the Navy, Washington, D.C., 1952.

37 Stein, R.: Liquid Metal Heat Transfer, *Adv. Heat Transfer*, **3**: (1966).

38 McAdams, W. H.: "Heat Transmission," 3d ed., McGraw-Hill Book Company, New York, 1954.

39 Fand, R. M.: Heat Transfer by Forced Convection from a Cylinder to Water in Crossflow, *Int. J. Heat Mass Transfer*, **8**:995–1010 (1965).

40 Whitaker, S.: Forced Convection Heat Transfer Calculations for Flow in Pipes, Past Flat Plates, Single Cylinders, and for Flow in Packed Beds and Tube Bundles, *AIChE J.*, **18**:361–371 (1972).

41 Eckert, E. R. G., and E. Soehngen: Distributions of Heat-Transfer Coefficients Around Circular Cylinders in Cross Flow at Reynolds Numbers from 20 to 500, *Trans ASME*, **74**:343–347 (1952).

42 Giedt, W. H.: Investigation of Variation of Point Unit-Heat-Transfer Coefficient Around a Cylinder Normal to an Air Stream, *Trans. ASME*, **71**:375–381 (1949).

43 Jakob, Max: "Heat Transfer," vol. 1, John Wiley & Sons, Inc., New York, 1949.

44 Kramers, H.: Heat Transfer from Spheres to Flowing Media, *Physica*, **12**:61–80 (1946).

45 Vliet, G. C., and G. Leppert: Forced Convection Heat Transfer from an Isothermal Sphere to Water, *J. Heat Transfer*, **83C**:163–175 (1961).

46 Witte, L. C.: An Experimental Study of Forced-Convection Heat Transfer from a Sphere to Liquid Sodium, *J. Heat Transfer*, **90C**:9–12 (1968).

47 Bergelin, O. P., G. A. Brown, and S. C. Doberstein: Heat Transfer and Fluid Friction During Flow Across Banks of Tubes, *Trans. ASME*, **74**:953–960 (1952).

48 Meece, W. E.: "The Effect of the Number of Tube Rows Upon Heat Transfer and Pressure Drop During Viscous Flow Across In-Line Tube Banks," M.S. thesis, University of Delaware, Newark, Del., 1949.

49 Grimison, E. D.: Correlation and Utilization of New Data on Flow Resistance and Heat Transfer for Cross Flow of Gases Over Tube Banks, *Trans. ASME*, **59**:583–594 (1937).

50 Kays, W. M., and R. K. Lo: Basic Heat Transfer and Fluid Friction Data for Gas Flow Normal to Banks of Staggered Tubes—Use of Transient Technique, *Stanford Univ. Dep. Mech. Eng., Tech. Rep.* 15, Navy Contract NG-ONR-251, T.O. 6, 1952.

51 Kays, W. M., and A. L. London: "Compact Heat Exchanger Design," McGraw-Hill Book Company, New York, 1958.

52 Hoe, R. J., D. Dropkin, and O. E. Dwyer: Heat Transfer Rates to Cross Flowing Mercury in a Staggered Tube Bank, I, *Trans. ASME*, **79**:899–908 (1957).

53 Richards, C. L., O. E. Dwyer, and D. Dropkin: Heat Transfer Rates to Cross Flowing Mercury in a Staggered Tube Bank, II, *ASME-AIChE Heat Transfer Conf. Paper* 57-HT-11, 1957.

54 Kalish, S., and O. E. Dwyer: Heat Transfer to NaK Flowing Through Unbuffled Rod Bundles, *Int. J. Heat Mass Transfer*, **10**:1533–1558 (1967).

55 Johnson, H. A., and M. W. Rubesin: Aerodynamic Heating and Convective Heat Transfer—Summary of Literature Survey, *Trans. ASME*, **71**:447–456 (1949).

56 Kaye, J.: Survey of Friction Coefficients, Recovery Factors, and Heat Transfer Coefficients for Supersonic Flow, *J. Appl. Sci.*, **21**:117–119 (1954).

57 Eckert, E. R. G.: Engineering Relations for Friction and Heat Transfer to Surfaces in High Velocity Flow, *J. Appl. Sci.*, **22**:585–587 (1955).

58 Eckert, E. R. G.: Engineering Relations for Heat Transfer and Friction in High-Velocity Laminar and Turbulent Boundary Layer Flow Over a Surface with Constant Pressure and Temperature, *Trans. ASME*, **78**:1273–1284 (1956).

59 Lin, C. C. (ed.): Turbulent Flows and Heat Transfer, in "High Speed Aerodynamics and Jet Propulsion," vol. 5, Princeton University Press, Princeton, N.J., 1954.

60 Eckert, E. R. G.: Survey of Boundary Layer Heat Transfer at High Velocities and High Temperatures, *WADC Tech. Rep.* 59–624, 1960.

NOTES

1 The left-hand side of Eq. (9-8a) is combined with the continuity equation (9-5) in the following manner:

$$u_i \frac{\partial u_i}{\partial x} + v_i \frac{\partial u_i}{\partial y} = \left(u_i \frac{\partial u_i}{\partial x} + v_i \frac{\partial u_i}{\partial y} \right) + u_i \left(\frac{\partial u_i}{\partial x} + \frac{\partial v_i}{\partial y} \right)$$

$$= 2u_i \frac{\partial u_i}{\partial x} + \left(v_i \frac{\partial u_i}{\partial y} + u_i \frac{\partial v_i}{\partial y} \right)$$

$$= \frac{\partial u_i^2}{\partial x} + \frac{\partial}{\partial y} \left(u_i v_i \right)$$

which is the same as the left-hand side of Eq. (9-8b).

2 The left-hand side of Eq. (9-11a) is combined with the continuity equation (9-5) in the following manner:

$$u_i \frac{\partial T_i}{\partial x} + v_i \frac{\partial T_i}{\partial y} = \left(u_i \frac{\partial T_i}{\partial x} + v_i \frac{\partial T_i}{\partial y} \right) + T_i \left(\frac{\partial u_i}{\partial x} + \frac{\partial v_i}{\partial y} \right)$$

$$= \left(u_i \frac{\partial T_i}{\partial x} + T_i \frac{\partial u_i}{\partial x} \right) + \left(v_i \frac{\partial T_i}{\partial y} + T_i \frac{\partial v_i}{\partial y} \right)$$

$$= \frac{\partial}{\partial x} (u_i T_i) + \frac{\partial}{\partial y} (v_i T_i)$$

which is the left-hand side of Eq. (9-11b).

3 The mean velocity u_m over the pipe cross section is given by

$$u_m = \frac{\int_0^R 2\pi r u(r)\, dr}{\int_0^R 2\pi r\, dr} = \frac{2}{R^2} \int_0^R r u(r)\, dr \tag{1}$$

where r is the radial variable and R is the tube radius. If y is the space variable measured from the tube wall, it is related to r by

$$y = R - r \tag{2a}$$

Hence,

$$dy = -dr \tag{2b}$$

By changing the variable from r to y, Eq. (1) becomes

$$u_m = \frac{2}{R^2} \int_0^R (R - y) u(y)\, dy \tag{3}$$

The logarithmic velocity distribution for turbulent flow is given by

$$u^+ = 2.5 \ln y^+ + 5.5 \tag{4}$$

and the velocity at the tube center $u^+ \equiv u_{max}^+$ is obtained from Eq. (4) by setting $y^+ = R^+$:

$$u_{max}^+ = 2.5 \ln R^+ + 5.5 \tag{5}$$

where

$$u^+ = \frac{u}{\sqrt{\tau_0/\rho}} \qquad y^+ = \frac{y}{\nu} \sqrt{\frac{\tau_0}{\rho}} \qquad R^+ = \frac{R}{\nu} \sqrt{\frac{\tau_0}{\rho}} \tag{6}$$

By combining Eqs. (4) and (5) we find

$$u_{max}^+ - u^+ = 2.5 \ln \frac{R^+}{y^+}$$

or

$$\frac{u_{max} - u(y)}{\sqrt{\tau_0/\rho}} = 2.5 \ln \frac{R}{y} \tag{7}$$

The substitution of $u(y)$ from Eq. (7) into Eq. (1) and performing the integration give

$$u_m = u_{max} - 3.75 \sqrt{\frac{\tau_0}{\rho}}$$

or

$$u_m^+ = u_{max}^+ - 3.75 \tag{8}$$

The elimination of u_{max}^+ between Eqs. (5) and (8) gives

$$u_m^+ = 2.5 \ln R^+ + 1.75$$

or

$$u_m^+ = 2.5 \ln \left(\frac{R}{\nu} \sqrt{\frac{\tau_0}{\rho}} \right) + 1.75 \tag{9}$$

which is the result given by Eq. (9-47).

Ten

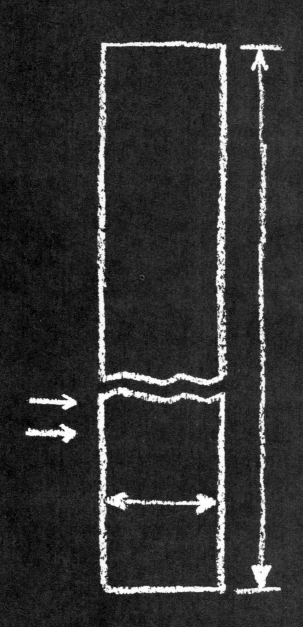

Free Convection

If fluid motion is generated predominantly by body forces caused by density variation in the fluid resulting from temperature gradients, the mechanism of heat transfer is called *free* or *natural convection*.   For example, when a hot object is placed in a fluid which is at rest at a uniform temperature less than that of the hot object, a temperature gradient is established within the fluid; as the density of the fluid increases with distance from the object, a velocity field is set up within the fluid as a result of the buoyancy forces.   Energy transfer by free convection arises in many engineering applications, such as heat transfer from steam radiators, refrigeration coils, transmission lines, electric transformers, heating elements, etc. Free-convection currents can also be set up at high-speed rotation if the temperature gradients decrease the density in the direction of the centrifugal force, because the magnitude of the body force due to centrifugal effects is proportional to the fluid density.   Thus free-convection heat transfer is possible in high-speed rotations with temperature gradients in the fluid.   In this chapter we examine energy transfer by free convection for situations in which convection currents are set up by the buoyancy forces.   The general analysis of heat transfer in free convection is a complicated matter.   However, to provide some insight into the physical significance of various factors affecting heat transfer in free convection, the dimensionless parameters of free convection are discussed and the integral method of solution is presented for laminar free convection from a vertical plate.   In many engineering applications, however, the geometry and the boundary conditions are complicated, and the flow is in the turbulent region.   For such situations an analytical method of solution becomes extremely tedious or impractical; hence the experimental approach is the only alternative to determine the heat-transfer coefficient.   Various empirical and semiempirical correlations are presented for free convection from typical geometries.

10-1   BOUNDARY-LAYER EQUATIONS OF FREE CONVECTION

Free convection is caused by density changes resulting from temperature gradients in the fluid, and the physical situation implies that the mathematical formulation of the heat-transfer problem should be based on the equations for a compressible fluid.   However, when the temperature difference between the surface of the body and the fluid is not large, the analysis can be simplified significantly by introducing the assumptions for an incompressible, constant-property fluid, while the effect of the density change resulting from temperature gradients is included in the body-force term in the momentum equation.   With this consideration, the equations of motion and energy derived in Chap. 6 are applicable for the analysis of free-convection problems, provided that an appropriate body-force term is included in the momentum equation.   For simplicity in the analysis, we consider the steady, laminar free convection from a vertical plate as illustrated in Fig. 10-1 and derive the dimensionless parameters directly from the governing differential equations.   Figure 10-1a and b shows the selection of the coordinate

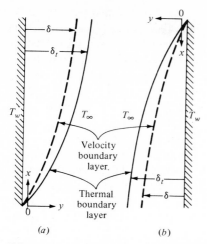

FIG. 10-1   Coordinates for boundary-layer equations for free convection from a hot and a cold vertical plate.

system for the cases of hot- and cold-plate conditions, respectively. We focus our attention on the situation shown in Fig. 10-1a in which a hot vertical plate at a uniform temperature T_w is immersed in a quiescent fluid (that is, $u_\infty = 0$) at a constant temperature $T_\infty(T_w > T_\infty)$. The body forces resulting from the buoyancy give rise to free-convection currents upward along the surface of the plate; hence a velocity and a thermal boundary layer are established as illustrated in Fig. 10-1a.   In the analysis of free convection the flow behaves as incompressible flow and thus density is considered constant in the governing equations except in the body-force term resulting from the buoyancy.   For free convection from a flat plate considered here the boundary-layer equations are applicable. Assuming that the boundary layer is laminar, the governing equations of continuity momentum, and energy are obtained from Eqs. (6-65) to (6-67), respectively, by introducing an appropriate body-force term into the momentum equation and neglecting the viscous-dissipation term in the energy equation.   We obtain

Continuity: $$\frac{\partial u}{\partial x} + \frac{\partial v}{\partial y} = 0 \qquad\qquad (10\text{-}1)$$

x Momentum: $$\rho\left(u\frac{\partial u}{\partial x} + v\frac{\partial u}{\partial y}\right) = -\rho g - \frac{\partial P}{\partial x} + \mu\frac{\partial^2 u}{\partial y^2} \qquad (10\text{-}2)$$

Energy: $$\rho c_p\left(u\frac{\partial T}{\partial x} + v\frac{\partial T}{\partial y}\right) = k\frac{\partial^2 T}{\partial y^2} \qquad (10\text{-}3)$$

Here the term $-\rho g$ on the right-hand side of the momentum equation represents the body force exerted on the fluid element in the negative x direction. *For small temperature differences the density ρ in the buoyancy term is considered to vary with temperature* whereas the density appearing elsewhere in these equations is considered constant. The pressure-gradient term $-\partial P/\partial x$ in the x direction results from the change in elevation; it is not zero. The viscous-dissipation term in the energy equation is omitted because the flow velocity is very small in free convection. To determine the pressure-gradient term the momentum equation is evaluated at the edge of the boundary layer where $\rho \to \rho_\infty$ and $u \to 0$. We obtain

$$\frac{\partial P}{\partial x} = -\rho_\infty g \tag{10-4}$$

where ρ_∞ is the fluid density outside the boundary layer. Then the term $-\rho g - \partial P/\partial x$ appearing in the momentum equation becomes

$$-\rho g - \frac{\partial P}{\partial x} = (\rho_\infty - \rho)g \tag{10-5}$$

If β is the *volumetric coefficient of thermal expansion* of the fluid, the change of density with temperature is related to β by

$$-\frac{1}{\rho}\left(\frac{\partial \rho}{\partial T}\right)_p = \beta \tag{10-6}$$

By expressing the derivative term in this relation with finite difference, Eq. (10-6) is approximated by

$$\Delta \rho = -\beta \rho \, \Delta T$$

or

$$\rho_\infty - \rho = -\beta \rho (T_\infty - T) \tag{10-7}$$

Then Eq. (10-5) becomes

$$-\rho g - \frac{\partial P}{\partial x} = -\beta \rho (T_\infty - T) \tag{10-8}$$

Equation (10-8) is now substituted into the momentum equation (10-2).

We now summarize the resulting equations for free convection from a vertical plate:

$$\frac{\partial u}{\partial x} + \frac{\partial v}{\partial y} = 0 \tag{10-9}$$

$$u \frac{\partial u}{\partial x} + v \frac{\partial u}{\partial y} = g\beta(T - T_\infty) + v \frac{\partial^2 u}{\partial y^2} \tag{10-10}$$

$$u \frac{\partial T}{\partial x} + v \frac{\partial T}{\partial y} = \alpha \frac{\partial^2 T}{\partial y^2} \tag{10-11}$$

If the fluid is considered to be an ideal gas we have

$$\rho = \frac{P}{\mathscr{R}T} \qquad (10\text{-}12)$$

Then the coefficient of expansion β in Eq. (10-7) becomes

$$\beta = \frac{\rho_\infty/\rho - 1}{T - T_\infty} = \frac{T/T_\infty - 1}{T - T_\infty} = \frac{1}{T_\infty} \qquad (10\text{-}13)$$

An examination of Eqs. (10-9) to (10-11) reveals that, for free convection, the continuity and momentum equations cannot be solved independently of the energy equation because the buoyancy term $g\beta(T - T_\infty)$ couples the momentum equation to the energy equation. Thus these three equations must be solved simultaneously. Without the buoyancy term these equations are identical to the boundary-layer equations for forced convection over a flat plate.

To determine the dimensionless groups for heat transfer in free convection, we now transform Eqs. (10-9) to (10-11) into dimensionless form by introducing the following dimensionless parameters:

$$X = \frac{x}{L} \qquad Y = \frac{y}{L} \qquad U = \frac{u}{U_0} \qquad V = \frac{v}{U_0} \qquad \theta = \frac{T - T_\infty}{T_w - T_\infty} \qquad (10\text{-}14)$$

where L is a characteristic length, U_0 is a reference velocity, T_w is the wall-surface temperature, and T_∞ is the fluid temperature at distances away from the hot plate. When these new variables are introduced into Eqs. (10-9) to (10-11), the resulting nondimensional equations become

$$\frac{\partial U}{\partial X} + \frac{\partial V}{\partial Y} = 0 \qquad (10\text{-}15)$$

$$U\frac{\partial U}{\partial X} + V\frac{\partial U}{\partial Y} = \frac{g\beta(T_w - T_\infty)L}{U_0{}^2}\theta + \frac{1}{\mathrm{Re}}\frac{\partial^2 U}{\partial Y^2} \qquad (10\text{-}16)$$

$$U\frac{\partial \theta}{\partial X} + V\frac{\partial \theta}{\partial Y} = \frac{1}{\mathrm{Re}\,\mathrm{Pr}}\frac{\partial^2 \theta}{\partial Y^2} \qquad (10\text{-}17)$$

Here the Reynolds and Prandtl numbers are defined as

$$\mathrm{Re} = \frac{U_0 L}{\nu} \qquad \mathrm{Pr} = \frac{\nu}{\alpha} \qquad (10\text{-}18)$$

The dimensionless group in the momentum equation can be rearranged as

$$\frac{g\beta(T_w - T_\infty)L}{U_0{}^2} = \frac{g\beta L^3(T_w - T_\infty)/\nu^2}{(LU_0/\nu)^2} \equiv \frac{\mathrm{Gr}}{\mathrm{Re}^2} \qquad (10\text{-}19)$$

where the *Grashof number* Gr is defined as

$$Gr = \frac{g\beta L^3(T_w - T_\infty)}{\nu^2} \tag{10-20}$$

As apparent from Eq. (10-19), the dimensionless group $g\beta(T_w - T_\infty)L/U_0^2$ is a ratio of the *buoyancy forces* to *inertial forces*.   In free convection, the buoyant forces are the only driving forces that generate the flow field because there is no external-flow field.   For such a case the Reynolds number appearing in the above equations cannot be an independent parameter because no external-flow velocity exists.   Then the Nusselt number characterizing the heat-transfer coefficient for free convection from the fluid to the surface is a function of the Prandtl and Grashof numbers and can be correlated by an equation of the form

$$Nu = f(Gr, Pr) \tag{10-21a}$$

For gases, $Pr \cong 1$; hence the Nusselt number is a function of the Grashof number only:

$$Nu = f(Gr) \qquad \text{for gases} \tag{10-21b}$$

Sometimes another dimensionless parameter called the *Rayleigh number*, Ra, is defined as

$$Ra = Gr \; Pr = \frac{g\beta L^3(T_w - T_\infty)}{\nu\alpha} \tag{10-22}$$

and is used instead of the Grashof number to correlate heat transfer in free convection.

In the foregoing formulation Eq. (10-9) to (10-11) are for free convection from a hot vertical plate with x, y coordinates chosen as shown in Fig. 10-1a.   The same equations are applicable for the case of free convection from a cold vertical plate (that is, $T_w < T_\infty$) if x, y coordinates are chosen as shown in Fig. 10-1b. This is apparent from the fact that, in Fig. 10-1a, g acts in the negative x direction; thus the product $g\beta(T - T_\infty)$ is a negative quantity because $T > T_\infty$ for the hot plate; in Fig. 10-1b, g acts in the positive x direction but the product $g\beta(T - T_\infty)$ is still a negative quantity because $T < T_\infty$ for the cold plate.

10-2   APPROXIMATE SOLUTION OF LAMINAR FREE CONVECTION FROM A VERTICAL PLATE

Consider that a vertical plate at a uniform temperature T_w is immersed in a large body of quiescent fluid at a uniform temperature T_∞.   The governing equations of motion and energy for free convection are given by Eqs. (10-9) to (10-11):

$$\frac{\partial u}{\partial x} + \frac{\partial v}{\partial y} = 0 \tag{10-9}$$

$$u\frac{\partial u}{\partial x} + v\frac{\partial u}{\partial y} = g\beta(T - T_\infty) + v\frac{\partial^2 u}{\partial y^2} \tag{10-10}$$

$$u\frac{\partial T}{\partial x} + v\frac{\partial T}{\partial y} = \alpha\frac{\partial^2 T}{\partial y^2} \tag{10-11}$$

The boundary conditions are taken as

$$u = 0 \qquad v = 0 \qquad T = T_w \qquad \text{at } y = 0 \text{ (wall)}$$

$$u = 0 \qquad T = T_\infty \qquad \text{as } y \to \infty \text{ (or outside boundary layers)} \tag{10-23}$$

The integral method described in Chap. 8 will now be used to solve this heat-transfer problem approximately. To simplify the analysis it is assumed that the Prandtl number is close to unity so that the thicknesses of the velocity and the thermal boundary layers are the same, that is, $\delta = \delta_t$. In this method the momentum equation (10-10) is integrated over the boundary-layer thickness δ, the velocity component v appearing in the resulting equation is eliminated by means of the continuity equation (10-9), and the boundary conditions on u as given above are utilized. Then the *momentum integral equation* becomes

$$\frac{d}{dx}\left(\int_0^\delta u^2\, dy\right) = -v\frac{\partial u}{\partial y}\bigg|_{y=0} + g\beta\int_0^\delta (T - T_\infty)\, dy \tag{10-24}$$

The energy equation (10-11) is integrated over the boundary-layer thickness, the velocity component v appearing in the resulting expression is eliminated by means of the continuity equation (10-9), and the boundary condition for T as given above is utilized to obtain the *energy integral equation* in the form

$$\frac{d}{dx}\left[\int_0^\delta u(T - T_\infty)\, dy\right] = -\alpha\frac{\partial T}{\partial y}\bigg|_{y=0} \tag{10-25}$$

We note that Eqs. (10-24) and (10-25) are coupled because temperature appears in the momentum integral equation; hence they must be solved simultaneously.

To solve these equations, suitable profiles must be chosen to represent the velocity and temperature distributions in the boundary layer. For free convection from a vertical hot plate the physical nature of the problem is such that the velocity and temperature profiles in the boundary layers are similar to those illustrated in Fig. 10-2. A polynomial representation is used to approximate these profiles within the boundary layer, as now described.

A second-degree polynomial is chosen for the temperature profile in the form

$$T(x, y) = a_0 + b_0 y + c_0 y^2 \tag{10-26}$$

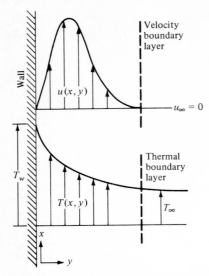

FIG. 10-2 Temperature and velocity profiles for free convection from a hot vertical plate.

where the coefficients a_0, b_0, and c_0 can be determined from the following three conditions:

$$T = T_w \qquad \text{at } y = 0 \tag{10-27}$$

$$T = T_\infty \qquad \text{at } y = \delta \tag{10-28}$$

$$\frac{\partial T}{\partial y} = 0 \qquad \text{at } y = \delta \tag{10-29}$$

Here we assume $\delta = \delta_t$; the resulting temperature profile becomes

$$\frac{T(x, y) - T_\infty}{T_w - T_\infty} = \left(1 - \frac{y}{\delta}\right)^2 \tag{10-30}$$

The velocity profile as illustrated in Fig. 10-2 implies that the velocity component u is zero both at the wall and at the edge of the velocity boundary layer but has a peak inside the boundary layer. A cubic polynomial is chosen to represent this velocity profile in the form

$$u(x, y) = u_0'(a_1 + b_1 y + c_1 y^2 + d_1 y^3) \tag{10-31}$$

where the coefficients a_1, b_1, c_1, d_1 and a reference velocity u_0' are considered to be functions of x and are yet to be determined. Four conditions are needed on the velocity to determine the four coefficients. Three of these conditions are taken as

$$u = 0 \quad \text{at } y = 0 \tag{10-32}$$

$$u = 0 \quad \text{at } y = \delta \tag{10-33}$$

$$\frac{\partial u}{\partial y} = 0 \quad \text{at } y = \delta \tag{10-34}$$

where δ is the edge of the velocity boundary layer. A fourth condition is obtained by evaluating the momentum equation (10-10) at $y = 0$ and noting that $u = v = 0$ and $T = T_w$ at $y = 0$. We find

$$\frac{\partial^2 u}{\partial y^2} = -\frac{g\beta}{\nu}(T_w - T_\infty) \tag{10-35}$$

With the application of these four conditions given by Eqs. (10-32) to (10-35), the velocity profile Eq. (10-31) becomes

$$u(x, y) = \left[u_0' \frac{\beta \delta^2 g(T_w - T_\infty)}{4\nu} \right] \frac{y}{\delta} \left(1 - \frac{y}{\delta}\right)^2 \tag{10-36}$$

which can be written more compactly

$$u(x, y) = u_0 \frac{y}{\delta} \left(1 - \frac{y}{\delta}\right)^2 \tag{10-37}$$

where $u_0 \equiv u_0(x) = u_0' \beta \delta^2 g(T_w - T_\infty)/4\nu$ is an arbitrary function of x with the dimension of velocity. By differentiating Eq. (10-37) with respect to y it can be shown that the maximum value of $u(x, y)$ occurs at a distance $y = \delta/3$.

The temperature profile Eq. (10-30) and the velocity profile Eq. (10-37) are introduced into the momentum integral equation (10-24) and the energy integral equation (10-25), and the indicated operations are performed. The momentum and the energy integral equations become, respectively,

$$\frac{1}{105} \frac{d}{dx} (u_0^2 \delta) = \frac{1}{3} g\beta(T_w - T_\infty)\delta - \nu \frac{u_0}{\delta} \tag{10-38}$$

$$\frac{1}{30} (T_w - T_\infty) \frac{d}{dx} (u_0 \delta) = 2\alpha \frac{T_w - T_\infty}{\delta} \tag{10-39}$$

which are two coupled ordinary differential equations for the two unknowns

$u_0 \equiv u_0(x)$ and $\delta = \delta(x)$; they can be solved as described in note 1 at the end of this chapter. The solution gives the boundary-layer thickness as (*see note 1*)

$$\delta(x) = 3.93(0.952 + Pr)^{1/4} \left[\frac{g\beta(T_w - T_\infty)}{v^2} \right]^{-1/4} Pr^{-1/2} x^{1/4} \tag{10-40}$$

or

$$\frac{\delta(x)}{x} = 3.93 Pr^{-1/2}(0.952 + Pr)^{1/4} Gr_x^{-1/4} \tag{10-41}$$

where the local Grashof number Gr_x is defined as

$$Gr_x = \frac{g\beta(T_w - T_\infty)x^3}{v^2} \tag{10-42}$$

Since we assumed $\delta = \delta_t$, Eq. (10-41) also gives the thickness of the thermal boundary layer.

The local Nusselt number Nu_x characterizing the local heat-transfer coefficient h_x at the plate surface is of interest in free-convection heat transfer. It is defined as

$$Nu_x \equiv \frac{h_x x}{k} = \frac{q_w}{T_w - T_\infty} \frac{x}{k} = \frac{-k(\partial T/\partial y |_{y=0}) x}{T_w - T_\infty} \frac{x}{k} = - \frac{x}{T_w - T_\infty} \frac{\partial T}{\partial y} \Big|_{y=0} \tag{10-43}$$

From the temperature profile Eq. (10-30) we find

$$\frac{\partial T}{\partial y} \Big|_{y=0} = - \frac{2(T_w - T_\infty)}{\delta} \tag{10-44}$$

Substitution of Eq. (10-44) into Eq. (10-43) gives

$$Nu_x = 2 \frac{x}{\delta} \tag{10-45}$$

and by combining Eqs. (10-41) and (10-45) the *local Nusselt number for free convection from a vertical plate becomes*

$$Nu_x \equiv \frac{h_x x}{k} = 0.508 Pr^{1/2} (0.952 + Pr)^{-1/4} Gr_x^{1/4} \tag{10-46}$$

where the Grashof number is defined by Eq. (10-42). The local heat-transfer coefficient h_x becomes

$$h_x = 0.508 Pr^{1/2} (0.952 + Pr)^{-1/4} \left[\frac{g\beta(T_w - T_\infty)}{v^2} \right]^{1/4} k x^{-1/4} \tag{10-47}$$

We note that h_x is *inversely proportional to the fourth root of x.*

In engineering applications the mean heat-transfer coefficient h_m over a distance $x = 0$ to $x = L$ along the plate is of interest. It is determined from

$$h_m = \frac{1}{L} \int_0^L h_x \, dx = \frac{4}{3} h_x \Big|_{x=L} \tag{10-48}$$

Then the mean Nusselt number Nu_m over the length $x = 0$ to $x = L$ becomes

$$\mathrm{Nu}_m = 0.677 \mathrm{Pr}^{1/2} (0.952 + \mathrm{Pr})^{-1/4} \mathrm{Gr}_L^{1/4} \tag{10-49}$$

where

$$\mathrm{Nu}_m \equiv \frac{hL}{k} \quad \text{and} \quad \mathrm{Gr}_L \equiv \frac{g\beta(T_w - T_\infty)L^3}{\nu^2} \tag{10-50}$$

In the case of a perfect gas, β is taken as [see Eq. (10-13)]

$$\beta = \frac{1}{T_\infty} \tag{10-51}$$

where T_∞ is absolute temperature.

For air at moderate temperatures $\mathrm{Pr} = 0.714$, and Eq. (10-46) and (10-49) simplify, respectively, to

$$\mathrm{Nu}_x = 0.378 \mathrm{Gr}_x^{1/4} \quad \text{for air} \tag{10-52}$$

$$\mathrm{Nu}_L = 0.504 \mathrm{Gr}_L^{1/4} \quad \text{for air} \tag{10-53}$$

These results for air obtained by the approximate method of solution of the free-convection problem as discussed above is only about 5 percent higher than the exact solution of the problem calculated numerically by Schmidt and Beckmann [1] given by the relation

$$\mathrm{Nu}_x = 0.360 \mathrm{Gr}_x^{1/4} \quad \text{for air} \tag{10-54}$$

In the foregoing analysis the velocity and thermal boundary-layer thicknesses are assumed to be the same. The problem is solved by Sparrow [2] for the case of velocity and thermal boundary-layer thicknesses not being equal, and by Ostrach [3] for a range of Prandtl numbers. The approximate solution given above for $\delta \cong \delta_t$ is within 10 percent of the exact solution for $0.01 < \mathrm{Pr} < 1000$.

Sparrow and Gregg [4] investigated free convection for very small Prandtl numbers, and LeFevre [5] studied the limiting cases $\mathrm{Pr} \to 0$ and $\mathrm{Pr} \to \infty$.

Finally we present in Table 10-1 the results for the mean Nusselt number Nu_m obtained from the *exact solution* of laminar free convection from a vertical plate, compiled by Schlichting [6] from several different references [3,4,7,8].

TABLE 10-1   Exact Solutions for the
Mean Nusselt Number in Laminar Free
Convection from a Vertical Plate, Compiled by Schlichting [6] from Refs. [3,4,
7,8]

Pr	$Nu_m/(Gr\ Pr)^{1/4}$
0.003	0.182
0.008	0.228
0.01	0.242
0.02	0.280
0.03	0.305
0.72	0.516
0.73	0.518
1	0.535
2	0.568
10	0.620
100	0.653
1000	0.665
∞	0.670

The heat-transfer relations for free convection from a vertical plate as given above is strictly applicable to situations in which the boundary layer is laminar. For a vertical plate the transition from laminar to turbulent flow begins for the values of the dimension parameter Gr Pr exceeding about 10^9. Therefore, the foregoing results are applicable for Gr Pr < 10^9.

The heat transfer in turbulent free convection is a more complicated matter, and empiricism is involved in the analysis. An early investigation of turbulent free convection from a vertical plate was performed by Eckert and Jackson [9] for a Prandtl number close to unity by using the integral method of solution. We shall not discuss here any of the theoretical approaches in this area but rather present some of the pertinent empirical correlations in the next section.

10-3   EMPIRICAL RELATIONS FOR FREE CONVECTION

For more practical applications the geometries involved are complicated and the free convection takes place in the turbulent-flow regime so that the analysis becomes very difficult or impractical. A wealth of experimental data and empirical correlations for free convection from bodies of different geometries are available in the literature for both laminar- and turbulent-flow regimes. In this section we present some of these correlations for typical geometries of practical interest.

Vertical Plates and Cylinders

The average Nusselt number for free convection from a vertical plate or cylinder (provided that the radius is much larger than the boundary-layer thickness) under *uniform surface-temperature* conditions can be correlated by the relation recommended by McAdams [10] in the form

$$Nu_m = c(Gr_L \, Pr)^n \tag{10-55}$$

where

$$Nu_m \equiv \frac{hL}{k} \qquad Gr_L \equiv \frac{g\beta(T_w - T_\infty)L^3}{\nu^2} \tag{10-56}$$

Here L is the height of the vertical plate or cylinder, and the fluid properties are evaluated at the arithmetic mean of T_w and T_∞. The values of the constant c and the exponent n are tabulated in Table 10-2 for the range $10^4 < Gr_L \, Pr < 10^{13}$. In the turbulent-flow range we note that the average heat-transfer coefficient is independent of the plate height since $Gr \sim L^3$ and $h \sim (1/L) \, Gr^{1/3}$. Figure 10-3 shows a correlation of the experimental data obtained by numerous investigators for the average heat-transfer coefficient in free convection from a vertical plate; the data extends to a value of $Gr_L \, Pr$ as low as about 0.5.

Free convection from a vertical plate for *uniform heat flux* at the wall surface was investigated by Sparrow and Gregg [13], Vliet [14a], and Vliet and Liu [14b]. The experiments reported in Refs. [14a,b] indicate that the local heat-transfer coefficient from a vertical plate under constant heat-flux boundary conditions can be correlated by the relations

Laminar: $\quad Nu_x = 0.60(Gr_x^* \, Pr)^{1/5} \qquad$ for $10^5 < Gr_x^* \, Pr < 10^{11} \quad$ (10-57)

Turbulent: $\quad Nu_x = 0.568(Gr_x^* \, Pr)^{0.22} \qquad$ for $2 \times 10^{13} < Gr_x^* \, Pr < 10^6 \quad$ (10-58)

where $Nu_x \equiv xh_x/k$ and the *modified local Grashof number* Gr_x^* is defined as

$$Gr^* = Gr_x \, Nu_x = \frac{g\beta(T_w - T_\infty)x^3}{\nu^2} \frac{q_w x}{T_w - T_\infty} = \frac{g\beta q_w x^4}{k\nu^2} \tag{10-59}$$

and q_w is the constant heat flux at the wall. All physical properties are to be evaluated at the arithmetic mean of wall-surface and fluid temperatures. The

TABLE 10-2   Constant c and n of Eq. (10-55) for Free Convection from a Vertical Plate or Cylinder at Uniform Temperature

Type of Flow	Range of $Gr_L \, Pr$	c	n	Reference
Laminar	10^4 to 10^9	0.59	1/4	[10]
Turbulent	10^9 to 10^{13}	0.10	1/3	[11,12]

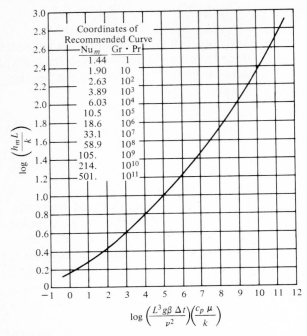

The coordinates of Recommended Curve table:

Nu$_m$	Gr · Pr
1.44	1
1.90	10
2.63	10^2
3.89	10^3
6.03	10^4
10.5	10^5
18.6	10^6
33.1	10^7
58.9	10^8
105.	10^9
214.	10^{10}
501.	10^{11}

Vertical axis: $\log\left(\dfrac{h_m L}{k}\right)$

Horizontal axis: $\log\left(\dfrac{L^3 g \beta \, \Delta t}{\nu^2}\right)\left(\dfrac{c_p \, \mu}{k}\right)$

FIG. 10-3   Average heat-transfer coefficient for free convection from a vertical plate at uniform temperature. (*From McAdams [10].*)

results of investigation in Ref. [14b] indicate that the transition from laminar to turbulent range occurs in the range $10^{12} < \text{Gr}_x^* \, \text{Pr} < 10^{14}$, that is, fully developed flow occurs by $\text{Gr}_x^* \, \text{Pr} = 10^{14}$ and it can be as low as 2×10^{13}.

In the case of *liquid metals* (that is, $\text{Pr} < 0.03$), the mean Nusselt number for free convection from a *vertical plate at uniform temperature* can be determined from the following relation:

Laminar: $\text{Nu}_m = \dfrac{h_m L}{k} = 0.68(\text{Gr}_L \, \text{Pr}^2)^{1/4}$ (10-60)

Recently Cebeci [15] studied the effects of curvature on heat transfer in *laminar* free convection from the outer surface of a vertical circular cylinder. Figure 10-4 shows a plot of the ratio of the Nusselt number for a vertical cylinder to that for a vertical flat plate, $(\text{Nu}_x)_{\text{cyl}}/(\text{Nu}_x)_{\text{f.p.}}$, against the parameter $\xi = (2\sqrt{2}/\text{Gr}_x^{1/4})(x/R)$, where R is the wire radius, for several different values of the Prandtl number. Here the local Nusselt number is defined as $\text{Nu}_x = hx/k$ and the local Grashof number as $\text{Gr}_x = g\beta(T_w - T_\infty)x^3/\nu^2$. Included in this figure are the results of Sparrow and Gregg [16] for two values of the Prandtl number.

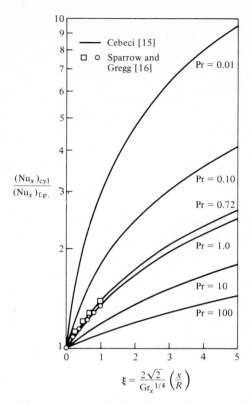

$$\xi = \frac{2\sqrt{2}}{Gr_x^{1/4}} \left(\frac{x}{R}\right)$$

FIG. 10-4   Ratio of the local Nusselt number for a vertical cylinder to that for a vertical flat plate for laminar free convection at a uniform wall temperature. (*From Cebeci* [15].)

Example 10-1   A cylindrical heating element 1 in $(2.54 \times 10^{-2}$ m) OD and 1.5 ft (0.457 m) long is inserted vertically into a body of water at 70°F (21.1°C). If the surface of the heating element is at a uniform temperature of 130°F (54.4°C), determine the mean heat-transfer coefficient over the entire length of the element and the rate of the heat transfer from the entire surface of the element to the water.

Solution   The physical properties of water at a mean temperature of $(70 + 130)/2 = 100$°F are

$k = 0.364$ Btu/h·ft·°F (0.630 W/m·°C)

$c_p = 1.0$ Btu/lb·°F (4187 W·s/kg·°C)

$v = 0.0266$ ft²/h (0.687 × 10⁻⁶ m²/s)

$\rho = 62$ lb/ft³ (993.1 kg/m³)

$\beta = 2 \times 10^{-4}$ °R⁻¹ (3.6 × 10⁻⁴ K⁻¹)

$Pr = 4.52$

Then the Grashof number is calculated as follows:
By using the English system of units, we find

$$\text{Gr}_L = \frac{g\beta(T_w - T_\infty)L^3}{v^2} = \frac{(4.17 \times 10^8)(2 \times 10^{-4})(130 - 70)(1.5^3)}{0.0266^2} = 2.38 \times 10^{10}$$

By using the SI system of units, we obtain the same result:

$$\text{Gr}_L = \frac{(9.8)(3.6 \times 10^{-4})(54.4 - 21.1)(0.457^3)}{(0.687 \times 10^{-6})^2} = 2.38 \times 10^{10}$$

Then,

$$\text{Gr}_L \text{ Pr} = (2.38 \times 10^{10})(4.52) = 1.08 \times 10^{11}$$

Therefore free convection is in the turbulent-flow range because $\text{Gr}_L \text{ Pr} > 10^9$. The mean Nusselt number is obtained by Eq. (10-55) and Table 10-2 as

$$\text{Nu}_m = \frac{h_m L}{k} = 0.10(\text{Gr}_L \text{ Pr})^{1/3} = (0.10)(1.08 \times 10^{11})^{1/3} = 476.2$$

and the mean heat-transfer coefficient is determined as follows:
By using the English system of units, we find

$$h_m = \frac{k}{L} 0.10(\text{Gr}_L \text{ Pr})^{1/3} = \frac{0.364}{1.5}(0.10)(476.2) = 11.56 \text{ Btu/h} \cdot \text{ft}^2 \cdot {}^\circ\text{F}$$

or, by using the SI system of units, we obtain

$$h_m = \frac{0.630}{0.457}(0.10)(476.2) = 6.56 \text{ W/m}^2 \cdot {}^\circ\text{C}$$

The heat-transfer rate Q from the entire surface of the element to the water is determined as follows:
By using the English system of units, we find

$$Q = (\pi DL)(h_m)(T_w - T_\infty) = (\pi \tfrac{1}{12} \tfrac{3}{2})(11.56)(130 - 70) = 272.3 \text{ Btu/h} \ (79.8 \text{ W})$$

or, by using the SI system of units, we obtain

$$Q = \pi(2.54 \times 10^{-2})(0.457)(65.6)(54.4 - 21.1) = 79.8 \text{ W}$$

Example 10-2 A vertical wall at a uniform temperature of 150°F is in contact with air at 50°F and atmospheric pressure. Determine the average free-convection heat-transfer coefficient over a distance 1 ft from the lower edge of the wall.

Solution The physical properties of air at a mean temperature of $(150 + 50)/2 = 100°\text{F}$ and atmospheric pressure are

$$\rho = 0.070 \text{ lb/ft}^3 \qquad v = 0.184 \times 10^{-3} \text{ ft}^2/\text{s}$$

$$k = 0.01698 \text{ Btu/h} \cdot \text{ft} \cdot {}^\circ\text{F} \qquad \beta = 1.79 \times 10^{-3} \ {}^\circ\text{R}^{-1}$$

$$\text{Pr} = 0.70 \qquad \text{and} \qquad \frac{g\beta}{v^2} = 1.705 \times 10^6 \ ({}^\circ\text{R} \cdot \text{ft}^3)^{-1}$$

The Grashof number for $L = 1$ ft is

$$\text{Gr}_L = \frac{g\beta(T_w - T_\infty)L^3}{\nu^2} = (1.70 \times 10^6)(150 - 50)(1^3) = 1.705 \times 10^8$$

and

$$\text{Gr}_L \, \text{Pr} = (1.705 \times 10^8)(0.70) \cong 1.19 \times 10^8$$

The flow is in the laminar range since $\text{Gr}_L \, \text{Pr} < 10^9$. The mean Nusselt number over the length $L = 1$ ft, by Eq. (10-55) and Table 10-2, becomes

$$\text{Nu}_m = \frac{h_m L}{k} = 0.59(\text{Gr}_L \, \text{Pr})^{1/4} = 61.7$$

and the mean heat-transfer coefficient h_m is determined as

$$h_m = \frac{k}{L} \, \text{Nu}_m = \left(\frac{0.0169}{1}\right)(61.7) = 1.05 \text{ Btu/h} \cdot \text{ft}^2 \cdot {}^\circ\text{F}$$

Horizontal Cylinders

The mean Nusselt number for free convection from a *horizontal cylinder at a uniform temperature* was correlated by McAdams [10] by a relation in the form

$$\text{Nu}_m = c(\text{Gr}_D \, \text{Pr})^n \tag{10-61}$$

where

$$\text{Nu}_m \equiv \frac{hD}{k} \qquad \text{Gr}_D \equiv \frac{g\beta(T_w - T_\infty)D^3}{\nu^2} \tag{10-62}$$

Here D is the outside diameter of the cylinder, and fluid properties are evaluated at the arithmetic mean of the tube wall temperature T_w and the fluid temperature T_∞. The values of the constant c and the exponent n are tabulated in Table 10-3 for the range $10^4 < \text{Gr}_D \, \text{Pr} < 10^{12}$. We note that in the turbulent-flow range the mean Nusselt number is independent of the tube diameter. Figure 10-5 shows a correlation of experimental data obtained by various investigators for the mean heat-transfer coefficient for free convection from a horizontal cylinder at uniform temperature. The data in this figure extend to a range as low as $\text{Gr}_D \, \text{Pr} = 10^{-4}$.

In the case of *liquid metals* (that is, $\text{Pr} < 0.03$) the mean Nusselt number for

TABLE 10-3 Constants c and n of Eq. (10-61) for Free Convection from a Horizontal Cylinder at Uniform Temperature

Type of Flow	Range of $\text{Gr}_D \, \text{Pr}$	c	n	Reference
Laminar	10^4 to 10^9	0.53	$\frac{1}{4}$	[10]
Turbulent	10^9 to 10^{12}	0.13	$\frac{1}{3}$	[10]

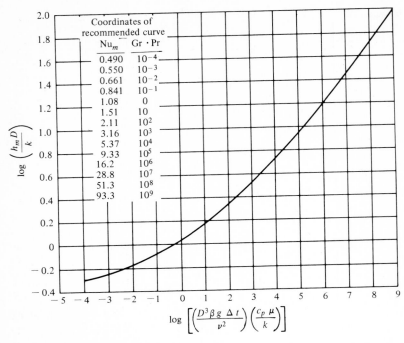

FIG. 10-5 Average heat-transfer coefficient for free convection from a horizontal cylinder at uniform temperature. (*From McAdams [10]*.)

free convection from a horizontal cylinder can be determined from the following relation [17]:

$$\text{Laminar:} \quad \text{Nu}_m = 0.53(\text{Gr}_D \, \text{Pr}^2)^{1/4} \tag{10-63}$$

where Nu_m and Gr_D are as defined by Eqs. (10-62).

It is to be noted that the foregoing relations are not applicable for free convection from horizontal wires for which the diameter is of the same order of magnitude as the boundary-layer thickness. Gebhart et al. [18] presented empirical correlations for free convection from horizontal wires for fluids having Prandtl numbers of 0.7, 6.3, and 63. These results indicate that for small wires the effect of Prandtl number on the heat-transfer coefficient is more involved than represented by the simple expressions given above.

S. Nakai and T. Okazaki [19] studied free convection from horizontal wires at uniform temperature for the case when the thickness of the boundary layer around the cylinder is sufficiently large compared with its diameter. A comparison of their proposed equation with experiments appears to give good correlation.

Horizontal Plates

The heat-transfer coefficient in free convection from a horizontal plate depends on the orientation of the heat-transfer surface, namely, whether the surface is facing up or down. The mean Nusselt number (i.e., average over the entire surface) for free convection from a horizontal surface is correlated by McAdams [10] with a relation in the form

$$Nu_m = c(Gr_L \, Pr)^n \qquad (10\text{-}64)$$

where

$$Nu_m \equiv \frac{h_m L}{k} \qquad Gr_L \equiv \frac{g\beta(T_w - T_\infty)L^3}{\nu^2} \qquad (10\text{-}65)$$

Here L is a *characteristic dimension* of the surfaces, and the coefficients c and n are tabulated in Table 10-4. The characteristic dimension L for a square plate is taken as the length of the side of the square, for a rectangular plate as the arithmetic mean of the lengths of two dimensions, and for a circular disk as $0.9 \times$ disk diameter. An examination of the values of exponents n in Table 10-4 reveals that for the case of a lower surface heated or an upper surface cooled the turbulent-flow condition is not reached even at $Gr_L \, Pr \simeq 3 \times 10^{10}$.

Example 10-3 Consider a 3 by 3-ft (0.915 by 0.915-m) panel with one surface insulated and the other kept at a uniform temperature 150°F (65.6°C). Calculate the average free-convection heat-transfer coefficient between the heated surface of the panel and the atmospheric air at 50°F (10°C) for the following arrangements:

(a) Heated surface is vertical.
(b) Panel is horizontal with the hot surface facing up.
(c) Panel is horizontal with the hot surface facing down.

TABLE 10-4 Constants c and n of Eq. (10-64) for Free Convection from a Horizontal Plate at Uniform Temperature

Type of Flow	Orientation of Plate	Range of $Gr_L \, Pr$	c	n	Reference
Laminar	Upper surface heated or Lower surface cooled	10^5 to 2×10^7	0.54	$\frac{1}{4}$	[10]
Turbulent	Upper surface heated or Lower surface cooled	2×10^7 to 3×10^{10}	0.14	$\frac{1}{3}$	[10]
Laminar	Lower surface heated or Upper surface cooled	3×10^5 to 3×10^{10}	0.27	$\frac{1}{4}$	[10]

Solution   The physical properties of atmospheric air at a mean temperature $(150 + 50)/2 = 100°F$ were given in Example 10-2. The characteristic length of the considered square panel is $L = 3$ ft (0.915 m); then the Grashof number becomes

$$Gr_L = \frac{g\beta(T_w - T_\infty)L^3}{v^2} = 4.6 \times 10^9$$

and

$$Gr_L \, Pr = (4.6 \times 10^9) \times 0.70 = 3.22 \times 10^9$$

The average heat-transfer coefficients for the cases considered above are now calculated as follows:

(a) When the heat-transfer surface is vertical the mean Nusselt number is obtained from Eq. (10-55) and Table 10-2; for the turbulent-flow condition it becomes

$$Nu_m = \frac{h_m L}{k} = 0.10(Gr_L \, Pr)^{1/3}$$

or

$$h_m = 0.10(Gr_L \, Pr)^{1/3} \frac{k}{L} = 0.10 \times (3.22 \times 10^9)^{1/3} \frac{0.0169}{3}$$

$$= 0.83 \text{ Btu/h} \cdot \text{ft}^2 \cdot °F \ (4.71 \text{ W/m}^2 \cdot °C)$$

(b) For a horizontal panel with a hot surface facing up, the mean Nusselt number is determined from Eq. (10-64) and Table 10-4; for the turbulent-flow condition it becomes

$$Nu_m = \frac{h_m L}{k} = 0.14(Gr_L \, Pr)^{1/3}$$

or

$$h_m = 0.14(Gr_L \, Pr)^{1/3} \frac{k}{L} = 0.14 \times (3.22 \times 10^9)^{1/3} \frac{0.0169}{3}$$

$$= 1.16 \text{ Btu/h} \cdot \text{ft}^2 \cdot °F \ (6.59 \text{ W/m}^2 \cdot °C)$$

(c) For a horizontal panel with a hot surface facing down, from Eq. (10-64) and Table 10-4 we obtain

$$Nu_m = \frac{h_m L}{k} = 0.27(Gr_L \, Pr)^{1/4}$$

or

$$h_m = 0.27(Gr_L \, Pr)^{1/4} \frac{k}{L} = 0.27 \times (3.22 \times 10^9)^{1/4} \times \frac{0.0169}{3}$$

$$= 0.362 \text{ Btu/h} \cdot \text{ft}^2 \cdot °F \ (2.06 \text{ W/m}^2 \cdot °C)$$

Spheres

The average Nusselt number for free convection from a sphere at uniform temperature to *air* can be calculated from the relation recommended by Yuge [20]:

$$Nu_m = \frac{h_m D}{k} = 2 + 0.392 Gr_D^{1/4} \qquad \text{for } 1 < Gr_D < 10^5 \qquad (10\text{-}66a)$$

where

$$\mathrm{Gr}_D \equiv \frac{g\beta(T_w - T_\infty)D^3}{\nu^2} \tag{10-66b}$$

where D is the diameter of the sphere. The physical properties of the fluid are evaluated at the arithmetic mean of wall temperature T_w and fluid temperature T_∞.

Equation (10-66a) is for air. To generalize this relation to other fluids having a Prandtl number different from unity, the term Gr_D is now replaced by Gr_D Pr and the constant 0.392 is modified by taking the Prandtl number for air as 0.72. We find

$$\mathrm{Nu}_m = 2 + 0.43(\mathrm{Gr}_D\ \mathrm{Pr})^{1/4} \tag{10-67}$$

and this expression should be used for cases with Prandtl number close to unity.

Enclosed Spaces

We now present empirical correlations for free convection in vertical and horizontal spaces and for a spherical space.

For the case of a *vertical enclosed space* as illustrated in Fig. 10-6a, Jakob [21] correlated a large number of experimental data for free convection with *air* with a length-to-gap ratio L/b from about 3 to 42 with following formulas:

Laminar: $\quad \mathrm{Nu}_m = 0.18\mathrm{Gr}_b^{1/4}\left(\dfrac{L}{b}\right)^{-1/9} \qquad$ for $2 \times 10^3 < \mathrm{Gr}_b < 2 \times 10^4$

$$\tag{10-68a}$$

Turbulent: $\quad \mathrm{Nu}_m = 0.064\mathrm{Gr}_b^{1/3}\left(\dfrac{L}{b}\right)^{-1/9} \qquad$ for $2 \times 10^4 < \mathrm{Gr}_b < 11 \times 10^6$

$$\tag{10-68b}$$

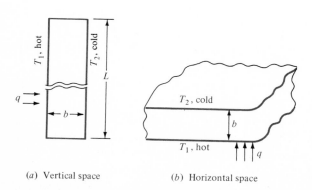

(a) Vertical space   (b) Horizontal space

FIG. 10-6   Nomenclature for free convection in enclosed vertical and horizontal spaces.

For the case of a *horizontal space* with the lower plate at a higher temperature than the upper plate, as illustrated in Fig. 10-6b, Jakob [21] correlated the free convection for air with the following relations:

Laminar: $\text{Nu}_m = 0.195\text{Gr}_b{}^{1/4}$ for $10^4 < \text{Gr}_b < 4 \times 10^5$ (10-69a)

Turbulent: $\text{Nu}_m = 0.068\text{Gr}_b{}^{1/3}$ for $4 \times 10^5 < \text{Gr}_b$ (10-69b)

where the Nusselt number and the Grashof number are based on the distance b between the two parallel surfaces, that is,

$$\text{Nu}_m \equiv \frac{qb}{k(T_1 - T_2)} \qquad \text{Gr}_b \equiv \frac{g\beta(T_1 - T_2)b^3}{v^2} \tag{10-70}$$

Here T_1 and T_2 are the temperatures of the hot and cold plates, respectively. Then, the heat flux q across the space is determined once the Nusselt number is known.

When the Grashof number is small, the convection is suppressed, and heat transfer across the air space takes place by pure conduction; for such cases the Nusselt number tends to unity.   That is,

$$\text{Nu}_m \to 1 \qquad \begin{cases} \text{for } \text{Gr}_b < 2000 \text{ in vertical space} \\ \text{for } \text{Gr}_b < 1700 \text{ in horizontal space} \end{cases} \tag{10-71}$$

It is to be noted that for a horizontal space with the upper plate hotter than the lower plate, the fluid is stratified because the lower-density fluid is above the higher-density fluid; then there is no convection.

For the case of a *liquid contained in a horizontal space* as shown in Fig. 10-6b with the lower plate hotter than the upper plate, the experiments by Globe and Dropkin [22] show that the Nusselt number can be correlated by the following expression:

$$\text{Nu}_m = 0.069\text{Gr}_b{}^{1/3}\,\text{Pr}^{0.407} \qquad \text{for } 3 \times 10^5 < \text{Gr}_b\,\text{Pr} < 7 \times 10^9 \tag{10-72}$$

The free convection inside a *spherical cavity* of diameter D is correlated by Kreith [23] with the following empirical expression:

$$\text{Nu}_m = \frac{hD}{k} = c(\text{Gr}_D\,\text{Pr})^n \tag{10-73}$$

where Gr_D is the Grashof number based on the sphere diameter D, and the constants c and n are given in Table 10-5.

TABLE 10-5   Constants c and n of Eq. (10-73) for Free Convection in a Spherical Cavity

Type of Flow	Range of Gr_D Pr	c	n	Reference
Laminar	10^4 to 10^9	0.59	$\frac{1}{4}$	[23]
Turbulent	10^9 to 10^{12}	0.13	$\frac{1}{3}$	[23]

Example 10-4   Atmospheric air is contained between two horizontal parallel panels separated by a distance 1 in (2.54×10^{-2} m).   The lower and upper plates are maintained at $140°F$ ($60°C$) and $60°F$ ($15.6°C$), respectively.   Determine the rate of heat transfer across the air space per square foot (square meter) of the panel surface.

Solution   The physical properties of atmospheric air at a temperature $(140 + 60)/2 = 100°F$ ($37.8°C$) are

$k = 0.0169$ Btu/h·ft·°F (0.0292 W/m·°C)

$\rho = 0.070$ lb/ft^3 (1.121 kg/m^3)

$v = 0.662$ ft^2/h (0.171×10^{-4} m^2/s)

$\beta = 1.79 \times 10^{-3}$ °R^{-1} (3.22×10^{-3} K^{-1})

Pr $= 0.70$

The Grashof number for a spacing $b = 1$ in (2.54×10^{-2} m) is calculated as follows: By using the English system of units, we find

$$Gr_b = \frac{g\beta(T_1 - T_2)b^3}{v^2} = \frac{(4.17 \times 10^8)(1.79 \times 10^{-3})(140 - 60)}{0.662^2 \times 12^3} = 7.9 \times 10^4$$

or, by using the SI system of units, the same result is obtained:

$$Gr_b = \frac{(9.8)(3.22 \times 10^{-3})(60 - 15.6)(2.54 \times 10^{-2})^3}{(0.171 \times 10^{-4})^2} \cong 7.9 \times 10^4$$

The Nusselt number is determined from Eq. (10-69a) as

$$Nu_m \equiv \frac{qb}{k(T_1 - T_2)} = 0.195Gr_b^{1/4} = (0.195)(7.9 \times 10^4)^{1/4} = 3.27$$

Then the heat flux q across the air space is given by

$$q = Nu_m \frac{k(T_1 - T_2)}{b}$$

If the English system of units is used, the heat flux becomes

$$q = (3.27)(0.0169)(140 - 60)(12) = 53 \text{ Btu/h·ft}^2 \text{ (167 W/m}^2)$$

or, by using the SI system of units, we obtain

$$q = 3.27 \frac{(0.0292)(60 - 15.6)}{2.54 \times 10^{-2}} = 167 \text{ W/m}^2$$

Simplified Equations for Free Convection of Air at Atmospheric Pressure

We present in Table 10-6 simplified expressions for a rapid but approximate estimation of the average heat-transfer coefficient from various geometries to *air at atmospheric pressure and moderate temperatures.* For more accurate results, previously given, more precise expressions should be used. The heat-transfer coefficients h_m considered in Table 10-6 refer to the mean value over the surface and apply reasonably well to air at 100 to 1500°F, and to CO, CO_2, O_2, N_2, and the flue gases. The Grashof number is defined as

$$Gr_L \equiv \frac{g\beta \, \Delta T L^3}{\nu^2} \tag{10-74}$$

where L is the characteristic dimension and ΔT is the temperature difference between the surface and the ambient air.

Example 10-5 Determine the free-convection heat-transfer coefficients for the cases considered in Example 10-3 by using the simplified relations given in Table 10-6.

Solution Taking $L = 3$ ft (0.915 m) and $\Delta T = 100°F$ (55.6°C), the heat-transfer coefficient for each of these three cases is determined by using the appropriate expressions from Table 10-6.

(a) *Vertical surface.* In Example 10-3a, it is shown that the flow is turbulent. Then, by using the English system of units, we obtain

$$h = 0.19\Delta T^{1/3} = 0.19 \times 100^{1/3} = 0.88 \text{ Btu/h} \cdot \text{ft}^2 \qquad (5.0 \text{ W/m}^2 \cdot °C)$$

or by using the SI system of units we obtain the same result:

$$h = 1.31\Delta T^{1/3} = 1.31 \times 55.6^{1/3} = 5.0 \text{ W/m}^2 \cdot °C$$

We note that the heat-transfer coefficient determined by the simplified expression is approximately 6 percent higher than that obtained in Example 10-3a by using Eq. (10-55).

(b) *Horizontal panel, hot surface facing up.* In this case the flow is also turbulent; by using the English system of units, we find

$$h = 0.22\Delta T^{1/3} = 0.22 \times 100^{1/3} = 1.02 \text{ Btu/h} \cdot \text{ft}^2 \cdot °F \qquad (5.8 \text{ W/m}^2 \cdot °C)$$

or, by using the SI system of units, we obtain

$$h = 1.52\Delta T^{1/3} = 1.52 \times 55.6^{1/3} = 5.8 \text{ W/m}^2 \cdot °C)$$

The result obtained with the simplified formula is about 12 percent lower than that determined in Example 10-3b by using Eq. (10-64).

(c) *Horizontal panel, hot surface facing down.* By using the English system of units, we find

$$h = 0.12\left(\frac{\Delta T}{L}\right)^{1/4} = 0.12\left(\frac{100}{3}\right)^{1/4} = 0.288 \text{ Btu/h} \cdot \text{ft}^2 \cdot °F \qquad (1.64 \text{ W/m}^2 \cdot °C)$$

TABLE 10-6 Simplified Equations for Free Convection to Air at Atmospheric Pressure and Moderate Temperatures†

Geometry	Characteristic Dimension, L	Type of Flow	Range of Gr_L Pr	Heat-Transfer Coefficient, h_m	
				Btu/h·ft²·°F‡	W/m²·°C§
Vertical plates and cylinders	Height	Laminar	10^4 to 10^9	$h_m = 0.29(\Delta T/L)^{1/4}$	$h_m = 1.42(\Delta T/L)^{1/4}$
		Turbulent	10^9 to 10^{13}	$h_m = 0.19\Delta T^{1/3}$	$h_m = 1.31\Delta T^{1/3}$
Horizontal cylinders	Outside diameter	Laminar	10^4 to 10^9	$h_m = 0.27(\Delta T/L)^{1/4}$	$h_m = 1.32(\Delta T/L)^{1/4}$
		Turbulent	10^9 to 10^{12}	$h_m = 0.18\Delta T^{1/3}$	$h_m = 1.24\Delta T^{1/3}$
Horizontal plates					
(a) Upper surface heated or lower surface cooled	As defined in the text	Laminar	10^5 to 2×10^7	$h_m = 0.27(\Delta T/L)^{1/4}$	$h_m = 1.32(\Delta T/L)^{1/4}$
		Turbulent	2×10^7 to 3×10^{10}	$h_m = 0.22\Delta T^{1/3}$	$h_m = 1.52\Delta T^{1/3}$
(b) Lower surface heated or upper surface cooled	As defined in the text	Laminar	3×10^5 to 3×10^{10}	$h_m = 0.12(\Delta T/L)^{1/3}$	$h_m = 0.59(\Delta T/L)^{1/4}$

† From McAdams [10].
‡ L is in feet, $\Delta T \equiv T_w - T_\infty$ is in degrees fahrenheit. These relations may be extended to higher or lower pressures about atmospheric pressure by multiplying by the following factors: $(P/14.7)^{1/2}$ for laminar, $(P/14.7)^{2/3}$ for turbulent, where P is in lb$_f$/in² abs.
§ L is in meters, $\Delta T \equiv T_w - T_\infty$ is in degrees Celsius. These relations may be extended to higher or lower pressures about atmospheric pressure by multiplying by the following factors: $(P/1.0132)^{1/2}$ for laminar, $(P/1.0132)^{2/3}$ for turbulent, where P is the pressure in bars.

or, by using the SI system of units, we obtain

$$h = 0.59 \left(\frac{\Delta T}{L} \right)^{1/4} = 0.59 \left(\frac{55.6}{0.915} \right)^{1/4} = 1.64 \text{ W/m}^2 \cdot {}^\circ\text{C}$$

In this case the result obtained with the simplified formula is approximately 20 percent lower than that obtained in Example 10-3c using Eq. (10-64).

10-4 COMBINED FREE AND FORCED CONVECTION

Free convection occurs whenever density gradients exist in a fluid due to the presence of temperature gradients. Therefore, in forced-convection heat-transfer problems some free convection always takes place; but when the externally applied velocity field is dominant, the free-convection effects are neglected. Conversely, if the velocity field generated by the buoyancy effects are dominant, the forced-convection effects, if any, are neglected and the problem is treated as a pure free-convection one. There are also a number of practical situations in which forced and free convection are of the same order of magnitude and neither of them can be neglected. The analysis of combined free- and forced-convection problems is a complicated matter; but it is of interest to know under what conditions the free- and forced-convection heat transfer are of the same order of magnitude or when the free-convection effects become negligible in forced-convection problems. This matter is now examined.

We recall the dimensionless equations (10-15) to (10-17) for free convection from a vertical plate and note that the dimensionless group

$$\frac{g\beta(T_w - T_\infty)L}{U_0^2} \equiv \frac{\text{Gr}}{\text{Re}^2} \tag{10-75}$$

appearing in these equations represents the ratio of the buoyancy forces to inertial forces. When this ratio is of the order of unity, that is, $\text{Gr} \cong \text{Re}^2$, the free convection cannot be ignored in comparison with forced convection; therefore, they should be analyzed simultaneously.

Heat transfer in combined free and forced convection has been studied by several investigators. Figure 10-7 shows the results of analysis by Lloyd and Sparrow [24] for an isothermal vertical plate with $T_w > T_\infty$ when u_∞ is upward (or $T_w < T_\infty$ when u_∞ is downward) for Prandtl numbers 0.003, 0.01, 0.03, 0.72, 10, and 100, thereby providing information for liquid metals, gases, and ordinary liquids. In these figures the local Nusselt number expressed as $\text{Nu}_x/\sqrt{\text{Re}_x}$ is plotted as a function of the dimensionless group $\text{Gr}_x/\text{Re}_x^2$, and the solid lines are the results of calculations for combined free and forced convection. The figures also contain the results for pure forced convection and for pure free convection. Examination of these figures reveals that for all the Prandtl numbers

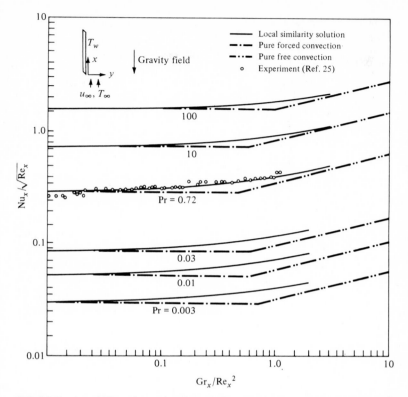

FIG. 10-7   Local Nusselt number for combined free and forced convection from an isothermal vertical plate. (*From Lloyd and Sparrow* [*24*].)

investigated the solid lines (i.e., combined free- and forced-convection results) merge smoothly with the pure forced-convection results at small values of Gr_x/Re_x^2. The solid lines tend to approach the asymptotes for pure free convection at larger values of Gr_x/Re_x^2. Included in this figure are the experimental data of Kliegel [25] for $Pr = 0.72$; the agreement between the experiments and the analysis is very good. Referring to the analysis in Fig. 10-7 we note that if the free-convection effects are neglected in forced convection from a vertical plate the greatest deviations are approximately 15 percent for $Pr = 100$ and 10, 19 percent for $Pr = 0.72$, and 22 percent for $Pr = 0.03$, 0.01, and 0.003. Free-convection effects in forced convection become more important at low Prandtl numbers. Table 10-7 shows the threshold values of Gr_x/Re_x^2 for 5 percent deviation in the Nusselt number due to neglecting the free-convection effects in forced convection from a vertical plate.

Mori's [26] investigations of forced convection from an isothermal horizontal plate showed that the effects of free convection upon the local heat-transfer coefficient on either side of the plate is less than 10 percent if

$$Gr_x \leq 0.083 Re_x^{0.25} \tag{10-76}$$

TABLE 10-7   Threshold Values of Gr_x/Re_x^2 for 5 Percent Deviation of the Nusselt Number Resulting from Neglecting Free Convection in Forced Flow Along a Vertical Plate for $T_w > T_\infty$ When u_∞ Is Upwards†

Pr	100	10	0.72	0.03–0.003
Gr_x/Re_x^2	0.24	0.13	0.08	0.056–0.05

† Based on data from Ref. [24].

A number of experimental results have been reported on combined free and forced convection for flow across spheres and cylinders. Yuge [20] studied flow of air across a sphere; Collis and Williams [27] and Oosthuizen and Madan [28], flow over horizontal cylinders; and Gebhart et al. [18], flow over horizontal wires.

A summary of free- and forced-convection effects for flow over horizontal and vertical tubes is given by Metais and Eckert [29], and a correlation has been developed by Brown and Gauvin [30] for the average Nusselt number for combined forced and free convection in the laminar-flow regime for flow over horizontal tubes. The reader should consult the original references for details in this matter.

Example 10-6   Consider air at atmospheric pressure and 70°F flowing upward along a heated vertical plate 1 ft high, and kept at a uniform temperature 130°F. Determine the minimum flow velocity of air below which the effect of free convection on the heat-transfer coefficient is more than 5 percent.

Solution   The physical properties of atmospheric air at a mean temperature $(70 + 130)/2 = 100°F$ are the same as those given in Example 10-4. Then, at $x = 1$ ft we have

$$Gr_x = \frac{g\beta(T_w - T_\infty)x^3}{\nu^2} = (1.705 \times 10^6) \times (130 - 70) \times 1 = 1.023 \times 10^8$$

According to Table 10-7 the threshold value of Gr_x/Re_x^2 for 5 percent deviation of the Nusselt number for $Pr = 0.72$ (i.e., air) is 0.08. Then,

$$\frac{Gr_x}{Re_x^2} = 0.08$$

$$Re_x^2 = \frac{1.023 \times 10^8}{0.08} = 12.79 \times 10^8$$

$$Re_x = \frac{u_\infty x}{\nu} = 3.58 \times 10^4$$

$$u_\infty = \frac{3.58 \times 10^4 \times \nu}{x} = \frac{(3.58 \times 10^4) \times (0.184 \times 10^{-3})}{1}$$

$$\cong 6.6 \text{ ft/s}$$

Thus the velocity should be greater than about 6.6 ft/s.

REFERENCES

1 Schmidt, E., and W. Beckmann: Das Temperatur und Geschwindigkeitsfeld von einer licher Wandtemperatur, *Forsch. Geb. Ingenieures.*, **1**:391 (1930).

2 Sparrow, E. M.: *NACA Tech. Note* 3508, 1955; also, E. M. Sparrow and J. L. Gregg, *Trans. ASME*, **78**:435–440 (1956).

3 Ostrach, S.: An Analysis of Laminar Free Convection Flow and Heat Transfer about a Flat Parallel to the Direction of the Generating Body Force, *NACA Rep.* 1111, 1953.

4 Sparrow, E. M., and J. L. Gregg: Details of Exact Low Prandtl Number Boundary Layer Solutions for Forced and Free Convection, *NACA Tech. Memo.* 2-27-59E, 1959.

5 LeFevre, E. J.: Laminar Free Convection from a Vertical Surface, *Mech. Eng. Res. Lab., Heat 113* (Gt. Britain), 1956.

6 Schlichting, H.: "Boundary Layer Theory," McGraw-Hill Book Company, New York, 1968.

7 Pohlhausen, E.: Der Wärmeaustrausch zwishen festen Körpen und Flüssigkeiten mit kleiner Reiburg und kleiner Warmeleitung, *Z. Angew. Math. Mech.*, **1**:115 (1921).

8 Schuh, H.: Einige Probleme bei Freirer Strömung Zaher Flüssigkeiten, *Göttinger Monogr. Bd. B., Grenzschichten*, 1946.

9 Eckert, E. R. G., and T. W. Jackson: Analysis of Turbulent Free Convection Boundary Layer on a Flat Plate, *NACA Rep.* 1015, 1951.

10 McAdams, W. H.: "Heat Transmission," 3d ed., McGraw-Hill Book Company, New York, 1954.

11 Bayley, F. J.: An Analysis of Turbulent Free Convection Heat Transfer, *Proc., Inst. Mech. Eng., London*, **169**:361 (1955).

12 Warner, C. Y., and V. S. Arpaci: An Investigation of Turbulent Natural Convection in Air at Low Pressure Along a Vertical Heated Flat Plate, *Int. J. Heat Mass Transfer*, **11**:397–406 (1968).

13 Sparrow, E. M., and J. L. Gregg: Laminar Free Convection from a Vertical Plate, *Trans. ASME*, **78**:435–440 (1956).

14a Vliet, G. C.: Natural Convection Local Heat Transfer on Constant-Heat-Flux Inclined Surfaces, *J. Heat Transfer*, **91C**:511–516 (1969).

14b Vliet, G. C., and C. K. Liu: An Experimental Study of Natural Convection Boundary Layers, *J. Heat Transfer*, **91C**:517–531 (1969).

15 Cebeci, T.: Laminar–Free-Convective-Heat Transfer from the Outer Surface of a Vertical Slender Circular Cylinder, *Fifth Int. Heat Transfer Conf.*, vol. 3, NC1.4, pp. 15–19, 1975.

16 Sparrow, E. M., and J. L. Gregg: Laminar-Free-Convection Heat Transfer from the Outer Surface of a Vertical Circular Cylinder, *Trans. ASME*, **78**:1823–1829 (1956).

17 Hyman, S. C., C. F. Borilla, and S. W. Ehrlich: Heat Transfer to Liquid Metals and Non-Metals at Horizontal Cylinders, *AIChE Symp. Heat Transfer*, Atlantic City, pp. 21–33, 1953.

18 Gebhart, B., T. Audunson, and L. Pera: Forced, Mixed and Natural Convection from Long Horizontal Wires, Experiments at Various Prandtl Numbers, *Fourth Int. Heat Transfer Conf.* Paris, vol. IV, sec. 3.2, August 1970.

19 Nakai, S., and T. Okazaki: Heat Transfer from a Horizontal Circular Wire at Small Reynolds and Grashof Numbers, Pts. I and II, *Int. J. Heat Mass Transfer*, **18**:387–413 (1975).

20 Yuge, T.: Experiments on Heat Transfer from Spheres Including Combined Natural and Free Convection, *J. Heat Transfer*, **82C**:214–220 (1960).

21 Jakob, M.: "Heat Transfer," vol. 1, John Wiley & Sons, Inc., New York, 1949.

22 Globe, S., and D. Dropkin: Natural Convection Heat Transfer in Liquids Confined by Two Horizontal Plates and Heated from Below, *J. Heat Transfer*, **81C**:24–28 (1959).

23 Kreith, F.: Thermal Design of High Altitude Balloons and Instrument Packages, *J. Heat Transfer*, **92C**:307–332 (1970).

24 Lloyd, J. R., and E. M. Sparrow: Combined Forced and Free Convection Flow on Vertical Surfaces, *Int. J. Heat Mass Transfer*, **13**:434–438 1970.

25 Kliegel, J. R.: "Laminar Free and Forced Convection Heat Transfer from a Vertical Flat Plate," Ph.D. thesis, University of California, Berkeley, Calif., 1959.

26 Mori, Y.: Buoyancy Effects in Forced Laminar Convection Flow Over a Horizontal Flat Plate, *J. Heat Transfer*, **83C**:479–482 (1961).

27 Collis, D. C., and M. J. Williams; Two Dimensional Convection from Heated Wires at Low Reynolds Numbers, *J. Fluid Mech.*, **6**:357–384 (1959).

28 Oosthuizen, P. H., and S. Madan: Combined Convective Heat Transfer from Horizontal Cylinders in Air, *J. Heat Transfer*, **92C**:194–196 (1970).

29 Metais, B., and E. R. G. Eckert: Forced, Mixed and Free Convection Regimes, *J. Heat Transfer*, **86C**:295–296 (1964).

30 Brown, C. K., and W. H. Gauvin: Combined Free and Forced Convection, Pts. I and II, *Can. J. Chem. Eng.*, **43**:306, 313 (1965).

NOTES

1 We look for a *similarity solution* of Eqs. (10-38) and (10-39), that is, we should find functional forms of $u_0(x)$ and $\delta(x)$ such that when they are substituted into Eqs. (10-38) and (10-39) the resulting expressions will be independent of x. To achieve this we assume that $u_0(x)$ and $\delta(x)$ depend on x in the form

$$u_0(x) = c_1 x^m \quad \text{and} \quad \delta(x) = c_2 x^n \tag{1}$$

where c_1, c_2, m, and n are constants. Equations (1) are substituted into Eqs. (10-38) and (10-39) to yield

$$\frac{(2m + n)c_1{}^2 c_2}{105} x^{2m+n-1} = \frac{1}{3} g\beta(T_w - T_\infty)c_2 x^n - \frac{vc_1}{c_2} x^{m-n} \tag{2}$$

$$\frac{(m + n)c_1 c_2}{30} x^{m+n-1} = \frac{2\alpha}{c_2} x^{-n} \tag{3}$$

If similarity solutions exist, both sides of these equations should be independent of x; this is possible if the exponents of x have the same value everywhere for each of these equations. Then, by equating the exponents of x, Eqs. (2) and (3) give, respectively,

$$2m + n - 1 = n = m - n \tag{4a}$$

$$m + n - 1 = -n \tag{4b}$$

These two relations establish the values of m and n as

$$m = \tfrac{1}{2} \quad \text{and} \quad n = \tfrac{1}{4} \tag{5}$$

To determine the unknown coefficients c_1 and c_2, the numerical values of m and n as given by Eqs. (5) are introduced into Eqs. (2) and (3).   Of course the x variable cancels out and we obtain

$$\frac{1}{105}\frac{5}{4}c_1{}^2 c_2 = \frac{1}{3} g\beta(T_w - T_\infty)c_2 - \frac{vc_1}{c_2} \tag{6a}$$

$$\frac{1}{30}\frac{3}{4} = \frac{2\alpha}{c_2} \tag{6b}$$

A simultaneous solution of Eqs. (6) gives the values of the coefficients c_1 and c_2 as

$$c_1 = 5.17v\left(\frac{20}{21} + \frac{\alpha}{v}\right)^{-1/2}\left[\frac{g\beta(T_w - T_\infty)}{v^2}\right]^{1/2} \tag{7a}$$

$$c_2 = 3.93\left(\frac{20}{21} + \frac{v}{\alpha}\right)^{1/4}\left[\frac{g\beta(T_w - T_\infty)}{v^2}\right]^{-1/4}\left(\frac{v}{\alpha}\right)^{-1/2} \tag{7b}$$

Substituting the above values of n and c_2 into Eq. (1), the boundary-layer thickness becomes

$$\delta(x) = 3.93(0.952 + \text{Pr})^{1/4}\left[\frac{g\beta(T_w - T_\infty)}{v^2}\right]^{-1/4}\text{Pr}^{-1/2} x^{1/4} \tag{8}$$

or

$$\frac{\delta(x)}{x} = 3.93\text{Pr}^{-1/2}(0.952 + \text{Pr})^{1/4}\,\text{Gr}_x{}^{-1/4} \tag{9}$$

where the local Grashof number Gr_x is defined as

$$\text{Gr}_x = \frac{g\beta(T_w - T_\infty)x^3}{v^2} \tag{10}$$

Eleven

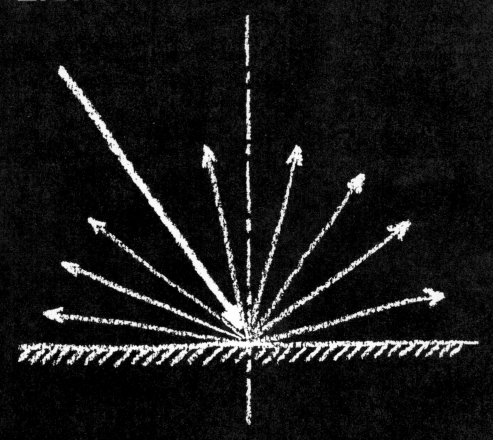

Radiation—
Basic Relations

In the preceding chapters we discussed heat transfer by the mechanisms of conduction and convection, both of which require an intervening carrier for the transport of thermal energy from the region of high temperature to the region of low temperature.   The energy transfer by thermal radiation, however, does not require an intervening carrier because the radiative energy is considered transported as photons or as electromagnetic waves.   If the space between a hot and a cold body is a vacuum, the radiative-energy transfer between them takes place with the speed of light in a vacuum and there is no attenuation of radiative energy as it travels through such a medium.   On the other hand, the presence of a partially absorbing medium in the space between them impedes the transport process.   With this consideration, the analysis of radiative transfer in this book will be separated into two distinct parts: (1) radiative transfer in a *nonparticipating medium* that deals with the problem of radiative-heat exchange between surfaces that are separated by a vacuum or by a medium that does not intervene with the propagation of radiation, and (2) the radiative transfer in a *participating medium* that deals with the situation in which the medium in the path of radiation intervenes with its propagation.   In the former case the radiative exchange is considered a surface phenomenon, and the analysis of such problems is relatively straightforward.   In the latter case the radiative transfer is analyzed as a bulk process because the intervening medium may absorb, emit, or scatter radiation; the scattering effects will not be considered in the present analysis because the analysis of radiative transfer with scattering is a very complicated matter and beyond the scope of this book.

In this chapter we present the various concepts for energy transport by radiation and discuss the basic laws for the emission and absorption of radiation by matter.   This information is needed in Chaps. 12 and 13 which will deal with the actual process of radiative-heat exchange in a nonparticipating and a participating medium, respectively.   The reader interested in a more comprehensive treatment of the subject is referred to the books by Hottel and Sarofim [1], Love [2], Özışık [3], Siegel and Howell [4], and Sparrow and Cess [5].

11-1   NATURE OF THERMAL RADIATION

Thermal radiation refers to radiation energy emitted by bodies because of their own temperature.   Other types of radiation include, for example, x-rays, gamma rays, cosmic rays, and so on.   X-rays are produced by the bombardment of a metal with high-frequency electrons and gamma rays by the fission of nuclei or by radioactive disintegration.   The actual mechanism of the emission and the propagation of radiation, however, is not fully understood.   Radiation is sometimes treated as electromagnetic waves that propagate according to Maxwell's classic electromagnetic theory, or it is treated as photons that derive its basis from Max Planck's concept of the quantum of energy.   Both of these concepts have been utilized in describing the emission of radiation and its propagation

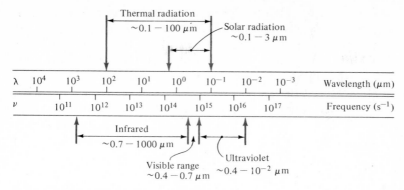

FIG. 11-1 Electromagnetic-wave spectrum.

in a medium. For example, the results obtained from the electromagnetic theory have been utilized to predict the radiative properties of materials, whereas the results from the quantum theory are useful to predict the amount of radiative energy emitted by a body because of its temperature. Figure 11-1 shows the electromagnetic-wave spectrum and typical subdivision of the spectrum where the major proportion of the energy in different types of radiation occurs. Although the range of thermal radiation extends, theoretically, from a wavelength of zero to infinity, for practical purposes the major portion of the energy of thermal radiation lies in the range from about 0.1 to 100 μm. Therefore, in Fig. 11-1 this portion of the electromagnetic-wave spectrum is indicated as *thermal radiation*. The visible part of the thermal radiation, for example, ranges from 0.4 to 0.7 μm. When radiation is treated as an electromagnetic wave, its propagation in a medium takes place with the speed of light, c; then the wavelength and the frequency of radiation are related by

$$c = \lambda v$$

where c = speed of light in medium
λ = wavelength
v = frequency

When the medium in which radiation travels is a vacuum, the speed of propagation is $c = 2.9979 \times 10^8$ m/s.

11-2   RADIATION INTENSITY AND BLACKBODY RADIATION

Radiation propagates as a beam, and the wavelength λ (or the frequency v) at which it is generated is important in the characterization of the energy carried by the beam. A fundamental quantity that is used to describe the amount of radiation energy propagating in a given direction $\hat{\Omega}$, at a wavelength λ, at a position $\mathbf{r}$ is the *spectral radiation intensity* $I_\lambda(\mathbf{r}, \hat{\Omega})$. It represents the amount *of energy streaming through a unit area perpendicular to the direction $\hat{\Omega}$, per unit time, per unit solid angle about the direction $\hat{\Omega}$ and per unit wavelength about the wavelength λ.* The *radiation intensity* $I(\mathbf{r}, \hat{\Omega})$ is used to characterize the amount of energy emitted over the entire wavelength spectrum from $\lambda = 0$ to ∞ in a beam and is related to the spectral radiation intensity $I_\lambda(\mathbf{r}, \hat{\Omega})$ by

$$I(\mathbf{r}, \hat{\Omega}) = \int_{\lambda=0}^{\infty} I_\lambda(\mathbf{r}, \hat{\Omega})\, d\lambda \tag{11-1}$$

Clearly, the radiation intensity I represents the *amount of radiant energy streaming through a unit area perpendicular to the direction of propagation $\hat{\Omega}$, per unit time, per unit solid angle about the direction $\hat{\Omega}$.*

We now consider a radiation intensity $I(\mathbf{r}, \hat{\Omega})$ contained within a solid angle $d\Omega$ to (or from) a surface element dA, propagating in the direction $\hat{\Omega}$ making an angle θ with the normal $\hat{n}$ to the surface element, as illustrated in Fig. 11-2. The quantity dq,

$$dq = I(\mathbf{r}, \hat{\Omega}) \cos \theta \, d\Omega \tag{11-2}$$

represents the amount of radiant energy per unit time, to (or from) per unit area of the surface due to radiation contained within a solid angle $d\Omega$. Here, $\cos \theta$ is included because, in the definition of the intensity $I(\mathbf{r}, \hat{\Omega})$, the energy is considered streaming per unit area perpendicular to the direction of propagation. The *radiative-energy flux* q to (or from) a surface due to radiation contained in a solid angle over an entire hemisphere is obtained by the integration of Eq. (11-2) as

$$q = \int_{\Omega} I \cos \theta \, d\Omega \tag{11-3a}$$

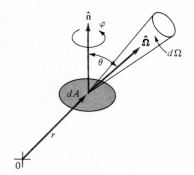

FIG. 11-2   Symbols for radiation intensity.

where the symbol $\cap$ denotes integration with respect to a solid angle over an entire hemisphere. If θ is the polar angle, φ is the azimuthal angle as illustrated in Fig. 11-2, then the above integration is written as

$$q = \int_{\varphi=0}^{2\pi} \int_{\theta=0}^{\pi/2} I(\mathbf{r}, \theta, \varphi) \cos \theta \sin \theta \, d\theta \, d\varphi \qquad (11\text{-}3b)$$

since $d\Omega = \sin \theta \, d\theta \, d\varphi$. It is apparent that the *radiative-energy flux* q *has dimensions in energy per unit time, per unit area of the surface* (that is, $\text{Btu/h} \cdot \text{ft}^2$).

Blackbody Radiation

There is a maximum amount of radiant energy that can be emitted at a given absolute temperature T at a wavelength λ by a body. The terminology *spectral blackbody radiation intensity* $I_{\lambda b}(T)$ is used to denote this maximum amount of radiation emission, and the emitter of such radiation is called a *blackbody*. The spectral blackbody radiation intensity $I_{\lambda b}(T)$ is independent of direction, but it is a function of the wavelength and the absolute temperature T of the blackbody. For a blackbody at an absolute temperature T and emitting radiation into a vacuum, $I_{\lambda b}(T)$ is determined from the expression given by Planck [6] in the form

$$I_{\lambda b}(T) = \frac{2hc^2}{\lambda^5[\exp{(hc/\lambda kT)} - 1]} \qquad (11\text{-}4)$$

where h ($= 6.6256 \times 10^{-34}$ J·s) and k ($= 1.38054 \times 10^{-23}$ J·K) are the Planck and Boltzmann constants, respectively, c is the speed of light in a vacuum, T is the absolute temperature, and λ is the wavelength.

In engineering applications the *spectral blackbody emissive flux* $q_{\lambda b}(T)$ at a surface is of interest; it is determined from the definition

$$q_{\lambda b}(T) = \int_{\cap} I_{\lambda b}(T) \cos \theta \, d\Omega \qquad (11\text{-}5a)$$

and the integration can be performed since $I_{\lambda b}(T)$ is independent of direction.

$$q_{\lambda b}(T) = I_{\lambda b}(T) \int_{\varphi=0}^{2\pi} \int_{\theta=0}^{\pi/2} \cos \theta \sin \theta \, d\theta \, d\varphi = \pi I_{\lambda b}(T) \qquad (11\text{-}5b)$$

Here $q_{\lambda b}(T)$ represents the *amount of radiative energy emitted by a blackbody at temperature T per unit of its surface, per unit time, per unit wavelength in all directions in the hemispherical space.*

The substitution of Eq. (11-4) into Eq. (11-5b) yields

$$q_{\lambda b}(T) = \frac{c_1}{\lambda^5[\exp{(c_2/\lambda T)} - 1]} \tag{11-6}$$

where $q_{\lambda b}(T)$ = spectral blackbody emissive flux at surface, Btu/h·ft²·μm
                   (W/m²·μm)
           T = absolute temperature of blackbody, °R (K)
           λ = wavelength, μm

$$c_1 = 2\pi hc^2 = 1.1870 \times 10^8 \text{ Btu } \mu m^4/ft^2 \cdot h = 3.743 \times 10^8 \text{ W} \cdot \mu m^4/m^2$$

$$c_2 = \frac{hc}{k} = 2.5896 \times 10^4 \ \mu m \cdot °R = 1.4387 \times 10^4 \ \mu m \cdot K$$

Figure 11-3 shows a plot of the spectral blackbody emissive flux as a function of wavelength at various temperatures. It is evident from this figure that at any given wavelength the radiative energy emitted by a blackbody increases as the absolute temperature of the body increases. Each curve shows a peak, and the peaks tend to shift toward smaller wavelengths as the temperature rises.

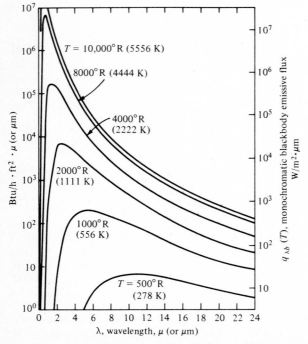

FIG. 11-3   Spectral blackbody emissive flux at different temperatures.

The locus of the peaks determined analytically by *Wien's displacement* rule is given as

$$(\lambda T)_{max} = 5215.6 \ \mu m \cdot {}^\circ R = 0.28976 \ cm \cdot K = 28997.6 \ \mu m \cdot K$$

The *blackbody radiation intensity* $I_b(T)$ is determined by the integration of $I_{\lambda b}(T)$ over all wavelengths $\lambda = 0$ to $\lambda = \infty$.

$$I_b(T) = \int_{\lambda=0}^{\infty} I_{\lambda b}(T) \ d\lambda \tag{11-7a}$$

Substituting Eq. (11-4) into Eq. (11-7a) and performing the integration yield

$$I_b(T) = \frac{\sigma T^4}{\pi} \tag{11-7b}$$

where σ is called the *Stefan-Boltzmann* constant; its numerical value is

$$\sigma = 0.1714 \times 10^{-8} \ Btu/h \cdot ft^2 \cdot {}^\circ R^4 = 0.56697 \times 10^{-8} \ W/m^2 \cdot K^4$$

The *blackbody emissive flux* $q_b(T)$ at a temperature T is obtained by integrating $q_{\lambda b}(T)$ over all wavelengths:

$$q_b(T) = \int_{\lambda=0}^{\infty} q_{\lambda b}(T) \ d\lambda = \pi \int_{\lambda=0}^{\infty} I_{\lambda b}(T) \ d\lambda = \pi I_b(T) = \sigma T^4 \tag{11-8a}$$

which has the dimensions of energy per unit time, per unit area.

TABLE 11-1   Numerical Values of Radiation Constants†

Quantity	$(cgs)_1$	$(cgs)_2$	Btu	SI
$q_{\lambda b}$	$erg/s \cdot cm^2 \cdot cm$	$W/cm^2 \cdot \mu m$	$Btu/h \cdot ft^2 \cdot \mu m$	$W/m^2 \cdot \mu m$
q_b	$erg/s \cdot cm^2$	W/cm^2	$Btu/h \cdot ft^2$	W/m^2
λ	cm	μm	μm	μm
c_1	3.704×10^{-5} $erg \cdot cm^2/s$	$37404 \ W \cdot \mu m^4/cm^2$	1.1870×10^8 $Btu \cdot \mu m^4/h \cdot ft^2$	3.7438×10^8 $W \cdot \mu m^4/m^2$
c_2	$1.4387 \ cm \cdot K$	$14387 \ \mu m \cdot K$	$25,896 \ \mu m \cdot {}^\circ R$	$14,387 \ \mu m \cdot K$
σ	5.6699×10^{-5} $erg/s \cdot cm^2 \cdot K^4$	5.6699×10^{-12} $W/cm^2 \cdot K^4$	1.714×10^{-9} $Btu/h \cdot ft^2 \cdot {}^\circ R^4$	5.6697×10^{-9} $W/m^2 \cdot K^4$
$(\lambda T)_{max}$	$0.28976 \ cm \cdot K$	$2897.6 \ \mu m \cdot K$	$5215.6 \ \mu m \cdot {}^\circ R$	$2897.6 \ \mu m \cdot K^4$

† From Snyder [7].

TABLE 11-2 Fractional Function of the First Kind[†]

λT $\mu m \cdot K$	λT $\mu m \cdot °R$	$f_{0-\lambda T} = \dfrac{q_{0-\lambda, b}}{q_b(T)}$	λT $\mu m \cdot K$	λT $\mu m \cdot °R$	$f_{0-\lambda T} = \dfrac{q_{0-\lambda, b}}{q_b(T)}$	λT $\mu m \cdot K$	λT $\mu m \cdot °R$	$f_{0-\lambda T} = \dfrac{q_{0-\lambda, b}}{q_b(T)}$
555.6	1000	0	4000.0	7200	0.4809	7444.4	13400	0.8317
666.7	1200	0	4111.1	7400	0.5007	7555.6	13600	0.8370
777.7	1400	0	4222.2	7600	0.5199	7666.7	13800	0.8421
888.9	1600	0.0001	4333.3	7800	0.5381	7777.8	14000	0.8470
1000.0	1800	0.0003	4444.4	8000	0.5558	7888.9	14200	0.8517
1111.1	2000	0.0009	4555.6	8200	0.5727	8000.0	14400	0.8563
1222.2	2200	0.0025	4666.7	8400	0.5890	8111.1	14600	0.8606
1333.3	2400	0.0053	4777.8	8600	0.6045	8222.2	14800	0.8648
1444.4	2600	0.0098	4888.9	8800	0.6195	8333.3	15000	0.8688
1555.6	2800	0.0164	5000.0	9000	6.6337	8888.9	16000	0.8868
1666.6	3000	0.0254	5111.1	9200	0.6474	9444.4	17000	0.9017
1777.7	3200	0.0368	5222.2	9400	0.6606	10000.0	18000	0.9142
1888.9	3400	0.0506	5333.3	9600	0.6731	10555.6	19000	0.9247
2000.0	3600	0.0667	5444.4	9800	0.6851	11111.1	20000	0.9335
2111.1	3800	0.0850	5555.6	10000	0.6966	11666.7	21000	0.9411
2222.2	4000	0.1051	5666.7	10200	0.7076	12222.2	22000	0.9475

2333.3	4200	0.1267	5777.8	10400	0.7181	12777.8	23000	0.9531
2444.4	4400	0.1496	5888.9	10600	0.7282	13333.3	24000	0.9589
2555.6	4600	0.1734	6000.0	10800	0.7378	13888.9	25000	0.9621
2666.7	4800	0.1979	6111.1	11000	0.7474	14444.4	26000	0.9657
2777.7	5000	0.2229	6222.2	11200	0.7559	15000.0	27000	0.9689
2888.9	5200	0.2481	6333.3	11400	0.7643	15555.6	28000	0.9718
3000.0	5400	0.2733	6444.4	11600	0.7724	16111.1	29000	0.9742
3111.1	5600	0.2983	6555.6	11800	0.7802	16666.7	30000	0.9765
3222.2	5800	0.3230	6666.7	12000	0.7876	22222.2	40000	0.9881
3333.3	6000	0.3474	6777.8	12200	0.7947	27777.7	50000	0.9941
3444.4	6200	0.3712	6888.9	12400	0.8015	33333.3	60000	0.9963
3555.6	6400	0.3945	7000.0	12600	0.8081	38888.9	70000	0.9981
3666.7	6600	0.4171	7111.1	12800	0.8144	44444.4	80000	0.9987
3777.8	6800	0.4391	7222.2	13000	0.8204	50000.0	90000	0.9990
3888.9	7000	0.4604	7333.3	13200	0.8262	55555.5	100000	0.9992
						∞	∞	1.0000

† From Dunkle [8].

It is apparent from Eq. (11-8a) that the blackbody radiation intensity $I_b(T)$ and the blackbody emissive flux $q_b(T)$ are related by

$$I_b(T) = \frac{q_b(T)}{\pi} \tag{11-8b}$$

Table 11-1 summarizes in different systems of units the numerical values of the constants c_1, c_2 of Eq. (11-6), the Stefan-Boltzmann constant σ, Wien's displacement rule, and the corresponding units for $q_{\lambda b}$ and q_b.

A quantity of interest in engineering applications is the blackbody emission in the range $\lambda = 0$ to λ, $q_{0-\lambda, b}(T)$, given by the expression

$$q_{0-\lambda, b}(T) = \int_{\lambda=0}^{\lambda} q_{\lambda b}(T)\, d\lambda \tag{11-9}$$

To compute this quantity readily, a dimensionless *fractional function of the first kind*, $f_{0-\lambda}(T)$, is defined as

$$f_{0-\lambda}(T) = \frac{q_{0-\lambda, b}(T)}{q_b(T)} = \frac{q_{0-\lambda, b}(T)}{\sigma T^4} \tag{11-10}$$

and the function $f_{0-\lambda}(T)$ has been calculated by Dunkle [8] and tabulated in Table 11-2 as a function of λT.

The concept of blackbody as discussed above refers to an idealized body that possesses the property of allowing all incident radiation to enter the medium without surface reflection and without allowing it to leave the medium again. Hence a blackbody absorbs all incident radiation from all directions at all frequencies without reflecting, transmitting, or scattering it out. The radiation-emission characteristics of a blackbody are understood better if one envisions a blackbody inside an isothermal enclosure whose boundaries absorb and emit radiation. When the blackbody and the enclosure walls are in thermal equilibrium, the blackbody emits as much radiation as it absorbs. Then, the emission of radiation by a blackbody must be the maximum possible at the considered temperature since the blackbody absorbs the maximum possible radiation from all directions and at all frequencies. In reality there is no such object as a blackbody. However, for experimental purposes, a blackbody can be approximated by a cavity, such as a hollow sphere whose interior surfaces are maintained at a uniform temperature T. If an opening which is very small compared with the size of the cavity is provided, then any radiation entering this cavity through this opening is almost entirely absorbed since it has very little chance to escape through the hole. Therefore, such a cavity is considered to approximate closely a blackbody. Similarly, radiation coming out of the hole of the cavity is considered almost a blackbody radiation at temperature T.

11-3   RADIATION TO AND FROM REAL SURFACES

In the analysis of radiative-heat exchange between surfaces that will be discussed in the next chapter, the radiation to and from a surface is of interest. Strictly speaking, the surface of a body alone never emits and absorbs radiation, because the emission and absorption of radiation by a body constitutes a bulk process.   That is, radiation coming from the interior of the body passes through the surface, or radiation incident on the surface penetrates into the medium where it is gradually attenuated.   However, in situations where a large portion of the incident radiation is attenuated within a very short distance from the surface, we speak, for the sake of simplicity, of radiation being absorbed by the surface.   For example, thermal radiation incident on a metal surface will not travel more than a few hundred angstroms before it is completely absorbed. Similarly thermal radiation originating from the interior of this body never reaches the surface.   In such cases radiation to and from a body is treated as a surface process and the surface is said to be *opaque* to radiation.   Furthermore, most surfaces encountered in engineering applications do not behave like black-bodies.   With this limitation to the concept of surface radiation, we now discuss the radiative properties of surfaces.

Reflectivity, Absorptivity, and Emissivity of Opaque Surfaces

Consider that a beam of radiation is incident on an opaque, real surface element. Part of this radiation is reflected and the rest is absorbed by the surface.   If I'_λ is the spectral radiation intensity incident on the surface, the spectral radiative heat flux incident on the surface is given by

$$q'_\lambda = \int_\Omega I'_\lambda \cos\theta' \, d\Omega' \qquad \text{energy/(time} \times \text{area} \times \text{wavelength)} \qquad (11\text{-}11)$$

where θ' is the polar angle between the direction of the incident radiation and the normal to the surface.   The *spectral hemispherical reflectivity* ρ_λ is then defined as

$$\rho_\lambda = \frac{\text{radiant energy reflected/(time} \times \text{area} \times \text{wavelength)}}{q_\lambda} \qquad (11\text{-}12a)$$

and the *spectral hemispherical absorptivity* α_λ as

$$\alpha_\lambda = \frac{\text{radiant energy absorbed/(time} \times \text{area} \times \text{wavelength)}}{q_\lambda} \qquad (11\text{-}12b)$$

For an opaque surface, ρ_λ and α_λ are related by

$$\rho_\lambda + \alpha_\lambda = 1 \tag{11-13}$$

In many engineering applications the *reflectivity* ρ and the *absorptivity* α averaged over the entire wavelengths is of interest.   The terminology *total reflectivity* and *total absorptivity* is also used to designate the averaging over the entire wavelengths; however, for brevity we shall omit the word *total* in the present definitions.   Then, the *hemispherical reflectivity* ρ is defined as

$$\rho = \frac{\int_0^\infty \rho_\lambda q_\lambda'\, d\lambda}{\int_0^\infty q_\lambda'\, d\lambda} \tag{11-14a}$$

and the *hemispherical absorptivity* α as

$$\alpha = \frac{\int_0^\infty \alpha_\lambda q_\lambda'\, d\lambda}{\int_0^\infty q_\lambda'\, d\lambda} \tag{11-14b}$$

For an opaque surface, α and ρ are related by

$$\alpha + \rho = 1 \tag{11-15}$$

Thus, if ρ is known, α is regarded known, and vice versa.

To characterize the properties of a surface for the emission of radiation is important in radiative-transfer applications.   The radiative energy emitted by a real surface at a temperature T is always less than that emitted by a black surface at the same temperature.   If $q_\lambda(T)$ is the *spectral emissive flux* [i.e., radiative energy emitted/(time × area × wavelength)] from a real surface at an absolute temperature T, and $q_{\lambda b}(T)$ is the *spectral blackbody emissive flux* for a black surface at the same temperature, the *spectral hemispherical emissivity* ε_λ of the surface is defined as

$$\varepsilon_\lambda = \frac{q_\lambda(T)}{q_{\lambda b}(T)} \tag{11-16}$$

and the *hemispherical emissivity* ε over the entire wavelengths is determined as

$$\varepsilon = \frac{\int_{\lambda=0}^\infty \varepsilon_\lambda q_{\lambda b}(T)\, d\lambda}{\int_{\lambda=0}^\infty q_{\lambda b}(T)\, d\lambda} = \frac{q(T)}{q_b(T)} \tag{11-17}$$

where $q(T)$ is the emissive flux from the real surface at temperature T [i.e., radiant energy/(time × area)].

In the foregoing discussion we defined the hemispherical reflectivity of a surface. However, depending on the manner and the direction that the incident radiation

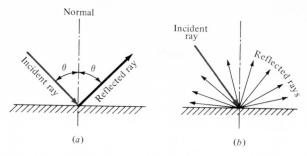

FIG. 11-4   Specular and diffuse reflection from surfaces. (a) Specular reflection; (b) diffuse reflection.

strikes the surface and on the manner the reflected energy is collected, several different reflectivity definitions have been used in the literature to characterize the reflection characteristics of surfaces.   The reader should consult Refs. [3–5] for the discussion of various types of reflectivity definitions.   However, for most engineering applications the hemispherical reflectivity as defined above is frequently used and considered satisfactory, even though it does not provide any information as to the angular distribution of the reflected energy.   On the other hand for most practical applications, it is not feasible to require such detailed information on reflection.

Finally, two idealized limiting types of reflections should be distinguished. The reflection is called *specular* if the incident and the reflected rays lie symmetrically with respect to the normal at the point of incidence, as illustrated in Fig. 11-4a.   The reflection is called *diffuse* if the intensity of the reflected radiation is constant for all angles of reflection and is independent of the direction of incident radiation.   Figure 11-4b illustrates a diffuse reflection from a surface. A real surface encountered in engineering applications is neither a perfectly diffuse nor a perfectly specular reflector.

Transmissivity of a Semitransparent Body

When radiation is incident on a semitransparent body of finite thickness, for example, on a glass plate as illustrated in Fig. 11-5, part of the incident radiation is reflected, part is absorbed, and the remainder is transmitted through the glass. With this consideration, we speak of the *spectral reflectivity* ρ_λ, *spectral absorptivity* α_λ, and *spectral transmissivity* τ_λ of the glass plate such that

$$\alpha_\lambda + \rho_\lambda + \tau_\lambda = 1 \tag{11-18}$$

When the averaged radiative properties over the entire wavelengths are considered, this relation is written

$$\alpha + \rho + \tau = 1 \tag{11-19}$$

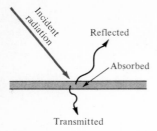

FIG. 11-5 Reflection, absorption, and transmission of incident radiation by a semitransparent material.

The reflectivity, absorptivity, and transmissivity of a semitransparent material depend not only on the surface conditions and the wavelength of the radiation but also on the composition of the material and the thickness of the body, since the radiation incident on the surface penetrates into the depths of the matter. The determination of the reflectivity and transmissivity of a semitransparent material is more involved because the attenuation of radiation within the body should be treated as a bulk process.

The terms "reflectivity, absorptivity, emissivity" are used by some authors to characterize radiation properties of *ideal surfaces* (i.e., those optically smooth and perfectly uncontaminated), and the terms "reflectance, absorptance, emittance" are used to characterize the radiation properties of real surfaces. We believe there is no need for a change in nomenclature just because the surface condition is changed; therefore we prefer to use the terms "reflectivity, absorptivity, and emissivity" for all cases in this book.

Kirchhoff's Law of Radiation

The absorptivity and the emissivity of a body can be related by Kirchhoff's law of radiation, as now discussed, by thermodynamic considerations. The reader should consult Ref. [6] for a rigorous derivation of this law.

Consider a body placed inside a perfectly black, closed container whose walls are maintained at a uniform temperature T and allowed to reach an equilibrium with the walls of the container. Let $q'_\lambda(T)$ be the spectral radiative-heat flux from the walls at temperature T *incident* upon the body. The spectral radiative-heat flux $q_\lambda(T)$ *absorbed* by the body at the wavelength λ is

$$q_\lambda(T) = \alpha_\lambda(T) q'_\lambda(T) \tag{11-20}$$

where $\alpha_\lambda(T)$ is the spectral absorptivity of the body. The quantity $q_\lambda(T)$ also represents the spectral radiative flux *emitted* by the body at the wavelength λ since the body is in radiative equilibrium. We note that the incident radiation $q'_\lambda(T)$ is coming from the perfectly black walls of the enclosure at temperature T

and the emission by the walls is unaffected whether the body introduced into the enclosure is a blackbody or not; with this consideration, we have

$$q_{\lambda b}(T) = q'_\lambda(T) \tag{11-21}$$

where $q_{\lambda b}(T)$ is the spectral blackbody emissive flux at temperature T. From Eqs. (11-20) and (11-21) we write

$$\frac{q_\lambda(T)}{q_{\lambda b}(T)} = \alpha_\lambda(T) \tag{11-22}$$

The spectral emissivity $\varepsilon_\lambda(T)$ of the body for radiation at temperature T is defined as the ratio of the spectral emissive flux $q_\lambda(T)$ of the body to the spectral black-body emissive flux $q_{\lambda b}(T)$ at the same temperature, that is,

$$\frac{q_\lambda(T)}{q_{\lambda b}(T)} = \varepsilon_\lambda(T) \tag{11-23}$$

From Eqs. (11-22) and (11-23) we find

$$\varepsilon_\lambda(T) = \alpha_\lambda(T) \tag{11-24}$$

which is the Kirchhoff law of radiation stating that *spectral emissivity for the emission of radiation at temperature T is equal to the spectral absorptivity for radiation coming from a blackbody at the same temperature T.*

Care must be exercised in the generalization of Eq. (11-24) for the spectral values of ε_λ and α_λ to the values averaged over the entire wavelength. That is, the relation given by Eq. (11-24) is always valid, but the relation

$$\varepsilon(T) = \alpha(T) \tag{11-25}$$

is applicable when the incident and emitted radiation have the same spectral distribution or when the body is *gray*, that is, the radiative properties are independent of wavelength.

Graybody

To simplify the analysis of radiative-heat transfer the *graybody* assumption is frequently made in many applications; that is, the radiative properties α_λ, ε_λ, and ρ_λ are assumed to be uniform over the entire wavelength spectrum. Such bodies are referred to as *graybodies*, and under the graybody assumption the absorptivity and the emissivity are related by the Kirchhoff law as $\alpha = \varepsilon$.

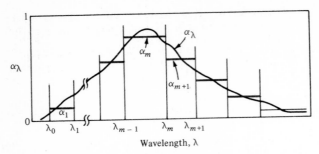

FIG. 11-6   Band approximation of spectral absorptivity α_λ over the entire wavelength.

Example 11-1   The spectral absorptivity α_λ for an incident spectral blackbody radiation flux $q'_{\lambda b}(T)$ at temperature T is approximated in bands, as illustrated in Fig. 11-6, in such a manner that over a wavelength band from $\lambda = \lambda_{m-1}$ to $\lambda = \lambda_m$ the absorptivity α_m, $m = 1, 2, 3, \ldots, M$, is taken to be uniform.   Show that the fractional functions of the first kind can be used to determine the average hemispherical absorptivity α over the entire wavelength by using the band absorptivity data as given in Fig. 11-6.

Solution   The average hemispherical absorptivity α over the entire wavelength is determined from the spectral absorptivity α for an incident radiation $q'_{\lambda b}(T)$ by Eq. (11-14b) as

$$\alpha = \frac{\int_0^\infty \alpha_\lambda q'_{\lambda b}(T)\, d\lambda}{\int_0^\infty q'_{\lambda b}(T)\, d\lambda} = \frac{\int_0^\infty \alpha_\lambda q'_{\lambda b}(T)\, d\lambda}{q'_b(T)} \tag{11-26}$$

When the spectral absorptivity is specified in the form of bands as illustrated in Fig. 11-6, the integral in Eq. (11-26) is replaced by a summation over the entire wavelengths as

$$
\begin{aligned}
\alpha &= \frac{\displaystyle\sum_{m=1}^{M} \alpha_m \int_{\lambda_{m-1}}^{\lambda_m} q'_{\lambda b}(T)\, d\lambda}{q'_b(T)} \\[2mm]
&= \sum_{m=1}^{M} \alpha_m \left[\frac{\int_0^{\lambda_m} q'_{\lambda b}(T)\, d\lambda}{q'_b(T)} - \frac{\int_0^{\lambda_{m-1}} q'_{\lambda b}(T)\, d\lambda}{q'_b(T)} \right] \\[2mm]
&= \sum_{m=1}^{M} \alpha_m \left(f_{0-\lambda_m T} - f_{0-\lambda_{m-1} T} \right)
\end{aligned}
\tag{11-27}
$$

where we utilized the definition of the fractional functions of the first kind as given by Eqs. (11-9) and (11-10), and M is the number of wavelength bands.

The summation in Eq. (11-27) is readily computed since α_m's are given in Fig. 11-6 and the functions $f_{0-\lambda_m T}$ are obtainable from Table 11-2 once the temperature T and the wavelength λ_m are specified.

Example 11-2   Transmissivity of a plate glass for incident solar radiation at $T = 10{,}000°R$ (5555.6 K) is given in the band approximation as

$\tau_1 = 0$ for $\lambda_0 = 0$ to $\lambda_1 = 0.4\ \mu m$

$\tau_2 = 0.8$ for $\lambda_1 = 0.4\ \mu m$ to $\lambda_2 = 3.0\ \mu m$

$\tau_3 = 0$ for $\lambda_2 = 3.0\ \mu m$ to $\lambda_3 \rightarrow \infty\ \mu m$

Determine the average hemispherical transmissivity of the glass over the entire wavelengths from the above transmissivity data by utilizing the fractional functions of the first kind.

Solution The average hemispherical transmissivity τ over the entire wavelengths is determined from the spectral transmissivity τ_λ for an incident radiation $q'_{\lambda b}(T)$ by the following relation:

$$\tau = \frac{\int_0^\infty \tau_\lambda q'_{\lambda b}(T)\, d\lambda}{\int_0^\infty q'_{\lambda b}(T)\, d\lambda} = \frac{\int_0^\infty \tau_\lambda q'_{\lambda b}(T)\, d\lambda}{q_b(T)} \tag{11-28}$$

The integral in this equation is now transformed into a summation involving the band values of the transmissivity τ_m and the fractional functions of the first kind by following a similar approach described in the previous example. We find

$$\tau = \sum_{m=1}^{M} \tau_m \left(f_{0-\lambda_m T} - f_{0-\lambda_{m-1} T} \right) \tag{11-29}$$

In the present example we have $M = 3$, $\tau_1 = 0$, $\tau_2 = 0.8$, $\tau_3 = 0$, and $T = 10,000°R$ (5555.6 K). Then Eq. (11-29) becomes

$$\tau = 0.8(f_{0-\lambda_2 T} - f_{0-\lambda_1 T}) \tag{11-30}$$

where the functions $f_{0-\lambda_1 T}$ and $f_{0-\lambda_2 T}$ are obtained from Table 11-2 for the values of

$\lambda_1 T = 0.4 \times 10,000 = 4000\ \mu\text{m} \cdot °\text{R}$ (or $0.4 \times 5555.6 = 2222.2\ \mu\text{m} \cdot \text{K}$)

$\lambda_2 T = 3 \times 10,000 = 30,000\ \mu\text{m} \cdot °\text{R}$ (or $3 \times 5555.6 = 16666.7\ \mu\text{m} \cdot \text{K}$)

We find $f_{0-\lambda_1 T} = 0.1051$ and $f_{0-\lambda_2 T} = 0.9765$. Then, the average hemispherical transmissivity of the glass for the solar radiation becomes

$$\tau = 0.8(0.9765 - 0.1051) \cong 0.70$$

11-4 RADIATION TO AND FROM A VOLUME ELEMENT

A beam of radiation incident upon a semitransparent body penetrates into the depth of the medium where it is attenuated as a result of absorption and in some cases scattering by the material. The scattering occurs if the medium contains inhomogeneities, such as extremely small particles distributed over the body. For example, dust particles or water droplets in the atmosphere scatter light traveling through it, or very small air bubbles inside a semitransparent plastic material scatter radiation traveling through it. Radiation is also emitted in a body because of its temperature; as the emitted radiation travels through the medium it is attenuated by absorption and scattering by the material. In this section we discuss the absorption and the emission of radiation by a volume element, but the scattering is not considered because it is a complicated matter and its analysis is beyond the scope of this work. Fortunately, for a large class of engineering applications the scattering of thermal radiation by the matter is negligible.

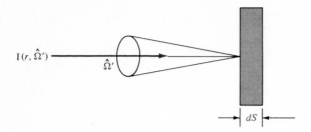

FIG. 11-7   Symbols for the absorption of radiation by a volume element.

Absorption of Radiation by a Volume Element

Consider that a beam of radiation of intensity $I(\mathbf{r}, \boldsymbol{\Omega})$ propagating in the direction of $\boldsymbol{\Omega}$ passes through a volume element at position $\mathbf{r}$ as illustrated in Fig. 11-7. As radiation propagates through the volume element, part of it is absorbed by the material.   To determine the amount of absorption of radiation by the material, a *volumetric absorption coefficient* κ is defined such that the quantity

$$\kappa I(\mathbf{r}, \boldsymbol{\Omega}) \qquad \text{energy/(time} \times \text{volume} \times \text{solid angle)} \qquad (11\text{-}31)$$

represents the *amount of radiation energy absorbed by the matter per unit time, per unit volume, and per unit solid angle.*   Noting that the radiation intensity $I(\mathbf{r}, \boldsymbol{\Omega})$ has dimensions energy/(time $\times$ area $\times$ solid angle), we conclude that the absorption coefficient κ has a dimension length^{-1}.

If the absorption of the spectral radiation of intensity $I_\lambda(\mathbf{r}, \boldsymbol{\Omega})$ by the matter is considered, a *spectral volumetric absorption coefficient* κ_λ is introduced such that

$$\kappa_\lambda I_\lambda(\mathbf{r}, \boldsymbol{\Omega}) \qquad \text{energy/(time} \times \text{volume} \times \text{solid angle} \times \text{wavelength)} \qquad (11\text{-}32)$$

represents the *amount of radiation energy absorbed by the matter per unit time, per unit volume, per unit solid angle, and per unit wavelength.*   Also in this case, κ_λ has a dimension length^{-1} since the spectral radiation intensity is given in energy/(time $\times$ area $\times$ solid angle $\times$ wavelength)

Emission of Radiation by a Volume Element

We now consider a medium which is in *local thermodynamic equilibrium,* that is, each point of the medium is characterized by a local temperature T.   Assuming that the Kirchhoff law is valid, the emission of radiation energy by the matter at temperature T *per unit time, per unit volume, per unit solid angle* is given by

$$\kappa I_b(T) \qquad \text{energy}/(\text{time} \times \text{volume} \times \text{solid angle}) \qquad (11\text{-}33)$$

where $I_b(T)$ is the blackbody radiation intensity at temperature T; the absorption coefficient κ in length^{-1} is used to represent emission because the Kirchhoff law is assumed to be valid.

11-5   RADIATIVE PROPERTIES OF SURFACES

The reflectivity and absorptivity of opaque, ideal surfaces (i.e., a surface that is optically smooth and perfectly clean) can be determined by the electromagnetic-wave theory; a discussion of the results obtained in this manner are given in several references [1–5]. The real surfaces encountered in engineering greatly deviate from ideality as a result of roughness, oxidation, contamination, etc.; therefore, there is no reliable way to predict theoretically the radiative properties of real surfaces, and the experimental approach is the only way to determine the reflectivity and absorptivity of real surfaces. Clearly, the surface condition is among the most important parameters that affect the results; unfortunately most of the experimental data available in the literature suffer from not being able to describe the exact surface conditions during experiments, because no standardized, sufficiently accurate method is yet available to characterize the actual surface conditions. Hence care must be exercised in interpreting the experimental data on radiative properties in relation to the actual surface condition during the tests.

The literature contains a vast amount of data on radiative properties of surfaces. For example, Gubareff, Janssen, and Torborg [9] give a comprehensive compilation of emissivity and reflectivity from nearly 320 references. Singham [10], Hottel [11], Svet [12], and Wood, Deem, and Lucks [13] have presented tabulated data on radiative properties of surfaces. The most comprehensive tabulation of radiative properties has been given more recently by Touloukian and DeWitt [14] as a function of the wavelength of radiation, together with detailed information describing surface conditions. To illustrate the effects of various parameters on radiative properties of surfaces, typical experimental data on reflectivity and emissivity of surfaces are now presented.

Figure 11-8 illustrates the effects of temperature and oxidation on the hemispherical emissivity of metals. We note that oxidation increases emissivity. In the case of copper, for example, as shown in Fig. 11-9, the degree of oxidation is very important for emissivity. Figure 11-10 shows the effects of source temperature of the incident radiation on the absorptivity of several different materials at room temperature as determined by Sieber [15]. We note that the absorptivity of good conductors such as aluminum and graphite increases as source temperature increases, whereas the trend appears to be reversed for nonconductors. Finally, we present in Table 11-3 a compilation of emissivities for typical surfaces.

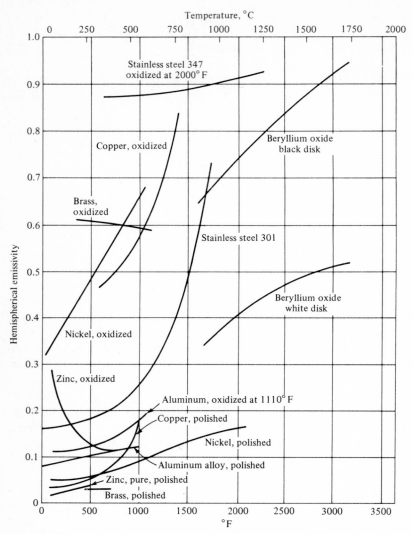

FIG. 11-8   Effects of temperature and oxidation on hemispherical emissivity of metals. (*Based on data from Gubareff et al.* [*9*].)

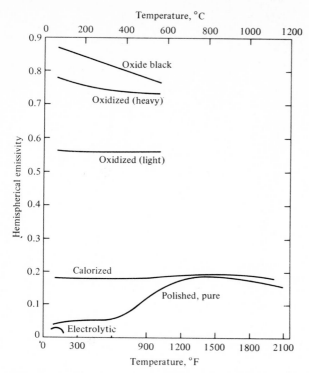

FIG. 11-9   Effects of various degrees of oxidation on the hemispherical emissivity of copper. (*From Gubareff et al.* [*9*].)

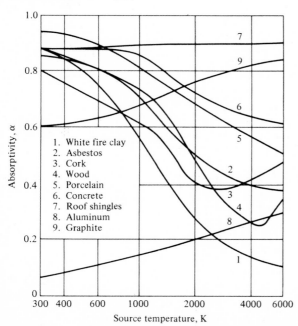

FIG. 11-10   Variation of absorptivity with source temperature of incident radiation for various common materials at room temperature. (*From Sieber* [*15*].)

TABLE 11-3 Emissivities of Various Surfaces †

Material	Temperature	ε
Asbestos:		
Board	100°F	0.96
Cloth	93°C	0.90
Paper	100°F	0.93
Brick:		
Red	0–93°C	0.93
Red, rough	0–200°C	0.93–0.95
Ceramic:		
Earthenware, glazed	20°C	0.90
Earthenware, matte	20°C	0.93
Porcelain	22°C	0.92
Refractory, black	93°C	0.94
Clay, fired	158°C	0.91
Concrete, rough	0–93°C	0.94
Glass	100°F	0.90
Glass, smooth	100°F	0.94
Ice	32°F	0.92–0.96
Lacquers:		
Black	93°C	0.96
Black, on iron	0–93°C	0.875
Clear, one thin coat on		
tarnished copper	93°C	0.64
Clear, two coats on tarnished		
copper	93°C	0.72
White	20°C	0.95
White, heavy coat on bright		
copper	93°C	0.93
Lampblack, 0.003 in or thicker	100°F	0.95
Marble, light-grey polished	22°C	0.93
Metals:		
Aluminum, polished	100°F	0.04
Aluminum, oxidized	100°F	0.20
Brass, polished	100°F	0.10
Brass, oxidized	100°F	0.46
Chromium, polished	100°F	0.08
Copper, polished	100°F	0.04
Copper, calorized	100°F	0.18
Copper, oxidized	100°F	0.73
Copper, black oxidized	100°F	0.87
Gold, polished	100°F	0.02
Iron, polished	100°F	0.06
Iron, oxidized	100°F	0.74
Lead, pure polished	100°F	0.05
Lead, gray, oxidized	100°F	0.28
Lead, unoxidized, rough	100°F	0.43
Mercury	100°F	0.10
Molybdenum, polished	100°F	0.06

Table 11-3   *continued*.

Material	Temperature	ε
Nickel, electrolytic	100°F	0.04
Nickel, polished	100°F	0.06
Nickel, matte	100°F	0.11
Nickel, oxidized	100°F	0.31
Platinum, pure polished	100°F	0.04
Platinum, black	100°F	0.93
Silver, polished	100°F	0.01
Silver, oxidized	100°F	0.02
Steel, polished	100°F	0.07
Steel, calorized	100°F	0.52
Steel, oxidized at 1100°F	100°F	0.68
Steel plate, rough	100°F	0.94
Zinc, polished	100°F	0.02
Zinc, matte	100°F	0.21
Zinc, polished, oxidized	100°F	0.28
Mica	100°F	0.75
Oil on polished iron:		
Very thin	0–100°F	0.06
0.0008 in thick	0–100°F	0.22
0.004 in thick	0–100°F	0.61
Very thick	0–100°F	0.83
Paints:		
Aluminum	100°C	0.27–0.67
Aluminum on galvanized iron	20°C	0.52
Aluminum with lacquer body	0–200°C	0.34–0.42
Enamel, vitreous, white	20°C	0.90
Flat white on polished aluminum	18–30°C	0.91
Flat black on polished aluminum	14–24°C	0.88
Oil, all colors	0–93°C	0.92–0.96
Paper, any color	0–93°C	0.92–0.94
Plaster of paris	100°F	0.92
Quartz	100°F	0.89
Sand	100°F	0.76
Soot (coal)	68°F	0.95
Tiles:		
Black	2500°F	0.94
Brown, rough	2500°F	0.92
Brown	2500°F	0.87
Uncolored	2500°F	0.63
Wood:		
Oak planed	100°F	0.91
Sawdust	100°F	0.75
Spruce sanded	100°F	0.82
Walnut sanded	100°F	0.83

† Data excerpted from Gubareff et al. [9] and Singham (data from Fishenden) [10].

11-6 RADIATIVE PROPERTIES OF MATERIALS

The absorption and emission characteristics of gases are quite different from those of solids. The absorption (or emission) of radiation by gases does not take place continuously over the entire wavelength spectrum; rather, it occurs over a large number of relatively narrow strips of intense absorption (or emission). Figure 11-11 shows the absorption spectrum for carbon dioxide from the data by Edwards [16] for which the density times the thickness (that is, ρL) of the gas layer was $\rho L = 0.5$ lb/ft^2. The spectrum is composed of four absorption bands approximately positioned at wavelengths 15, 4.3, 2.7, and 1.9 μm; these absorption bands are caused by changes in the energy levels of the molecules. The prediction of radiative properties of gases is a very complicated matter. The reader should consult the text by S. S. Penner [17] for comprehensive treatment of various models on gas emission, and Refs. [1,3,5] for a discussion of results on the macroscopic characteristics of various gases for absorption.

Gases such as air and nitrogen are relatively transparent to thermal radiation unless the temperature is extremely high; hence for most engineering applications their presence in the path of radiation as a participating medium is ignored. On the other hand, gases such as CO_2, H_2O, CO, SO_2, and various hydrocarbons absorb and emit radiation; hence their presence in the path of radiation signif-

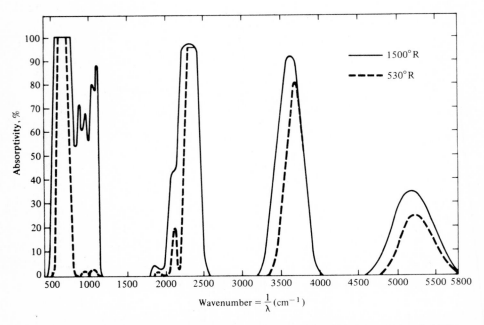

FIG. 11-11 Absorption spectra for carbon dioxide for $\rho L = 0.5$ lb/ft^2. (*From Edwards [16]*.)

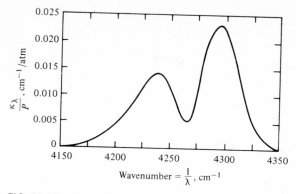

FIG. 11-12   Spectral-absorption coefficient of carbon monoxide at room temperature. (*Based on data from Penner* [*17*].)

icantly influences the transport of energy by radiation.   Figures 11-12 and 11-13 show, respectively, the spectral-absorption coefficient κ_λ for CO at room temperature and for water vapor at 1000 K obtained from measurements by Goldstein [18] as a function of the wavelength.   The figure for water vapor is for absorption characteristics of the 2.7-μm band; other absorption bands for water vapor occur at wavelengths 1.38, 1.87, and 6.3 μm.

Figure 11-14 shows the spectral-absorption coefficient κ_λ for *liquid water* in the near-infrared spectrum at temperatures 27, 89, 159, and 209°C obtained from measurements by Goldstein and Penner [19].   We note that strong absorptions are localized between 4600 and 5900 cm^{-1} and between 5900 and 7800 cm^{-1}.

To   illustrate   the   absorption   characteristics   of   glass,   Fig.   11-15   shows   the

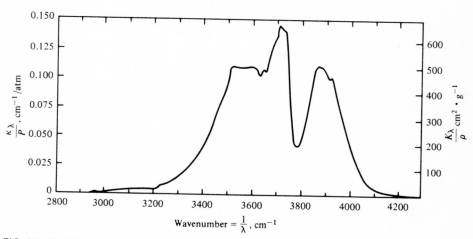

FIG. 11-13   Spectral-absorption coefficient of water vapor at 1000 K for the 2.7-μm region. (*From Goldstein* [*18*].)

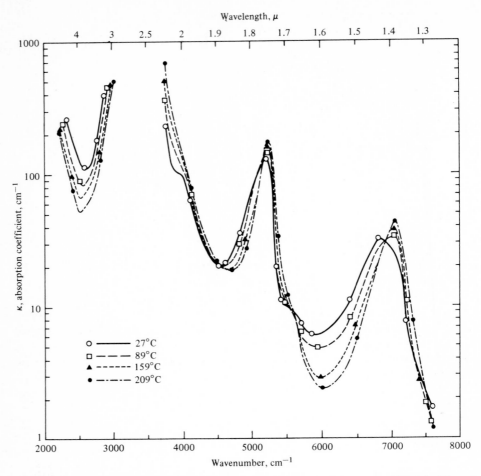

FIG. 11-14 Spectral-absorption coefficient of liquid water at temperatures of 27, 89, 159, and 209°C. (*From Goldstein and Penner [19].*)

spectral-absorption coefficient for ordinary window glass as a function of wavelength for several different temperatures as obtained from measurements by Neuroth [20]. It is apparent from this figure that ordinary window glass transmits radiation well in the visible range of the spectrum but in the infrared range (i.e., beyond about 2.6 μm) it is considered opaque to radiation. This behavior of transmittance is the reason for the so-called greenhouse effect, that is, the trapping of solar radiation by a glass enclosure.

Example 11-3 Compare the emissivity and reflectivity of polished and oxidized copper at 500 and 1000°F by using the data in Fig. 11-8.

Solution We obtain from this figure for *polished copper* surface $\varepsilon = 0.05$ and 0.18 at 500 and 1000°F, respectively, and for the *oxidized copper* $\varepsilon \cong 0.46$ and 0.57 at 500 and

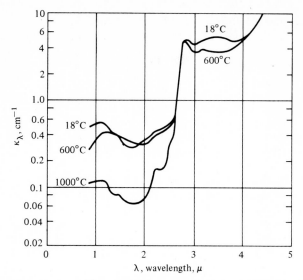

FIG. 11-15   Spectral-absorption coefficient of window glass. (*From Neuroth* [20].)

1000°F, respectively.   We note that the emissivity increases with increasing temperature and oxidation for the copper surface.   Taking $\varepsilon = \alpha$, the reflectivity is determined from $\rho = 1 - \alpha = 1 - \varepsilon$.   In this case reflectivity decreases with increasing temperature and oxidation.

REFERENCES

1   Hottel, H. C., and A. F. Sarofim: "Radiative Transfer," McGraw-Hill Book Company, New York, 1967.
2   Love, T. J.: "Radiative Heat Transfer," Charles E. Merrill Books, Inc., Columbus, Ohio, 1968.
3   Özişık, M. N.: "Radiative Transfer and Interactions with Conduction and Convection," John Wiley & Sons, Inc., New York, 1973.
4   Siegel, R., and R. Howell: "Thermal Radiation Heat Transfer," vol. 1, 1968; vol. 2, 1969; and vol. 3, 1971, NASA Publications, Government Printing Office, Washington, D.C.
5   Sparrow, E. M., and R. D. Cess: "Radiation Heat Transfer," Brooks/Cole Publishing Company, Belmont, Calif., 1970.
6   Planck, M.: "The Theory of Heat Radiation," Dover Publications, Inc., New York, 1959.
7   Snyder, N. W.: A Review of Thermal Radiation Constants, *Trans. ASME*, **76**:537–540 (1954).
8   Dunkle, R. V.: Thermal Radiation Tables and Applications, *Trans. ASME*, **76**:549–552 (1954).

9 Gubareff, G. G., J. E. Janssen, and R. H. Torborg: "Thermal Radiation Properties Survey," Honeywell Research Center, Honeywell Regulator Company, Minneapolis, 1960.

10 Singham, J. R.: Tables of Emissivity of Surfaces, *Int. J. Heat Mass Transfer*, **5**:67–76 (1962).

11 Hottel, H.: Radiant Heat Transmission, in W. H. McAdams (ed.), "Heat Transmission," 3d ed., McGraw-Hill Book Company, New York, 1954.

12 Svet, D. Y.: Thermal Radiation, Metals, Semiconductors, Ceramics, Partly Transparent Bodies and Films, Consultants Bureau, New York, 1965.

13 Wood, D. H., H. W. Deem, and C. F. Lucks: "Thermal Radiative Properties," vol. 3, Plenum Press, New York, 1964.

14 Touloukian, Y. S., and D. P. DeWitt: Metallic Elements and Alloys, in "Thermal Radiative Properties," vol. 7, 1F1/Plenum, New York, 1970.

15 Sieber, W.: *Z. Tech. Phys.*, **22**:130–135 (1941).

16 Edwards, D. K.: Radiation Interchange in a Nongray Enclosure Containing an Isothermal Carbon Dioxide–Nitrogen Gas Mixture, *J. Heat Transfer*, **84C**:1–11 (1962).

17 Penner, S. S.: "Quantitative Molecular Spectroscopy and Gas Emissivities," Addison-Wesley Publishing Company, Inc., Reading, Mass., 1959.

18 Goldstein, R.: Measurements of Infrared Absorption of Water Vapor at Temperatures to 1000°K, *J. Quant. Spectrosc. Radiat. Transfer*, **4**:343–352 (1964).

19 Goldstein, R., and S. S. Penner; The Near-Infrared Absorption of Liquid Water at Temperatures Between 27 and 209°C, *J. Quant. Spectrosc. Radiat. Transfer*, **4**:441–451 (1964).

20 Neuroth, N.: Das Einfluss der Temperatur auf die Spektrale Absorption von Blasern in Ultraroter, I, *Glastech Ber.*, **25**:242–249 (1952).

Twelve

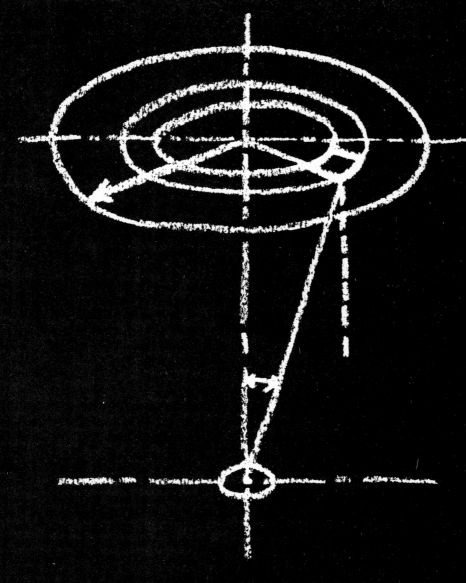

Radiation—
Energy Exchange by Radiation
in a Nonparticipating Medium

In this chapter we consider radiative-heat exchange among the surfaces of an *enclosure* which is filled with a nonparticipating medium, that is, a medium that does not absorb, emit, or scatter radiation, hence has no effect on its propagation through the medium.   A vacuum is a perfect nonabsorbing medium, or air at low and moderate temperatures is considered for all practical purposes a non-participating medium.   The term *enclosure* is used to designate a region surrounded by a set of surfaces that are characterized by their radiative properties and temperatures (or heat fluxes) so that a full account can be made of the incoming and outgoing radiation to any of these surfaces.   With this definition of an enclosure, an opening can be envisioned as an *imaginary surface* such that, if radiative energy is streaming into the enclosure through the opening, the incoming flux is characterized as the emissive flux at this imaginary surface.   The problem of radiative-heat exchange in an enclosure is concerned with the determination of the net radiative-heat flux at the surfaces for which temperature distribution is specified, or with the determination of temperature distribution at the surfaces for which heat flux is prescribed.   The general analysis of such problems is complicated because the radiative properties vary with the frequency, direction, and position, and the distribution of the temperature or the heat flux is not uniform over the surfaces.   Several simplified methods of analysis have been developed in the literature to solve the problem of radiative-heat exchange in enclosures under very restrictive assumptions.   For example, Hottel [1] introduced the so-called script $\mathscr{F}$ method, Eckert and Drake [2] applied the radiosity approach, Gebhart [3,4] used the concept of absorption factor, Oppenheim [5] developed the method of electric-network analogy, Sparrow [6] proposed a method which has computational advantages, and Clark and Korybalski [7] presented an approach based on the concept of radiosity similar to that of Hottel [1].   However, a close scrutiny of all these methods reveals that there is no significant difference among them, since all of them utilize the same simplifying assumptions, and that for a given system they provide the same answer.   The principal difference between these different approaches lies in the method of formulation of the problem; hence the advantages of one method over the other may be assessed by its computational merits.   The method described by Sparrow [6] appears to be more straightforward and offers computational advantages.

In this chapter we present a simplified method of analysis of the radiative-heat-exchange problem for enclosures; therefore care must be exercised in the interpretation of the results obtained from such a simplified analysis by ensuring that all the assumed conditions are strictly satisfied in the practical situation.   The reader is referred to the texts by Hottel and Sarofim [8], Love [9], Özışık [10], Siegel and Howell [11], and Sparrow and Cess [12] for a discussion of more general, sophisticated methods of analysis of the enclosure problems under less restrictive assumptions.

12-1   CONCEPT OF VIEW FACTOR

In the problem of radiative-heat exchange for an enclosure filled with a non-participating medium, the geometric orientation of surfaces with respect to each other affects the radiative interchange.   For convenience in the analysis, the *view factor* is introduced to characterize the effects of geometry and orientation of surfaces on the radiative-heat exchange between them.   The terms *shape factor, configuration factor*, and *angle factor* have also been used in the literature for the view factor.   A distinction should also be made between a *diffuse view factor* and a *specular view factor*.   The former refers to the situation in which the surfaces are diffuse reflectors and diffuse emitters, whereas the latter refers to the situation in which the surfaces are diffuse emitters and specular reflectors.   In this chapter we consider only the cases in which surfaces are diffuse emitters and diffuse reflectors; therefore the term *view factor* that is used in this chapter actually refers to the *diffuse view factor*.   The physical significance of view factor is *the fraction of the radiative energy leaving one surface element that strikes the other surface directly*.   The symbol $F_{A_i - A_j}$, or written more compactly as F_{i-j}, will be used to denote the *view factor* from a surface A_i to surface A_j; the symbol $dF_{dA_i - dA_j}$ will be used to denote the *elemental view* factor from an elemental surface dA_i to an elemental surface dA_j.   The view factor $F_{A_i - A_j}$ represents the fraction of radiative energy leaving surface A_i diffusely that strikes the surface A_j directly with no intervening reflections.   The view factor as defined in this manner is a function of the size, geometry, relative position, and orientation of the two surfaces.   We first derive the basic expressions defining the view factors *between two elemental surfaces, between an elemental surface and a finite surface, and between two finite surfaces*, and then discuss the properties of view factors for an enclosure.

View Factor Between Two Elemental Surfaces

Consider two elemental surfaces dA_1 and dA_2 as illustrated in Fig. 12-1.   Let r be the distance between these two surfaces, θ_1 be the polar angle between the normal $\hat{n}_1$ to the surface element dA_1 and the line r joining dA_1 to dA_2, and θ_2 be the polar angle between the normal $\hat{n}_2$ to the surface element dA_2 and the line r.   The elemental view factor $dF_{dA_1 - dA_2}$ from the elemental surface dA_1 to the elemental surface dA_2 is now determined.

Let $d\Omega_{12}$ be the solid angle under which an observer at dA_1 sees the surface element dA_2, and I_1 be the intensity of radiation leaving the surface element diffusely in all directions in the hemispherical space.   The rate of radiative energy dQ_1 leaving dA_1 that strikes dA_2 is determined by Eq. (11-2) as

$$dQ_1 = dA_1 I_1 \cos \theta_1 \, d\Omega_{12} \tag{12-1}$$

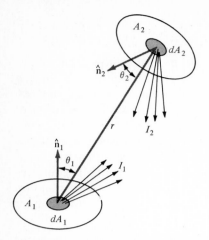

FIG. 12-1 Coordinates for the definition of view factor.

where the solid angle $d\Omega_{12}$ is given by

$$d\Omega_{12} = \frac{dA_2 \cos \theta_2}{r^2}$$ (12-2)

The substitution of Eq. (12-2) into Eq. (12-1) results in

$$dQ_1 = dA_1 I_1 \frac{\cos \theta_1 \cos \theta_2 \, dA_2}{r^2}$$ (12-3)

The rate of radiation energy Q_1 leaving the surface element dA_1 in all directions over the hemispherical space is given, according to Eq. (11-3b), by

$$Q_1 = dA_1 \int_{\varphi=0}^{2\pi} \int_{\theta_1=0}^{\pi/2} I_1 \cos \theta_1 \sin \theta_1 \, d\theta_1 \, d\varphi$$ (12-4a)

where φ is the azimuthal angle. For a diffusely reflecting and diffusely emitting surface the radiation intensity leaving the surface is independent of direction. Then, for constant I_1, Eq. (12-4a) is integrated to give

$$Q_1 = \pi I_1 \, dA_1$$ (12-4b)

The elemental view factor $dF_{dA_1-dA_2}$, by definition, is the ratio of the radiative energy leaving dA_1 that strikes dA_2 directly to the radiative energy leaving dA_1 in all directions into the hemispherical space; hence, it is obtained by dividing Eq. (12-3) by Eq. (12-4b) to yield

$$dF_{dA_1-dA_2} = \frac{dQ_1}{Q_1} = \frac{\cos \theta_1 \cos \theta_2 \, dA_2}{\pi r^2}$$ (12-5)

The elemental view factor $dF_{dA_2-dA_1}$ from dA_2 to dA_1 is now immediately obtained from Eq. (12-5) by interchanging the subscripts 1 and 2. We find

$$dF_{dA_2-dA_1} = \frac{\cos\theta_1 \cos\theta_2 \, dA_1}{\pi r^2} \tag{12-6}$$

The *reciprocity relation* between the view factors $dF_{dA_1-dA_2}$ and $dF_{dA_2-dA_1}$ follows from Eqs. (12-5) and (12-6) as

$$dA_1 \, dF_{dA_1-dA_2} = dA_2 \, dF_{dA_2-dA_1} \tag{12-7}$$

This relation implies that, for two elemental surfaces dA_1 and dA_2, when one of the view factors is known the other is readily computed by the reciprocity relation.

View Factor Between an Elemental Surface dA_1 and a Finite Surface A_2

Consider an elemental surface dA_1 and a finite surface A_2 as illustrated in Fig. 12-1. The view factor $F_{dA_1-A_2}$, from dA_1 to A_2, is immediately determined by integrating the elemental view factor $dF_{dA_1-dA_2}$ given by Eq. (12-5) over the area A_2 as

$$F_{dA_1-A_2} = \int_{A_2} dF_{dA_1-dA_2} = \int_{A_2} \frac{\cos\theta_1 \cos\theta_2}{\pi r^2} dA_2 \tag{12-8}$$

Let q_2 be the radiative-heat flux [i.e., radiative energy/(time × area)] leaving the surface A_2. When q_2 is uniform over the surface, the radiative energy leaving A_2 that strikes dA_1 directly is given by

$$q_2 \int_{A_2} dF_{dA_2-dA_1} \, dA_2$$

and the radiative energy leaving A_2 into the hemispherical space by

$$q_2 \, A_2$$

Then, the view factor $F_{A_2-dA_1}$ from A_2 to dA_1 becomes

$$F_{A_2-dA_1} = \frac{q_2 \int_{A_2} dF_{dA_2-dA_1} \, dA_2}{q_2 \, A_2} = \frac{1}{A_2} \int_{A_2} dF_{dA_2-dA_1} \, dA_2 \tag{12-9a}$$

The substitution of $dF_{dA_2 - dA_1}$ from Eq. (12-6) into Eq. (12-9a) yields

$$F_{A_2 - dA_1} = \frac{dA_1}{A_2} \int_{A_2} \frac{\cos \theta_1 \cos \theta_2}{\pi r^2} dA_2 \qquad (12\text{-}9b)$$

The *reciprocity relation between the view factors* $F_{dA_1 - A_2}$ and $F_{A_2 - dA_1}$ follows from Eq. (12-8) and (12-9b):

$$dA_1 F_{dA_1 - A_2} = A_2 F_{A_2 - dA_1} \qquad (12\text{-}10)$$

View Factor Between Two Finite Surfaces A_1 and A_2

Consider two finite surfaces A_1 and A_2 as shown in Fig. 12-1. Let the radiative-heat flux q_1 leaving the surface area A_1 be uniform over the surface. The radiative energy leaving A_1 that strikes A_2 directly is given by

$$q_1 \int_{A_1} F_{dA_1 - A_2} \, dA_1$$

and the radiative energy leaving A_1 into the hemispherical space is given by

$$q_1 A_1$$

Then, the view factor $F_{A_1 - A_2}$ from A_1 to A_2 becomes

$$F_{A_1 - A_2} = \frac{q_1 \int_{A_1} F_{dA_1 - A_2} \, dA_1}{q_1 A_1} = \frac{1}{A_1} \int_{A_1} dF_{dA_1 - A_2} \, dA_2 \qquad (12\text{-}11a)$$

Substituting $dF_{dA_1 - A_2}$ from (Eq. 12-8) into (Eq. 12-11a) we find

$$F_{A_1 - A_2} = \frac{1}{A_1} \int_{A_1} \int_{A_2} \frac{\cos \theta_1 \cos \theta_2}{\pi r^2} dA_2 \, dA_1 \qquad (12\text{-}11b)$$

The view factor $F_{A_2 - A_1}$ is obtained from Eq. (12-11b) by interchanging the subscripts 1 and 2:

$$F_{A_2 - A_1} = \frac{1}{A_2} \int_{A_2} \int_{A_1} \frac{\cos \theta_1 \cos \theta_2}{\pi r^2} dA_1 \, dA_2 \qquad (12\text{-}12)$$

The reciprocity relation between the view factors $F_{A_1 - A_2}$ and $F_{A_2 - A_1}$ follows from Eqs. (12-11b) and (12-12):

$$A_1 F_{A_1 - A_2} = A_2 F_{A_2 - A_1} \qquad (12\text{-}13)$$

Properties of View Factors for an Enclosure

Consider an enclosure consisting of N surfaces (or zones), each of area A_i, $i = 1, 2, \ldots, N$. It is assumed that each zone is *isothermal, a diffuse reflector, a diffuse emitter, and the radiative-heat flux incident upon each zone is uniform over the surface of the zone.* The view factors between the two surfaces A_i and A_j of the enclosure obeys the following *reciprocity relation:*

$$A_i F_{A_i - A_j} = A_j F_{A_j - A_i} \tag{12-14}$$

The view factors from the surface, say A_i of the enclosure to all surfaces of the enclosure, including to itself, when summed up should be equal to unity by the definition of the view factor. This is called the *summation relation* among the view factors for an enclosure, and it is written

$$\sum_{k=1}^{N} F_{A_i - A_k} = 1 \tag{12-15}$$

where N is the number of zones in the enclosure. In this summation the term $F_{A_i - A_i}$ is the view factor from the surface A_i to itself; it represents the fraction of radiative energy leaving the surface A_i that strikes itself directly. Clearly, $F_{A_i - A_i}$ vanishes if A_i is flat or convex, and it is nonzero if A_i is concave; this is stated as

$$F_{A_i - A_i} = 0 \qquad \text{if } A_i \text{ plane or convex} \tag{12-16a}$$

$$F_{A_i - A_i} \neq 0 \qquad \text{if } A_i \text{ concave} \tag{12-16b}$$

The reciprocity and summation rules as given above are useful in providing additional simple relations to calculate view factors for an enclosure from the knowledge of others. That is, to determine all possible view factors for an enclosure, one need not compute every one of them directly but should make use of the reciprocity and summation relations whenever possible. This situation is envisioned better if all possible view factors for an N-zone enclosure are expressed in the matrix notation as

$$F_{ij} \equiv \begin{bmatrix} F_{11} & F_{12} & \cdots & F_{1N} \\ F_{21} & F_{22} & \cdots & F_{2N} \\ \multicolumn{4}{c}{\dotfill} \\ F_{N1} & F_{N2} & \cdots & F_{NN} \end{bmatrix} \tag{12-17}$$

It is apparent from this relation that there are N^2 view factors to be determined for an N-zone enclosure. However, the reciprocity rule given by Eq. (12-14) provides $N(N - 1)/2$ relations, and the summation rule Eq. (12-15) provides N additional relations among the view factors. Then, the total number of view factors that are to be calculated for an N-zone enclosure from the view-factor expressions becomes

$$N^2 - \tfrac{1}{2}N(N - 1) - N = \tfrac{1}{2}N(N - 1) \qquad (12\text{-}18)$$

If the surfaces are convex or flat, by Eq. (12-16a), N of these view factors, from a surface to itself, vanish and the total number of view factors to be calculated directly from the geometric arrangement of surfaces reduces to

$$\frac{1}{2} N(N - 1) - N = \frac{N(N - 3)}{2} \qquad (12\text{-}19)$$

For example, for an $(N = 5)$-zone enclosure with a flat surface at each zone, out of all the possible $N^2 = 25$ view factors, the number of view factors that are to be determined from the geometric arrangement of surfaces is only $\tfrac{1}{2}N(N - 3) = 5$.

If the geometry possesses symmetry, some of the view factors are known from the symmetry condition, thus reducing further the number of view factors to be calculated.

12-2   METHODS OF DETERMINATION OF VIEW FACTORS

The computation of the view factor between two elemental surfaces as given by Eq. (12-6) poses no problem; but to determine the view factor between an elemental and a finite surface as given by Eq. (12-8) or between two finite surfaces as given by Eq. (12-11b), the indicated surface integrals should be evaluated.   Such integrals are very difficult to perform analytically for arbitrary geometrical arrangements; only for very simple geometries are analytical results possible.   Hamilton and Morgan [13] evaluated view factors for simple configurations involving rectangles, triangles, cylinders, etc., and presented the results in the form of tables and charts.   A systematic tabulation of references on view factors has been given in Ref. [11, vol. II], and view factors for various simple geometries can be found in Ref. [12,14–18].   The reader should consult Refs. [10,11,12,19] for a discussion of various analytical techniques for the determination of view factors; a discussion of various experimental techniques is also given in Ref. [19].   We present in Table 12-1 some of the analytical expressions for the view factors for simple geometrical arrangements.

For convenience in practical applications, view factors for various geometrical arrangements are presented in the literature in the form of charts.   For example, Fig. 12-2 shows the view factor $F_{dA_1 - A_2}$ from an elemental surface dA_1 to a rectangular finite surface A_2 with dA_1 parallel to A_2 and positioned as shown in the figure.   Figure 12-3 shows the view factor $F_{A_1 - A_2}$ from a rectangular surface A_1 to a rectangular surface A_2 which are adjacent along the edge and

TABLE 12-1   Analytical Expressions for View Factors for Simple Geometrical Arrangements†

Geometrical Arrangement	Analytical Expression for the View Factor

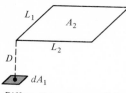

Differential surface parallel to a finite rectangular surface

$$F_{dA_1 - A_2} = \frac{1}{2\pi} \left(\frac{X}{\sqrt{1 + X^2}} \tan^{-1} \frac{Y}{\sqrt{1 + X^2}} + \frac{Y}{\sqrt{1 + Y^2}} \tan^{-1} \frac{X}{\sqrt{1 + Y^2}} \right)$$

where $X = \dfrac{L_1}{D}$ and $Y = \dfrac{L_2}{D}$

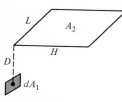

Differential surface perpendicular to a rectangular finite surface

$$F_{dA_1 - A_2} = \frac{1}{2\pi} \left[\tan^{-1} \frac{1}{X} - \frac{1}{\sqrt{1 + (Y/X)^2}} \tan^{-1} \frac{1}{\sqrt{X^2 + Y^2}} \right]$$

where $X = \dfrac{D}{L}$ and $Y = \dfrac{H}{L}$

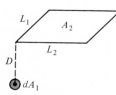

A differential spherical surface and a finite rectangular surface

$$F_{dA_1 - A_2} = \frac{1}{4\pi} \sin^{-1} \frac{XY}{\sqrt{1 + X^2 + Y^2 + X^2 Y^2}}$$

where $X = \dfrac{L_1}{D}$ and $Y = \dfrac{L_2}{D}$

Plane circular surfaces with a common central normal

$$F_{A_1 - A_2} = \frac{1 + B^2 + C^2 - \sqrt{(1 + B^2 + V^2)^2 - 4B^2 V^2}}{2B^2}$$

where $B = \dfrac{b}{a}$ and $C = \dfrac{c}{a}$

† From Refs. [16,19].

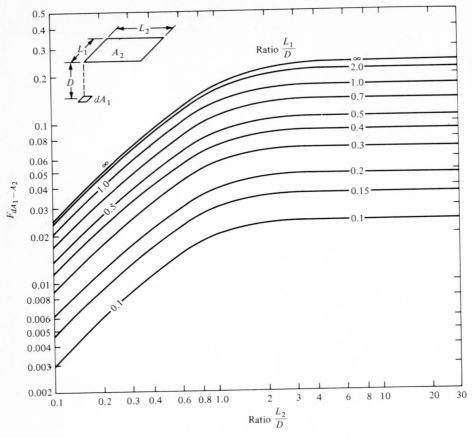

FIG. 12-2 View factor $F_{dA_1 - A_2}$ from an elemental surface dA_1 to a rectangular surface A_2. (*From Mackey et al. [16].*)

perpendicular to each other. Figure 12-4 is the view factor $F_{A_1 - A_2}$ from a rectangular surface A_1 to a rectangular surface A_2 which are directly opposite.

There are numerous geometric arrangements for which view-factor charts are not available, but the view factor can be computed by the method of *view-factor algebra*. That is, the geometrical arrangement is separated by the principle of arithmetical addition and subtraction of view factors into other simple arrangements for which view-factor charts are available. To illustrate the application of view-factor algebra, we consider the determination of the view factor $F_{dA_1 - A_2'}$ from an elemental surface dA_1 to a finite rectangular surface A_2' which are parallel to each other and positioned as illustrated in Fig. 12-5. A view-factor chart is not available for this particular geometrical arrangement, but it is possible to express the area A_2' as the algebraic sum of the four areas A_3, A_4, A_5, A_6 as illustrated in Fig. 12-6, that is,

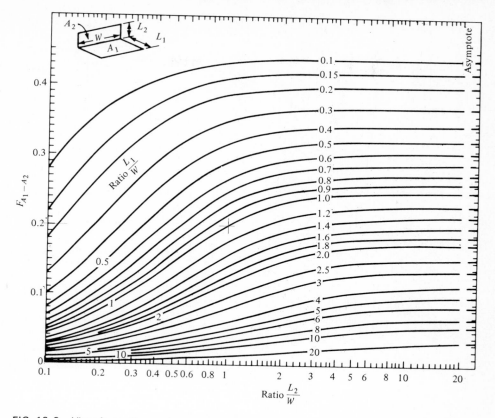

FIG. 12-3 View factor $F_{A_1-A_2}$ from a rectangular surface A_1 to a rectangular surface A_2 which are adjacent and in perpendicular planes. (*From Mackey, et al.* [16].)

$$A_2' = A_3 - A_4 - A_5 + A_6 \tag{12-20}$$

Then the view factor $F_{dA_1-A_2'}$ can be expressed as the algebraic sum of the view factors from dA_1 to the areas A_i ($i = 3, 4, 5, 6$) as

$$F_{dA_1-A_2'} = F_{dA_1-A_3} - F_{dA_1-A_4} - F_{dA_1-A_5} + F_{dA_1-A_6} \tag{12-21}$$

Here the view factors on the right-hand side of this equation are obtainable from the view-factor chart given by Fig. 12-2, hence the view factor $F_{dA_1-A_2'}$ is determined.

Example 12-1 Determine analytically the view factor from an elemental surface dA_1 to a circular disk A_2 of radius R which are parallel to each other and positioned at a distance L as shown in Fig. 12-7.

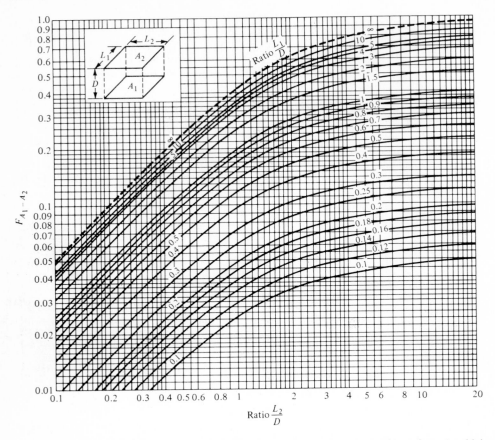

FIG. 12-4   View factor $F_{A_1 - A_2}$ from a rectangular surface A_1 to a rectangular surface A_2 which are parallel and directly opposite to each other.

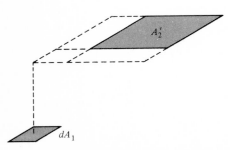

FIG. 12-5   Arrangement of an elemental surface dA_1 and a large rectangular surface A_2 which are parallel to each other.

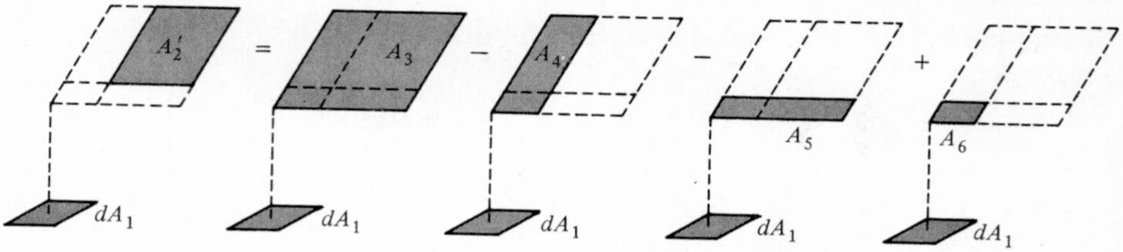

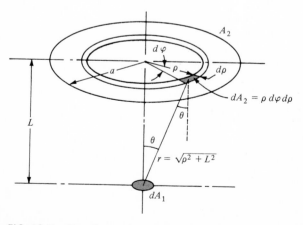

FIG. 12-6   Determination of view factor $F_{dA_1-A_2}$ by view-factor algebra.

Solution   The view factor $F_{dA_1-A_2}$ from an elemental surface dA_1 to a finite surface A_2 is given by Eq. (12-8) as

$$F_{dA_1-A_2} = \int_{A_2} \frac{\cos\theta_1 \cos\theta_2}{\pi r^2} \, dA_2 \qquad\qquad (12\text{-}22)$$

Referring to the geometrical arrangement of Fig. 12-7, the elemental area dA_2 is expressed in the ρ and φ coordinates as

$$dA_2 = \rho \, d\varphi \, d\rho \qquad\qquad (12\text{-}23a)$$

where φ is the azimuthal angle.   The distance r between the elemental surfaces dA_1 and dA_2 is given by

$$r^2 = \rho^2 + L^2 \qquad\qquad (12\text{-}23b)$$

and the angles θ_1 and θ_2 are equal to the angle θ shown in Fig. 12-7 and given as

$$\cos\theta_1 = \cos\theta_2 = \cos\theta = \frac{L}{r} = \frac{L}{(\rho^2 + L^2)^{1/2}} \qquad\qquad (12\text{-}23c)$$

FIG. 12-7   Coordinates for the determination of view factor $F_{dA_1-A_2}$ from an elemental surface dA_1 to a large disk A_2 which is parallel and directly above.

Substituting Eqs. (12-23) into Eq. (12-22) and noting that the limits of integration are $0 \leq \rho \leq R$ and $0 \leq \varphi \leq 2\pi$, we obtain the view factor

$$F_{dA_1 - A_2} = \frac{1}{\pi} \int_{\rho=0}^{R} \int_{\rho=0}^{2\pi} \frac{L^2}{(\rho^2 + L^2)^2} \rho \, d\rho \, d\varphi$$

$$= 2 \int_{\rho=0}^{R} \frac{L^2}{(\rho^2 - L^2)^2} \rho \, d\rho = \frac{R^2}{R^2 + L^2} \tag{12-24}$$

Example 12-2   View factors are to be determined for all possible combinations of the surfaces of an enclosure consisting of six different zones each having a flat surface. If the system does not possess symmetry, how many of these view factors are to be computed from the individual view-factor relations?

Solution   The enclosure contains $N = 6$ zones; hence the maximum possible combination of view factors is $N^2 = 36$.   However, the reciprocity relation between view factors provides $\frac{1}{2}N(N-1) = (6 \times 5)/2 = 15$ relations, the summation rule $\sum_{i=1}^{6} F_{i-j} = 1$ provides six additional relations, and the fact that surfaces are flat, hence $F_{i-i} = 0$, $i = 1$ to 6, provides another additional six relations.   Then, the number of individual view factors that are to be evaluated from the view-factor relations becomes

$$36 - 15 - 6 - 6 = 9$$

This result could also be obtained by Eq. (12-19), that is,

$$\frac{N(N-3)}{2} = \frac{6 \times 3}{2} = 9$$

Example 12-3   A rectangular room, as illustrated in Fig. 12-8, has dimensions $a = 25$ ft in depth, $b = 20$ ft in width, and $c = 10$ ft in height.   Using the view-factor charts given by Figs. 12-3 and 12-4, determine the view factors from the floor to the ceiling, and from the floor to the side walls.   Using the values of view factors obtained in this manner, show that the view factors from the floor to the entire surfaces of the enclosure sum up to unity.

Solution   We use Fig. 12-3 to determine view factors from the floor to the side walls, and Fig. 12-4 from the floor to the ceiling.   The results are summarized below.

Geometrical Arrangement	From Figure	Characteristic Dimensions		View Factor
Floor to wall 1	12-3	$\frac{L_1}{W} = \frac{b}{a} = 0.8$	$\frac{L_2}{W} = \frac{c}{a} = 0.4$	$F_{\text{floor-1}} = 0.15$
Floor to wall 2	12-3	$\frac{L_1}{W} = \frac{a}{b} = 1.25$	$\frac{L_2}{W} = \frac{c}{b} = 0.5$	$F_{\text{floor-2}} = 0.125$
Floor to ceiling	12-4	$\frac{L_1}{D} = \frac{a}{c} = 2.5$	$\frac{L_2}{D} = \frac{b}{c} = 2$	$F_{\text{floor-ceiling}} = 0.45$

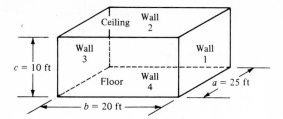

FIG. 12-8   Dimensions of a rectangular enclosure.

We note that, by symmetry, $F_{\text{floor-1}} = F_{\text{floor-3}}$ and $F_{\text{floor-2}} = F_{\text{floor-4}}$. Then, the sum of the view factors from the floor to all surfaces of the enclosure is

$$2 \times 0.15 + 2 \times 0.125 + 0.45 = 1.00$$

Thus they add up to unity.

12-3   RADIATIVE-HEAT EXCHANGE IN ENCLOSURES

In the problems of radiative-heat exchange inside enclosures the interest is in the determination of the net radiative-heat flux at the surfaces for which temperature is prescribed, and in the determination of temperature at the surfaces for which the net heat flux is specified.   For enclosures encountered in real physical situations the radiative properties of the surfaces may vary with direction, frequency, and position, and the temperature or the heat flux may vary from point to point over the surfaces.   The heat-transfer analysis that takes into consideration the effects of all these variables is a very complicated matter. On the other hand, the mathematical formulation of the problem can be simplified significantly and the heat-transfer analysis can be reduced to the solution of a set of coupled algebraic equations if the following assumptions are made.

The entire surface of the enclosure is divided into N number of zones, and the following conditions are assumed to be satisfied at the surface of each zone.

1 Radiative properties (i.e., reflectivity, emissivity, absorptivity) are uniform and independent of direction and frequency.

2 Surfaces are diffuse emitters and diffuse reflectors.

3 The radiative-heat flux leaving the surface is uniform over the surface of each zone.

4 Surfaces are opaque (that is, $\alpha + \rho = 1$).

5 Either a uniform temperature or a uniform heat flux is prescribed over the surface of each zone.

6 The enclosure is filled with a nonparticipating medium.

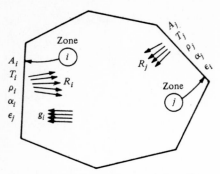

FIG. 12-9 *N*-zone enclosure filled with a nonparticipating medium.

Figure 12-9 illustrates such an enclosure divided into N zones each having a surface A_i, $i = 1, 2, \ldots, N$. Let α_i, ρ_i, and T_i be the absorptivity, the reflectivity, and the temperature, respectively, of the zone A_i. For the graybody assumption made here, the emissivity is taken equal to the absorptivity, that is, $\varepsilon_i = \alpha_i$. The governing equations of the radiative-heat exchange among the surfaces of the enclosure are now derived.

Let R_i and g_i denote the radiative-heat fluxes (i.e., Btu/h·ft²) *leaving* and *incident* on the surface A_i, respectively. The quantity R_i is generally referred to in the engineering literature as the *radiosity* of the surface A_i. The radiative-heat flux g_i incident on the surface A_i should be uniform over zone i to ensure the uniformity of the radiative-heat flux leaving the surface as specified in assumption 3 above. The net *radiative-heat flux* q_i at the surface A_i is equal to the difference between the leaving and incident fluxes and is given as

$$q_i = R_i - g_i \tag{12-25}$$

According to this definition, *the net radiative-heat flux is from the surface into the enclosure when q_i is positive, and vice versa.*

The radiative-heat flux R_i leaving the surface A_i is composed of the radiative-heat flux emitted by the surface due to its temperature T_i and the incident radiative-heat flux that is reflected by the surface. Therefore,

$$R_i = \begin{pmatrix} \text{radiative flux} \\ \text{emitted at } T_i \end{pmatrix} + \begin{pmatrix} \text{incident radiative flux} \\ \text{reflected by the surface} \end{pmatrix} \tag{12-26a}$$

or

$$R_i = \varepsilon_i \sigma T_i^4 + \rho_i g_i \tag{12-26b}$$

The incident radiative-heat flux g_i can be related to the radiosities at the surfaces of the enclosure as now described. The radiative energy leaving the surface A_j of zone j that strikes the surface A_i of zone i directly is given by

$$R_j A_j F_{j-i} \tag{12-27a}$$

where F_{j-1} is the view factor from the surface A_j to the surface A_i and the quantity $R_j A_j$ represents the total radiative energy leaving the surface A_j. By utilizing the reciprocity relation $A_j F_{j-i} = A_i F_{i-j}$, the expression (12-27a) becomes

$$R_j A_i F_{i-j} \qquad (12\text{-}27b)$$

The total radiative energy leaving all the zones of the enclosure and incident upon the surface A_i is determined by summing up the expression (12-27b) for $j = 1$ to N; we obtain

$$\sum_{j=1}^{N} R_j A_i F_{i-j} \qquad (12\text{-}27c)$$

The radiative-heat flux g_i incident upon the surface A_i of zone i is obtained by dividing the total radiative energy given in (12-27c) by A_i, that is,

$$g_i = \sum_{j=1}^{N} R_j F_{i-j} \qquad (12\text{-}28)$$

Substituting g_i from Eq. (12-28) into Eqs. (12-25) and (12-26), we find

$$q_i = R_i - \sum_{j=1}^{N} R_j F_{i-j} \qquad i = 1, 2, \ldots, N \qquad (12\text{-}29)$$

$$R_i = \varepsilon_i \sigma T_i^4 + \rho_i \sum_{j=1}^{N} R_j F_{i-j} \qquad i = 1, 2, \ldots, N \qquad (12\text{-}30)$$

Equations (12-29) and (12-30) are the basic relations for the analysis of radiative-heat exchange in enclosures. Two situations are of particular interest: (1) an N-zone enclosure temperature prescribed for all zones and (2) an N-zone enclosure temperature prescribed for some of the zones and the heat flux prescribed for the others. We now examine the application of Eqs. (12-29) and (12-30) for the formulation of radiative-heat exchange in enclosures for these two situations.

An N-Zone Enclosure Temperature Prescribed for All Zones

Consider an N-zone enclosure in which temperatures T_i are prescribed for all the surfaces A_i, $i = 1, 2, \ldots, N$, of the enclosure. The net radiative-heat fluxes q_i, $i = 1, 2, \ldots, N$, are to be determined for the surfaces. In this problem Eqs. (12-30),

$$R_i = \varepsilon_i \sigma T_i^4 + \rho_i \sum_{j=1}^{N} R_j F_{i-j} \qquad i = 1, 2, \ldots, N \qquad (12\text{-}31)$$

provide N simultaneous algebraic equations for the N unknown radiosities R_i, $i = 1, 2, \ldots, N$, since the quantities ε_i, ρ_i, F_{i-j}, and T_i are all considered to be known for all surfaces of the enclosure. Once R_i's are determined from the solution of Eqs. (12-31), the net radiative-heat flux q_i at any surface A_i is determined from Eq. (12-29), that is,

$$q_i = R_i - \sum_{j=1}^{N} R_j F_{i-j} \qquad i = 1, 2, \ldots, N \tag{12-32}$$

An alternative form of Eq. (12-32) is obtainable by the elimination of the summation term between Eqs. (12-31) and (12-32); we find

$$q_i = \frac{\varepsilon_i \sigma T_i^4 - (1 - \rho_i) R_i}{\rho_i} = \frac{\varepsilon_i}{\rho_i} (\sigma T_i^4 - R_i) \qquad \text{for } \rho_i \neq 0 \tag{12-33}$$

where we have substituted $\varepsilon_i = 1 - \rho_i$.

Equations (12-31) can be solved by hand computation if the number of zones are not many. If the enclosure contains a large number of zones, Eqs. (12-31) are readily solved with a digital computer. When the solution is to be performed with a digital computer it is desirable to rearrange the equations in the matrix form as now described.

Equations (12-31) are written

$$\sum_{j=1}^{N} \frac{\delta_{ij} - \rho_i F_{i-j}}{\varepsilon_i} R_j = \sigma T_i^4 \qquad i = 1, 2, \ldots, N \tag{12-34}$$

where

$$\delta_{ij} = \begin{cases} 1 & \text{for } i = j \\ 0 & \text{for } i \neq j \end{cases}$$

Equations (12-34) can now be expressed in the matrix form

$$\mathbf{MR} = \mathbf{T} \tag{12-35a}$$

where the matrix $\mathbf{M}$ is defined as

$$\mathbf{M} \equiv \begin{bmatrix} m_{11} & m_{12} & m_{13} & \cdots & m_{1N} \\ m_{21} & m_{22} & m_{23} & \cdots & m_{2N} \\ \cdots\cdots\cdots\cdots\cdots\cdots\cdots \\ m_{N1} & m_{N2} & m_{N3} & \cdots & m_{NN} \end{bmatrix} \tag{12-35b}$$

with the elements m_{ij} of the matrix given by

$$m_{ij} \equiv \frac{\delta_{ij} - \rho_i F_{i-j}}{\varepsilon_i} \tag{12-35c}$$

and the vectors $\mathbf{R}$ and $\mathbf{T}$ are defined as

$$\mathbf{R} \equiv \begin{bmatrix} R_1 \\ R_2 \\ \vdots \\ R_N \end{bmatrix} \quad \text{and} \quad \mathbf{T} \equiv \begin{bmatrix} \sigma T_1{}^4 \\ \sigma T_2{}^4 \\ \vdots \\ \sigma T_N{}^4 \end{bmatrix} \tag{12-35d}$$

The matrix equations (12-35a) are now solved for the radiosities as

$$\mathbf{R} = \mathbf{M}^{-1}\mathbf{T} \tag{12-36a}$$

where the inverse matrix $\mathbf{M}^{-1}$ is defined as

$$\mathbf{M}^{-1} \equiv \begin{bmatrix} m'_{11} & m'_{12} & m'_{13} & \cdots & m'_{1N} \\ m'_{21} & m'_{22} & m'_{23} & \cdots & m'_{2N} \\ \cdots\cdots\cdots\cdots\cdots\cdots\cdots\cdots \\ m'_{N1} & m'_{N2} & m'_{N3} & \cdots & m'_{NN} \end{bmatrix} \tag{12-36b}$$

Here the elements m'_{ij} of the inverse matrix $\mathbf{M}^{-1}$ are readily determined by a digital computer using a standard matrix-inversion subroutine from the known elements m_{ij} of the matrix $\mathbf{M}$. The solutions for the radiosities R_i given in the matrix form by Eqs. (12-36) are written explicitly in the form

$$R_i = m'_{i1}\sigma T_1{}^4 + m'_{i2}\sigma T_2{}^4 + \cdots + m'_{iN}\sigma T_N{}^4 \qquad i = 1, 2, \ldots, N \tag{12-37}$$

where m'_{ij}'s are considered to be known from the matrix inversion.

Thus, once the radiosities R_j, $j = 1, 2, \ldots, N$, are known, the net radiative-heat flux q_i at the surface A_i of any one of the zones is determined by Eq. (12-32) or (12-33).

An N-Zone Enclosure Temperature Prescribed for Some Zones and Heat Flux Prescribed for the Others

We now examine a situation in which temperatures are prescribed at the surfaces of some zones and heat fluxes are prescribed at the surfaces of the remaining zones in an N-zone enclosure. In this problem the interest is in the determination of the net radiative-heat fluxes at the surfaces for which temperatures are prescribed, and in the determination of temperatures at the surfaces for which heat fluxes are prescribed. We now assume that the temperatures T_i are prescribed for the zones $i = 1, 2, \ldots, r$, and the net heat fluxes q_i are prescribed for the remaining zones $i = r + 1, r + 2, \ldots, N$ of the enclosure. The N equations needed for the determination of N unknown radiosities R_i are now obtained from Eqs. (12-29) and (12-30) with the following considerations.

Equations for the zones $i = 1, 2, \ldots, r$ are obtained from Eqs. (12-30) since the temperatures T_i are prescribed for these zones, and the equations for the zones $i = r + 1, r + 2, \ldots, N$ are obtained from Eqs. (12-29) because the net radiative-heat fluxes q_i are given for these zones. Then, the N equations for the determination of N unknown radiosities R_i are given as

$$R_i = \varepsilon_i \sigma T_i^4 + \rho_i \sum_{j=1}^{N} R_j F_{i-j} \qquad i = 1, 2, \ldots, r \tag{12-38a}$$

$$R_i = q_i + \sum_{j=1}^{N} R_j F_{i-j} \qquad i = r + 1, r + 2, \ldots, N \tag{12-38b}$$

Once N simultaneous algebraic equations (12-38) are solved and the N unknown radiosities R_i are determined, the net radiative-heat fluxes for the zones with prescribed temperatures are obtained from Eqs. (12-38b), and the temperatures for the zones with prescribed heat fluxes are obtained from Eqs. (12-38a).

When the net radiative-heat flux at the surface of a zone, say zone j, is zero (that is, $q_j = 0$), that zone is called a *reradiating* or an *adiabatic* zone because the surface of that zone does not participate in the net radiative-heat exchange. A reradiating surface emits as much energy as it receives by radiation from the surrounding zones of the enclosure; or it is said to behave as a perfectly reflecting surface (that is, $\rho_j = 1$ or $\varepsilon_j = 0$).

The limitations to the range of applicability of the simplified analysis of radiative-heat transfer in enclosures as described above should be recognized. The analysis is strictly applicable to situations in which the assumptions 1 to 6 are satisfied. Especially assumption 3 is a very restrictive one because of the limitation to be imposed on the geometrical arrangement between the surfaces. For example, if the spacing between the surfaces of the two zones is very small compared with the characteristic dimension of the surfaces, as in the case of closely spaced fins, assumption 3 becomes invalid because the radiosity is not uniform over the surfaces of the zones for such geometrical arrangements.

Example 12-4 Consider an equilateral-triangular enclosure with sides $a = 2$ ft, as shown in Fig. 12-10, and infinitely long in the direction perpendicular to the plane of the figure. The surfaces are opaque, gray, diffuse emitters and diffuse reflectors; the air contained in the enclosure is considered to be a nonparticipating medium. For the purpose of analysis of radiative-heat exchange, the surfaces of the enclosure are separated into three distinct zones as illustrated in the figure. The temperatures T_1, T_2, T_3, the emissivities ε_1, ε_2, ε_3, and the reflectivities ρ_1, ρ_2, ρ_3 are uniform over the surface of each zone and their values are prescribed. Write the governing equations for the determination of three unknown radiosities R_1, R_2, and R_3 over the surfaces of the three zones and describe how the net radiative-heat fluxes at the surfaces are computed.

Solution The equations for the determination of the three unknown radiosities R_1, R_2, and R_3 are obtainable from Eq. (12-31) or (12-34). We prefer the latter because it is more convenient to express the results in the matrix form. By setting $i = 1, 2,$ and 3 in Eq. (12-34) we obtain, respectively,

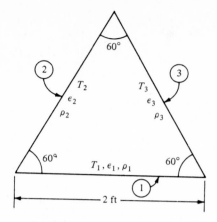

FIG.12-10   Equilateral-triangular enclosure infinitely long in the direction normal to the plane of this figure.

$$\frac{1 - \rho_1 F_{11}}{\varepsilon_1} R_1 - \frac{\rho_1 F_{12}}{\varepsilon_1} R_2 - \frac{\rho_1 F_{13}}{\varepsilon_1} R_3 = \sigma T_1{}^4 \qquad (12\text{-}39a)$$

$$-\frac{\rho_2 F_{21}}{\varepsilon_2} R_1 + \frac{1 - \rho_2 F_{22}}{\varepsilon_2} R_2 - \frac{\rho_2 F_{23}}{\varepsilon_2} R_3 = \sigma T_2{}^4 \qquad (12\text{-}39b)$$

$$-\frac{\rho_3 F_{31}}{\varepsilon_3} R_1 - \frac{\rho_3 F_{32}}{\varepsilon_3} R_2 + \frac{1 - \rho_3 F_{33}}{\varepsilon_3} R_3 = \sigma T_3{}^4 \qquad (12\text{-}39c)$$

These equations are written in the matrix form as

$$\begin{bmatrix} \dfrac{1 - \rho_1 F_{11}}{\varepsilon_1} & -\dfrac{\rho_1 F_{12}}{\varepsilon_1} & -\dfrac{\rho_1 F_{13}}{\varepsilon_1} \\[2ex] -\dfrac{\rho_2 F_{21}}{\varepsilon_2} & \dfrac{1 - \rho_2 F_{22}}{\varepsilon_2} & -\dfrac{\rho_2 F_{23}}{\varepsilon_2} \\[2ex] -\dfrac{\rho_3 F_{31}}{\varepsilon_3} & -\dfrac{\rho_3 F_{32}}{\varepsilon_3} & \dfrac{1 - \rho_3 F_{33}}{\varepsilon_3} \end{bmatrix} \begin{bmatrix} R_1 \\[2ex] R_2 \\[2ex] R_3 \end{bmatrix} = \begin{bmatrix} \sigma T_1{}^4 \\[2ex] \sigma T_2{}^4 \\[2ex] \sigma T_3{}^4 \end{bmatrix} \qquad (12\text{-}40)$$

The view factors F_{ij} for the geometry in Fig. 12-10 are readily determined; $F_{11} = F_{22} = F_{33} = 0$ because the walls are plane surfaces, and the geometry is such that each of the remaining view factors is equal to 0.5.   Then Eqs. (12-40) become

$$\begin{bmatrix} \dfrac{1}{\varepsilon_1} & -\dfrac{\rho_1}{2\varepsilon_1} & -\dfrac{\rho_1}{2\varepsilon_1} \\[2ex] -\dfrac{\rho_2}{2\varepsilon_2} & \dfrac{1}{\varepsilon_2} & -\dfrac{\rho_2}{2\varepsilon_2} \\[2ex] -\dfrac{\rho_3}{2\varepsilon_3} & -\dfrac{\rho_3}{2\varepsilon_3} & \dfrac{1}{\varepsilon_3} \end{bmatrix} \begin{bmatrix} R_1 \\[2ex] R_2 \\[2ex] R_3 \end{bmatrix} = \begin{bmatrix} \sigma T_1{}^4 \\[2ex] \sigma T_2{}^4 \\[2ex] \sigma T_3{}^4 \end{bmatrix} \qquad (12\text{-}41)$$

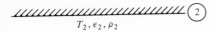

T_2, ϵ_2, ρ_2

T_1, ϵ_1, ρ_1

FIG.12-11   Enclosure consisting of two infinite parallel plates.

The radiosities R_1, R_2, and R_3 are determined from the solution of Eqs. (12-41) since the emissivities, the reflectivities, and the temperatures of the surfaces are known.   When the radiosities are known, the net radiative-heat fluxes q_i at the surfaces are determined from Eqs. (12-32) or Eqs. (12-33).

Example 12-5   Consider an enclosure consisting of two parallel, infinite, opaque plates as shown in Fig. 12-11.   The surfaces 1 and 2 are gray, kept at uniform temperatures T_1 and T_2 ($T_1 > T_2$), have emissivities ε_1 and ε_2, and reflectivities ρ_1 and ρ_2, respectively.   The medium between the plates is nonparticipating.   Determine a relation for the net radiative-heat flux at the surfaces of the plates.

Solution   The considered problem has two zones, hence $N = 2$.   The equations for the two unknown radiosities R_1 and R_2 are obtained from Eqs. (12-31) because the temperatures are specified at the surfaces.   We obtain

$$R_1 = \varepsilon_1 \sigma T_1{}^4 + \rho_1 (R_1 F_{11} + R_2 F_{12}) \tag{12-42a}$$

$$R_2 = \varepsilon_2 \sigma T_2{}^4 + \rho_2 (R_1 F_{21} + R_2 F_{22}) \tag{12-42b}$$

For the geometry considered here, the view factors $F_{11} = F_{22} = 0$ because the walls are plane surfaces, and $F_{12} = F_{21} = 1$ because the surfaces are of infinite extent.   Equations (12-42) become

$$R_1 = \varepsilon_1 \sigma T_1{}^4 + (1 - \varepsilon_1) R_2 \tag{12-43a}$$

$$R_2 = \varepsilon_2 \sigma T_2{}^4 + (1 - \varepsilon_2) R_1 \tag{12-43b}$$

where we set $\rho_1 = 1 - \varepsilon_1$ and $\rho_2 = 1 - \varepsilon_2$ since the surfaces are opaque.
A simultaneous solution of Eqs. (12-43) gives the radiosities as

$$R_1 = \frac{\varepsilon_2 (1 - \varepsilon_1) \sigma T_2{}^4 + \varepsilon_1 \sigma T_1{}^4}{1 - (1 - \varepsilon_1)(1 - \varepsilon_2)} \tag{12-44a}$$

$$R_2 = \frac{\varepsilon_1 (1 - \varepsilon_2) \sigma T_1{}^4 + \varepsilon_2 \sigma T_2{}^4}{1 - (1 - \varepsilon_1)(1 - \varepsilon_2)} \tag{12-44b}$$

The net radiative-heat fluxes at the surfaces are determined by Eq. (12-32); by setting $F_{11} = F_{22} = 0$ and $F_{12} = F_{21} = 1$ this equation gives for $i = 1$ and 2, respectively,

$$q_1 = R_1 - R_2 \tag{12-45a}$$

$$q_2 = R_2 - R_1 \tag{12-45b}$$

The substitution of radiosities from Eqs. (12-44) into Eq. (12-45a) gives the net radiative-heat flux q_1 at the surface 1 as

$$q_1 = \frac{\sigma T_1{}^4 - \sigma T_2{}^4}{1/\varepsilon_1 + 1/\varepsilon_2 - 1} \tag{12-46}$$

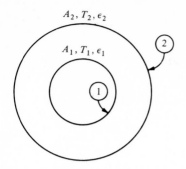

FIG. 12-12   Enclosure consisting of two concentric spheres or two coaxial, very long cylinders.

and we note from Eq. (12-45b) that

$$q_2 = -q_1 \tag{12-47}$$

The result in Eq. (12-47) is to be expected from the conservation of energy, because these are only two zones and the net radiative-heat loss at one of the zones becomes the net radiative-heat gain at the other.

For *black surfaces* we set $\varepsilon_1 = \varepsilon_2 = 1$, and Eq. (12-46) simplifies to

$$q_1 = \sigma T_1{}^4 - \sigma T_2{}^4 \tag{12-48}$$

Example 12-6   Consider an enclosure composed of two concentric, very long cylinders (or two concentric spheres) as illustrated in Fig. 12-12.   The surfaces are opaque, gray, diffuse emitters and diffuse reflectors.   Let A_1 and A_2 be the surface areas, ε_1 and ε_2 be the emissivities, ρ_1 and ρ_2 be the reflectivities, and T_1 and T_2 be the temperatures of the inner and outer surfaces, respectively.   Determine the net radiative-heat fluxes at the surfaces.

Solution   In this problem the enclosure contains two zones for which the temperatures T_1 and T_2 are prescribed.   The mathematical formulation of this problem is the same as that given by Eqs. (12-42) except the view factors are different.   Then the equations for the radiosities are

$$R_1 = \varepsilon_1 \sigma T_1{}^4 + \rho_1(R_1 F_{11} + R_2 F_{12}) \tag{12-49a}$$

$$R_2 = \varepsilon_2 \sigma T_2{}^4 + \rho_2(R_1 F_{21} + R_2 F_{22}) \tag{12-49b}$$

The view factors for the geometry shown in Fig. 12-12 are determined in the following manner.

The inner surface is convex, hence

$$F_{11} = 0 \tag{12-50a}$$

Introducing this result into the summation rule $F_{11} + F_{12} = 1$ gives

$$F_{12} = 1 \tag{12-50b}$$

Utilizing Eq. (12-50b) in the reciprocity relation $A_1 F_{12} = A_2 F_{21}$ we find

$$F_{21} = \frac{A_1}{A_2} \tag{12-50c}$$

and introducing Eq. (12-50c) into the summation rule $F_{21} + F_{22} = 1$ gives

$$F_{22} = 1 - \frac{A_1}{A_2} \tag{12-50d}$$

With the substitution of the view factors given by Eqs. (12-50) into Eqs. (12-49), and setting $\rho = 1 - \varepsilon$, the equations for the radiosities R_1 and R_2 become

$$R_1 = \varepsilon_1 \sigma T_1{}^4 + (1 - \varepsilon_1) R_2 \tag{12-51a}$$

$$R_2 = \varepsilon_2 \sigma T_2{}^4 + (1 - \varepsilon_2)\left[R_1 \frac{A_1}{A_2} + R_2\left(1 - \frac{A_1}{A_2}\right) \right] \tag{12-51b}$$

Once the radiosities are obtained from the solutions of Eqs. (12-51), the net radiative-heat fluxes at the surfaces 1 and 2 are obtained by means of Eqs. (12-32), which now become

$$q_1 = R_1 - R_2 \tag{12-52a}$$

$$q_2 = (R_2 - R_1) \frac{A_1}{A_2} \tag{12-52b}$$

The solution of Eqs. (12-51) and the substitution of the radiosities in Eq. (12-52a) give the net radiative-heat flux at surface 1 as

$$q_1 = \frac{\sigma T_1{}^4 - \sigma T_2{}^4}{1/\varepsilon_1 + (A_1/A_2)(1/\varepsilon_2 - 1)} \tag{12-53}$$

This result is applicable for both coaxial long cylinders and concentric spheres. The area ratio A_1/A_2 can be replaced by r_1/r_2 for coaxial cylinders and by $(r_1/r_2)^2$ for concentric spheres, where r_1 and r_2 are the inner and outer radius, respectively. When the spacing between the surfaces is small compared with the inner radius, we have $A_1/A_2 \simeq 1$, and Eq. (12-53) reduces to the result given by Eq. (12-46) for two parallel infinite plates.

REFERENCES

1 Hottel, H. C.: Radiant Heat Transmission, in W. H. McAdams, "Heat Transmission," 3d ed., pp. 72–79, McGraw-Hill Book Company, New York, 1954.
2 Eckert, E. R. G., and R. M. Drake: "Analysis of Heat and Mass Transfer," pp. 619–646, McGraw-Hill Book Company, New York, 1972.
3 Gebhart, B.: A New Method for Calculating Radiant Exchanges, *Heat. Piping/Air Cond.*, **30**:131–135 (July 1958).
4 Gebhart, B.: Surface Temperature Calculations in Radiant Surroundings of Arbitrary Complexity for Gray, Diffuse Radiation, *Int. J. Heat Mass Transfer*, **3**:341–346 (1961).
5 Oppenheim, A. K.: Radiation Analysis by the Network Method, *Trans. ASME*, **78**:725–735 (1956).
6 Sparrow, E. M.: Radiation Heat Transfer Between Surfaces, in James P. Hartnett and T. F. Irvine, Jr. (eds.), "Advances in Heat Transfer," pp. 407–411, Academic Press, Inc., New York, 1965.
7 Clark, J. A., and E. Korybalski: Radiation Heat Transfer in an Enclosure Having Surfaces Which Are Adiabatic or of Known Temperature, *First. Natl. Heat Mass Transfer Conf.*, Madras, India, December 1971.

8   Hottel, H. C., and A. F. Sarofim: "Radiative Transfer," McGraw-Hill Book Company, New York, 1967.

9   Love, T. J.: "Radiation Heat Transfer," Charles E. Merrill Books, Inc., Columbus, Ohio, 1968.

10 Özışık, M. N.: "Radiative Transfer and Interactions with Conduction and Convection," John Wiley & Sons, Inc., New York, 1973.

11 Siegel, R., and R. Howell: "Thermal Radiation Heat Transfer," vol. 1, 1968; vol. 2, 1969, NASA Publication, Government Printing Office, Washington, D.C.

12 Sparrow, E. M., and R. D. Cess: "Radiation Heat Transfer," Brooks/Cole Publishing Company, Belmont, Calif., 1970.

13 Hamilton, D. C., and W. R. Morgan: Radiant Interchange Configuration Factors, *NACA Tech. Note,* 2836, 1952.

14 Leuenberger, H., and R. A. Pearson: Compilation of Radiant Shape Factors for Cylindrical Assemblies, *ASME Paper* 56-A-144, 1956.

15 Kreith, F.: "Radiation Heat Transfer for Spacecraft and Solar Power Design," International Textbook Company, Scranton, Pa., 1962.

16 Mackey, C. O., L. T. Wright, R. E. Clark, and N. R. Gray: Radiant Heating and Cooling, Pt. I, *Cornell Univ., Eng. Exp. Sta. Bull.* 22, 1943.

17 Feingold, A., and K. G. Gupta: New Analytical Approach to the Evaluation of Configuration Factors in Radiation from Spheres and Infinitely Long Cylinders, *J. Heat Transfer,* **92C:**69–76 (1970).

18 Singer, G. L.: Viewpin: FORTRAN Program to Calculate View Factors for Cylindrical Pins, Aerojet Nuclear Co., Rept. ANCR-1054, Idaho Falls, Idaho, March 1972.

19 Jakob, M.: "Heat Transfer," vol. 2, John Wiley & Sons, Inc., New York, 1957.

Thirteen

Radiation—
Energy Exchange by Radiation
in an Absorbing, Emitting Medium

In the preceding chapter we discussed radiative exchange between surfaces separated by a nonparticipating medium. In many engineering applications the radiative-heat transfer through a medium that absorbs and/or emits radiation is of interest. For example, gases such as carbon dioxide, carbon monoxide, sulfur dioxide, ammonia, hydrogen chloride, water vapor, and many others emit radiation, and these gases also absorb radiation passing through them; a molten glass emits radiation and it also absorbs radiation propagating through it. In this chapter we present a simplified analysis of radiative-heat transfer in an absorbing, emitting medium for the case of a simple slab geometry and discuss the approximate, semiempirical methods of prediction of the absorption and emission of radiation by a body of gas having an arbitrary shape. The reader should consult the texts by Chandrasekhar [1], Hottel and Sarofim [2], Kourganoff [3], Love [4], Özışık [5], Siegel and Howell [6], Sobolev [7], Sparrow and Cess [8], and Viskanta [9] for detailed discussion of the analysis of radiative-heat transfer in participating media, including the effects of scattering of radiation.

13-1 EQUATION OF RADIATIVE TRANSFER FOR A PLANE-PARALLEL MEDIUM

The equation of radiative transfer is now derived for an absorbing, emitting medium in the one-dimensional plane-parallel geometry as illustrated in Fig. 13-1. The plane-parallel geometry implies that the radiative properties, temperature, and pressure (if the medium is gas) are uniform in planes perpendicular to the $0y$ axis in Fig. 13-1, but they may vary in the y direction. For simplicity in the analysis it is assumed that the radiation propagating in the medium is characterized by its intensity $I(y, \theta)$, that is, considered to be a function of only the space variable y and the polar angle θ between the direction of the radiation beam and the positive y axis. Let a beam of radiation of intensity $I(y, \theta)$ propagate through a differential element of thickness dS and of unit area normal to the direction of propagation, as illustrated in Fig. 13-1, and S be the distance measured along the direction of propagation. The radiation beam enters this

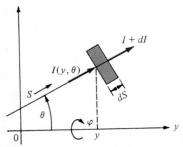

FIG. 13-1 Nomenclature for the propagation of radiation in a plane-parallel medium.

volume element with an intensity I and, after traveling a distance dS in the medium, leaves it with an intensity $I + dI$. Then the quantity dI/dS represents the increase in the intensity of radiation per unit length along the direction of propagation through the volume element. This increase in the intensity of radiation must be the result of a net balance between the rate of emission and the rate of absorption of radiation by the volume element; hence the energy-balance equation is written

$$\frac{dI}{dS} = W_e - W_a \tag{13-1}$$

where the terms W_e and W_a represent the emission and the absorption of radiation, respectively, by the volume element in radiative energy/(time × volume × solid angle). The emission and absorption of radiation by a volume element have been discussed in Chap. 11; therefore the quantities on the right-hand side of Eq. (13-1) are immediately obtainable from the results given in that chapter. The absorption term is obtained from Eq. (11-31) as

$$W_a = \kappa I \tag{13-2}$$

Assuming that the medium is in local thermodynamic equilibrium and that the Kirchhoff law is valid, the emission term is given by Eq. (11-33) as

$$W_e = \kappa I_b(T) \tag{13-3}$$

where κ is the absorption coefficient and $I_b(T)$ is the blackbody radiation intensity of the temperature T of the body.

The substitution of Eqs. (13-2) and (13-3) into (13-1) yields

$$\frac{dI}{dS} = \kappa I_b(T) - \kappa I \tag{13-4}$$

In this equation d/dS is the derivative with respect to the S variable measured along the direction of propagation of the radiation beam; but it can be expressed in terms of the derivative with respect to the y variable in the following manner:

$$\frac{d}{dS} = \frac{\partial}{\partial y}\frac{dy}{dS} = \cos\theta\,\frac{\partial}{\partial y} \tag{13-5}$$

since $dy/dS = \cos\theta$, as apparent from the geometry in Fig. 13-1. The substitution of Eq. (13-5) into Eq. (13-4) yields

$$\cos\theta\,\frac{\partial I(y,\theta)}{\partial y} + \kappa I(y,\theta) = \kappa I_b(T) \tag{13-6}$$

where the polar angle θ varies from 0 to π. It is convenient to define a new variable μ as

$$\mu = \cos\theta \tag{13-7}$$

Then Eq. (13-6) becomes

$$\mu\frac{\partial I(y,\mu)}{\partial y} + \kappa I(y,\mu) = \kappa I_b(T) \qquad \text{for } -1 \le \mu \le 1 \tag{13-8a}$$

or

$$\mu\frac{\partial I(y,\mu)}{\partial y} + \kappa I(y,\mu) = \kappa\frac{\sigma T^4}{\pi} \qquad \text{for } -1 \le \mu \le 1 \tag{13-8b}$$

since the blackbody radiation intensity $I_b(T)$ is related to the temperature T of the medium by [see Eq. (11-7b)]

$$I_b(T) = \frac{\sigma T^4}{\pi} \tag{13-9}$$

In Eqs. (13-8) the range of the μ variable is taken from -1 to 1 since θ varies from 0 to π.

Equation (13-8) is called the *equation of radiative transfer* for an absorbing, emitting, plane-parallel medium.

Sometimes it is convenient to define an optical variable τ as

$$\tau = \kappa y \tag{13-10}$$

where κ has the dimension of length^{-1}, hence τ is a *dimensionless* variable. Then, the *equation of radiative transfer* (13-8) in the τ variable becomes

$$\mu\frac{\partial I(\tau,\mu)}{\partial \tau} + I(\tau,\mu) = I_b(T) \cdot \qquad \text{for } -1 \le \mu \le 1 \tag{13-11a}$$

where

$$I_b(T) = \frac{\sigma T^4}{\pi} \tag{13-11b}$$

In the problems of radiative-heat transfer the net radiative-heat flux q (Btu/h·ft^2) in the τ (or y) direction is a quantity of interest; it is determined from the definition

$$q(\tau) = \int_{\varphi=0}^{2\pi}\int_{\theta=0}^{\pi} I(\tau,\theta)\cos\theta\sin\theta\,d\theta\,d\varphi \tag{13-12}$$

where φ is the azimuthal angle in the plane perpendicular to the y axis. Equation (13-12) implies that the net radiative-heat flux is obtained by integrating the quantity $I(\tau, \theta) \cos \theta$ over the entire spherical space. Here $I(\tau, \theta)$ is independent of the φ variable, hence the integration with respect to φ is readily performed; in addition, if the variable θ is changed into the variable μ by the transformation $\mu = \cos \theta$ (that is, $d\mu = -\sin \theta \, d\theta$), Eq. (13-12) is written

$$q(\tau) = 2\pi \int_{-1}^{1} I(\tau, \mu)\mu \, d\mu \qquad (13\text{-}13)$$

Thus, once the angular distribution of the radiation intensity $I(\tau, \mu)$ is determined from the solution of the equation of radiative transfer (13-11) subject to appropriate boundary conditions, the net radiative-heat flux $q(\tau)$ anywhere in the medium is determined by Eq. (13-13).

13-2   INTEGRATION OF THE EQUATION OF RADIATIVE TRANSFER

We now examine the integration of the equation of radiative transfer (13-11). The independent variable μ varies from -1 to $+1$ (that is, θ varies from 0 to π), but for $\mu = 0$ the equation becomes *singular*; that is, the derivative $\partial I / \partial \tau$ is multiplied by zero for $\mu = 0$. To circumvent the difficulty associated with the singularity of the differential equation, the solution of this equation is performed by separating it into two parts: one for the range of μ from 0 to $+1$ and the other from -1 to 0. Here, the positive range of μ in $0 < \mu \le 1$ characterizes the radiation intensity propagating in the *forward direction* or in the increasing τ direction, and the negative range of μ in $-1 \le \mu < 0$ characterizes the radiation intensity propagating in the *backward direction* or in the decreasing τ direction. With this consideration, Eq. (13-11a) is separated into two equations, one for the radiation intensity $I^{+}(\tau, \mu)$ in the *forward direction* (that is, $0 < \mu \le 1$) and the other for the radiation intensity $I^{-}(\tau, \mu)$ in the *backward direction* (that is, $-1 \le \mu < 0$) in the following manner:

$$\mu \frac{\partial I^{+}(\tau, \mu)}{\partial \tau} + I^{+}(\tau, \mu) = I_{b}(T) \qquad \text{for } 0 < \mu \le 1 \qquad (13\text{-}14a)$$

$$\mu \frac{\partial I^{-}(\tau, \mu)}{\partial \tau} + I^{-}(\tau, \mu) = I_{b}(\tau) \qquad \text{for } -1 \le \mu < 0 \qquad (13\text{-}14b)$$

Then, Eq. (13-13) defining the net radiative-heat flux in the medium is written in terms of the forward and backward radiation intensities as

$$q(\tau) = 2\pi \int_{-1}^{0} I^{-}(\tau, \mu)\mu \, d\mu + 2\pi \int_{0}^{1} I^{+}(\tau, \mu)\mu \, d\mu \qquad (13\text{-}15)$$

Thus the net radiative-heat flux anywhere in the medium can be calculated by Eq. (13-15) if the radiation intensities $I^-(\tau, \mu)$ and $I^+(\tau, \mu)$ are available from the solution of the above differential equations.

Boundary Conditions

To solve Eqs. (13-14a) and (13-14b) boundary conditions are needed; we discuss the boundary conditions with reference to a slab geometry of optical thickness τ_0 as illustrated in Fig. 13-2. Each of these equations requires one boundary condition. The boundary condition for Eq. (13-14a) is specified at the surface $\tau = 0$ because the radiation intensity $I^+(\tau, \mu)$ is propagating in the *forward* direction. The boundary condition for Eq. (13-14b) is specified at the surface $\tau = \tau_0$ because the radiation intensity $I^-(\tau, \mu)$ is propagating in the *backward* direction. Let $I^+(0, \mu)$ and $I^-(\tau_0, \mu)$ be the intensities at the boundaries $\tau = 0$ and $\tau = \tau_0$, respectively. The relations defining the radiation intensities at the boundary surfaces depend on the type of boundary surfaces. The boundaries at $\tau = 0$ and $\tau = \tau_0$ may be (1) *transparent* to radiation, (2) *black surfaces* which absorb and emit radiation, or (3) *reflective surfaces* which absorb, emit, and reflect radiation. We now discuss the physical significance of these three types of boundary conditions.

Transparent Boundaries When the boundaries at $\tau = 0$ and $\tau = \tau_0$ are both transparent, there is no attenuation of radiation at the boundary surfaces. If both boundaries at $\tau = 0$ and $\tau = \tau_0$ are irradiated by an externally incident *isotropic radiation* (i.e., radiation intensity is independent of direction) of known intensities f_1 and f_2, respectively, the boundary conditions for Eqs. (13-14a) and (13-14b) are given, respectively, as

$$I^+(0, \mu) = f_1 \qquad \text{for } \mu > 0 \tag{13-16a}$$

$$I^-(\tau_0, \mu) = f_2 \qquad \text{for } \mu < 0 \tag{13-16b}$$

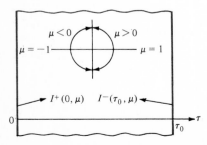

FIG. 13-2 Coordinates for the solution of the equation of radiative transfer for a plane-parallel slab.

If there is no externally incident radiation these boundary conditions become

$$I^+(0, \mu) = 0 \qquad \text{for } \mu > 0 \tag{13-17a}$$

$$I^-(\tau_0, \mu) = 0 \qquad \text{for } \mu < 0 \tag{13-17b}$$

Black Boundaries   If the boundaries at $\tau = 0$ and $\tau = \tau_0$ are both black surfaces maintained at uniform temperatures T_1 and T_2, respectively, they emit radiation into the medium with intensities equal to the blackbody radiation intensity at their respective temperatures.   Then the boundary conditions for Eqs. (13-14a) and (13-14b) become, respectively,

$$I^+(0, \mu) = \frac{\sigma T_1^{\,4}}{\pi} \equiv f_1 \qquad \text{for } \mu > 0 \tag{13-18a}$$

$$I^-(\tau_0, \mu) = \frac{\sigma T_2^{\,4}}{\pi} \equiv f_2 \qquad \text{for } \mu < 0 \tag{13-18b}$$

We note that transparent boundaries with externally incident isotropic radiation given by Eqs. (13-16) and the black boundaries as discussed above are characterized with boundary conditions of a similar form.

Reflective Boundaries   If the boundaries at $\tau = 0$ and $\tau = \tau_0$ are reflective, the radiation intensity from the interior of the slab is partially reflected when it strikes the surface.   The boundary conditions at $\tau = 0$ and $\tau = \tau_0$ for such a situation become a function of the radiation intensity coming from the interior of the medium; hence they are unknown quantities.   Then the analysis of radiative-heat transfer in a participating medium subject to reflection at the boundaries becomes involved.   Therefore the reflective boundaries are not discussed here; the reader interested in this subject should consult Refs. [4,5].

Integration of the Equation of Radiative Transfer   We now examine the integration of the equations of radiative transfer (13-14) for a slab $0 \leq \tau \leq \tau_0$ having *black* boundaries at $\tau = 0$ and $\tau = \tau_0$ at uniform temperatures T_1 and T_2, respectively.   The differential equation and the boundary condition for the *forward radiation intensity* $I^+(\tau, \mu)$ are obtained from Eqs. (13-14a) and (13-18a), respectively, as

$$\mu \frac{\partial I^+(\tau, \mu)}{\partial \tau} + I^+(\tau, \mu) = I_b(T) \qquad \text{for } 0 < \mu \leq 1 \tag{13-19a}$$

$$I^+(\tau, \mu) = \frac{\sigma T_1^{\,4}}{\pi} \equiv f_1 \qquad \text{at } \tau = 0, \text{ for } \mu > 0 \tag{13-19b}$$

The equation and the boundary condition for the *backward radiation intensity* $I^-(\tau, \mu)$ are obtained from Eqs. (13-14b) and (13-18b) as

$$\mu \frac{\partial I^-(\tau, \mu)}{\partial \tau} + I^-(\tau, \mu) = I_b(T) \qquad \text{for } -1 \leq \mu < 0 \tag{13-20a}$$

$$I^-(\tau, \mu) = \frac{\sigma T_2^{\,4}}{\pi} \equiv f_2 \qquad \text{at } \tau = \tau_0, \text{ for } \mu < 0 \tag{13-20b}$$

The solutions of Eqs. (13-19) and (13-20) are given, respectively, as

$$I^+(\tau, \mu) = f_1 e^{-\tau/\mu} + \int_0^\tau \frac{1}{\mu} I_b(T) e^{-(\tau-\tau')/\mu} \, d\tau' \qquad \text{for } \mu > 0 \tag{13-21}$$

$$I^-(\tau, \mu) = f_2 \, e^{(\tau_0 - \tau)/\mu} - \int_\tau^{\tau_0} \frac{1}{\mu} I_b(T) e^{-(\tau-\tau')/\mu} \, d\tau' \qquad \text{for } \mu < 0 \tag{13-22}$$

The net radiative-heat flux $q(\tau)$ anywhere in the slab is obtained by substituting the intensities given by Eqs. (13-21) and (13-22) into Eq. (13-15). We obtain

$$q(\tau) = 2\pi \left(f_1 \int_0^1 \mu e^{-\tau/\mu} \, d\mu + f_2 \int_{-1}^0 \mu e^{(\tau_0 - \tau)/\mu} \, d\mu \right)$$
$$+ 2\pi \left[\int_0^\tau I_b(T) \int_0^1 e^{-(\tau-\tau')/\mu} \, d\mu \, d\tau' - \int_\tau^{\tau_0} I_b(T) \int_{-1}^0 e^{-(\tau-\tau')/\mu} \, d\mu \, d\tau' \right] \tag{13-23}$$

Various integrals with respect to the μ variable are performed:

$$\int_0^1 \mu e^{-\tau/\mu} \, d\mu \equiv E_3(\tau) \tag{13-24a}$$

$$\int_{-1}^0 \mu e^{(\tau_0 - \tau)/\mu} \, d\mu = -\int_0^1 \mu e^{-(\tau_0 - \tau)/\mu} \, d\mu \equiv -E_3(\tau_0 - \tau) \tag{13-24b}$$

$$\int_0^1 e^{-(\tau-\tau')/\mu} \, d\mu \equiv E_2(\tau - \tau') \tag{13-24c}$$

$$\int_{-1}^0 e^{-(\tau-\tau')/\mu} \, d\mu = -\int_1^0 e^{-(\tau'-\tau)/\mu} \, d\mu = \int_0^1 e^{-(\tau'-\tau)/\mu} \, d\mu = E_2(\tau' - \tau) \tag{13-24d}$$

where the functions $E_2(z)$ and $E_3(z)$ are called the *exponential integrals* of the second and the third kind, respectively. In general, the *exponential integral* of the nth kind is defined as

$$E_n(z) = \int_0^1 \mu^{n-2} e^{-z/\mu} \, d\mu \tag{13-25}$$

and $E_n(z)$ functions are well tabulated in the literature. In Appendix C we present a tabulation of the functions E_1, E_2, E_3, E_4 and give a brief discussion of the properties of $E_n(x)$ functions.

When the integrals in Eqs. (13-24) are substituted into Eq. (13-23), the expression for the net radiative-heat flux $q(\tau)$ anywhere in the slab becomes

$$q(\tau) = 2\pi[f_1 E_3(\tau) - f_2 E_3(\tau_0 - \tau)]$$

$$+ 2\pi\left[\int_0^\tau I_b(T)E_2(\tau - \tau')\,d\tau' - \int_\tau^{\tau_0} I_b(T)E_2(\tau' - \tau)\,d\tau'\right] \qquad (13\text{-}26)$$

where

$$f_1 \equiv \frac{\sigma T_1{}^4}{\pi} \qquad f_2 = \frac{\sigma T_2{}^4}{\pi} \qquad \text{and} \qquad I_b(T) = \frac{\sigma T(\tau)^4}{\pi}$$

Thus, knowing the temperature distribution $T(\tau)$ in the medium and the boundary surface temperatures T_1 and T_2, the net radiative-heat flux $q(\tau)$ at any position τ in the slab is determined from Eq. (13-26). It is to be noted that *the heat flow is in the positive τ direction when $q(\tau)$ is positive*, and vice versa. We illustrate the application of Eq. (13-26) in the next section to predict the net radiative-heat flux in a medium at uniform temperature.

13-3   NET RADIATIVE-HEAT FLUX IN AN ABSORBING, EMITTING SLAB AT UNIFORM TEMPERATURE

There are numerous engineering applications in which the net radiative-heat flux from an emitting, absorbing medium is needed. For example, consider the high-velocity flow of a hot exhaust gas at a uniform temperature T_0 through a parallel-plate channel having walls which are black surfaces and kept at uniform temperatures T_1 and T_2. In this problem the heat flux at the walls is of interest. The exhaust gases absorb and emit radiation, and for sufficiently high temperatures the convection is neglected and radiation is considered to be the dominant mode of heat transfer. Then, this heat-transfer problem can be analyzed as a pure radiative heat-transfer problem from an absorbing and emitting slab as now described.

Let κ be the average absorption coefficient for the exhaust gas and L the distance between the parallel plates. The gas layer can be treated as an absorbing, emitting slab of optical thickness $\kappa L = \tau_0$, at a uniform temperature T_0, contained between two black boundary surfaces at $\tau = 0$ and $\tau = \tau_0$ which are kept at uniform temperatures T_1 and T_2, respectively, as illustrated in Fig. 13-3. This heat-transfer problem is exactly the same as the radiative-heat-transfer problem for an absorbing and emitting slab of optical thickness τ_0 and at a

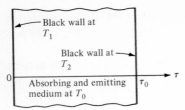

FIG. 13-3 Absorbing and emitting medium at uniform temperature T_0 between two parallel black walls at temperatures T_1 and T_2.

specified temperature T_0 considered in the previous section. The net radiative-heat flux $q(\tau)$ anywhere in the medium is calculated from Eq. (13-26) by setting in that equation

$$I_b(T) = \frac{\sigma T_0^4}{\pi} \tag{13-27}$$

where T_0 is the medium temperature which is now constant. When $I_b(T)$ is given as above, various integrals in Eq. (13-26) are evaluated as

$$\int_0^\tau I_b(T)E_2(\tau - \tau')\, d\tau' = \frac{\sigma T_0^4}{\pi} \int_0^\tau E_2(\tau - \tau')\, d\tau'$$

$$= \frac{\sigma T_0^4}{\pi}\left[E_3(\tau - \tau')\right]_{\tau'=0}^{\tau'=\tau}$$

$$= \frac{\sigma T_0^4}{\pi}\left[\frac{1}{2} - E_3(\tau)\right] \tag{13-28a}$$

$$\int_\tau^{\tau_0} I_b(T)E_2(\tau' - \tau)\, d\tau' = \frac{\sigma T_0^4}{\pi} \int_\tau^{\tau_0} E_2(\tau' - \tau)\, d\tau'$$

$$= \frac{\sigma T_0^4}{\pi}\left[-E_3(\tau' - \tau)\right]_{\tau'=\tau}^{\tau'=\tau_0}$$

$$= \frac{\sigma T_0^4}{\pi}\left[\frac{1}{2} - E_3(\tau_0 - \tau)\right] \tag{13-28b}$$

since $E_0(0) = \frac{1}{2}$.

The substitution of Eqs. (13-28) into Eq. (13-26) gives an explicit expression for the net radiative-heat flux in the medium as

$$q(\tau) = 2\pi[f_1 E_3(\tau) - f_2 E_3(\tau_0 - \tau)] + 2\sigma T_0^4[E_3(\tau_0 - \tau) - E_3(\tau)] \tag{13-29}$$

where

$$f_1 \equiv \frac{\sigma T_1^4}{\pi} \qquad f_2 = \frac{\sigma T_2^4}{\pi}$$

In this expression the first bracket on the right-hand side is the contribution of the boundary conditions and the second is the contribution of the medium temperature to the net radiative-heat flux in the medium.

The net radiative-heat fluxes at the boundary surfaces are quantities of practical interest; they are obtained by setting in Eq. (13-29) $\tau = 0$ and $\tau = \tau_0$. Then the net radiative-heat fluxes at the boundaries $\tau = 0$ and $\tau = \tau_0$ become, respectively,

$$q(0) = 2[\tfrac{1}{2}\sigma T_1^{\,4} - \sigma T_2^{\,4} E_3(\tau_0)] + 2\sigma T_0^{\,4}[E_3(\tau_0) - \tfrac{1}{2}] \tag{13-30a}$$

$$q(\tau_0) = 2[\sigma T_1^{\,4} E_3(\tau_0) - \tfrac{1}{2}\sigma T_2^{\,4}] + 2\sigma T_0^{\,4}[\tfrac{1}{2} - E_3(\tau_0)] \tag{13-30b}$$

since $E_3(0) = \tfrac{1}{2}$.

If the gas between the plates is nonparticipating (i.e., transparent to radiation), then the absorption coefficient κ is taken as zero. For this special case the optical thickness $\tau_0 = \kappa L$ of the gas layer is zero; by setting $\tau_0 = 0$ in Eqs. (13-30) the net radiative-heat fluxes at the boundary surfaces become

$$q(0) = \sigma T_1^{\,4} - \sigma T_2^{\,4} \tag{13-31a}$$

$$q(\tau_0) = \sigma T_1^{\,4} - \sigma T_2^{\,4} \tag{13-31b}$$

We note that the result in Eqs. (13-31) is similar to that given in the previous chapter for the net radiative-heat flux between two parallel, black, infinite plates at uniform temperatures T_1 and T_2 separated by a nonparticipating medium. However, care must be exercised in establishing the direction of the heat flow from the expressions as given by Eqs. (13-31) and from those given in the previous chapter for a nonparticipating medium. In *the above expressions the heat flow is in the positive τ direction if q is positive*, whereas *in the expressions given in the previous chapter the heat flow is from the surface into the enclosure space if q is positive*. For example, when $T_1 > T_2$, in Eqs. (13-31) both $q(0)$ and $q(\tau_0)$ are positive, hence the heat must flow in the positive τ direction; now, the arrangement of surfaces as shown in Fig. 13-3, implies that heat flows in a direction from the wall into the gas at the surface $\tau = 0$ and from the gas into the wall at the surface $\tau = \tau_0$.

Example 13-1   An absorbing, emitting, gray gas at temperature $T_0 = 2000°R$ (1111.1 K) flows between two parallel, black, infinite plates kept at temperatures $T_1 = T_2 = 1000°R$ (555.6 K). Determine the net radiative-heat flux at the walls if the optical spacing between the plates is $\tau_0 = 0.5$.

Solution   Taking the geometry for the considered problem as the same as that shown in Fig. 13-3, the net radiative-heat flux $q(0)$ at the boundary surface at $\tau = 0$ is given by Eq. (13-30a) as

$$q(0) = 2[\tfrac{1}{2}\sigma T_1^{\,4} - \sigma T_2^{\,4} E_3(\tau_0)] + 2\sigma T_0^{\,4}[E_3(\tau_0) - \tfrac{1}{2}]$$

and the numerical values of various quantities are $T_1 = T_2 = 1000°R$ (555.6 K), $T_0 = 2000°R$ (1111.1 K), $\tau_0 = 0.5$, $E_3(0.5) = 0.2216$ (from Appendix C), and $\sigma = 0.1714 \times 10^{-8}$ Btu/h·ft²·°R⁴ (0.56697 × 10⁻⁸ W/m²·K⁴). Substituting these values in the above expression and using the English system of units, we find

$$q(0) = 2 \times (0.1714 \times 10^4) \times (\tfrac{1}{2} - 0.2216) + 2 \times (0.1714 \times 20^4) \times (0.2216 - \tfrac{1}{2})$$

$$q(0) = 2 \times (0.1714 \times 10^4) \times (\tfrac{1}{2} - 0.2216) \times (1^4 - 2^4) = -1.432 \times 10^4 \text{ Btu/h·ft}^2$$

$$(-4.51 \times 10^4 \text{ W/m}^2)$$

or, by using the SI system of units, we obtain

$$q(0) = 2 \times (0.56697 \times 10^4) \times (\tfrac{1}{2} - 0.2216) \times (0.5556^4 - 1.1111^4) = -4.51 \times 10^4 \text{ W/m}^2$$

Here, $q(0)$ being negative, heat flows from the gas into the wall at a rate of 1.432×10^4 Btu/h·ft² (or 4.51×10^4 W/m²).

The radiative-heat flux $q(\tau_0)$ at the wall $\tau = \tau_0$ is determined by Eq. (13-30b). Because of symmetry, we find

$$q(\tau_0) = 1.432 \times 10^4 \text{ Btu/h·ft}^2 \ (4.51 \times 10^4 \text{ W/m}^2)$$

$q(\tau_0)$ being positive, the heat flow is in the positive τ direction or from the gas stream into the wall at $\tau = \tau_0$.

13-4 EMPIRICAL DATA FOR ABSORPTION AND EMISSION OF RADIATION BY NONLUMINOUS GASES

Elementary gases such as H_2, N_2, O_2, and dry air are transparent to thermal radiation at temperatures met in most engineering practice, except at extremely high temperatures when gases become ionized plasma. On the other hand, gases such as water vapor, carbon dioxide, carbon monoxide, sulfur dioxide, ammonia, hydrogen chloride, alcohol, and hydrocarbons absorb and emit radiation significantly at certain wavelengths. For example, if a beam of radiation passes through a gas mass containing, say, carbon dioxide, absorption takes place in certain regions of the infrared spectrum, as illustrated in Fig. 11-11. Conversely, if such a gas is heated it radiates in the same wavelength bands. In the previous chapter we described an analytical method of determining radiative-heat exchange between an absorbing and emitting gas body and its enclosure for the case of a very simple slab geometry. The shape and the size of the gas body affect the radiative exchange, and for bodies of arbitrary shape the exact analysis is extremely difficult. Therefore, approximate and semiempirical methods have been developed for the prediction of the absorption or emission of radiation by a body of *nonluminous gas;* for engineering purposes such methods may yield results of sufficient accuracy. The term *nonluminous gas* is used here to indicate that the results are for the absorption or emission by the gas mass itself, but not by the matter such as carbon particles, etc., which may exist in the combustion products and radiate in the visible-light spectrum at high temperatures.

To provide some background information for the understanding of the semiempirical results that will be given in this section, we recall the equation of radiative transfer (13-4), and write this equation for the spectral-radiation intensity $I_\lambda(S)$ in the form

$$\frac{dI_\lambda(S)}{dS} + \kappa_\lambda I_\lambda(S) = 0 \tag{13-32}$$

where the emission term $\kappa_\lambda I_{\lambda b}(T)$ is neglected and the distance S is measured along the path of the propagation of radiation. We now consider the integration of this ordinary differential equation with respect to S along the path of propagation of the beam. A boundary condition is needed to perform the integration; let the spectral-radiation intensity be specified at $S = 0$ as

$$I_\lambda(0) \equiv I_{\lambda 0} \qquad \text{at } S = 0 \tag{13-33}$$

The integration of Eq. (13-32) with respect to S for constant κ_λ and subject to the condition Eq. (13-33) yields

$$I_\lambda(S) = I_{\lambda 0} e^{-\kappa_\lambda S} \tag{13-34}$$

Then the spectral intensity at $S = L$ becomes

$$I_\lambda(L) = I_{\lambda 0} e^{-\kappa_\lambda L} \tag{13-35}$$

The difference between the radiation intensities at $S = 0$ and $S = L$ given by Eqs. (13-33) and (13-35), respectively, represents the amount of attenuation of radiation in traveling through an absorbing gas layer of thickness L. We find

$$I_\lambda(0) - I_\lambda(L) = I_{\lambda 0}(1 - e^{-\kappa_\lambda L}) \tag{13-36}$$

where the quantity

$$\alpha_\lambda \equiv 1 - e^{-\kappa_\lambda L} \tag{13-37}$$

is called the *spectral absorptivity of the gas layer*. When the Kirchhoff law is applicable, the spectral absorptivity α_λ is taken to be equal to the *spectral emissivity* ε_λ of the gas layer. In practice, the gas absorptivity α_g or the gas emissivity ε_g over all wavelengths is of interest. Attempts have been made to determine the gas absorptivity or emissivity over all wavelength bands by the integration of Eq. (13-37) over all radiation bands; but the analysis is quite involved because the spectral absorption coefficient κ_λ for gases exhibit very complex variations with the wavelength, and their determination is very difficult theoretically. The reader should consult Ref. [10] for a discussion of various analytical techniques for the determination of gas absorptivities. Therefore, experimental approaches have been made to predict gas emissivity and absorptivity.

Hottel [11,12] and Hottel and Egbert [13] measured gas emissivity ε_g and presented emissivity charts for gases such as CO_2, H_2O, CO, ammonia, SO_2, etc., as a function of temperature and a product term $P_i L$, where P_i is the partial pressure (in atmospheres) of the gas i in the gas mass and L is the beam length (in feet). The gas-emissivity charts prepared by them are for a hemispherical gas mass of radius L, at a temperature T_g, radiating to a black surface located at the center of the base of the hemisphere; the emission of radiation from the hemispherical gas mass to a unit area at the center of the base of the hemisphere is expressed as $\varepsilon_g \sigma T_g^4$, where ε_g is the gas emissivity (i.e., the ratio of radiation from the gas to the surface to the radiation from a blackbody at the same temperature to the surface). Most of these experiments are conducted at a total pressure of 1 atm; as the total pressure is increased the absorption lines are broadened, hence the gas emissivity is affected. Therefore approximate correction factors are also presented to adjust such emissivity data for a total pressure differing from 1 atm. To generalize the utility of these charts for the prediction of gas emissivities for gas masses having shapes other than a hemisphere, corrections are applied by introducing the *equivalent mean path length* which are determined by graphical or analytical means. We now present some of these charts and discuss their application in the prediction of radiative-heat exchange between a gas mass and a surface element.

Figure 13-4a gives the emissivity ε_c for carbon dioxide in a gas mass at a total pressure of 1 atm, plotted as a function of gas temperature T_g for several different values of the product $P_c L$, where P_c is the partial pressure in atmospheres of carbon dioxide and L is the radius in feet of the hemispherical gas mass. Figure 13-4b gives an approximate correction factor C_c to adjust the emissivity data of Fig. 13-4a for a total pressure of gas mass differing from 1 atm; that is, the emissivity of carbon dioxide ε_c read from Fig. 13-4a is multiplied by the factor C_c from the correction chart, Fig. 13-4b, to adjust the emissivity for other total pressures.

Figure 13-5a shows the emissivity ε_w for water vapor in a gas mass at a total pressure of 1 atm plotted as a function of gas temperature T_g for several different values of the product term $P_w L$. The data presented in this figure were obtained by Hottel and Egbert [13] by reducing all measured gas emissivities to values corresponding to an idealized case of $P_w \rightarrow 0$. By this procedure they were able to correlate the data of various other experiments at a total pressure of 1 atm. Figure 13-5b gives an approximate correction factor C_w for converting the emissivity of water vapor to values of P_w other than almost zero and to values of total pressure P_T other than 1 atm; that is, the emissivity of water vapor ε_w from Fig. 13-5a is multiplied by the correction factor C_w from Fig. 13-5b to adjust it to values of P_w and P_T other than 0 and 1 atm, respectively.

When carbon dioxide and water vapor are present together in a gas mass, the emissivity of the gas mixture ε_m is somewhat less by an amount $\Delta\varepsilon$ than that determined by adding separately calculated emissivities ε_c and ε_w for carbon dioxide and water vapor from Figs. 13-4 and 13-5, respectively. The reason for this is the mutual absorption of radiation. The emissivity ε_m of the mixture of carbon

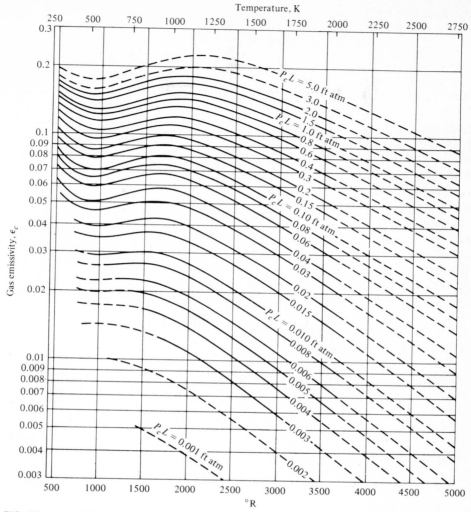

FIG. 13-4a Emissivity ε_c of carbon dioxide at a total pressure of $P_T = 1$ atm. (*From Hottel* [*12*].)

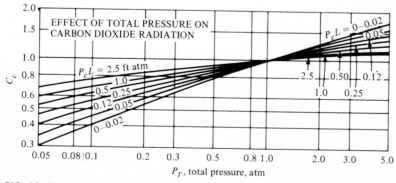

FIG. 13-4b Correction factor C_c for converting the emissivity of CO_2 at 1 atm pressure to emissivity at P_T atm. (*From Hottel* [*12*].)

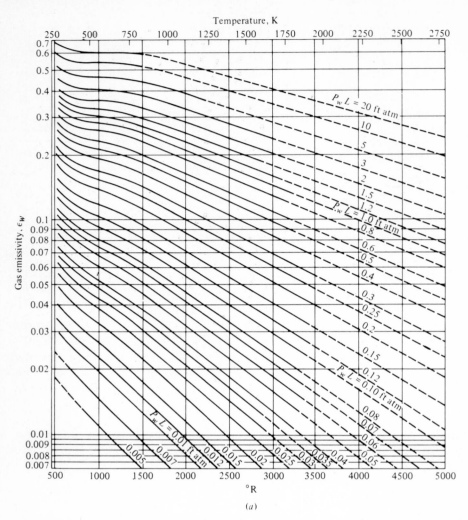

Temperature, K

FIG. 13-5a   Emissivity ε_w of water vapor at a total pressure of $P_T = 1$ atm and corresponding to an idealized case of $P_w \to 0$. (*From Hottel* [12].)

dioxide and water vapor can be determined approximately from the relation

$$\varepsilon_m = \varepsilon_c + \varepsilon_w - \Delta\varepsilon \tag{13-38}$$

where ε_c and ε_w are emissivities of carbon dioxide and water vapor given in Figs. 13-4 and 13-5, respectively, and the emissivity correction $\Delta\varepsilon$ for mutual absorption can be obtained from Fig. 13-6.   In this figure P_c and P_w denote, respectively, the partial pressures of carbon dioxide and water vapor.

The reader should consult Ref. [12] for similar emissivity charts for various other gases.

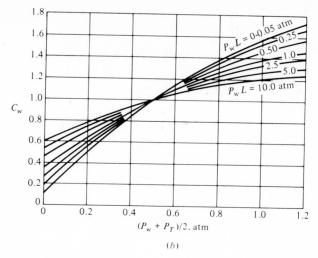

(b)

FIG. 13-5b Correction factor C_w for converting emissivity of H_2O to values of P_w and P_T other than 0 and 1 atm, respectively. (*From Hottel* [12].)

Recently Penner and Varanasi [14], by using basic spectroscopic data on line spacing and simplified procedures, reproduced Hottel's emissivity charts for carbon dioxide and water vapor.   Figure 13-7 shows a comparison of Penner and Varanasi's [14] calculated emissivities for carbon dioxide and water vapor and Hottel's [12] empirical data.   It is apparent from this figure that the agreement between the calculated data and the empirical measurement for CO_2 lies well within the estimated reliability of either set of data.   However, in the case of water vapor, there are some discrepancies at large optical depths and at low temperatures.

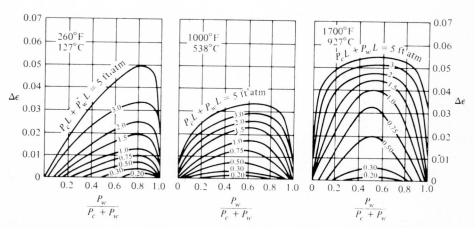

FIG. 13-6   Emissivity correction $\Delta\epsilon$ for mutual absorption of water vapor and carbon dioxide. (*From Hottel* [12].)

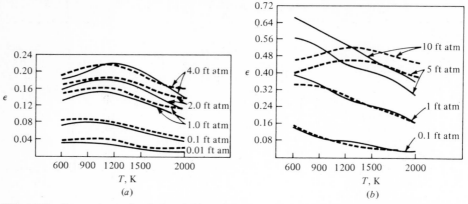

FIG. 13-7 Comparison of calculated (dotted lines) and observed (solid lines) gas emissivity for (a) carbon dioxide and (b) water vapor at atmospheric pressure as a function of temperature. (*From Penner and Varanasi* [14].)

Effects of Geometry

The foregoing emissivity charts are for gas masses having a hemispherical geometry and the path length L that refer to the radius of the hemisphere. To generalize the application of these charts for gas masses having other shapes, an *equivalent path length L* may be used in the product PL appearing in the emissivity charts of Figs. 13-4 to 13-6. Table 13-1 gives the equivalent path lengths for several simple geometries.

Radiative-Heat Exchange Between a Gas Mass and the Enclosure Surface

We now consider the determination of net radiative-heat exchange between a gas mass at a temperature T_g and its surrounding black walls at a temperature T_w. Let ε_g be the emissivity of the gas mass for the considered geometry, partial pressure, total pressure, and temperature. The radiation emitted Q_e by the gas mass to the surrounding black walls of the container is

$$Q_e = A\varepsilon_g \sigma T_g^4 \tag{13-39}$$

where A is the surface area of the walls. The radiation emitted by the surrounding black walls of temperature T_w is $A\sigma T_w^4$; if α_g is the absorptivity of the gas for radiation emitted at temperature T_w, then the radiation absorbed Q_a by the gas mass is

$$Q_a = A\alpha_g \sigma T_w^4 \tag{13-40}$$

The net radiative-heat exchange Q between the gas mass at temperature T_g and its black surroundings at temperature T_w is then determined from

$$Q = Q_e - Q_a = \sigma A(\varepsilon_g T_g^4 - \alpha_g T_w^4) \qquad \text{Btu/h (W)} \tag{13-41a}$$

TABLE 13-1   Equivalent Path Length for Various Gas Body Geometries†

Geometry	L
1 Sphere	$\frac{2}{3}$ × diameter
2 Infinite cylinder	1 × diameter
3 Space between infinite parallel plates	1.8 × distance between planes
4 Space outside infinite bank of tubes with centers on equilateral triangles; clearance between the tubes equal to tube diameter	2.8 × clearance
5 Same as preceding except clearance is twice the tube diameter	3.8 × clearance
6 Same, except tube centers on squares; clearance equal to tube diameter	3.5 × clearance

† From Hottel [12].

or

$$q \equiv \frac{Q}{A} = \sigma\left(\varepsilon_g T_g^4 - \alpha_g T_w^4\right) \qquad \text{Btu/h·ft}^2 \ (\text{W/m}^2) \tag{13-41b}$$

where the gas emissivity ε_g of the gas mass is obtained from the emissivity chart for the gas temperature T_g, and the gas absorptivity α_g is obtained from the same chart for the wall temperature T_w.

Example 13-2   Determine the emissivity of a gas mass at a temperature $T_g = 1000°R$, at a total pressure of $P_T = 1.5$ atm, and containing 10% water vapor for the geometry of the gas mass having an equivalent path length $L = 2$ ft.

Solution   The partial pressure P_w of water vapor in the gas mass is $P_w = 1.5\frac{10}{100} = 0.15$ atm. Then, $P_w L = 0.15 \times 2 = 0.3$ ft·atm. From Fig. 13-5a, for $T_g = 1000°R$ and $P_w L = 0.3$ ft·atm, the emissivity of water vapor at a total pressure of 1 atm is $\varepsilon_g = 0.165$. This result is to be corrected for the desired total pressure and the partial pressure. From Fig. 13-5b, for $(P_w + P_T)/2 = (0.15 + 1.5)/2 = 0.825$ and $P_w L = 0.3$ ft·atm, the correction factor is found to be $C_w \cong 1.4$. Then, the corrected emissivity of the gas mass becomes $\varepsilon_{g,\text{corrected}} = 0.165 \times 1.4 \cong 0.23$.

Example 13-3   A flue gas at temperature $T_g = 2000°R$, total pressure 1 atm, and containing 10% water vapor, flows over a bank of tubes arranged in an equilateral-triangular pitch with tubes of 3-in OD and 5-in spacing. Assuming the tube surfaces are black and at a uniform temperature $T_w = 1000°R$, determine the net radiative-heat exchange between the gas and the tube walls.

Solution   For the considered tube-bundle geometry the equivalent mean path length for the gas is obtained from Table 13-1 as

$$L = 2.8 \times \text{tube diameter} = 2.8 \times \tfrac{3}{12} = 0.7 \text{ ft}$$

The partial pressure of the water vapor is $P_w = 0.1$ atm, and the product $P_w L$ becomes $P_w L = 0.1 \times 0.7$ ft·atm. Then from Fig. 13-5a the emissivity ε_g of the gas mass for $T_g = 2000°R$ and $P_w L = 0.07$ ft·atm is obtained as

$$\varepsilon_g = 0.041$$

The absorptivity α_g of the gas mass is also obtained from Fig. 13-5a for $T_w = 1000°R$ and $P_w L = 0.07$ ft·atm as

$$\alpha_g = 0.075$$

Figure 13-5b is now used to correct these results for the effects of pressure. However, for $(P_w + P_T)/2 = (0.1 + 1)/2 = 0.55$ atm and $P_w L = 0.07$ ft·atm, the correction factor is $C_w \cong 1$; thus no correction is needed for ε_g and α_g.

The net radiative-heat flux q between the gas mass and the surrounding walls is determined, according to Eq. (13-41b), as

$$q = \sigma(\varepsilon_g T_g^4 - \alpha_g T_w^4)$$

$$= (0.171 \times 10^{-8}) \times (0.041 \times 2000^4 - 0.07 \times 1000^4)$$

$$= 0.171 \times (0.041 \times 20^4 - 0.07 \times 10^4)$$

$$= (0.171 \times 10^4) \times (0.041 \times 16 - 0.07) \cong 1002 \text{ Btu/h·ft}^2 \ (3532 \text{ W/m}^2)$$

REFERENCES

1 Chandrasekhar, S.: "Radiative Transfer," Oxford University Press, London, 1950; also, Dover Publications, Inc., New York, 1960.

2 Hottel, H. C., and A. F. Sarofim: "Radiative Transfer," McGraw-Hill Book Company, New York, 1967.

3 Kourganoff, V.: "Basic Methods in Transfer Problems," Dover Publications, Inc., New York, 1963.

4 Love, T. J.: "Radiation Heat Transfer," Charles E. Merrill Books, Inc., Columbus, Ohio, 1968.

5 Özışık, M. N.: "Radiative Transfer and Interactions with Conduction and Convection," John Wiley & Sons, Inc., New York, 1973.

6 Siegel, R., and R. Howell: "Thermal Radiation Heat Transfer," vol. 3, 1971, NASA Publication, Government Printing Office, Washington, D.C.

7 Sobolev, V. V.: "A Treatise on Radiative Transfer," D. Van Nostrand Company, Inc., Princeton, N.J., 1963.

8 Sparrow, E. M., and R. D. Cess: "Radiation Heat Transfer," Brooks/Cole Publishing Company, Belmont, Calif., 1970.

9 Viskanta, R.: Radiation Transfer and Interaction of Convection with Radiation Heat Transfer, in T. F. Irvine and J. P. Hartnett (eds.), "Advances in Heat Transfer," Academic Press, Inc., New York, 1966.

10 Penner, S. S.: "Quantitative Molecular Spectroscopy and Gas Emissivities," Addison-Wesley Publishing Company, Inc., Reading, Mass., 1959.

11 Hottel, H. C.: Heat Transmission by Radiation from Non-luminous Gases, *Trans. AIChE*, **19**:173–205 (1927).

12 Hottel, H. C.: Radiant Heat Transmission, in W. H. McAdams, "Heat Transmission," McGraw-Hill Book Company, New York, 1954.

13 Hottel, H. C., and R. B. Egbert: Radiant Heat Transmission from Water Vapor, *Trans. AIChE*, **38**:531–565 (1942).

14 Penner, S. S., and P. Varanasi: Simplified Procedures for Estimating Band and Total Emissivity of Polyatomic Molecules, *Eleventh Symp. (Int.) Combustion*, pp. 569–576, The Combustion Institute, Pittsburgh, Pa., 1967.

Fourteen

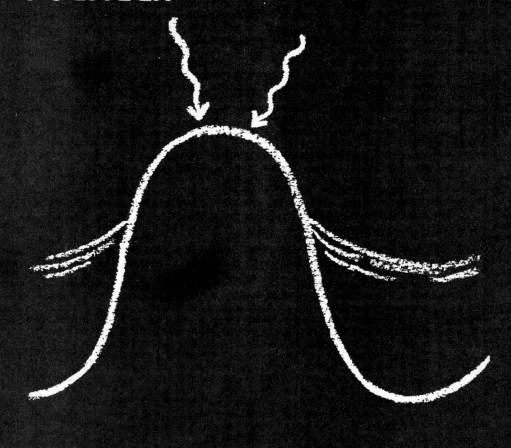

Heat Transfer with Change
in Phase

Condensers and boilers constitute an important and widely used type of heat exchangers with unique characteristics of heat-transfer mechanism on the condensing and boiling side.   If a vapor strikes a surface that is at a temperature slightly below the corresponding saturation temperature, the vapor will immediately condense into the liquid phase.   If the condensation takes place continuously over the surface which is kept cooled by some cooling process and the condensed liquid is removed from the surface by the motion resulting from gravity, the condensing surface is usually covered with a thin layer of liquid and the situation is known as *filmwise condensation*.   Under certain conditions, for example, if traces of oil are present during the condensation of steam on a highly polished surface, the film of condensate is broken into droplets and the situation is known as *dropwise condensation*.   The presence of condensate acts as a barrier to heat transfer from the vapor to the metal surface, and the dropwise condensation offers much less resistance to heat flow on the vapor side than the filmwise condensation.   If vapor contains some noncondensable gas, this gas will collect on the condensing side while condensation takes place.   The presence of noncondensable gas acts as resistance to heat flow on the condensing side because the vapor must diffuse through the noncondensable gas before it comes into contact with the cool surface of the condensate.   Therefore, an understanding of the mechanism of heat transfer and an accurate prediction of the heat-transfer coefficient for condensing vapors with and without the presence of noncondensable gas are important in the design of condensers.

When a liquid is in contact with a surface maintained at a temperature above saturation temperature of the liquid, boiling may occur.   The phenomenon of heat transfer in boiling is extremely complicated because of a large number of variables involved and very complex hydrodynamic developments occurring in the process.   Therefore, considerable work has been directed toward gaining a better understanding of the boiling mechanism, but because of the extremely complex nature of the problem most of the heat-transfer correlations in boiling still remain of empirical and semiempirical nature.

In this chapter we present a simple analysis for the determination of the heat-transfer coefficient during filmwise condensation on a plane, vertical surface; discuss the effects of noncondensable gas on heat transfer; and give various correlations on the heat-transfer coefficient during condensation of vapor.   Different boiling regimes and their heat-transfer characteristics are described, and pertinent heat-transfer relations are presented.

14-1   FILMWISE CONDENSATION OF PURE VAPORS

The first fundamental analysis leading to the determination of the heat-transfer coefficient during filmwise condensation of pure vapors (i.e., without the presence of noncondensable gas) on a flat plate and a circular tube was given by Nusselt

[1] in 1916. Subsequently many modifications of the original Nusselt theory appeared in the literature with various relaxations of the restrictive assumptions, and numerous experimental studies were made to verify the validity of various theories. For example, Refs. [2–14] deal with the condensation of pure vapors, and Refs. [15–19] consider the effects of noncondensable gas on condensation heat transfer. In this section we present a derivation of Nusselt's theory of condensation of pure vapors on a vertical plate in order to provide some insight into the physical significance of various parameters affecting heat transfer during condensation. Various correlations and semiempirical relations that are useful in practice are given.

Consider the condensation of vapor on a vertical plane surface and the condensate film draining down under the influence of gravity, as illustrated in Fig. 14-1. Let x be the coordinate axis from the top of the plate measured downward and y the coordinate axis normal to the condensing surface. Nusselt in his analysis of film condensation made the following simplifying assumptions: (1) the plate is maintained at a uniform temperature T_w which is less than the saturation temperature T_v of the vapor, (2) the vapor is stationary and exerts no drag on the motion of the condensate, (3) the flow of condensate is laminar, (4) the fluid acceleration within the condensate layer may be neglected, (5) the fluid properties are constant, and (6) heat transfer across the condensate layer is by pure conduction, and a linear temperature distribution is assumed.

Nusselt's analysis begins with the determination of the velocity distribution $u(y)$ across the condensate layer from the equation of motion obtained by balancing the gravitational force and the viscous shear force acting on a differential volume element of the condensate at any axial position x. For the small shaded area

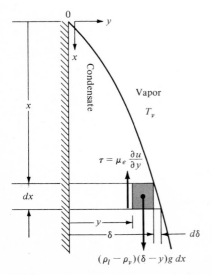

FIG. 14-1 Nomenclature for filmwise condensation on a vertical plane surface.

shown in Fig. 14-1 a force balance gives

$$\mu_l \frac{du}{dy} dx = (\rho_l - \rho_v)(\delta - y)g \, dx$$

or

$$\frac{du}{dy} = \frac{g(\rho_l - \rho_v)}{\mu_l} (\delta - y) \tag{14-1}$$

where δ is the thickness of the condensate film at the position x, μ is viscosity, and the subscripts l and v refer to the liquid and vapor phases, respectively. The boundary condition for Eq. (14-1) is taken as the liquid velocity zero at the wall surface

$$u = 0 \quad \text{at } y = 0 \tag{14-2}$$

The integration of Eq. (14-1) subject to the boundary condition (14-2) gives the velocity distribution in the condensate layer:

$$u(y) = \frac{g(\rho_l - \rho_v)}{\mu_l} \left(\delta y - \frac{1}{2} y^2 \right) \tag{14-3}$$

The mass-flow rate of condensate, M, per unit width of the plate (i.e., lb/h·ft) is given by

$$M = \int_0^\delta \rho_l u \, dy \tag{14-4}$$

Substituting u from Eq. (14-3) into (14-4) and performing the integration yield the mass-flow rate of condensate per unit width of the plate at the position x as

$$M = \frac{g\rho_l(\rho_l - \rho_v)\delta^3}{3\mu_l} \tag{14-5}$$

As the condensate flows from a position x to $x + dx$, the thickness of the layer increases from δ to $\delta + d\delta$. Then dM represents the amount of condensate added over the length dx about the position x per unit width of the plate and is obtained from Eq. (14-5) by differentiation:

$$dM = \frac{g\rho_l(\rho_l - \rho_v)\delta^2}{\mu_l} d\delta \tag{14-6}$$

The rate of heat released, Q, during the condensation of mass dM is given by

$$Q = h_{fg} dM = h_{fg} \frac{g\rho_l(\rho_l - \rho_v)\delta^2}{\mu_l} d\delta \tag{14-7}$$

where h_{fg} is the latent heat of condensation. This amount of heat Q must be transferred by conduction across the condensate layer to the plate through an area $1dx$ with a temperature difference $(T_v - T_w)$ as

$$Q = \frac{k_l}{\delta}(T_v - T_w)\, dx \qquad (14\text{-}8)$$

where k_l is the thermal conductivity of liquid, and T_v and T_w are the vapor-saturation and wall-surface temperatures, respectively. Equating Eqs. (14-7) and (14-8) we obtain the following differential equation for the thickness of the condensate layer:

$$\frac{d\delta}{dx} = \frac{\mu_l k_l (T_v - T_w)}{g\rho_l(\rho_l - \rho_v)h_{fg}}\frac{1}{\delta^3} \qquad (14\text{-}9)$$

The integration of Eq. (14-9) with the condition $\delta = 0$ for $x = 0$ yields the thickness of the condensate layer as a function of x along the plate:

$$\delta(x) = \left[\frac{4\mu_l k_l (T_v - T_w)x}{g(\rho_l - \rho_v)\rho_l h_{fg}}\right]^{1/4} \qquad (14\text{-}10)$$

The local heat-transfer coefficient h_x from the vapor side to the plate surface is defined as

$$h_x = \frac{k_l}{\delta} \qquad (14\text{-}11)$$

Then, by Eqs. (4-10) and (4-11), we obtain the local heat-transfer coefficient:

$$h_x = \left[\frac{g\rho_l(\rho_l - \rho_v)h_{fg}k_l^3}{4\mu_l(T_v - T_w)x}\right]^{1/4} \qquad (14\text{-}12)$$

When $\rho_v \ll \rho_l$, Eq. (14-12) simplifies to

$$h_x = \left[\frac{g\rho_l^2 h_{fg}k_l^3}{4\mu_l(T_v - T_w)x}\right]^{1/4} \qquad (14\text{-}13)$$

or the local Nusselt number Nu_x is defined as

$$\mathrm{Nu}_x \equiv \frac{h_x x}{k_l} = \left[\frac{g\rho_l^2 h_{fg} x^3}{4\mu_l k_l(T_v - T_w)}\right]^{1/4} \qquad (14\text{-}14)$$

The average value of the heat-transfer coefficient h_m over the length $0 \le x \le L$ of the plate is determined from

$$h_m = \frac{1}{L} \int_0^L h_x \, dx = 0.943 \left[\frac{g\rho_l^2 h_{fg} k_l^3}{\mu_l(T_v - T_w)L} \right]^{1/4} \tag{14-15a}$$

$$\text{or} \quad h_m = \tfrac{4}{3} h_x \big|_{x=L} \tag{14-15b}$$

In the above expressions various quantities are defined as follows:

g = gravitational acceleration, ft/h^2 (m/s^2)
h_{fg} = latent heat of condensation, Btu/lb (J/kg or W·s/kg)
h_x, h_m = local and average heat-transfer coefficient, respectively, Btu/h·ft^2·°F (W/m^2·°C)
k_l = thermal conductivity of liquid, Btu/h·ft·°F (W/m·°C)
L = length of surface, ft (m)
T_v, T_w = vapor and wall-surface temperatures, respectively, °F (°C)
μ_l = viscosity of liquid, lb/ft·h (kg/m·s)
ρ_l, ρ_v = density of liquid and vapor, respectively, lb/ft^3 (kg/m^3)

The heat-transfer coefficient derived above for a vertical plate is also applicable for condensation on the outside or inside surface of a vertical tube provided that the tube radius is large compared with the thickness of the condensate film.

Condensation on Inclined Plates

Nusselt's analysis of filmwise condensation given above for a vertical surface can readily be extended for condensation on an inclined plane surface making an angle φ with the horizontal as illustrated in Fig. 14-2. The results for the local and the average heat-transfer coefficients are given, respectively, as

$$h_x = \left[\frac{g\rho_l(\rho_l - \rho_v)h_{fg} k_l^3}{4\mu_l(T_v - T_w)x} \sin \varphi \right]^{1/4} \tag{14-16}$$

$$h_m = 0.943 \left[\frac{g\rho_l(\rho_l - \rho_v)h_{fg} k_l^3}{\mu_l(T_v - T_w)L} \sin \varphi. \right]^{1/4} \tag{14-17}$$

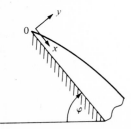

FIG. 14-2   Nomenclature for filmwise condensation on an inclined plane surface.

Condensation on Horizontal Tubes

The analysis of heat transfer for condensation on the outside surface of a horizontal tube is more complicated than the one given above for a vertical surface. Nusselt's analysis for laminar filmwise condensation on the surface of a horizontal tube gives the average heat-transfer coefficient as

$$h_m = 0.725 \left[\frac{g\rho_l(\rho_l - \rho_v)h_{fg}k_l^3}{\mu_l(T_v - T_w)D} \right]^{1/4}$$

(14-18)

where D is the outside diameter of the tube. A comparison of Eqs. (14-15) and (14-18) for a given $T_v - T_w$ shows that the average heat-transfer coefficients for a vertical tube of length L and a horizontal tube of outside diameter D become equal when $L = 2.87D$. For example, when $L = 100D$, theoretically h_m for the horizontal position would be 2.44 times that for the vertical position. With this consideration, horizontal tube arrangements are generally preferred to vertical tube arrangements in condenser design.

Condensation on Horizontal Tube Banks

Condenser design generally involves horizontal tubes arranged in vertical tiers in such a way that the condensate from one tube drains onto the tube just below. If it is assumed that the drainage from one tube flows smoothly onto the tube below, it can be shown that, for a vertical tier of N tubes each of diameter D, the average heat-transfer coefficient h_m for the N tubes is given by

$$h_m = 0.725 \left[\frac{g\rho_l(\rho_l - \rho_v)h_{fg}k_l^3}{\mu_l(T_v - T_w)ND} \right]^{1/4}$$

(14-19)

This relation generally yields a conservative result since some turbulence and disturbance of condensate film are unavoidable during drainage and increase the heat-transfer coefficient.

Reynolds Number for Condensate Flow

Although the flow hardly changes to turbulent during condensation on a single horizontal tube, turbulence may start at the lower portions of a vertical tube. When turbulence occurs in the condensate film the average heat-transfer coefficient begins to increase with the length of the tube in contrast to its decrease with the length for laminar film condensation. To establish a criterion for transition from laminar to turbulent flow, a *Reynolds number for condensate flow* is defined as

$$\text{Re} = \frac{D_e u_m \rho_l}{\mu_l}$$

(14-20a)

where u_m is the average velocity of condensate film and D_e is the hydraulic diameter for condensate flow given by

$$D_e \equiv \frac{4A}{p} = \frac{4 \times \text{cross-sectional area for condensate flow}}{\text{wetted perimeter}} \qquad (14\text{-}20b)$$

For example, for *condensate flow over a vertical plate* of width b if δ is the thickness of the condensate film at a particular location, we have $p = b$, $A = b \times \delta$, and $D_e = 4\delta$; then the Reynolds number for condensate flow over a vertical plate becomes

$$\text{Re} = \frac{4\delta u_m \rho_l}{\mu_l} \equiv \frac{4\Gamma}{\mu_l} \qquad (14\text{-}21a)$$

where

$$\Gamma = \delta u_m \rho_l = \text{mass-flow rate of condensate per unit width of plate lb/ft·h} \qquad (14\text{-}21b)$$

For *condensation over a vertical tube* of outside diameter D, the Reynolds number for condensate flow at the bottom of the tube is given as

$$\text{Re} = \frac{4}{\mu_l}\left(\frac{W}{\pi D}\right) \qquad (14\text{-}22a)$$

where

$$W = \text{mass-flow rate of condensate at bottom of vertical tube lb/h or kg/s} \qquad (14\text{-}22b)$$

For *condensation over a horizontal tube* the Reynolds number at the bottom of the tube is given as

$$\text{Re} = \frac{4\Gamma_H}{\mu_l} \qquad (14\text{-}23a)$$

where

$$\Gamma_H = \text{mass-flow rate of condensate per unit length at bottom of horizontal tube lb/ft·h or kg/m·s} \qquad (14\text{-}23b)$$

Experiments have shown that for condensation over a vertical plate or a vertical tube the transition from laminar to turbulent flow takes place at a Reynolds number of about 1800. In the case of condensation over a horizontal tube with Reynolds number at the bottom of the tube as defined above [i.e., Eq. (14-23)] the transition from laminar to turbulent flow takes place at a Reynolds number about 3600 because the condensate film from both sides of the tube join at the bottom of the tube to give the mass-flow rate Γ_H.

Correlation with Experiments

For the purpose of correlating the theory of condensation with experiments, the above relations for the average heat-transfer coefficient are generally rearranged as a function of the Reynolds number for the condensate flow. Nusselt's theory for the average heat-transfer coefficient for condensation on a *vertical surface* [Eq. (14-15), for example] is rearranged as

$$
h_m \left(\frac{\mu_l^2}{k_l^2 \rho_l^2 g} \right)^{1/3} = 1.47 \left(\frac{4\Gamma}{\mu_l} \right)^{-1/3}
\tag{14-24}
$$

where Γ is the mass-flow rate of condensate per unit width at the bottom of the condensing surface. A comparison of this relation with the experiments has shown that due to the occurrence of ripples on the surface of the condensate film the measured heat-transfer coefficient is about 20 percent higher than the theory. With this consideration, McAdams [20] recommends that Nusselt's theory for the average heat coefficient for condensation on a vertical surface should be multiplied by 1.2, that is, Eqs. (14-15) and Eq. (14-24) should be replaced, respectively, by

$$
h_m = 1.13 \left[\frac{g \rho_l^2 h_{fg} k_l^3}{\mu_l (T_v - T_w) L} \right]^{1/4}
\tag{14-25a}
$$

and

$$
h_m \left(\frac{\mu_l^2}{k_l^3 \rho_l^2 g} \right)^{1/3} = 1.76 \left(\frac{4\Gamma}{\mu_l} \right)^{-1/3}
\tag{14-25b}
$$

which is valid for $\mathrm{Re} = 4\Gamma/\mu_l < 1800$.

When the Reynolds number for condensate flow exceeds 1800, turbulence starts in the condensate film and the average heat-transfer coefficient begins to increase with increasing length of the surface. For condensation on a vertical surface after the start of turbulence the average heat-transfer coefficient h_m can be represented by the following empirical relation proposed by Kirkbride [21]:

$$
h_m \left(\frac{\mu_l^2}{k_l^2 \rho_l^2 g} \right)^{1/3} = 0.0076 \left(\frac{4\Gamma}{\mu_l} \right)^{0.4}
\tag{14-26}
$$

for $\mathrm{Re} = 4\Gamma/\mu_l > 1800$. Figure 14-3 shows a plot of Eqs. (14-25b) and (14-26) as a function of the Reynolds number for the condensate flow in the laminar- and turbulent-flow regions, respectively. Included in this figure in the laminar-flow region is Nusselt's theory given by Eq. (14-24).

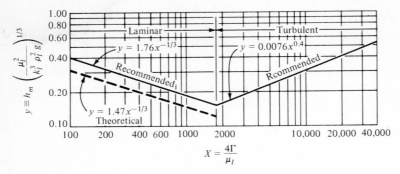

FIG. 14-3 Average heat-transfer coefficient for filmwise condensation on a vertical surface for laminar- and turbulent-flow regions.

In the foregoing relations the units of various quantities are as given below:

D, L	in ft (m)
g	in ft/h^2 (m/s^2)
h_{fg}	in Btu/lb (J/kg or W·s/kg)
h_m	in Btu/h·ft^2·°F (W/m^2·°C)
k_l	in Btu/h·ft·°F (W/m·°C)
T_v, T_w	in °F (°C)
ρ_l, ρ_v	in lb/ft^3 (kg/m^3)
μ_l	in lb/h·ft (kg/m·s)
Γ	in lb/h·ft (kg/m·s)

and the physical properties of the condensate should be evaluated at the arithmetic mean of the vapor and wall-surface temperatures.

A more accurate analysis of laminar filmwise condensation of vapors on vertical surfaces has been performed by various investigators [6–10] by using the mathematical techniques of boundary-layer theory. The results of these studies show that the Prandtl number of the condensing vapor is a factor that influences the condensation heat-transfer coefficient. The Prandtl numbers for steam and other common engineering fluids lie between 1 and 10; in this range the effect for practical purposes appears to be negligible for laminar film condensation. At low Prandtl numbers in the range of 0.003 to 0.03, which embrace the currently important liquid metals, these analyses show that the heat-transfer coefficient drops below the Nusselt prediction as the parameter $c_p(T_v - T_w)/h_{fg}$ increases (i.e., relatively thicker condensate flow). This matter will be discussed further later in this chapter.

Example 14-1 Air-free steam at 2.222 lb$_f$/in^2 abs. (i.e., at a saturation temperature $T_v = 130°F$ or 54.44°C) condenses on the outside surface of a 1-in (2.54 × 10^{-2}-m) OD, 12-ft (3.66-m) long vertical tube maintained at a uniform temperature $T_w = 110°F$ (43.33°C).

Assuming a filmwise condensation, determine the average condensation heat-transfer coefficient over the entire length of the tube and the rate of condensate flow at the bottom of the tube.

Solution The physical properties of condensate (i.e., water) at the average of wall and steam temperatures, $(110 + 130)/2 = 120°F$, are

$$h_{fg} = 1020 \text{ Btu/lb } (2372.4 \text{ kJ/kg})$$

$$k_l = 0.371 \text{ Btu/h·ft·°F } (0.642 \text{ W/m·°C})$$

$$\rho_l = 61.7 \text{ lb/ft}^3 \text{ (988.4 kg/m}^3)$$

$$\mu_l = 1.35 \text{ lb/h·ft } (0.558 \times 10^{-3} \text{ kg/m·s})$$

and

$$g = 4.17 \times 10^8 \text{ ft/h}^2 \text{ (9.8 m/s}^2) \quad \text{under normal conditions}$$

The average heat-transfer coefficient h_m for laminar filmwise condensation on a vertical surface is given by Eq. (14-25a):

$$h_m = 1.13 \left[\frac{g\rho_l^2 h_{fg} k_l^3}{\mu_l (T_v - T_w)L} \right]^{1/4}$$

Using the English system of units, we find

$$h_m = 1.13 \left[\frac{(4.17 \times 10^8) \times 61.7^2 \times 1020 \times 0.371^3}{1.35 \times (130 - 110) \times 12} \right]^{1/4} = 803 \text{ Btu/h·ft}^2·°F \text{ (4559 W/m}^2·°C)$$

or, by using the SI system of units, we obtain

$$h_m = 1.13 \left[\frac{9.8 \times 988.4^2 \times (2372.4 \times 10^3) \times 0.642^3}{(0.558 \times 10^{-3}) \times (54.44 - 43.33) \times 3.66} \right]^{1/4} = 4559 \text{ W/m}^2·°C$$

The mass-flow rate of condensate W at the bottom of the tube is determined from

$$W = \frac{\pi DL \times h_m \times (T_v - T_w)}{h_{fg}}$$

Using the English system of units, we obtain

$$W = \frac{(\pi \times \frac{1}{12} \times 12) \times 803 \times (130 - 110)}{1020} = 49.5 \text{ lb/h·tube } (6.24 \times 10^{-3} \text{ kg/s·tube})$$

or, using the SI system of units, we find

$$W = \frac{\pi \times (2.54 \times 10^{-2}) \times 3.66 \times 4559 \times (54.44 - 43.33)}{2372.4 \times 10^3} = 6.24 \times 10^{-3} \text{ kg/s·tube}$$

In the above calculations we assumed that the condensate flow was in the laminar range. To check the validity of this assumption, we now evaluate the Reynolds number for condensate flow at the bottom of the tube.

$$\text{Re} = \frac{4\Gamma}{\mu_l} = \frac{4 \ W}{\mu_l \ \pi D} = \frac{4 \times 49.5}{1.35 \times \pi \times \frac{1}{12}} = 560.2$$

Since the Reynolds number is less than 1800, the condensate flow is laminar, and the above calculations are valid.

Example 14-2 Air-free steam at 2.222 lb/in² abs. (i.e., at a saturation temperature $T_v = 130°F$ or 54.44°C) condenses on the outside surface of a 1-in (2.54 × 10⁻²-m) OD vertical tube. Determine the tube length L for a condensate flow rate at the bottom of the tube $W = 25$ lb/h (3.15 × 10⁻³ kg/s) per tube and a temperature difference of $T_v - T_w = 20°F$ (11.1°C) between the vapor and the tube surface temperatures.

Solution We note that this problem is essentially the inverse of the problem considered in Example 14-1. The physical properties of the condensate are the same as those given previously. We start the calculation by checking the Reynolds number for condensate flow.

$$\mathrm{Re} = \frac{4\Gamma}{\mu_l} = \frac{4}{\mu_l} \frac{W}{\pi D} = \frac{4 \times 25}{1.35 \times \pi \times \frac{1}{12}} = 283.1$$

The same result is obtainable by using the SI system of units, that is,

$$\mathrm{Re} = \frac{4}{0.558 \times 10^{-3}} \frac{3.15 \times 10^{-3}}{\pi \times (2.54 \times 10^{-2})} = 283$$

The flow is in the laminar range. In this problem it is convenient to use Eq. (14-25b) to determine the average heat-transfer coefficient.

$$h_m = 1.76 \left(\frac{k_l^3 \rho_l^2 g}{\mu_l^2} \right)^{1/3} \mathrm{Re}^{-1/3}$$

By using the English system of units, we find

$$h_m = 1.76 \left[\frac{0.371^3 \times 61.7^2 \times (4.17 \times 10^8)}{1.35^2} \right]^{1/3} \left(\frac{1}{283.1} \right)^{1/3} = 949 \ \mathrm{Btu/h \cdot ft^2 \cdot F}$$
$$(5391 \ \mathrm{W/m^2 \cdot °C})$$

or, by using the SI system of units, we obtain

$$h_m = 1.76 \left[\frac{0.642^3 \times 988.4^2 \times 9.8}{(0.558 \times 10^{-3})^2} \right]^{1/3} \left(\frac{1}{283.1} \right)^{1/3} = 5391 \ \mathrm{W/m^2 \cdot °C}$$

The tube length L is determined from

$$W h_{fg} = (\pi D L) h_m (T_v - T_w)$$

or

$$L = \frac{W h_{fg}}{\pi D h_m (T_v - T_w)}$$

By using the English system of units, we find

$$L = \frac{25 \times 1020}{\pi \times \frac{1}{12} \times 949 \times 20} = 5.1 \ \mathrm{ft} \ (1.56 \ \mathrm{m})$$

or, by using the SI system of units, we obtain

$$L = \frac{(3.15 \times 10^{-3}) \times 2,372,400}{\pi \times (2.54 \times 10^{-2}) \times 5391 \times (54.44 - 43.33)} = 1.56 \ \mathrm{m}$$

Effects of High Vapor Velocity.

The foregoing analysis and results for the filmwise condensation of vapors do not take into consideration the effects of high vapor velocity. There are many practical applications in which vapor velocity affects significantly the heat transfer in condensation. For example, with vapor condensing on the inside surface of a long vertical tube, the upward flow of vapor retards the condensate flow by causing the thickening of the condensate layer which in turn decreases the condensation heat-transfer coefficient. On the other hand, the downward flow of vapor decreases the thickness of the condensate film, hence increases the heat-transfer coefficient. The effects of vapor velocity on the heat-transfer coefficient for condensation inside tubes have been studied in the literature [4,5,22–27]. Tepe and Mueller [23] condensed benzene and methanol vapor inside a single copper tube, $\frac{7}{8}$ in in diameter and 3 ft in length, at several different inclinations. The average heat-transfer coefficient was shown to increase considerably with increasing vapor velocity and was much higher than that predicted by Nusselt's theory. Carpenter and Colburn [4] condensed steam, methanol, ethanol, toluene, and trichloroethylene inside a $\frac{1}{2}$-in ID, 8-ft-long vertical tube with inlet velocities up to 500 ft/s and covering a Prandtl number range of 2 to 5. They presented a correlation which appeared to be in fair agreement with the experiments over a limited range of data considered. A later work by Soliman et al. [27] discusses some of the limitations of the analysis in Ref. [4]. Goodykoontz and Dorsch [25] condensed steam ($Pr_l = 1.5$ to 2) in downflow inside a 0.293-in-ID, 8-ft-long vertical tube with inlet velocities ranging from approximately 300 to 1000 ft/s. Goodykoontz and Brown [26] extended the experimental work of Ref. [25] for the condensation of Freon 113 ($Pr_l = 7$ to 8) with inlet velocities greater than 200 ft/s. Soliman et al. [27] presented a correlation for condensation of vapors inside pipes which appears to agree well with the available experimental data [24–26,28] covering a range of Prandtl numbers from 1 to 10, a range of vapor velocities from 20 to 1000 ft/s, and a range of vapor qualities from 0.99 to 0.03. Their correlation for local condensing heat-transfer coefficient h inside the tubes is given in the form

$$\frac{h\mu_l}{k_l \rho_l^{1/2}} = 0.036 \, Pr_l^{0.65} \, F_0^{1/2} \tag{14-27a}$$

where the wall shear stress F_0 is composed of three components:

$$F_0 = F_f + F_m + F_a \tag{14-27b}$$

Here, F_f is for the effects of two-phase flow friction, F_m for the effects of momentum changes in the flow, and F_a for the effects of an axial gravitational field on the wall shear stress. The expressions defining F_f, F_m, and F_a being rather lengthy, they are not presented here; the reader should consult the original reference for details. However, as a special case, consider condensation in the region of sufficiently high vapor quality. In that case it may be assumed that $F_0 \cong F_f$

and in addition the expression defining F_f is simplified; then the heat-transfer relation given by Eqs. (14-27) is approximated by [27]

$$\text{Nu} \cong 0.0054 \, \text{Pr}_l{}^{0.65} \, \text{Re}_v{}^{0.9} \frac{\mu_v}{\mu_l} \left(\frac{\rho_l}{\rho_v}\right)^{0.5} \tag{14-28}$$

where

$$\text{Nu} = \frac{hD}{k_l} \qquad \text{Re}_v = \frac{4}{\mu_v} \frac{W_v}{\pi D}$$

and various quantities are evaluated in the units listed below:

h = heat-transfer coefficient, Btu/h·ft^2·°F (W/m^2·°C)

D = inside diameter, ft (m)

Pr_l = Prandtl number for liquid

W_v = mass-flow rate of vapor, lb/h (kg/s)

k_l = thermal conductivity of liquid, Btu/h·ft·°F (W/m·°C)

μ_l, μ_v = viscosity of liquid and vapor, respectively, lb/ft·h (kg/m·s)

ρ_l, ρ_v = density of liquid and vapor, respectively, lb/ft^3 (kg/m^3)

and all fluid properties are taken at the local saturation temperature. At intermediate and low vapor qualities, however, the momentum and gravity contributions to the wall shear stress cannot be neglected.

14-2 FILMWISE CONDENSATION OF LIQUID-METAL VAPORS

In the preceding section we considered the condensation of common liquids (i.e., $\text{Pr} \geq 0.5$) and noted that the theoretical predictions of the condensation heat-transfer coefficient agreed fairly well with experiments. In the case of condensation of liquid metals, which have a low value of Prandtl number (i.e., in the range of 0.003 to 0.03) there is a significant difference between the experiments and Nusselt's theory or modifications of it. Experimental data on liquid-metal condensation are scarce. Misra and Bonilla [29] and Cohn [30] investigated the condensation of mercury and sodium vapors on a vertical tube, Roth [31] studied the condensation of mercury and cadmium vapors, the data of General Electric Company [32] were for the condensation of potassium vapor inside a horizontal tube, and Sukhatme and Rohsenow [34] investigated the condensation of mercury vapor at low pressures on a vertical tube. Figure 14-4 shows a comparison in nondimensional form of the experimental data with Nusselt's theory and its modifications by various investigators [7–10]. Clearly, the measured values of the condensing-side heat-transfer coefficient for liquid metals are lower, as much

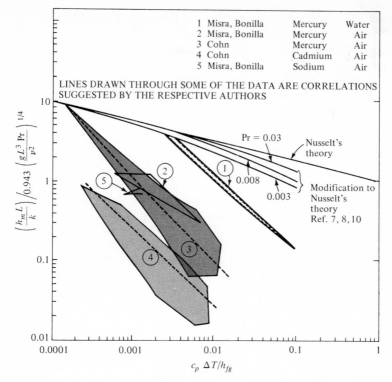

1	Misra, Bonilla	Mercury	Water
2	Misra, Bonilla	Mercury	Air
3	Cohn	Mercury	Air
4	Cohn	Cadmium	Air
5	Misra, Bonilla	Sodium	Air

LINES DRAWN THROUGH SOME OF THE DATA ARE CORRELATIONS
SUGGESTED BY THE RESPECTIVE AUTHORS

FIG. 14-4   Comparison of experimental data on condensation of liquid metals with Nusselt's theory and modifications of it. (*From Sukhatme and Rohsenow [34].*)

as an order of magnitude, than the theoretical predictions; the deviation appears to increase with the increasing value of the parameter $c_p \, \Delta T / h_{fg}$. In an attempt to explain this discrepancy Sukhatme and Rohsenow [34] measured the condensate thickness δ for condensing mercury by measuring the attenuation of gamma-ray radiation across the condensate layer. At the measured heat flux, the experimentally determined value of the condensate film thickness compared well with δ calculated from the Nusselt theory [Eq. (14-10)] based on a temperature difference $T_i - T_w$ instead of $T_v - T_w$, where T_i is the liquid-vapor interface temperature. This finding suggested that the true liquid-vapor interface temperature T_i must be lower than the saturation temperature T_v of the bulk vapor for the condensation of liquid-metal vapors, hence there must be an additional thermal resistance to heat flow between the outer surface of the condensate and the bulk vapor. An attempt was made to explain this thermal resistance by the hypothesis that during the condensation of liquid metals not all the vapor molecules striking the liquid surface actually condense. That is, only a fraction σ of the molecules striking the surface actually condense, and the fraction σ is called the *condensation coefficient*. Although the condensation coefficient σ compiled in Ref. [36] appeared to be much less than unity, later on a critical evaluation of these data in Ref. [37] suggested that the condensation coefficient compiled in Ref. [36] might have errors due to lack of adequate precise measurements and due to the presence of noncondensable gas; the true magnitude of σ might be close to unity or unity. Therefore, no satisfactory general correlation sufficiently consistent with the experimental data is yet available to predict heat-transfer coefficients with reasonable accuracy during the condensation of liquid metals. More theoretical and experimental investigations are needed in this area.

14-3 DROPWISE CONDENSATION

The condensation of steam is important in many industrial processes, and the rate at which condensation takes place depends on the rate of heat transfer from the vapor side to the coolant on the other side. It has been found in experiments that if traces of oil are present in steam and the condensing surface is highly polished the condensate film breaks into droplets; this type of condensation is called *dropwise condensation*. Figure 14-5 shows an example of an ideal dropwise condensation of steam on a vertical surface. The droplets grow, coalesce, and run off the surface, leaving a greater portion of the condensing surface freely exposed to incoming steam. Since the entire condensing surface is not covered with a continuous layer of liquid film, the heat transfer for ideal dropwise condensation of steam is much higher than that for filmwise condensation of steam. The heat-transfer coefficients may be five to ten times greater, but the overall heat-transfer coefficient between the steam and the coolant in a typical surface condenser may be about two to three times greater for dropwise than for filmwise condensation.

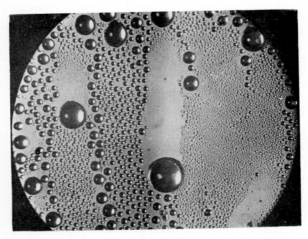

FIG. 14-5 Dropwise condensation of steam under ideal conditions. (*From Hampson and Özışık [39].*)

If dropwise condensation could be maintained under practical conditions in a surface condenser, it might result in a significant reduction in the size of heat-exchanger equipment. Therefore, the heat-transfer coefficient for dropwise condensation of steam and the conditions that produce dropwise condensation on a condensing surface have been studied extensively in the literature [39–46]. The results of these studies indicate that dropwise condensation is obtainable on a polished surface if the surface is coated with a thin film of promoter or if trace quantities of a promoter are injected into the steam at regular intervals. Promoters such as oleic, stearic, or linoleic acids, benzyl mercaptan, and many others have been used during earlier investigations. Several groups of promoters containing sulfur or selenium and long-chain hydrocarbons have been prepared [40] as possible promoters of dropwise condensation of steam on copper or copper-alloy surfaces, and some of these promoter compounds have been tested [41,42] with pure steam and industrial steam under laboratory conditions. The results of these tests indicate that the lives obtainable with different promoter compounds vary between 100 and 3000 h with pure steam and are shorter with industrial steam or with intermittent operations [42]. Failure occurs either because of fouling or oxidation of the surface or by the gradual removal of the promoter on the surface by the flow of condensate, or by a combination of these effects.

To prevent the failure of dropwise condensation due to the oxidation, noble-metal coating of the condensing surface has been tried [43], and coatings of gold, silver, rhodium, palladium, and platinum have been used. Although some of these coated surfaces could produce dropwise condensation under laboratory conditions for more than 10,000 h of continuous operation, the cost of coating the condensing surface with noble materials is so high that the economics of such an approach for industrial applications has yet to be proved. A series of field tests have also been carried out with a number of compound promoters under industrial-steam condensing conditions in a power-generating plant [42]. The results of these tests have shown that the condenser tubes treated with compound promoter gave dropwise condensation lasting from 2 weeks to 5 months under intermittent operation. However, as the number of stoppages and overall periods of operation increased, a longer time was required to establish dropwise condensation.

It appears that it is unlikely that long-lasting dropwise condensation can be produced under practical conditions by a single treatment of any of the promoters currently available. Although it may be possible to produce dropwise condensation for periods up to a year by the injection of a small quantity of promoter into the steam at regular intervals, the successful operation depends on the amount and the cost of the promoter and to what extent the cumulative effect of the injected promoter can be tolerated in the rest of the plant. Therefore, *in the analysis of a heat exchanger involving the condensation of steam, it is recommended that filmwise condensation should be assumed for the condensing surface.*

14-4 CONDENSATION ON FLUTED SURFACES

The average heat-transfer coefficient for filmwise condensation can be increased by fluting the surface so that surface tension concentrates the condensate in the grooves and leaves a thin film of liquid over the section of the surface between the grooves. Figure 14-6 shows filmwise condensation on a vertical tube with axial flutes. Although the local heat-transfer coefficient for the thicker conden-

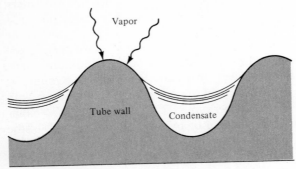

FIG. 14-6 Section through a vertical tube with axial flutes, showing the effects of surface tension on the configuration of the condensate film. (*From Lusterader et al.* [*47*].)

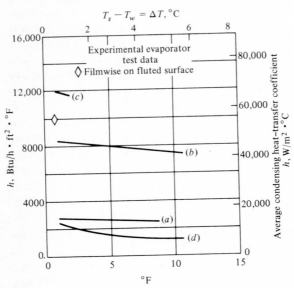

FIG. 14-7 Average heat-transfer coefficient for filmwise condensation of steam on a smooth and a fluted surface. (*a*) Filmwise on a smooth brass; (*b*) filmwise on a fluted surface at 212°F; (*c*) filmwise on a fluted surface at 86°F; (*d*) calculated from Nusselt's theoretical derivation for laminar flow in the condensate film for the conditions under which the data for parts *c* and *d* were obtained. (*From Lusterader et al.* [*47*].)

sate layer in the grooves is somewhat less, the major portion of the crests having a very thin film of condensate has greatly reduced resistance to heat flow through the crest area. As a result, the average heat-transfer coefficient is higher than that obtainable for filmwise condensation on a smooth tube. Some indication of the effectiveness of fluting the surface is given in Fig. 14-7 obtained from Ref. [47]. It is to be noted that fluting the surface increases the heat-transfer coefficient for condensation if the type of condensation is filmwise. With drop-wise condensation, experiments have shown that fluting the surface may lead to a drop in the overall heat-transfer coefficient.

14-5 CONDENSATION IN THE PRESENCE OF NONCONDENSABLE GAS

In the previous sections we considered the heat-transfer coefficient for condensing vapors that did not contain any noncondensable gas. If noncondensable gas such as air is present in the vapor, even in very small amounts, it has been observed that the heat-transfer coefficient for condensation is greatly reduced. The reason for this is that when a vapor containing noncondensable gas conden-ses, the noncondensable gas is left at the surface and the incoming condensable vapor must diffuse through this body of vapor-gas mixture collected in the vicinity of the condensate surface before it reaches the cold surface to condense. There-fore the presence of noncondensable gas adjacent to the condensate surface acts as a thermal resistance to heat transfer. The resistance to this diffusion process causes a drop in the partial pressure of the condensing vapor which in turn drops the saturation temperature; that is, the temperature of the outside surface of the condensate layer is lower than the saturation temperature at the bulk mixture. A simple condensation theory as discussed in the previous sections is not applicable because the condensate surface temperature is not known; the analysis is complicated in that heat, mass, and momentum transfer problems in the liquid and the vapor-gas mixture should be solved simultaneously. Various semiempirical computational methods were proposed by the early investigators for the prediction of the heat-transfer coefficient in the presence of noncondens-able gas [48–51], and later on several analytical investigations were presented for laminar film condensation in the presence of noncondensable gas [52–57]. For example, Othmer [48] showed that the condensation heat-transfer coeffi-cient is reduced by 50 percent or more with a few percent by volume of air in steam. Later it was demonstrated by the analysis that the presence of a very small amount of noncondensable gas in the bulk vapor can cause a large buildup of noncondensable gas at the liquid-vapor interface and may reduce the heat-transfer coefficient well over 50 percent [53]; also the influence of noncondensable gas is accentuated at lower pressures [54]. These analytical studies were based on laminar film condensation on a vertical surface for the limiting case of zero forced vapor flow, but the results are very much dependent upon vapor flow patterns in the vicinity of the condensing surface. The effects of forced vapor flow have been analyzed under idealized conditions for fluids having low Prandtl number [56] and high Prandtl number [57]. High velocities over the surface tend to reduce the accumulation of the noncondensable gas and to alleviate the problem.

In steam condensers for industrial applications, however, the vapor flow patterns are so complicated that neither an analysis performed under idealized conditions nor the data obtained from simple laboratory experiments can be representative of the actual situations, but they help to demonstrate the importance of various factors. Therefore, *the general practice in condenser design still remains to be the venting of noncondensable gas as much as possible.*

14-6   POOL BOILING

The simplest form of boiling is *pool boiling* in which a heated surface at a temperature above the saturation temperature of a liquid and immersed below the free surface of a liquid causes boiling. At this point it should be determined whether the main body of the liquid in the immediate vicinity of the heated surface is at (or slightly above) the saturation temperature or below the saturation temperature. The latter is called *subcooled (or local) boiling* because the vapor bubbles that are formed at the hot metal surface either collapse without leaving the surface or collapse immediately upon leaving the surface. The former is called *saturated (or bulk) boiling* because liquid is maintained at saturation temperature. Figure 14-8 illustrates the variation of the heat-transfer coefficient as a function of the temperature difference between the wall-surface T_w and the liquid-saturation T_s temperatures in the pool boiling of a liquid at saturation temperature. These results, obtained by Farber and Scorah [58] for pool boiling of water at atmospheric pressure and saturation temperature on the surface of a submerged, electrically heated, platinum wire, are typical of pool boiling and are applicable to other geometries besides a wire. If the heat flux q is also plotted on this figure as a function of the temperature difference $T_w - T_s$, the general shape of the heat-flux curve would be similar to that of the heat-transfer coefficient curve. The h curve in this figure is divided into six different regions, and the slope of the curve in each region is distinctly different from those in other regions. In region I, no vapor bubbles are formed because the energy transfer from the heated surface to the saturated liquid is by free-convection currents which produce sufficient circulation so that heat is removed by evaporation from the free surface. In region II, bubbles begin to form at the hot surface of the wire, but as soon as they are detached from the surface they are dissipated in the liquid. In region III, bubbles detached from the wire surface rise to the surface of the liquid where they are dissipated. In regions II and III the boiling process is called *nucleate boiling*. In region IV, the bubble formation is so rapid

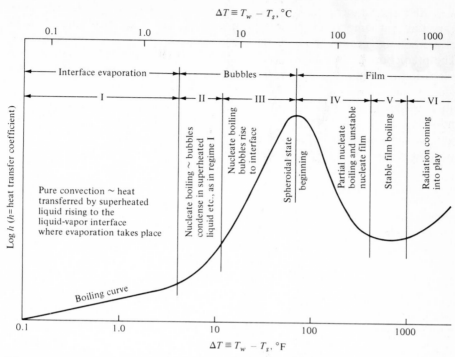

FIG. 14-8 Principal boiling regimes in pool boiling of water at atmospheric pressure and saturation temperature T_s from an electrically heated platinum wire. (*From Farber and Scorah* [*58*].)

that bubbles begin to coalesce before they are detached; as a result a large fraction of the heating surface is blanketed by an unstable film of vapor which causes an abrupt loss in the heat flux and in the boiling heat-transfer coefficient. This is an *unstable region* and it signifies transition from *nucleate boiling* to *film boiling*. In region V, the heat flux drops to a minimum and the wire surface is blanketed with a stable film of vapor; it is called the *stable film-boiling* region. In region VI, both the heat flux and the heat-transfer coefficient increase with $T_w - T_s$ because the wire surface temperature in this region is sufficiently high for thermal radiation effects to augment heat transfer through the vapor film. In this region the boiling also takes place as *stable film boiling* but radiation effects are dominant.

For the pool boiling of water shown in Fig. 14-8 the *peak heat flux* is of the order of 10^6 Btu/h·ft² (3.154×10^6 W/m² or 3.154 MW/m²) and in the nucleate-boiling region this peak heat flux is obtainable with a temperature difference of less than 100°F (55.6°C). If the same heat flux is to be obtained in the film-boiling region VI it requires a temperature difference well above the melting point of most materials. Clearly, while the boiling takes place in the nucleate-boiling regime, if the applied heat flux exceeds the peak heat flux, boiling is immediately shifted to film-boiling region VI where boiling continues to take place with large temperature differences, provided that the heater material does not melt at that temperature. For this reason the maximum heat-flux point in nucleate boiling is called the *burnout point*.

The prediction of the heat-transfer coefficient in various boiling regimes and the location of the peak heat flux in pool boiling are of practical interest. The reader should consult Refs. [59–64] for a discussion of various models and postulates toward the understanding of the mechanism of boiling and the determination of boiling heat transfer. Here we present a brief discussion of this matter for the nucleate-boiling and the film-boiling regimes which are most important in engineering applications.

Nucleate Boiling

The analysis of the nucleate-boiling process is extremely difficult because the exact mechanism of bubble formation and the motion of bubbles are not yet fully understood. Therefore, to date, there has been no satisfactory analysis of heat transfer in nucleate boiling. On the other hand, a considerable amount of experimental data are available on boiling heat flux as a function of the temperature

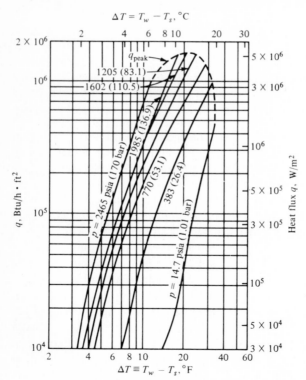

FIG. 14-9 Heat flux as a function of temperature difference between the wall and saturation temperatures of water boiling on a horizontal 0.024-in (0.061 × 10⁻²-m) diameter platinum wire. (*From Addoms* [67].)

difference between the wall-surface and saturation temperatures, $\Delta T \equiv T_w - T_s$, for a variety of liquids and over a wide range of pressure levels [65–67]. Figure 14-9 shows typical experimental data for the heat flux q plotted as a function of ΔT at various pressure levels for boiling of water on a horizontal 0.024-in-diameter electrically heated platinum wire. The dashed lines in this figure denote the location of the peak heat flux in nucleate boiling. Heat-flux correlations of the form shown in Fig. 14-9 were used by numerous early investigators for each different liquid by plotting q versus ΔT. Later on, attempts were made to obtain more general empirical correlations between q and ΔT that could be applicable to a group of liquids over a wide range of pressure levels. A discussion of such attempts is given in Refs. [62,68]. The basic concept behind such correlations was the fact that the principal mechanism of heat transfer in nucleate boiling is forced convection caused by the agitation produced by bubbles. Rohsenow [69], by analyzing the significance of various parameters in relation to forced-convection effects, proposed the following empirical relation to correlate the heat flux in the nucleate-boiling regime:

$$\frac{c_{pl}\,\Delta T}{h_{fg}\,\mathrm{Pr}_l^{1.7}} = C_{sf} \left[\frac{q}{\mu_l h_{fg}} \sqrt{\frac{g_c \sigma^*}{g(\rho_l - \rho_v)}} \right]^{0.33} \tag{14-29}$$

where
c_{pl} = specific heat of saturated liquid, Btu/lb·°F (W·s/kg·°C)
C_{sf} = constant to be determined from experimental data depending on "heating surface-fluid" combination
h_{fg} = latent heat of vaporization, Btu/lb (W·s/kg)
g, g_c = gravitational acceleration, ft/h² (m/s²), and gravitational acceleration conversion factor, lb·ft/lb$_f$·h² (kgm/N·s²), respectively
$\mathrm{Pr}_l = c_{pl}\mu_l/k_l$ = Prandtl number of saturated liquid
q = boiling heat flux, Btu/h·ft² (W/m²)
$\Delta T = T_w - T_s$, temperature difference between wall and saturation temperature, °F (°C)
μ_l = viscosity of saturated liquid, lb/h·ft (kg/m·s)
ρ_l, ρ_v = density of liquid and saturated vapor, respectively, lb/ft³ (kg/m³)
σ^* = surface tension of liquid-vapor interface, lb$_f$/ft (N/m)

The correlation given by Eq. (14-29) is found applicable for boiling of *single-component liquids* on *clean surfaces*. For *dirty* or *contaminated surfaces*, however, the exponent of Pr_l has been found to vary between 0.8 and 2.0 instead of 1.7 given in the above equation. Then the coefficient C_{sf} is the only provision in this equation to adjust the correlation for the liquid–heating-surface combination. Table 14-1 lists the experimentally determined values of the coefficient C_{sf} for a variety of liquid-surface combinations as obtained from various references [66,67,70,71]. Table 14-2 gives the values of the vapor-liquid surface tension σ^* for a variety of liquids.

Figure 14-10 gives some idea of the accuracy of Eq. (14-29) to correlate the experimental data on nucleate boiling. In this figure Addoms' results [67] at

TABLE 14-1   Values of the Coefficient C_{sf} of Eq. (14-29) for Various Liquid-Surface Combinations

Liquid-Surface Combination	C_{sf}	Reference
Water–copper	0.0130	[71]
Water–scored copper	0.0068	[70]
Water–emery-polished copper	0.0128	[70]
Water–emery-polished, paraffin-treated copper	0.0147	[70]
Water–chemically etched stainless steel	0.0133	[70]
Water–mechanically polished stainless steel	0.0132	[70]
Water–ground and polished stainless steel	0.0080	[70]
Water–Teflon pitted stainless steel	0.0058	[70]
Water–platinum	0.0130	[67]
Water–brass	0.0060	[70]
Benzene–chromium	0.0100	[66]
Ethyl alcohol–chromium	0.0027	[66]
Carbon tetrachloride–copper	0.0130	[71]
Carbon tetrachloride–emery-polished copper	0.0070	[70]
n-Pentane–emery-polished copper	0.0154	[70]
n-Pentane–emery-polished nickel	0.0127	[70]
n-Pentane–emery-rubbed copper	0.0074	[70]
n-Pentane–lapped copper	0.0049	[70]

TABLE 14-2   Values of Liquid-Vapor Surface Tension σ for Various Liquids

Liquid	Saturation Temperature		Surface Tension	
	°F	°C	$\sigma^* \times 10^4$ lb_f/ft	$\sigma^* \times 10^3$ N/m
Water	32	0	51.8	75.6
Water	60	15.56	50.2	73.2
Water	100	37.78	47.8	69.7
Water	200	93.34	41.2	60.1
Water	212	100	40.3	58.8
Water	320	160	31.6	46.1
Water	440	226.7	21.9	31.9
Water	560	293.3	11.1	16.2
Water	680	360	1.0	1.46
Water	705.4	374.11	0.0	0
Sodium	1618	881.1	77	11.2
Potassium	1400	760	43	62.7
Rubidium	1270	687.8	30	43.8
Cesium	1260	682.2	20	29.2
Mercury	675	357.2	27	39.4
Benzene (C_6H_6)	176	80	19	27.7
Ethyl alcohol (C_2H_6O)	173	78.3	15	21.9
Freon 11	112	44.4	5.8	8.5

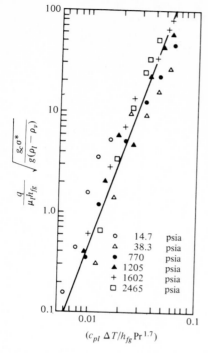

FIG. 14-10   Rohsenow's [69] correlation of Addoms' [67] data for pool-boiling heat transfer in nucleate boiling by Eq. (14-29).

various pressures from 14.7 to 2465 lb_f/in^2 abs., given previously in Fig. 14-9, are correlated by Eq. (14-29) for $C_{sf} = 0.013$. The spread of experimental data is approximately ± 20 percent about the calculated solid line.

Peak Heat Flux in Nucleate Boiling   The correlation given by Eq. (14-29) provides information for the heat flux in nucleate boiling, but it cannot predict the *peak heat flux*. The determination of peak heat flux in nucleate boiling is of interest because of burnout considerations; that is, if the applied heat flux exceeds the peak heat flux, the transition takes place from the nucleate to the stable film-boiling regime in which, depending on the kind of fluid, boiling may occur at temperature differences well above the melting point of the heating surface. Equations have been developed for the prediction of the peak heat flux [72–75]. Kutateladze [72] treated the failure of nucleate boiling as a purely hydrodynamic problem and derived an equation which required an experimentally determined constant. Zuber et al. [73,74] developed a relation analytically by the stability requirement of the liquid-vapor interface. Zuber's [73] equation for the maximum heat flux q_{max} in nucleate boiling is given as

$$q_{max} = \frac{\pi}{24} \rho_v h_{fg} \left[\frac{\sigma^* g g_c (\rho_l - \rho_v)}{\rho_v^2} \right]^{1/4} \left(\frac{\rho_l}{\rho_l + \rho_v} \right)^{1/2} \tag{14-30}$$

where $\sigma^* =$ surface tension of liquid-vapor interface, lb_f/ft (N/m)

        $g =$ gravitational acceleration, ft/h^2 (m/s^2)

        $g_c =$ gravitational acceleration conversion factor, $lb \cdot ft/lb_f \cdot h^2$ (kg·m/ N·s²)

   $\rho_l, \rho_v =$ density of liquid and vapor, respectively, lb/ft^3 (kg/m^3)

      $h_{fg} =$ latent heat of vaporization, Btu/lb (W·s/kg or J/kg)

    $q_{max} =$ peak heat flux, $Btu/h \cdot ft^2$ (W/m^2)

It is apparent from this equation that large values of h_{fg}, ρ_v, g, and σ^* are desirable for a large value of the peak heat flux. For example, water has a large value of h_{fg}; hence the peak heat flux obtainable with the boiling of water is high. This equation also shows that a reduced gravitational field decreases the peak heat flux. The validity of Eq. (14-30) for the effects of reduced g has been shown by Usiskin and Siegel [76]. The relation given by Eq. (14-30) is in reasonably good agreement with the experimental data.

Stable Film Boiling

As illustrated in Fig. 14-8, the nucleate-boiling region ends and the unstable film-boiling region begins after the peak heat flux is reached. No analysis is available for the prediction of heat flux as a function of the temperature difference $(T_w - T_s)$ in this unstable region until the minimum point in the boiling curve is reached and the stable film-boiling region starts. In stable film-boiling regions V and VI, the heating surface is separated from the liquid by a vapor layer across which heat must be transferred. Since the thermal conductivity of the vapor is low, large temperature differences are needed for heat transfer in this region; therefore heat transfer in this region is generally avoided when high temperatures are involved. On the other hand, stable film boiling has numerous applications in the boiling of cryogenic fluids. A theory was developed by Bromley [77] for the prediction of the heat-transfer coefficient for stable film boiling on the outside surface of a horizontal cylinder. The basic approach in the analysis is similar to Nusselt's theory for filmwise condensation on a horizontal tube. The resulting equation for the average heat-transfer coefficient h_0 for stable film boiling on the outside surface of a horizontal cylinder in the *absence of radiation* is given by

$$h_0 = 0.62 \left[\frac{k_v^3 \rho_v (\rho_l - \rho_v) g h_{fg}}{\mu_v D_o \, \Delta T} \left(1 + \frac{0.4 c_{pv} \, \Delta T}{h_{fg}} \right) \right]^{1/4} \tag{14-31}$$

where $h_0 =$ average boiling heat-transfer coefficient in absence of radiation, $Btu/h \cdot ft^2 \cdot {}^\circ F$ $(W/m^2 \cdot {}^\circ C)$

     $c_{pv} =$ specific heat of saturated vapor, $Btu/lb \cdot {}^\circ F$ (W·s/kg·°C)

      $D_o =$ outside diameter of tube, ft (m)

     $h_{fg} =$ latent heat of vaporization, Btu/lb (W·s/kg)

k_v = thermal conductivity of saturated vapor, Btu/h·ft·°F (W/m·°C)

$\Delta T = T_w - T_s$, temperature difference between wall and saturation temperatures, °F (°C)

It is recommended that the physical properties of the vapor are evaluated at the arithmetic mean of the wall and the saturation temperatures, while ρ_l and h_{fg} are evaluated at the saturation temperature.

Equation (14-31) has been derived by assuming that heat transfer across the vapor film is by pure conduction; therefore it does not include the radiation effects. Bromley [77] suggested that when the surface temperature is sufficiently high for the radiation effects to be important the average heat-transfer coefficient h_m can be determined from the following empirical relation:

$$h_m = h_0 \left(\frac{h_0}{h_m}\right)^{1/3} + h_r \qquad (14\text{-}32a)$$

where h_0 is the boiling heat-transfer coefficient given by Eq. (14-31) without the radiation effects and h_r is the radiation heat-transfer coefficient which can be estimated approximately from the following relation:

$$h_r = \frac{1}{1/\varepsilon + 1/\alpha - 1} \frac{\sigma(T_w^4 - T_s^4)}{T_w - T_s} \qquad (14\text{-}32b)$$

where α = absorptivity of liquid
ε = emissivity of hot tube
σ = Stefan-Boltzmann constant
T_w = wall temperature,
T_s = saturation temperature of liquid

Equation (14-32a) is difficult to use because a trial-and-error approach is needed to determine h_m. When h_r is smaller than h_0, Eq. (14-32a) may be replaced by the following relation:

$$h_m = h_0 + \tfrac{3}{4}h_r \qquad (14\text{-}33)$$

To check the validity of the above theory, Bromley [77] used Eqs. (14-31) and (14-32) to correlate film-boiling data at atmospheric pressure for a number of liquids such as water, benzene, carbon tetrachloride, n-pentane, and nitrogen boiling on carbon tubes of various diameters over a wide range of $T_w - T_s$ up to 2500°F. Figure 14-11 shows the comparison of calculated and experimental heat-transfer coefficients for nitrogen boiling on a 0.35-in-diameter tube. The agreement between the theory and experiment is fairly good. However, the correlation of film-boiling data for water on very small diameter wires (i.e., from 0.004 to 0.024 in) showed that the theory predicted lower values, the deviation

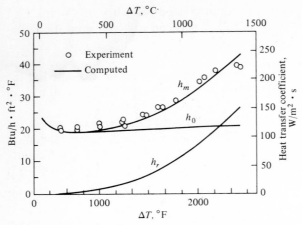

FIG. 14-11 Heat-transfer coefficient for stable film boiling of liquid nitrogen on an electrically heated 0.35-in-diameter carbon tube. (*From Bromley* [77].)

ranging from 30 to 100 percent as the wire size decreased from 0.024 to 0.004 in. It appears that for such cases when the vapor film thickness is comparable to the wire diameter the model is not applicable. Breen and Westwater [78] investigated experimentally the effects of diameter for film boiling on horizontal tubes and proposed a simple correlation. Berenson [79] derived an equation for film boiling on horizontal surfaces facing up, and Hsu and Westwater [80] and Bankoff [81] proposed relations for boiling on vertical surfaces. The reader should consult the original references for details of the analysis and the resulting expressions.

14-7 BOILING IN FORCED FLOW INSIDE TUBES

In the previous sections we considered the boiling on a heated surface immersed in a quiescent mass of liquid. If boiling takes place on the inside surface of a heated tube through which the liquid flows with some velocity, boiling is called *forced-convection boiling*. Boiling of liquids in forced flow inside heated tubes has numerous applications in the design of steam generators for nuclear power plants, space power plants, and various advanced power-generation systems. Since the velocity inside the tube affects the bubble growth and separation, the mechanism and hydrodynamics of boiling in forced convection are much more complex than in the pool boiling of a quiescent liquid. Therefore, no fundamental theory is yet available to predict the boiling heat-transfer coefficient for forced-convection boiling. Photographic studies of forced-convection boiling have been presented in the literature [82,83] to gain some understanding of the complex flow patterns in various boiling regimes. These observations have shown that flow behavior in forced-convection boiling is significantly different from that of pool boiling as velocity and vapor quality increased over the low values that prevail under pool-boiling conditions. Various flow characteristics and their effect on the heat-transfer coefficient for boiling in forced flow inside a uniformly heated tube is best visualized by examining Fig. 14-12 in which subcooled water is first heated, boiled, and then superheated as it progresses from one end of the tube to the other end. At the inlet region, heat transfer to the subcooled liquid is by forced convection, and this regime prevails until boiling starts. Boiling is accompanied by a sudden increase in the heat-transfer coefficient. In the boiling region, the bubbles appear on the heated surface, grow, and are washed away into the main stream, so that a *bubbly flow regime* prevails for some distance along the tube. As the individual bubbles coalesce, plugs or slugs of vapor are formed in the flow, and this regime, known as a *slug-flow regime*, prevails up to vapor qualities as much as 50 percent by volume. As the volume fraction of vapor in the fluid stream increases further, the nature of the flow changes markedly; that is, the vapor begins to move as a continuous stream down through the center of the tube while the liquid adheres to the wall and moves in an annular film. This regime is called an *annular film regime* in which the liquid film becomes progressively thinner with progress along the length of the tube and the vapor quality runs from 50 to 90 percent by volume. As long as the liquid film wets the wall surface the heat-transfer coefficient remains high. Then, depending on the surface condition, pressure, and flow rate, dry spots appear in the wall which is accompanied by a sharp decrease in the heat-transfer coefficient. This is called the

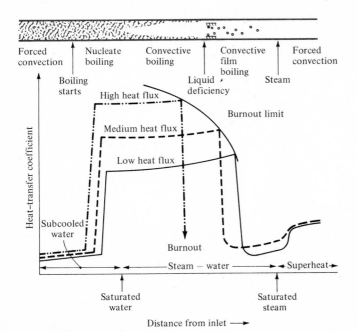

FIG. 14-12 Boiling of water in forced convection inside a once-through boiler tube and effects of heat flux on the heat-transfer coefficient. (*From Polomik et al.* [84].)

transition region from the annular flow to the mist flow. The dry spots continue to enlarge until all remaining liquid is in the form of fine droplets in the water. This is known as the *mist-flow regime* which persists until the vapor quality reaches 100 percent. Beyond 100 percent quality the flow is all vapor, and the super-heating starts.

The prediction of the heat-transfer coefficient for boiling of liquids in forced flow inside tubes is of interest in practice. The heat-transfer coefficient to the subcooled liquid before boiling starts and to the superheated vapor can be determined from the relations for forced convection inside tubes. But no general correlation is yet available to predict the heat-transfer coefficient for all the various boiling regimes that take place inside the tube. Davis and David [85] proposed the following empirical relation for the average heat-transfer coefficient h_m for the flow of vapor-liquid mixtures through tubes as long as liquid wets the wall:

$$\frac{h_m D}{k_l} = 0.06 \left(\frac{\rho_l}{\rho_v}\right)^{0.28} \left(\frac{DG\chi}{\mu}\right)^{0.87} Pr_l^{0.4} \tag{14-34}$$

where G = mass-flow rate, $lb/h \cdot ft^2$ ($kg/s \cdot m^2$)
D = tube inside diameter, ft (m)
k = thermal conductivity, $Btu/h \cdot ft \cdot °F$ ($W/m \cdot °C$)
μ = viscosity, $lb/h \cdot ft$ ($kg/m \cdot s$)
χ = mass fraction of vapor in mixture or quality
ρ = density, lb/ft^3 (kg/m^3)

and subscripts l and v refer to liquid and vapor, respectively.

The estimation of the *pressure drop for two-phase flow inside tubes* is also of interest in engineering applications but it is greatly complicated by the existence of variety of modes of flow. A summary index to the two-phase flow literature is given by Gouse [86]. Under bubbly flow conditions, Wallis [87] found that experimental data for the two-phase pressure drop are correlated well by the relation

$$\frac{\Delta P_{TP}}{\Delta P_l} = 1 + 3 \frac{\rho_l G_v}{\rho_v G_l} (G \times 10^{-6})^{0.33} \tag{14-35}$$

where the subscripts l and v refer to the liquid and vapor, respectively, ΔP_{TP} is the two-phase pressure drop, ΔP_l is the pressure drop for liquid flow only, and G is the mass-flow rate in $lb/ft^2 \cdot h$ ($kg/m^2 \cdot s$). Figure 14-13 shows the two-phase pressure-drop correlation according to Eq. (14-35) for the flow of air-water mixtures in tubes ranging from $\frac{3}{8}$ to $\frac{7}{8}$ in ID. Observations through the transparent walls of the tubes during the tests showed that, for the experimental points that fell to the right of the straight line in Fig. 14-13, the flow regime changed to annular flow. A discussion of the determination of the total pressure drop associated with boiling flow through tubes is given by Fraas and Özışık [88].

Example 14-3 A heated clean copper plate is immersed in a pool of water at saturation temperature and atmospheric pressure. Determine the rate of evaporation per unit area

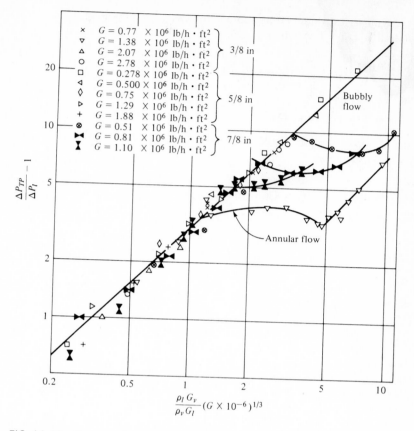

FIG. 14-13 Pressure-drop correlation for a bubbly flow regime. Data were obtained isothermally with an air-water mixture in tubes ranging from $\frac{3}{8}$ to $\frac{7}{8}$ in ID. (*From Wallis [87].*)

of the plate if the plate surface is kept at a temperature 25°F (13.9°C) above the saturation temperature (212°F or 100°C) of water.

Solution The heat-transfer data in Fig. 14-8 imply that for a temperature difference of 25°F (13.9°C) the boiling takes place in the nucleate-boiling regime. Therefore Eq. (14-29) can be used to determine the boiling heat flux q.

$$\frac{c_{pl}\,\Delta T}{h_{fg}\,\mathrm{Pr}_l^{1.7}} = C_{sf}\left[\frac{q}{\mu_l h_{fg}}\sqrt{\frac{g_c\sigma^*}{g(\rho_l-\rho_v)}}\right]^{0.33}$$

For a water-copper combination the coefficient C_{sf} is obtained from Table 14-1 as $C_{sf} = 0.013$. The physical properties of water and vapor and other quantities are taken as

$c_{pl} = 1.007$ Btu/lb·°F $= 4216$ W·s/kg·°C $\qquad \rho_l = 59.8$ lb/ft³ $= 961.88$ kg/m³

$h_{fg} = 970.3$ Btu/lb $= 2256.8 \times 10^3$ W·s/kg $\qquad \rho_v = 0.373$ lb/ft³ $= 5.975$ kg/m³

$\sigma^* = 40.3 \times 10^{-4}$ lb$_f$/ft $= 58.8 \times 10^{-3}$ N/m $\quad$ Pr$_l = 1.76$

$\mu_l = 0.686$ lb/ft·h $= 0.2836 \times 10^{-3}$ kg·m/s

Substituting these numerical values in the above expression, using the English system of units, and noting that $g_c/g \simeq 1$, we obtain

$$\frac{1.007 \times 25}{970.3 \times 1.76^{1.7}} = 0.013 \left(\frac{q}{0.686 \times 970.3} \sqrt{\frac{40.3 \times 10^{-4}}{59.8 - 0.373}} \right)^{0.33}$$

Then, the heat flux becomes

$$q = 35,700 \text{ Btu/h·ft}^2 \ (113,000 \text{ W/m}^2)$$

Now, using the SI system of units and noting that $g_c = 1 \text{ kg·m/N·s}^2$ and $g = 9.8 \text{ m/s}^2$, the above expression becomes

$$\frac{4216 \times 13.9}{(2256.8 \times 10^3) \times 1.76^{1.7}} = 0.013 \left[\frac{q}{(0.2836 \times 10^{-3}) \times (2256.8 \times 10^3)} \right.$$

$$\left. \times \sqrt{\frac{1 \times (58.8 \times 10^{-3})}{9.8 \times (961.88 - 5.975)}} \right]^{0.33}$$

and the heat flux becomes

$$q \cong 113,000 \text{ W/m}^2$$

The rate of evaporation per unit area of the heater surface is

$$W = \frac{q}{h_{fg}} = \frac{35,700}{970.3} = 36.8 \text{ lb/h·ft}^2 = 49.9 \times 10^{-3} \text{ kg/m}^2\text{·s}$$

Example 14-4   Water at saturation temperature and atmospheric pressure is boiled with an electrically heated horizontal platinum wire of diameter 0.05 in $(0.127 \times 10^{-2} \text{ m})$. Determine the boiling heat-transfer coefficient h_m if the boiling takes place in the stable film-boiling regime with a temperature difference $T_w - T_s = 1200°F$ (666.7°C).

Solution   The heat-transfer coefficient h_0 for stable film-boiling without the effects of radiation is evaluated by Eq. (14-31):

$$h_0 = 0.62 \left[\frac{k_v^3 \rho_v (\rho_l - \rho_v) g h_{fg}}{\mu_v D_o \Delta T} \left(1 + \frac{0.4 c_{pv} \Delta T}{h_{fg}} \right) \right]^{1/4}$$

The radiation heat-transfer coefficient h_r is determined from Eq. (14-32b):

$$h_r = \frac{1}{1/\varepsilon + 1/\alpha - 1} \frac{\sigma(T_w^4 - T_s^4)}{T_w - T_s}$$

where the physical properties of steam evaluated at $212 + 600 = 812°F$ (433.3°C), ρ_l and h_{fg} evaluated at 212°F (100°C), are given as

$$c_{pv} = 0.498 \text{ Btu/lb·°F} = 2085 \text{ W·s/kg·°C}$$

$$h_{fg} = 970.3 \text{ Btu/lb} = 2256.8 \times 10^3 \text{ W·s/kg}$$

$$k_v = 0.0295 \text{ Btu/h·ft·°F} = 0.0510 \text{ W/m·°C}$$

$$\mu_v = 0.0595 \text{ lb/ft·h} = 0.026 \times 10^{-3} \text{ kg/m·s}$$

$$\rho_v = 0.0196 \text{ lb/ft}^3 = 0.314 \text{ kg/m}^3$$

$$\rho_l = 59.8 \text{ lb/ft}^3 = 957.9 \text{ kg/m}^3$$

and other quantities are taken as

$$\Delta T = T_w - T_s = 1200°F = 648.9°C$$

$$D_o = \frac{0.05}{12} \text{ ft} = 0.127 \times 10^{-2} \text{ m}$$

$$g = 4.17 \times 10^8 \text{ ft/h}^2 = 9.8 \text{ m/s}^2$$

$$\sigma = 0.1714 \times 10^{-8} \text{ Btu/h·ft}^2·°R^4 = 0.55697 \times 10^{-8} \text{ W/m}^2·K^4$$

$$T_s = 460 + 212 = 672°R = 373.3 \text{ K}$$

$$T_w = 672 + 1200 = 1872°R = 1040 \text{ K}$$

$$\alpha = \varepsilon = 1 \text{ (assumed)}$$

Then, from the above relations we obtain

$$h_0 = 52.4 \text{ Btu/h·ft}^2·°F = 297.6 \text{ W/m}^2·°C$$

$$h_r = 17.2 \text{ Btu/h·ft}^2·°F = 32.26 \text{ W/m}^2·°C$$

Since $h_r < h_0$, we apply the simple equation (14-33) to determine the film-boiling heat-transfer coefficient h_m that includes the effects of radiation:

$$h_m = h_0 + \tfrac{3}{4}h_r = 65.3 \text{ Btu/h·ft}^2·°F = 370.9 \text{ W/m}^2·°C$$

REFERENCES

1 Nusselt, W.: Die Oberflachenkondensation des Wasserdampfes, *Z. Ver. Deut. Ing.*, **60**:541–569 (1916).

2 Colburn, A. P.: The Calculation of Condensation Where a Portion of the Condensate Layer Is the Turbulent Flow, *Trans. AIChE*, **30**:187–193 (1933).

3 Kirkbride, C. G.: Heat Transfer by Condensing Vapors, *Trans. AIChE*, **30**:170–186 (1934).

4 Carpenter, E. F., and A. P. Colburn: The Effect of Vapor Velocity on Condensation Inside Tubes, *Inst. Mech. Eng.–ASME, Proc. General Discussion Heat Transfer*, pp. 20–26, 1951.

5 Rohsenow, W. M., J. M. Weber, and A. T. Ling: Effect of Vapor Velocity on Laminar and Turbulent Film Condensation, *Trans. ASME*, **78**:1637–1644 (1956).

6 Sparrow, E. M., and J. L. Gregg: A Boundary Layer Treatment of Laminar Film Condensation, *J. Heat Transfer*, **81C**:13–18 (1959).

7 Chen, M. M.: An Analytical Study of Laminar Film Condensation: Pt. I, Flat Plates, *J. Heat Transfer*, **83C**:48–54 (1961).

8 Koh, J. C. Y., E. M. Sparrow, and J. P. Hartnett: The Two Phase Boundary Layer in Laminar Film Condensation, *Int. J. Heat Mass Transfer*, **2**:69–82 (1961).

9 Koh, J. C. Y.: Film Condensation in a Forced-Convection Boundary Layer Flow, *Int. J. Heat Mass Transfer*, **5**:941–954 (1962).

10 Koh, J. C. Y.: An Integral Treatment of Two-Phase Boundary Layer in Film Condensation, *J. Heat Transfer*, **83C**:359–362 (1961).

11 Chung, P. M.: Unsteady Laminar Film Condensation on Vertical Plate, *J. Heat Transfer*, **85C**: 63–70 (1963).

12 Madejski, J.: The Effect of Molecular-Kinetic Resistances on Heat Transfer in Condensation, *Int. J. Heat Mass Transfer*, **9**:35–39 (1966).

13 Jones, W. P., and U. Renz: Condensation from a Turbulent Stream onto a Vertical Surface, *Int. J. Heat Mass Transfer*, **17**:1019–1028 (1974).

14 Tamir, A., and I. Rachmilev: Direct Contact Condensation of an Immiscible Vapor on a Thin Film of Water, *Int. J. Heat Mass Transfer*, **17**:1241–1251 (1974).

15 Colburn, A. P., and T. B. Drew: The Condensation of Mixed Vapors, *Trans. AIChE*, **33**:197–208 (1937).

16 Sparrow, E. M., and E. R. G. Eckert: Effects of Superheated Vapor and Noncondensable Gas on Laminar Film Condensation, *AIChE J.*, **7**:473–477 (1961).

17 Sparrow, E. M., and S. H. Lin: Condensation Heat Transfer in the Presence of Noncondensable Gas, *J. Heat Transfer*, **86C**:430–436 (1964).

18 Minkowycz, W. J., and E. M. Sparrow: Condensation Heat Transfer in the Presence of Noncondensables, Interfacial Resistance, Superheating, Variable Properties, and Diffusion, *Int. J. Heat Mass Transfer*, **9**:1125–1144 (1966).

19 Mori, Yasuo, and Kunio Hijikata: Free Convective Condensation Heat Transfer with Noncondensable Gas on a Vertical Surface, *Int. J. Heat Mass Transfer*, **16**: 2229–2240 (1973).

20 McAdams, W. H.: "Heat Transmission," 3d ed., McGraw-Hill Book Company, New York, 1954.

21 Kirkbride, C. G.: Heat Transfer by Condensing Vapors, *Trans. AIChE*, **30**:170–186 (1934).

22 Colburn, A. P.: Problems in Design and Research on Condensers of Vapors and Mixtures, *Int. Mech. Eng.–ASME Proc. General Discussions Heat Transfer*, pp. 1–11, 1951.

23 Tepe, J. B., and A. C. Mueller: Condensation and Subcooling Inside an Inclined Tube, *Chem. Eng. Prog.*, **43**:267 (1947).

24 Altman, M., F. W. Staub, and R. H. Norris: Local Heat Transfer and Pressure Drop for Refrigerant-22 Condensing in Horizontal Tubes, *ASME–AIChE*, Storrs, Conn., August 1959.

25 Goodykoontz, J. H., and R. G. Dorsch: Local Heat-Transfer Coefficients and Static Pressures for Condensation of High-Velocity Steam within a Tube, *NASA Tech. Note* D-3953, May 1967.

26 Goodykoontz, J. H., and W. F. Brown: Local Heat Transfer and Pressure Distribution for Freon-113 Condensing Downward Flow in a Vertical Tube, *NASA Tech. Note* D-3952, May 1967.

27 Soliman, M., J. R. Schuster, and P. J. Berenson: A General Heat Transfer Correlation for Annular Flow Condensation, *J. Heat Transfer*, **90C**:267–276 (1968).

28 Carpenter, F. G.: "Heat Transfer and Pressure Drop for Condensing Pure Vapors Inside Vertical Tubes at High Vapor Velocities," Ph.D. dissertation, University of Delaware, Newark, Del., 1948.   Also available as American Documentation Institute, Document 3274, Washington, D.C.

29 Misra, B., and C. F. Bonilla: Heat Transfer in the Condensation of Liquid Metal Vapors: Mercury and Sodium at Atmospheric Pressure, *Chem. Eng. Prog.*, Ser. no. 8, **52**:7–21 (1956).

30 Cohn, P. D.: M.S. thesis, Oregon State College, Corvallis, Ore., 1960.

31 Roth, J. A.: Wright-Patterson AFB, Ohio, Rep. ASD-TDR-62-738, 1962.

32 General Electric Co., Missile and Space Div., Cincinnati, Alkali Metals Boiling and Condensing Investigations, *Proc. 1962 High Temperature Liquid Metal Heat Transfer Technology Meeting*, Brookhaven National Laboratory.

33 Singer, R. M.: The Control of Condensation Heat Transfer Rates Using an Electromagnetic Field, *ANL Rep.* 6861, 1964.

34 Sukhatme, S. P., and W. M. Rohsenow: Heat Transfer During Film Condensation of a Liquid Metal Vapor, *J. Heat Transfer*, **88C**:19–28 (1966).

35 Barry, R. E., and R. E. Balzhiser: Condensation of Sodium at High Heat Fluxes, *Proc. Third Int. Heat Transfer Conf.*, Chicago, **2**:318–328 (1966).

36 Kroger, D. G., and W. M. Rohsenow: Film Condensation of Saturated Potassium Vapor, *Int. J. Heat Mass Transfer*, **10**:1891–1894 (1967).

37 Rohsenow, W. M.: Film Condensation of Liquid Metals, in O. E. Dwyer (ed.), "Progress in Heat and Mass Transfer," vol. 7, "Heat Transfer in Liquid Metals," Pergamon Press, New York, 1973.

38 Denny, V. E., A. F. Mills, and J. R. Gardiner: Nonsimilar Solutions for Laminar Film Condensation of Liquid Metals, *Proc. Fourth Int. Heat Transfer Conf.*, The Netherlands, Paper Cs 2.1, 1970.

39 Hampson, H., and N. Özışık: An Investigation into the Condensation of Steam, *Proc. Inst. Mech. Eng., London*, **1B**:282–294 (1952).

40 Blackman, L. C. F., and M. J. S. Dewar: Promoters for Dropwise Condensation of Steam, Pts. I–IV, *J. Chem. Soc.*, pp. 162–176, January–March 1957.

41 Blackman, L. C. F., M. J. S. Dewar, and H. Hampson: An Investigation of Compounds Promoting Dropwise Condensation of Steam, *J. Appl. Chem.*, **7**:160–171 (1957).

42 Osment, B. D. J., D. Tudor, R. M. M. Speirs, and W. Rugman: Promoters for the Dropwise Condensation of Steam, *Trans. Inst. Chem. Eng.*, **40**:152–160 (1962).

43 Erb, R. A., and E. Thelen: Promoting Permanent Dropwise Condensation, *Ind. Eng. Chem.*, **57**:49–52 (1965).

44 Umur, A., and P. Griffith: Mechanism of Dropwise Condensation, *ASME Paper* 64-WA/HT-3, 1964.

45 Lefevre, E. S., and J. Rose: A Theory of Dropwise Condensation of Steam, *Proc. ASME–Inst. Mech. Eng. Heat Transfer Conf.*, 1966.

46 Çitakoğlu, E., and J. W. Rose: Dropwise Condensation—Some Factors Influencing the Validity of Heat Transfer Measurements, *Int. J. Heat Mass Transfer*, **11**:523–537 (1968).

47 Lusterader, E. L., R. Richter, and F. N. Neugebauer: The Use of Thin Films for Increasing Evaporation and Condensation Rates in Process Equipment, *J. Heat Transfer*, **81C**:297–307 (1959).

48 Othmer, D. F.: The Condensation of Steam, *Ind. Eng. Chem.*, **21**:576–583 (1929).

49 Colburn, A. P., and T. B. Drew: The Condensation of Mixed Vapors, *Trans. AIChE*, **33**:197–208 (1937).

50 Meisenburg, S. J., R. M. Boarts, and W. L. Badger: *Trans. AIChE*, **31**:622–637 (1935).

51 Hampson, H.: The Condensation of Steam on a Metal Surface, *Inst. Mech. Eng.– ASME, Proc. General Discussion Heat Transfer*, pp. 58–61, 1951.

52 Sparrow, E. M., and E. R. G. Eckert: Effects of Superheated Vapor and Noncondensable Gas on Laminar Film Condensation, *AIChE J.*, **7**:473–477 (1961).

53 Sparrow, E. M., and S. H. Lin: Condensation Heat Transfer in the Presence of Noncondensable Gas, *J. Heat Transfer*, **86C**:430–436 (1964).

54 Minkowycz, W. J., and E. M. Sparrow: Condensation Heat Transfer in the Presence

of Noncondensables, Interfacial Resistance, Superheating, Variable Properties, and Diffusion, *Int. J. Heat Mass Transfer*, **9**:1125–1144 (1966).

55   Mori, Y., and K. Hijikata: Free Convective Condensation Heat Transfer with Noncondensable Gas on a Vertical Surface, *Int. J. Heat Mass Transfer*, **16**:2229–2240 (1973).

56   Turner, R. H., A. F. Mills, and V. E. Denny: The Effect of Noncondensable Gas on Laminar Film Condensation of Liquid Metals, *J. Heat Transfer*, **95C**:6–11 (1973).

57   Denny, V. E., and V. J. Jusionis: Effects of Noncondensable Gas on Forced Flow on Laminar Film Condensation, *Int. J. Heat Mass Transfer*, **15**:315–326 (1972).

58   Farber, E. A., and R. L. Scorah: Heat Transfer to Water Boiling Under Pressure, *Trans. ASME*, **79**:369–384 (1948).

59   Jens, W. H., and G. Leppert: Recent Developments in Boiling Research, Pts, I, II, *J. Am. Soc. Nav. Eng.*, **67**:137–155 (1955); **66**:437–456 (1955).

60   Rohsenow, W. M.: Boiling Heat Transfer, in W. M. Rohsenow (ed.), "Developments in Heat Transfer," The M.I.T. Press, Cambridge, Mass., 1964.

61   Leppert, G., and C. C. Pitt: Boiling, in T. F. Irvine, Jr., and J. P. Hartnett (eds.), "Advances in Heat Transfer," vol. 1, Academic Press, Inc., New York, 1964.

62   Tong, L. S.: "Boiling Heat Transfer and Two-Phase Flow," John Wiley & Sons, Inc., New York, 1966.

63   Jordan, D. P.: Film and Transition Boiling, in T. F. Irvine, Jr., and J. P. Hartnett (eds.), "Advances in Heat Transfer," vol. 5, Academic Press, Inc., New York, 1968.

64   Wallis, G. B.: "One-dimensional Two-Phase Flow," McGraw-Hill Book Company, New York, 1969.

65   Cryder, D. S., and A. C. Finalbargo: Heat Transmission from Metal Surfaces to Boiling Liquids: Effects of Temperature of the Liquid on Film Coefficient, *Trans. AIChE*, **33**:346–362 (1937).

66   Chichelli, M. T., and C. F. Bonilla: Heat Transfer to Liquids Boiling Under Pressure, *Trans. AIChE*, **41**:755–787 (1945).

67   Addoms, J. N.: "Heat Transfer at High Rates to Water Boiling Outside Cylinders." D. Sc. thesis, Massachusetts Institute of Technology, Department of Chemical Engineering, Cambridge, Mass., 1948.

68   Rohsenow, W. M., and H. Y. Choi: "Heat, Mass and Momentum Transfer," Prentice-Hall, Inc., Englewood Cliffs, N. J., 1961.

69   Rohsenow, W. M.: A Method of Correlating Heat Transfer Data for Surface Boiling Liquids, *Trans. ASME*, **74**:969–975 (1952).

70   Vahon, R. I., G. H. Nix, and G. E. Tanger: Evaluation of Constants for the Rohsenow Pool-Boiling Correlation, *J, Heat Transfer*, **90C**:239–247 (1968).

71   Piret, E. L., and H. S. Isbin: Natural Circulation Evaporation Two-Phase Heat Transfer, *Chem. Eng. Prog.*, **50**:305–311 (1954).

72   Kutateladze, S. S.: A Hydrodynamic Theory of Changes in Boiling Process Under Free Convection, *Iz. Akad. Nauk SSSR, Otd. Tekh. Nauk*, no. 4, p. 524, 1951.

73   Zuber, N.: On the Stability of Boiling Heat Transfer, *J. Heat Transfer*, **80C**:711 (1958).

74   Zuber, N., and M. Tribus: Further Remarks on the Stability of Boiling Heat Transfer, *Univ. Calif.*, Los Angeles, *Dept. Eng.*, *Rep.* 58-5, 1958.

75   Moissis, R., and P. J. Berenson: On the Hydrodynamic Transitions in Nucleata Boiling, *J. Heat Transfer*, **85C**:221–229 (1963).

76   Usiskin, C. M., and R. Siegel: An Experimental Study of Boiling in Reduced and Zero Gravity Fields, *Trans. ASME*, **83C**:243–253 (1961).

77 Bromley, L. A.: Heat Transfer in Stable Film Boiling, *Chem. Eng. Prog.*, **46**:221–227 (1950).

78 Breen, B. P., and J. W. Westwater: Effect of Diameter of Horizontal Tubes on Film Boiling Heat Transfer, *Chem. Eng. Prog.*, **58**:67–72 (1962).

79 Berenson, P. J.: Film-Boiling Heat Transfer from a Horizontal Surface, *Trans. ASME*, **83C**:351–356 (1961).

80 Hsu, Y. Y., and J. W. Westwater: Approximate Theory for Film Boiling on Vertical Surfaces, *Chem. Eng. Prog., Symp. Ser., AIChE Heat Transfer Conf.*, Storrs, Conn., **56**:15–22 (1959).

81 Bankoff, S. G.: Discussion of Approximate Theory for Film Boiling on Vertical Surfaces, *Chem. Eng. Prog., Symp. Ser., AIChE Heat Transfer Conf.*, Storrs, Conn., pp. 22–24, 1959.

82 Berenson, P. J., and R. A. Stone: A Photographic Study of the Mechanism of Forced-Convection Evaporation, Symposium on Heat Transfer, San Juan, Puerto Rico, *AIChE Reprint* 21, 1963.

83 Konmutsos, K., R. Moissis, and A. Spyridonos: A Study of Bubble Departure in Forced Convection Boiling, *J. Heat Transfer*, **90C**:223–230 (1968).

84 Polomik, E. E., S. Levy, and S. G. Sawochka: Film Boiling of Steam-Water Mixtures in Annular Flow at 800, 1100, and 1400 Psi, *J. Heat Transfer*, **86C**:81–88 (1964).

85 Davis, E. J., and M. M. David: Two-Phase Gas-Liquid Convection Heat Transfer, *Ind. Eng. Chem., Fundam.*, **3**:111–118 (1964).

86 Gouse, S. W., Jr.: An Index to the Two-Phase Gas-Liquid Flow Literature, Pt. I, *M.I.T. Eng. Proj. Lab. Rep. DSR*-8734-1, 1963.

87 Wallis, G. B.: Some Hydrodynamic Aspects of Two-Phase Flow and Boiling, in "International Developments in Heat Transfer," p. 319, ASME, 1962.

88 Fraas, A. P., and M. N. Özışık: "Heat Exchanger Design," John Wiley & Sons, Inc., New York, 1965.

Fifteen

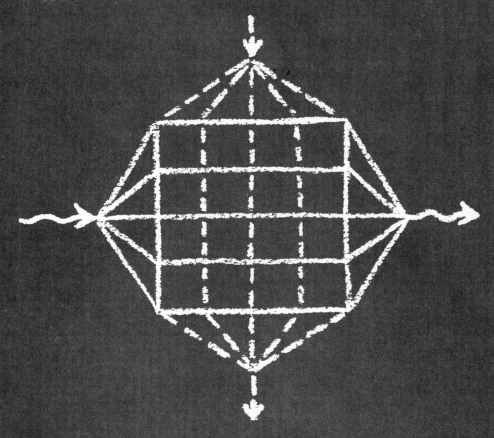

Heat Exchangers

Many types of heat exchangers have been developed for use at such varied levels of technological sophistication and sizes as steam power plants, chemical processing plants, building heating and air conditioning, household refrigerators, car radiators, radiators for space vehicles, and so on.   In the common types, such as shell-and-tube heat exchangers and car radiators, heat transfer is primarily by conduction and convection from a hot to a cold fluid which are separated by a metal wall.   In boilers and condensers, heat transfer by boiling and condensation is of primary importance.   In certain types of heat exchangers, such as cooling towers, hot fluid (i.e., water) is cooled by direct mixing with the cold fluid (i.e., air), that is, the water sprayed or falling down into an induced air draft is cooled both by convection and vaporization.   In radiators for space applications the waste heat carried by the coolant fluid is transported by convection and conduction to the fin surface and from there by thermal radiation into the atmosphere-free space.   Therefore, the thermal design of heat exchangers is an area where the principles of heat transfer discussed in the previous chapters find numerous applications.

The actual design of heat exchangers is a much more complicated problem than that of heat-transfer analysis alone because cost, weight, size, and economic considerations play an important role in the selection of the final design.   For example, although the cost considerations are very important for applications in large installations such as power plants and chemical processing plants, the weight and size considerations become the dominant factor in the choice of design for space and aeronautical applications.   A comprehensive treatment of heat-exchanger design is, therefore, beyond the scope of this chapter.   The reader should consult Refs. [1–13] for detailed consideration of various specific design problems.   In the limited space available here, we will present a brief discussion of thermal analysis of the more common types of heat exchangers and the basic design considerations.

15-1   TYPES OF HEAT EXCHANGERS

Most heat exchangers may be classified on the basis of (1) the configuration of the fluid flow paths through the heat exchanger and (2) the application for which they are intended. We now examine the classification of heat exchangers according to these two different considerations.

Classification by Fluid-Flow Arrangement

The four most common types of flow-path configurations are illustrated in Fig. 15-1.   In the *parallel-flow* arrangement shown in Fig. 15-1a, hot and cold fluids enter at one end of the heat exchanger, flow through in the same direction, and leave together at the other end.   In the *counterflow* arrangement in Fig. 15-1b, the hot and cold fluids enter in the opposite ends of the heat exchanger and

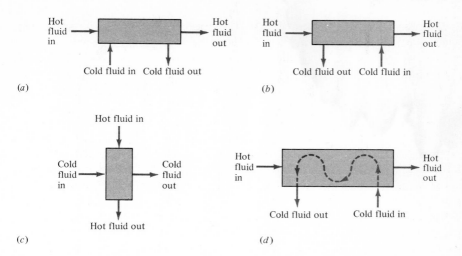

FIG. 15-1   Types of flow-path configuration through heat exchangers. (*a*) Parallel flow; (*b*) counterflow; (*c*) single-pass cross-flow; (*d*) multipass cross-flow.

flow through in opposite directions.   In the *single-pass cross-flow* arrangement shown in Fig. 15-1*c*, one fluid moves through the heat-transfer matrix at right angles to the flow path of the other fluid.   In the *multipass cross-flow* unit shown in Fig. 15-1*d*, one fluid stream shuttles back and forth across the flow path of the other fluid stream.

Classification by Types of Applications

Special terms are generally employed to characterize heat exchangers on the basis of their application.   The terms used for major types include *boilers* (or *steam generators*), *condensers, shell-and-tube heat exchangers, cooling towers, compact heat exchangers, radiators for space power plants,* and *regenerators*. Typical features of these principal types of heat exchangers are now described below.

Boilers   Steam boilers used to generate steam constitute one of the earliest application of heat exchangers.   The term *steam generator* is often applied to boilers in which the heat source is a hot fluid stream rather than the hot products of combustion.   A typical steam generator for a pressurized-water nuclear-reactor power plant is shown in Fig. 15-2.   High-pressure, high-temperature (approximately 2200 lb_f/in^2 abs. and 600°F or 149.7 atm and 316°C) water from the nuclear reactor enters the upper head and flows downward inside the tubes, giving up heat to produce steam in the space between the tubes.   The feedwater enters through the nozzles into the space between the lower tube-bundle shroud and the shell, is heated to the saturation temperature while flowing downward in the annulus, and then enters the tube bundle through the ports in the lower shroud.   The water begins to boil first by nucleate boiling, then film boiling

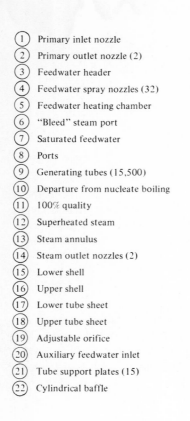

1. Primary inlet nozzle
2. Primary outlet nozzle (2)
3. Feedwater header
4. Feedwater spray nozzles (32)
5. Feedwater heating chamber
6. "Bleed" steam port
7. Saturated feedwater
8. Ports
9. Generating tubes (15,500)
10. Departure from nucleate boiling
11. 100% quality
12. Superheated steam
13. Steam annulus
14. Steam outlet nozzles (2)
15. Lower shell
16. Upper shell
17. Lower tube sheet
18. Upper tube sheet
19. Adjustable orifice
20. Auxiliary feedwater inlet
21. Tube support plates (15)
22. Cylindrical baffle

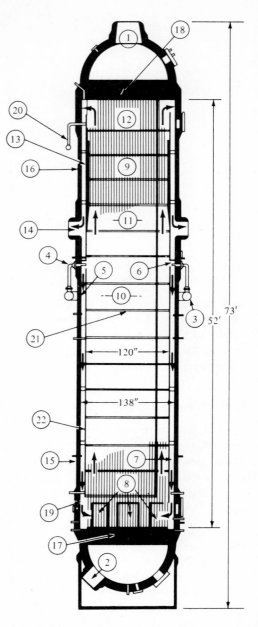

FIG. 15-2 Once-through steam generator for a pressurized-water nuclear-reactor power plant. (*From The Babcock and Wilcox Company.*)

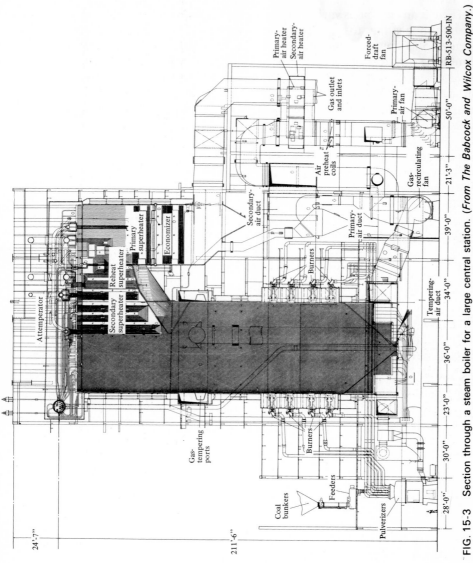

FIG. 15-3 Section through a steam boiler for a large central station. (*From The Babcock and Wilcox Company.*)

as it flows upward, finally becoming superheated steam (at about 900 lb_f/in^2 abs. and 594°F or 61.2 atm and 312°C) before leaving the tube bundle at the upper tube sheet. The superheated steam is routed downward in the annulus between the upper shroud and the shell, finally exiting through the steam outlet to drive the turbogenerator.

There is an enormous variety of boiler types ranging from small units for house heating applications to huge, complex, expensive units for modern power stations. Figure 15-3 shows a steam boiler for a large central station, in which the heat transmitted by thermal radiation from the oil or powdered-coal flame goes to *preheating* and *boiling* the water flowing in the banks of tubes that form the furnace walls. The gases then flow downward at reduced and more uniform temperature through the *superheater*, the *economizer*, and the *air heater* to the base of the stack.

Condensers   Condensers are used for such varied applications as steam power plants, chemical processing plants, and nuclear electric plants for space vehicles. The major types include the *surface condensers, jet condensers,* and *evaporative condensers.* The most common type is the surface condenser which has the advantage that the condensate is returned to the boiler through the feedwater system.   Figure 15-4 shows a section through a typical two-pass surface condenser for a large steam turbine in a power plant.   Since the steam pressure at the turbine exit is only 1.0 to 2.0 in Hg abs., the steam density is very low and the volume rate of flow is extremely large.   To minimize the pressure loss in

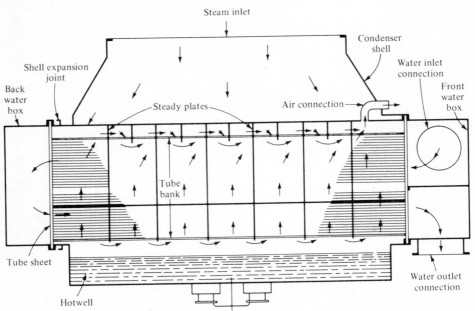

FIG. 15-4   Section through a typical two-pass surface condenser for a large steam power plant. (*From Allis-Chalmers Manufacturing Company.*)

transferring steam from the turbine to the condenser, the condenser is normally mounted beneath and attached to the turbine. Cooling water flows horizontally inside the tubes, while the steam flows vertically downward from the large opening at the top and passes transversely over the tubes. Note that provision is made to aspirate cool air from the regions just above the center of the hot well. This is important because the presence of noncondensable gas in the steam reduces the heat-transfer coefficient for condensation. The reader should consult Refs. [1,2,14] for a discussion of the design considerations.

Shell-and-Tube Heat Exchangers   The units known as shell-and-tube heat exchangers essentially consist of round tubes mounted in a cylindrical shell with their axes parallel to that of the shell. Liquid-to-liquid heat exchangers commonly fall in this group, and in some cases gas-to-gas heat exchangers are also of the shell-and-tube type. They are especially well suited for applications in which heat-transfer coefficients for the two fluids are within a factor of 2 or 3 of each other so that there is little incentive to employ extended surfaces. In the case of gas-to-gas applications the heat-transfer coefficients on the opposite sides of the heat-transfer surface are usually within a factor of 3 or 4 of each other, and the absolute values usually are lower than the corresponding values of liquid-to-liquid heat exchangers by a factor of 10 to 100; thus a much larger volume of heat-transfer matrix is required to transfer a given amount of heat. Many variations of this type of heat exchanger are available; the differences lie in the arrangement of flow configurations and in the detailed features of construction. Figure 15-5 shows some of the more commonly used flow configurations for liquid-to-liquid shell-and-tube heat exchangers. Figure 15-5a illustrates a baffled, single-pass tube-side, single-pass shell-side, counterflow arrangement, and Fig. 15-5b shows a baffled, two-pass tube-side, single-pass shell-side arrangement. Figure 15-5c illustrates a baffled, four-pass tube-side, two-pass shell-side arrangement. The allowable pressure drop for the tube-side fluid stream is a major factor in determining the number of flow passes chosen, while the tube bundles are baffled to give a reasonably uniform flow distribution across the tube matrix. The reader should consult Refs. [15–25] for heat-transfer and pressure-drop characteristics of shell-and-tube-type heat exchangers.

Cooling Towers   Cooling towers have been widely used to dispose waste heat from industrial processes by rejecting heat into the atmosphere rather than to water in a river, lake, or ocean. The most common types include the *natural-convection* and *forced-convection* cooling towers. In the natural-convection type of cooling tower shown in Fig. 15-6 the water is sprayed directly into the air stream that moves through the cooling tower by thermal convection. The falling water droplets are cooled both by ordinary convection and by the evaporation of water. The deck of *fills* positioned inside the cooling tower reduces the average velocity of the falling droplets and thus increases the time droplets are exposed to the cooling air stream in falling through the tower. Large natural-convection-type cooling towers over 300 ft high have been built to cool waste

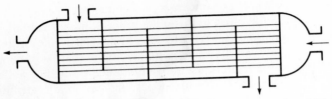

(a) Single-pass tube–side; baffled, single–pass shell–side; counterflow, full baffles

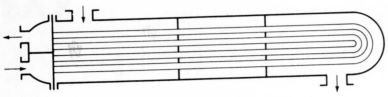

(b) U tube with single axial pass on the shell–side, annular orifice baffles

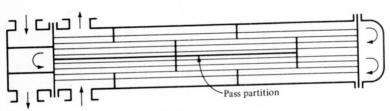

(c) Four-pass tube–side, baffled two–pass shell-side

FIG. 15-5 Typical flow arrangements for shell-and-tube heat exchangers.

heat from power plants. In a forced-convection type of cooling tower the water is sprayed into the air stream circulated through the tower by a fan which can be mounted either on top of the tower so that it draws air upward or just outside the base so that air flows directly inward. Figure 15-7 shows a section through a forced-circulation cooling tower with draft induced by a fan. The increased air circulation increases the heat-transfer capacity of the cooling tower. The reader should consult Refs. [26–30] for the characteristics and performance of cooling towers.

Compact Heat Exchangers The relative importance of the criteria such as pumping power, cost, weight, and size of a heat exchanger varies so much from one installation to another that it is not always possible to generalize such requirements with the type of application. However, when heat exchangers are to be employed for aircraft, marine, and aerospace vehicles, the weight and size consideration becomes important. In order to increase the effectiveness of the heat exchanger, fins are used on the surface where the heat-transfer coefficient is much lower. The detailed dimensions of the heat-transfer matrix and the

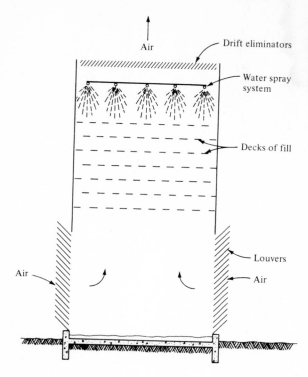

FIG. 15-6   Section through a natural-convection cooling tower with "fill" to increase the effective water-droplet surface area by multiple splashing.

type of fins, and the appropriate fin size and dimension vary with the specific application, and various types have been designed and used in numerous applications.   The reader should consult Refs. [3,31] for a discussion of heat-transfer and pressure-drop characteristics of various types of heat-transfer matrices for compact heat exchangers.

Radiators for Space Power Plants   The rejection of waste heat from the condenser of a power plant intended to produce electricity for the propulsion, guidance, or the communication equipment of a space vehicle poses serious problems even for a power plant producing only a few kilowatts of electricity. The only way the waste heat can be dissipated from a space vehicle is by thermal radiation by taking advantage of the fourth-power relationship between the absolute temperature of the surface and the radiative-heat flux.   Thus, in the operation of some power plants for space vehicles the thermodynamic cycle is at such high temperatures that the radiator runs red-hot.   Even so, it is difficult to keep the radiator size within a reasonable envelope for launch vehicles. Figure 15-8 shows some typical configurations that have appeared attractive as space radiators to dissipate waste heat by thermal radiation into the space. In this figure, the radiator configurations 1 and 2 make use of type A tubes

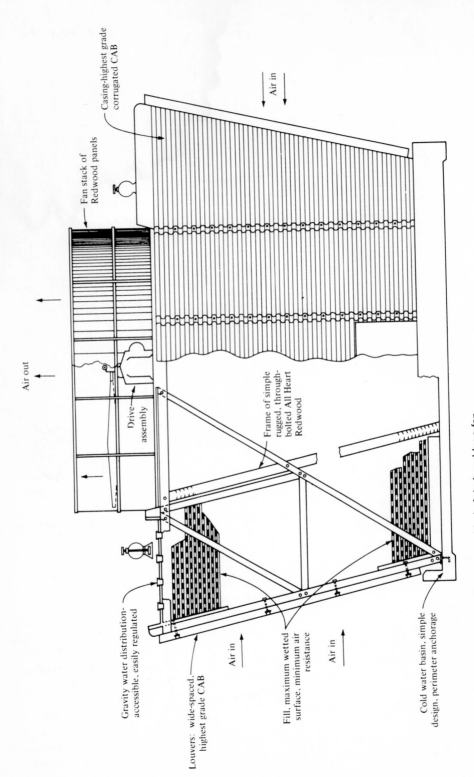

Casing-highest grade corrugated CAB

Air in

Fan stack of Redwood panels

Air out

Drive assembly

Frame of simple rugged, through-bolted All Heart Redwood

Gravity water distribution-accessible, easily regulated

Air in

Louvers: wide-spaced, highest grade CAB

Fill, maximum wetted surface, minimum air resistance

Air in

Cold water basin, simple design, perimeter anchorage

FIG. 15-7   Forced-convection cooling tower with draft induced by a fan.

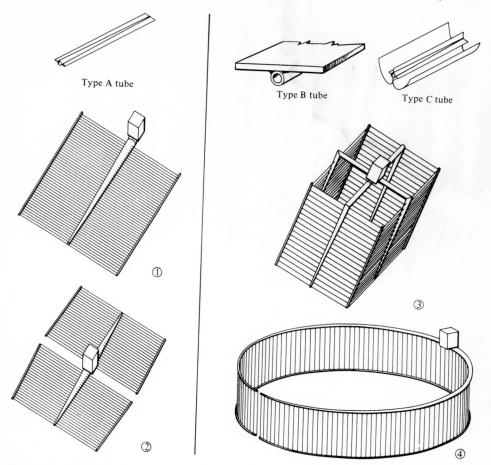

Type A tube

Type B tube

Type C tube

①

③

②

④

FIG. 15-8   Typical finned-tube and tube-manifold radiator configurations for radiators for space vehicles. The box indicates the position of the turbine. (*From Oak Ridge National Laboratory.*)

with two axial opposed fins.   Configurations 3 and 4 are designed to employ either type B or type C finned tubes.

Regenerators   In the various types of heat exchangers discussed above, the hot and cold fluids are separated by a solid wall, whereas the term *regenerator* is applied to the periodic-flow type of heat exchanger.   That is, the same space is alternately occupied by the hot and cold gases between which heat is exchanged. Regenerators are generally employed to preheat the air for steam power plants, open-hearth furnaces, blast furnaces, as well as in a host of other applications including oxygen production and the separation of gases at very low temperatures. A typical example of such a regenerator is the *Ljungstrom regenerative air pre-heater* shown in Fig. 15-9.   In this regenerator, the cylindrical heat-transfer

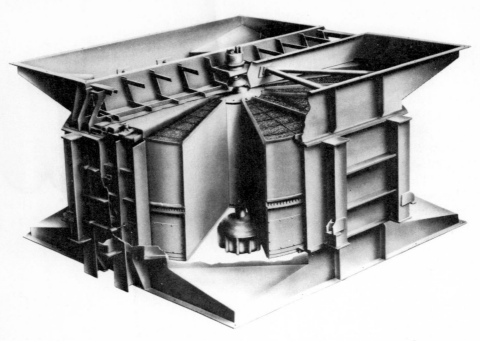

FIG. 15-9   A Ljungstrom® air preheater. (*By permission from C. E. Air Preheater, Combustion Engineering, Inc.*)

matrix is made of alternate layers of flat and corrugated sheets and is so designed that in each rotation the matrix passes through a warm and a cold gas stream. The heat stored in the matrix during its contact by the hot gas is given up to the cold gas on the other side. There is, of course, a certain amount of leakage between the two fluids. This type of unit is suitable only for gas-to-gas heat exchange, because only for gases the heat capacity of the heat-transfer matrix is much greater than the heat capacity of the gas contained in the flow passages. It is not suitable for liquid-to-liquid heat exchange, because the heat capacity of the heat-transfer matrix is much less than that of the liquid.

Since the heat-transfer matrix is rotating, the temperatures of the gases and the wall depend on space and time; as a result, the heat-transfer analysis of regenerators is involved, in that the periodic flow introduces several new variables. For conventional, stationary heat exchangers it is sufficient to define the inlet and the outlet temperatures, the flow rates, the heat-transfer coefficients for the two fluids, and the surface areas of the two sides of the heat exchangers. On the other hand, for the rotary heat exchanger, it is also necessary to relate the heat capacity of the rotor to that of the fluid streams, the fluid flow rates, and the rotation speed under consideration. The reader should consult Refs. [32–38] for detailed discussion of heat-transfer and design considerations for regenerative-type heat exchangers.

15-2   LOGARITHMIC MEAN TEMPERATURE DIFFERENCE

In the stationary types of heat exchangers described previously, the heat transfer from the hot to the cold fluid gives rise to a change of temperature of one or both of the fluids while flowing through the heat exchanger. Figure 15-10 illustrates how the temperature of the fluid varies along the path of the heat exchanger for a number of different cases. In each instance the temperature distribution in the heat exchanger is plotted as a function of the distance from the cold-fluid inlet end. Figure 15-10a, for example, characterizes a pure counterflow heat exchanger in which the temperature rise in the cold fluid is equal to the temperature drop in the hot fluid; thus the temperature difference ΔT between the hot and cold fluids is constant throughout. However, in all other cases (i.e., Fig. 15-10b to e) the temperature difference ΔT between the hot and cold fluids varies with position along the path of flow.

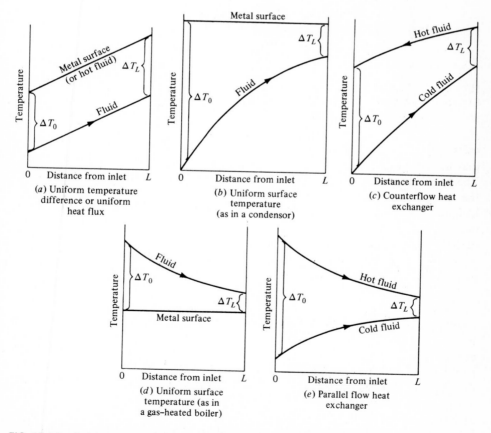

FIG. 15-10   Axial temperature distribution in typical heat-transfer matrices.

To simplify the heat-transfer analysis of heat exchangers, it is desirable to establish a mean temperature difference between cold and hot fluids. We now describe the determination of the mean temperature difference for stationary-type single-pass heat exchangers. Although the following analysis is presented with reference to the parallel-flow arrangement shown in Fig. 15-11, the result is applicable for all the cases illustrated in Fig. 15-10.

Referring to the parallel-flow arrangement shown in Fig. 15-11, let

$$A = \text{heat-transfer area measured from inlet end of heat exchanger,}$$
$$\text{ft}^2 \ (\text{m}^2)$$
$$m_c, m_h = \text{mass-flow rate of cold and hot fluids, respectively, lb/h (kg/h)}$$
$$\Delta T = T_h - T_c = \text{local temperature difference between hot and cold fluids at}$$
$$\text{position } A \text{ from inlet, } °F \ (°C)$$
$$U = \text{local overall heat-transfer coefficient between two fluids, Btu/}$$
$$\text{h·ft}^2·°F \ (W/m^2·°C)$$

The rate of heat transfer dQ from the hot to the cold fluid through an elemental area dA about the location A is given by

$$dQ = U \, dA \, \Delta T \tag{15-1}$$

On the other hand, this heat transfer dQ should be equal to the heat given up by the hot fluid flowing from position A to $A + dA$; with this consideration, we write

$$dQ = -m_h c_{ph} \, dT_h \tag{15-2}$$

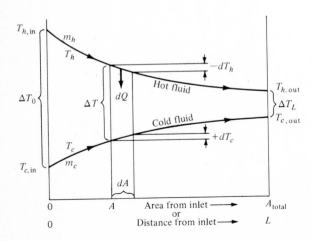

FIG. 15-11   Nomenclature for the derivation of the logarithmic mean temperature difference.

Similarly, for the heat gain by the cold fluid flowing from position A to $A + dA$, we write

$$dQ = m_c c_{pc} \, dT_c \tag{15-3}$$

where c_{pc} and c_{ph} are the specific heats, and dT_c and dT_h are the changes in the temperatures of the cold and hot fluids, respectively. We now consider the relation

$$\Delta T = T_h - T_c \tag{15-4}$$

where T_c and T_h are the temperatures of the cold and hot fluids, respectively, at position A, and by differentiation we obtain

$$d(\Delta T) = d(T_h - T_c) = dT_h - dT_c \tag{15-5}$$

The substitution of dT_h and dT_c from Eqs. (15-2) and (15-3) into Eq. (15-5) yields

$$d(\Delta T) = -\frac{dQ}{m_h c_{ph}} - \frac{dQ}{m_c c_{pc}} = -dQ\left(\frac{1}{m_h c_{ph}} + \frac{1}{m_c c_{pc}}\right) \tag{15-6}$$

which can be written more compactly in the form

$$d(\Delta T) = -M \, dQ \tag{15-7a}$$

where

$$M = \frac{1}{m_h c_{ph}} + \frac{1}{m_c c_{pc}} \tag{15-7b}$$

When dQ is substituted from Eq. (15-1) into Eq. (15-7a), we obtain

$$\frac{d(\Delta T)}{\Delta T} = -UM \, dA \tag{15-8}$$

Equation (15-7a) is now integrated over the entire length of the heat exchanger by assuming M is constant:

$$\int_{\Delta T_0}^{\Delta T_L} d(\Delta T) = -M \int_0^Q dQ \tag{15-9}$$

where ΔT_0 and ΔT_L are the temperature difference between the hot and cold fluid at the inlet and outlet ends, respectively, and Q is the total heat-transfer rate from the hot to the cold fluid in the heat exchanger. By performing the integration in Eq. (15-9) we obtain

$$\Delta T_0 - \Delta T_L = MQ$$

or

$$Q = \frac{\Delta T_0 - \Delta T_L}{M} \tag{15-10}$$

Equation (15-8) is integrated over the entire length of the heat exchanger to give

$$\int_{\Delta T_0}^{\Delta T_L} \frac{d(\Delta T)}{\Delta T} = -M \int_0^{A_t} U \, dA$$

or

$$\int_{\Delta T_0}^{\Delta T_L} \frac{d(\Delta T)}{\Delta T} = -MA_t \frac{\int_0^{A_t} U \, dA}{A_t} \tag{15-11}$$

where A_t is the total heat-transfer area of the heat exchanger. We now define the average overall heat-transfer coefficient U_m for the entire heat exchanger as

$$U_m = \frac{1}{A_t} \int_0^{A_t} U \, dA \tag{15-12}$$

Then Eq. (15-11) becomes, after the integration is performed,

$$\ln \frac{\Delta T_0}{\Delta T_L} = M U_m A_t \tag{15-13}$$

Eliminating M between Eqs. (15-10) and (15-13), we obtain

$$Q = A_t U_m \frac{\Delta T_0 - \Delta T_L}{\ln (\Delta T_0 / \Delta T_L)} \tag{15-14}$$

Equation (15-14) is now written more compactly in the form

$$Q = A_t U_m \Delta T_{\ln} \tag{15-15}$$

where we have defined the *logarithmic mean temperature difference* $\Delta T_{\ln}$ between the hot and cold fluids through the heat exchanger as

$$\Delta T_{\ln} = \frac{\Delta T_0 - \Delta T_L}{\ln (\Delta T_0 / \Delta T_L)} \tag{15-16}$$

The logarithmic mean temperature difference (LMTD) as defined by Eq. (15-16) is applicable for all flow arrangements shown in Fig. 15-10. When $\Delta T_0 = \Delta T_L$, as in the case of counterflow with $m_h c_{ph} = m_c c_{pc}$, then $\Delta T_{\text{ln}} = \Delta T_0 = \Delta T_L$. If ΔT_0 is not more than 50 percent greater than ΔT_L, the logarithmic mean temperature difference given by Eq. (15-16) can be approximated by the arithmetic mean temperature difference within about 1 percent. It is to be noted that *LMTD is always less than the arithmetic mean.*

15-3   CORRECTION FOR LMTD WITH CROSS-FLOW HEAT EXCHANGERS

The cross-flow heat exchangers involve more complicated temperature distribution patterns than the simple profiles shown in Fig. 15-10. Therefore, the LMTD derived in the preceding section for a single pass on the tube and shell sides should be modified if it is to be applied to cross-flow arrangement. The analysis of LMTD for such cases is more involved. Fortunately, correction charts have been developed [39–41] to convert the LMTD for counterflow condition to the LMTD for cross-flow conditions. Figure 15-12 shows correction charts for typical single- and multipass cross-flow conditions. To obtain the *corrected mean temperature difference* $\Delta T_{\text{corrected}}$ for any of these arrangements, the LMTD for the *counterflow* condition should be multiplied by the appropriate *correction factor*, that is,

$$\Delta T_{\text{corrected}} = (\text{LMTD})(\text{correction factor from Fig. 15-12}) \qquad (15\text{-}17)$$

In these figures the abscissa is the dimensionless temperature ratio P defined as

$$P = \frac{t_2 - t_1}{T_1 - t_1} \qquad (15\text{-}18)$$

where capital T refers to the *shell-side temperature*, lowercase t to the *tube-side temperature*, and the subscripts 1 and 2 refer, respectively, to the *inlet* and *outlet* conditions. The parameter R appearing on the curves is defined as

$$R = \frac{T_1 - T_2}{t_2 - t_1} = \frac{(mc_p)_{\text{tube side}}}{(mc_p)_{\text{shell side}}} \qquad (15\text{-}19)$$

It is to be noted that the correction factors in Fig. 15-12 are applicable whether the hot fluid is in the shell side or the tube side. Correction-factor charts for several other flow arrangements are available in Ref. [39].

Example 15-1 Consider an oil cooler for a large diesel engine to cool SAE-30 lubricating oil from 150°F (65.6°C) to 130°F (54.4°C) using seawater at an inlet temperature 80°F (26.7°C) with a temperature rise of 20°F (11.1°C). The design heat load is $Q = 650,000$ Btu/h (190,500 W). Assuming an average overall heat-transfer

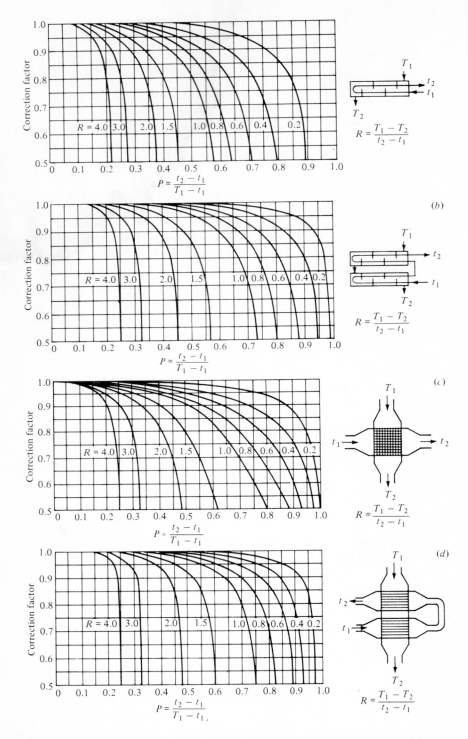

FIG. 15-12 Factors for computing the corrected logarithmic mean temperature difference for typical multipass cross-flow heat exchangers. (*From Bowman et al.* [*39*].)

coefficient $U_m = 130$ Btu/h·ft²·°F (738 W/m²·°C) based on the outer surface area of the tubes, determine the heat-transfer surface area required for a single-pass (a) counterflow and (b) parallel-flow arrangement.

Solution   The temperature profiles in the heat exchanger for the counterflow and parallel-flow arrangements are shown in Fig. 15-13. The heat-transfer areas required for each of these arrangements are evaluated as follows:

(a) Counterflow arrangement.
$\Delta T_0 = 50°F, \Delta T_L = 50°F$.   Then LMTD = 50°F (27.8°C) and the total heat-transfer area A_t is determined from

$$A_t = \frac{Q}{\text{LMTD} \times U_m}$$

By using the English system of units, we find

$$A_t = \frac{650,000}{50 \times 130} = 100 \text{ ft}^2 \ (9.29 \text{ m}^2)$$

or, by using the SI system of units, we obtain

$$A_t = \frac{190,500}{27.8 \times 738} = 9.29 \text{ m}^2$$

(b) Parallel-flow arrangement.
$\Delta T_0 = 70°F, \ \Delta T_L = 30°F$.   Then LMTD $= \dfrac{70 - 30}{\ln{(70/30)}} = 47.2°F \ (26.2°C)$ and the total heat-transfer area is

$$A_t = \frac{650,000}{47.2 \times 130} = 106 \text{ ft}^2 \ (9.85 \text{ m}^2) \quad \text{or} \quad A_t = \frac{190,500}{26.2 \times 738} = 9.85 \text{ m}^2$$

We note that slightly less area is required for the counterflow arrangement.

Example 15-2   Water is to be cooled from 65°F (18.33°C) to 44°F (6.67°C) by using brine at an inlet temperature of 30°F (-1.11°C) with a temperature rise of 7°F (3.89°C). The brine and water flows are on the tube and shell sides, respectively.   Determine the total heat-transfer area required for a cross-flow arrangement of the type shown in Fig. 15-12b by assuming an average overall heat-transfer coefficient $U_m = 150$ Btu/h·ft²·°F (851.5 W/m²·°C) and a design heat load $Q = 20,000$ Btu/h (5862 W).

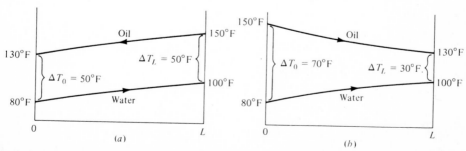

FIG. 15-13   Inlet and outlet temperatures for Example 15-1.   (a) Counterflow, (b) parallel flow.

Solution   The inlet and outlet temperatures of water and brine are:

Water on the shell side: inlet, $T_1 = 65°F$ (18.33°C)   outlet, $T_2 = 44°F$ (6.67°C)

Brine on the tube side: inlet, $t_1 = 30°F$ (−1.11°C)   outlet, $t_2 = 37°F$ (2.78°C)

Assuming a *counterflow* arrangement, the temperature differences at the two ends of the heat exchanger are

$$\Delta T_0 = 65 - 37 = 28°F \ (15.56°C)$$

$$\Delta T_L = 44 - 30 = 14°F \ (7.78°C)$$

and the corresponding logarithmic mean temperature difference becomes

$$\text{LMTD} = \frac{\Delta T_0 - \Delta T_L}{\ln (\Delta T_0/\Delta T_L)} = \frac{28 - 14}{\ln (28/14)} = 20.2°F \ (11.22°C)$$

or by using the SI system of units we obtain

$$\text{LMTD} = \frac{15.56 - 7.78}{\ln (15.56/7.78)} = 11.22°C$$

The parameters P and R are evaluated as

$$P = \frac{t_2 - t_1}{T_1 - t_1} = \frac{37 - 30}{65 - 30} = \frac{7}{35} = 0.2$$

$$R = \frac{T_1 - T_2}{t_2 - t_1} = \frac{65 - 44}{37 - 30} = \frac{21}{7} = 3$$

Then, the correction factor from Fig. 15-12b for $P = 0.2$ and $R = 3$ is

Correction factor $= 0.98$

and the corrected temperature difference becomes

$$\Delta T_{corrected} = 20.2 \times 0.98 = 19.8°F \ (11.0°C)$$

The required total heat-transfer area A_t is determined from

$$A_t = \frac{Q}{\Delta T_{corrected} \times U_m} = \frac{20,000}{19.8 \times 150} = 6.73 \text{ ft}^2 \ (0.625 \text{ m}^2)$$

The same result is obtained by using the SI system of units, i.e.,

$$A_t = \frac{5862}{11 \times 851.5} = 0.625 \text{ m}^2$$

15-4   HEAT-EXCHANGER EFFECTIVENESS

In the preceding section we have shown that the total heat-transfer rate Q through a single-pass heat exchanger is given by

$$Q = AU_m \, \Delta T_{\ln} \tag{15-20a}$$

where

$$\Delta T_{\text{ln}} = \frac{\Delta T_0 - \Delta T_L}{\ln \left(\Delta T_0 / \Delta T_L \right)} \tag{15-20b}$$

U_m is the average overall heat-transfer coefficient and A is the total heat-transfer surface area.   Clearly, Eqs. (15-20) are useful if the inlet and outlet temperatures of the hot and cold fluids are all specified so that ΔT_{ln} can be calculated. On the other hand, there are many practical design applications in which the inlet temperatures of the hot and cold fluids are known and U_m can be estimated, but the outlet temperatures of the fluids are to be determined.   If Eqs. (15-20) are used for such purposes, very tedious calculations are needed to determine the outlet temperatures and the rates of flow of the fluids.   To alleviate this difficulty the total heat-transfer rate Q is expressed in terms of the *inlet temperature difference* $(T_{h,\,\text{in}} - T_{c,\,\text{in}})$ of the hot and cold fluids and the *heat-exchanger effectiveness* ε, instead of the logarithmic mean temperature difference, as given by Eqs. (15-20).   We now describe the determination of Q in terms of heat-exchanger effectiveness.

The heat-exchanger effectiveness ε is defined as

$$\varepsilon = \frac{Q}{Q_{\text{max}}} = \frac{\text{actual heat-transfer rate}}{\text{maximum possible heat-transfer rate}} \tag{15-21a}$$

where the maximum possible heat-transfer rate Q_{max} is defined as the heat-transfer rate through a heat exchanger with infinite heat-transfer area and no losses and is given by

$$Q_{\text{max}} = (mc_p)_{\text{min}}(T_{h,\,\text{in}} - T_{c,\,\text{in}}) \tag{15-21b}$$

Here, $(mc_p)_{\text{min}}$ is the smaller of the $m_h c_{ph}$ and $m_c c_{pc}$ for the hot and cold fluids, and $T_{h,\,\text{in}}$ and $T_{c,\,\text{in}}$ are the inlet temperatures of the hot and cold fluid, respectively.   By combining Eqs. (15-21a) and (15-21b), we write

$$Q = \varepsilon(mc_p)_{\text{min}}(T_{h,\,\text{in}} - T_{c,\,\text{in}}) \tag{15-22}$$

This equation eliminates the need for LMTD to compute the total heat-transfer rate Q, but it requires that the heat-exchanger effectiveness ε be known.   The heat-exchanger effectiveness is determined as now described.

The actual heat-transfer rate Q through the heat exchanger is given by

$$Q = m_h c_{ph}(T_{h,\,\text{in}} - T_{h,\,\text{out}}) = m_c c_{pc}(T_{c,\,\text{out}} - T_{c,\,\text{in}}) \tag{15-23}$$

The substitution of Eqs. (15-21b) and (15-23) into Eq. (15-21a) yields

$$\varepsilon = \frac{C_h(T_{h,\,\text{in}} - T_{h,\,\text{out}})}{C_{\text{min}}(T_{h,\,\text{in}} - T_{c,\,\text{in}})} \tag{15-24a}$$

or

$$\varepsilon = \frac{C_c(T_{c,\,out} - T_{c,\,in})}{C_{min}(T_{h,\,in} - T_{c,\,in})} \tag{15-24b}$$

where we defined

$$C_h \equiv m_h c_{ph} \qquad C_c \equiv m_c c_{pc} \tag{15-25}$$

and $\quad C_{min} \equiv$ smaller of C_h and C_c

Equations (15-24) are now utilized to derive the desired expression for ε in terms of the heat-capacity rates C_c, C_h, the overall heat-transfer coefficient U_m, and the total heat-transfer area A of the heat exchanger. The analysis depends upon the flow arrangement, that is, whether it is a parallel-flow or a counterflow arrangement. To illustrate the basic approach in the analysis, we present a derivation of the heat-exchanger effectiveness ε with particular emphasis to the *parallel-flow arrangement* shown in Fig. 15-11.

We exponentiate Eq. (15-13) and obtain

$$\frac{\Delta T_L}{\Delta T_0} = e^{-MAU_m} \tag{15-26}$$

For the parallel-flow arrangement shown in Fig. 15-11, we have $\Delta T_L = T_{h,\,out} - T_{c,\,out}$ and $\Delta T_0 = T_{h,\,in} - T_{c,\,in}$. Then Eq. (15-26) becomes

$$\frac{T_{h,\,out} - T_{c,\,out}}{T_{h,\,in} - T_{c,\,in}} = e^{-MAU_m} \tag{15-27}$$

Equation (15-23) is solved for $T_{h,\,out}$:

$$T_{h,\,out} = T_{h,\,in} - \frac{C_c}{C_h}(T_{c,\,out} - T_{c,\,in}) \tag{15-28}$$

This result is substituted into Eq. (15-27) and $T_{h,\,out}$ is eliminated:

$$1 - \frac{T_{c,\,out} - T_{c,\,in}}{T_{h,\,in} - T_{c,\,in}}\left(1 + \frac{C_c}{C_h}\right) = e^{-MAU_m} \tag{15-29}$$

The temperature ratio in this equation is eliminated by means of Eq. (15-24b):

$$1 - \varepsilon\left(\frac{C_{min}}{C_c} + \frac{C_{min}}{C_h}\right) = e^{-MAU_m}$$

and solving for ε we obtain the relation for the *effectiveness of a parallel-flow heat exchanger* as

$$\varepsilon = \frac{1 - e^{-MAU_m}}{C_{min}/C_c + C_{min}/C_h} \qquad (15\text{-}30a)$$

where M is defined by Eq. (15-7b), i.e.,

$$M = \frac{1}{C_h} + \frac{1}{C_c} \qquad (15\text{-}30b)$$

Depending on whether C_{min} occurs on the hot- or cold-fluid side, Eq. (15-30a) is simplified as follows:

1   If C_{min} occurs on the cold-fluid side, we set $C_{min} = C_c$, and Eq. (15-30a) becomes

$$\varepsilon = \frac{1 - e^{-MAU_m}}{1 + C_c/C_h} \qquad (15\text{-}31a)$$

2   If C_{min} occurs on the hot-fluid side, we set $C_{min} = C_h$, and Eq. (15-30a) becomes

$$\varepsilon = \frac{1 - e^{-MAU_m}}{1 + C_h/C_c} \qquad (15\text{-}31b)$$

An examination of Eqs. (15-31) reveals that these results can be written more compactly in the form

$$\varepsilon = \frac{1 - e^{-MAU_m}}{1 + C_{min}/C_{max}} \qquad (15\text{-}31b)$$

where C_{min} and C_{max} are, respectively, the smaller and the larger of the two quantities C_h and C_c. The effectiveness given by Eq. (15-32) is a function of the parameters A, U_m, C_h, and C_c, but it does not depend on the LMTD.

The effectiveness for a counterflow heat exchanger can be derived in a similar manner. Kays and London [3] presented *effectiveness charts* for various types of heat exchangers. Figures 15-14 to 15-18 show the effectiveness charts for typical flow arrangements of practical interest. In these figures ε is plotted against AU_m/C_{min} for several different values of the ratio C_{min}/C_{max}. The parameter AU_m/C_{min} is sometimes called the *number of heat-transfer units* or briefly NTU. It is to be noted that the case $C_{min}/C_{max} = 0$ refers to an evaporator or a condenser, because when a fluid remains at a constant temperature throughout the heat exchanger its heat-capacity rate is considered infinite, hence $C_{max} \to \infty$

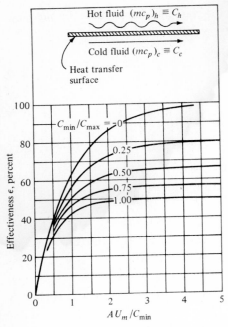

FIG. 15-14   Effectiveness for a parallel-flow heat exchanger. (*From Kays and London* [*3*].)

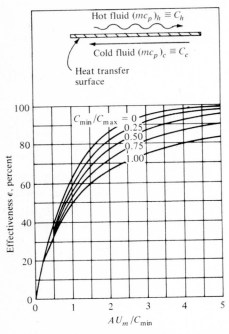

FIG. 15-15   Effectiveness for a counterflow heat exchanger. (*From Kays and London* [*3*].)

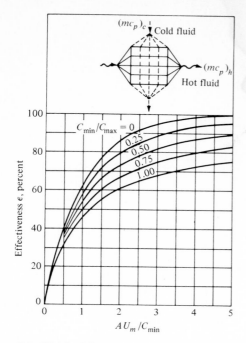

FIG. 15-16 Effectiveness for a cross-flow heat exchanger with fluids unmixed. (*From Kays and London* [*3*].)

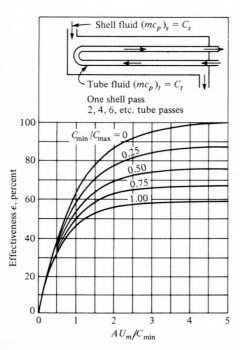

FIG. 15-17 Effectiveness for a single-shell pass heat exchanger with 2, 4, 6, etc. tube passes. (*From Kays and London* [*3*].)

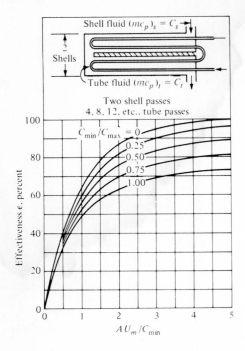

FIG. 15-18   Effectiveness of a two-shell pass heat exchanger with 4, 8, 12, etc. tube passes. (*From Kays and London* [3].)

and $C_{min}/C_{max} = 0$.   The reader should consult Ref. [3] for effectiveness charts for various other flow arrangements.

Example 15-3   A shell-and-tube type of heat exchanger is designed to cool 12,000 lb/h (1.512 kg/s) oil ($c_p = 0.5$ Btu/lb·°F or 2093 W·s/kg·°C) from 150°F (65.56°C) to 108°F (42.22°C) by using 8000 lb/h (1.008 kg/s) water ($c_p = 1.0$ Btu/lb·°F or 4187 W·s/kg·°C) at an inlet temperature of 80°F (26.67°C).   The average overall heat-transfer coefficient for the heat exchanger is estimated to be $U_m = 120$ Btu/h·ft²·°F (681.6 W/m²·°C).   By using the effectiveness method, determine the heat-transfer area required for the following flow arrangements: (*a*) a single-shell pass heat exchanger shown in Fig. 15-17, and (*b*) a two-shell pass heat exchanger shown in Fig. 15-18.

Solution   The values of mc_p for the hot and cold fluids, respectively, are

$$C_h = m_h c_{ph} = 12,000 \times 0.5 = 6000 \text{ Btu/h·°F (3165 W/°C)}$$

$$C_c = m_c c_{pc} = 8000 \times 1.0 = 8000 \text{ Btu/h·°F (4220 W/°C)}$$

Then

$$\frac{C_{min}}{C_{max}} = \frac{6000}{8000} = 0.75$$

The same results are obtainable by using the SI system of units, that is,

$$C_h = 1.512 \times 2093 = 3165 \text{ W/}^\circ\text{C}$$

$$C_c = 1.008 \times 4187 = 4220 \text{ W/}^\circ\text{C}$$

$$\frac{C_{\min}}{C_{\max}} = \frac{3165}{4220} = 0.75$$

The effectiveness ε is determined from Eq. (15-24a) by setting $C_{\min} = C_h$ since $C_h < C_c$. We obtain

$$\varepsilon = \frac{T_{h,\,\text{in}} - T_{h,\,\text{out}}}{T_{h,\,\text{in}} - T_{c,\,\text{in}}}$$

and substituting the temperatures as given above we find

$$\varepsilon = \frac{150 - 108}{150 - 80} = 0.6$$

Then, for the values of $\varepsilon = 0.6$ and $C_{\min}/C_{\max} = 0.75$, the total required heat-transfer area A for the two flow arrangements is determined as follows:

(a) For the single-shell pass arrangement, from Fig. 15-17 we find

$$\frac{AU_m}{C_{\min}} = 1.7$$

or

$$A = \frac{1.7 C_{\min}}{U_m} = \frac{1.7 \times 6000}{120} = 85 \text{ ft}^2 \ (7.9 \text{ m}^2)$$

(b) For the two-shell pass arrangement from Fig. 15-18 we obtain

$$\frac{AU_m}{C_{\min}} = 1.35$$

or

$$A = \frac{1.35 C_{\min}}{U_m} = \frac{1.35 \times 6000}{120} = 67.5 \text{ ft}^2 \ (6.27 \text{ m}^2)$$

15-5 FOULING AND SCALING OF HEAT EXCHANGERS

During operation heat exchangers tend to become fouled with deposits of one kind or another [42–46], and the deposits act as an additional thermal resistance to heat transfer. The thermal resistance due to fouling is generally referred to as the *fouling factor* or *unit fouling resistance* ($\text{h} \cdot \text{ft}^2 \cdot {}^\circ\text{F/Btu}$ or $\text{m}^2 \cdot {}^\circ\text{C/W}$), and its magnitude depends on the nature of the fouling. Table 15-1 lists typical magnitudes of *unit fouling resistances* resulting from the flow of water, oil, and condensation of vapors. It is apparent from this table that the fouling resulting from the flow of water depends on the type, temperature, and velocity of water.

TABLE 15-1   Unit Fouling Resistance F for Heat-Transfer Equipment†

Types of Water	Water Temperature 125 F (51.7 C) or less			
	Water Velocity 3 ft/s (1 m/s) and less		Water Velocity over 3 ft/s (1 m/s)	
	$h \cdot ft^2 \cdot F/Btu$	$m^2 \cdot C/W$	$h \cdot ft^2 \cdot F/Btu$	$m^2 \cdot C/W$
Seawater	0.0005	0.000088	0.0005	0.000088
Distilled	0.0005	0.000088	0.0005	0.000088
Treated boiler feedwater	0.001	0.00018	0.0005	0.000088
Engine jacket	0.001	0.00018	0.001	0.00018
Great Lakes	0.001	0.00018	0.001	0.00018
Cooling tower and spray pond:				
Treated makeup	0.001	0.00018	0.001	0.00018
Untreated	0.003	0.00053	0.003	0.00053
Boiler blowdown	0.002	0.00035	0.002	0.00035
Brackish water	0.002	0.00035	0.001	0.00018
River water:				
Minimum	0.002	0.00036	0.001	0.00018
Mississippi	0.003	0.00053	0.002	0.00035
Delaware, Schuylkill	0.003	0.00053	0.002	0.00035
East River and New York Bay	0.003	0.00053	0.002	0.00035
Chicago sanitary canal	0.008	0.00141	0.006	0.00106
Muddy or silty	0.003	0.00053	0.002	0.00035
Hard (over 15 grains/gal)	0.003	0.00053	0.003	0.00053

Types of Fluid	$h \cdot ft^2 \cdot F/Btu$	$m^2 \cdot C/W$
Industrial oils:		
Clean recirculating oil	0.001	0.00018
Machinery and transformer oils	0.001	0.00018
Vegetable oils	0.003	0.00053
Quenching oil	0.004	0.00070
Fuel oil	0.005	0.00088
Industrial gases and vapors:		
Organic vapors	0.0005	0.000088
Steam (non-oil bearing)	0.0005	0.000088
Alcohol vapors	0.0005	0.000088
Steam, exhaust	0.001	0.00018
Refrigerating vapors	0.002	0.00035
Air	0.002	0.00035
Industrial liquids:		
Organic	0.001	0.00018
Refrigerating liquids	0.001	0.00018
Brine (cooling)	0.001	0.00018

† Source: Tubular Exchanger Manufacturer Association [9].

To illustrate the use of unit fouling resistance in heat-transfer calculations, we formulate the overall heat-transfer coefficient for a circular tube with flow and fouling at both the inside and outside surfaces. The *total thermal resistance R* to heat flow across the tube from the inside to the outside flow is composed of the sum of the individual thermal resistances of various layers as

$$
R = \begin{pmatrix} \text{thermal} \\ \text{resistance} \\ \text{of inside} \\ \text{flow} \end{pmatrix} + \begin{pmatrix} \text{thermal} \\ \text{resistance} \\ \text{of fouling} \\ \text{at inside} \\ \text{surface} \end{pmatrix} + \begin{pmatrix} \text{thermal} \\ \text{resistance} \\ \text{of tube} \\ \text{material} \end{pmatrix} + \begin{pmatrix} \text{thermal} \\ \text{resistance} \\ \text{of fouling} \\ \text{at outside} \\ \text{surface} \end{pmatrix} + \begin{pmatrix} \text{thermal} \\ \text{resistance} \\ \text{of outside} \\ \text{flow} \end{pmatrix}
$$

$$(15\text{-}33a)$$

and various terms are given by

$$
R = \frac{1}{A_i h_i} + \frac{F_i}{A_i} + \frac{t}{k A_m} + \frac{F_o}{A_o} + \frac{1}{A_o h_o} \tag{15-33b}
$$

where A_o, A_i = outside and inside surface area of tube respectively, ft^2 (m^2)

$$
A_m = \frac{A_o - A_i}{\ln (A_o/A_i)} = \text{logarithmic mean area, ft}^2 \text{ (m}^2)
$$

F_i, F_o = unit fouling resistance at inside and outside surface of tube, respectively, h·ft^2·°F/Btu (m^2·°C/W)

h_i, h_o = heat-transfer coefficients for inside and outside flow, respectively, Btu/h·ft^2·°F (W/m^2·°C)

k = thermal conductivity of tube material, Btu/h·ft·°F (W/m·°C)

R = total thermal resistance from inside to outside flow, h·°F/Btu (°C/W)

t = thickness of tube, ft (m)

The *overall heat-transfer coefficient* U_o based on the outside surface of the tube is related to R by

$$
R = \frac{1}{A_o U_o} \qquad \text{or} \qquad U_o = \frac{1}{A_o R} \tag{15-34}
$$

The substitution of Eq. (15-33b) into Eq. (15-34) gives U_o as

$$
U_o = \frac{1}{\dfrac{A_o}{A_i}\dfrac{1}{h_i} + \left(\dfrac{A_o}{A_i}\right) F_i + \left(\dfrac{A_o}{A_m}\right)\dfrac{t}{k} + F_o + \dfrac{1}{h_o}} \qquad \text{Btu/h·ft}^2\text{·°F (W/m}^2\text{·°C)} \tag{15-35}
$$

The area ratios can be related to the diameter ratios as

$$\frac{A_o}{A_i} = \frac{D_o}{D_i} \tag{15-36a}$$

$$\frac{A_o}{A_m} = \frac{\pi D_o}{\pi(D_o - D_i)/\ln(D_o/D_i)} = \frac{D_o}{2t}\ln\frac{D_o}{D_i} \quad \text{where } D_o - D_i = 2t \tag{15-36b}$$

Substituting Eqs. (15-36) into Eq. (15-35) we obtain

$$U_o = \frac{1}{\dfrac{D_o}{D_i}\dfrac{1}{h_i} + \dfrac{D_o}{D_i}F_i + \dfrac{D_o}{2k}\ln\dfrac{D_o}{D_i} + F_o + \dfrac{1}{h_o}} \quad \text{Btu/h·ft}^2\cdot{}^\circ\text{F } (\text{W/m}^2\cdot{}^\circ\text{C}) \tag{15-37}$$

where D_o and D_i are the outside and inside diameter of the tube, respectively.

Example 15-4   Determine the overall heat-transfer coefficient U_o based on the outer surface of a 1.5-in (3.81×10^{-2}-m) OD, 1.25-in (3.175×10^{-2}-m) ID brass tube ($k = 60$ Btu/h·ft·°F or 103.8 W/m·°C) for the following flow and fouling conditions.   The heat-transfer coefficients for the inside and outside flow are, respectively, $h_i = 400$ Btu/h·ft^2·°F (2272 W/m^2·°C) and $h_o = 500$ Btu/h·ft^2·°F (2840 W/m^2·°C) and the unit fouling resistances at inside and outside surfaces are $F_i = F_o = 0.005$ h·ft^2·°F/Btu (8.8×10^{-3} m^2·°C/W).

Solution   Substituting these numerical values into Eq. (15-37) and using the English system of units, the overall heat-transfer coefficient based on the outside surface of the tube becomes

$$U_o = \frac{1}{\dfrac{1.5}{1.25}\dfrac{1}{400} + \dfrac{1.5}{1.25}0.005 + \dfrac{1.5}{2 \times 12 \times 60}\ln\dfrac{1.5}{1.25} + 0.005 + \dfrac{1}{500}}$$

$$= \frac{1}{(3 \times 10^{-3}) + (6 \times 10^{-3}) + (0.19 \times 10^{-3}) + (5 \times 10^{-3}) + (2 \times 10^{-3})} = \frac{10^3}{16.19}$$

$$= 61.8 \text{ Btu/h·ft}^2\cdot{}^\circ\text{F } (3510 \text{ W/m}^2\cdot{}^\circ\text{C})$$

REFERENCES

1   Afgan, N. H., and E. U. Schlünder: "Heat Exchangers: Design and Theory," McGraw-Hill Book Company, New York, 1974.
2   Fraas, A. P., and M. N. Özışık: "Heat Exchanger Design," John Wiley & Sons, Inc., New York, 1965.
3   Kays, W. M., and A. L. London: "Compact Heat Exchangers," 2d ed., McGraw-Hill Book Company, New York, 1964.

4 "Heat Exchangers," The Patterson-Kelley Co., East Stroudsburg, Pa., 1960.
5 Chilton, C. H. (ed.): "Cost Engineering in the Process Industries," McGraw-Hill Book Company, New York, 1960.
6 Hyrnisak, W.: "Heat Exchangers," Academic Press, Inc., New York, 1958.
7 Kern, D. Q.: "Process Heat Transfer," McGraw-Hill Book Company, New York, 1950.
8 Sim, J.: "Steam Condensing Plant—in Theory and Practice," Blackie and Son, Ltd., Glasgow, 1926.
9 Tubular Exchanger Manufacturers Association, "Standards," TEMA, New York, 1959.
10 Berman, L. D.: "Evaporative Cooling of Circulating Water," Pergamon Press, New York, 1961.
11 Fraas, A. P., and M. N. Özışık: "Steam Generators for High Temperature Gas-cooled Reactors," ORNL-3208, Oak Ridge National Laboratory, Oak Ridge, Tenn., April 1963.
12 Shields, C. D.: "Boilers: Types, Characteristics, and Functions," McGraw-Hill Informations Systems Company, McGraw-Hill, Inc., New York, 1961.
13 Fax, D. H., and R. R. Mills, Jr.: General Optimal Heat Exchanger Design, *Trans. ASME*, **79**:653–661 (1957).
14 Colburn, A. P.: Problems in Design and Research on Condensers of Vapors and Vapor Mixtures, *Inst. Mech. Eng.–ASME, Proc. General Discussion Heat Transfer*, pp. 1–11, 1951.
15 Ten Broeck, H.: Multipass Heat Exchanger Calculations, *Ind. Eng. Chem.*, **30**:1041–1042 (1938).
16 Gardner, H. S., and I. Siller: Shell-Side Coefficients of Heat Transfer in a Baffled Heat Exchanger, *Trans. ASME*, **69**:687–694 (1947).
17 Bergelin, O. P., G. A. Brown, H. L. Hull, and F. W. Sullivan: Heat Transfer and Fluid Friction During Viscous Flow Across Banks of Tubes, III. A Study of Tube Spacing and Tube Size, *Trans. ASME*, **72**:881–888 (1950).
18 Tinker, Townsend: Shell-Side Characteristics of Shell and Tube Heat Exchangers, *Inst. Mech. Eng.–ASME, Proc. General Discussion Heat Transfer*, pp. 89–116, 1951.
19 Bergelin, O. P., G. A. Brown, and S. C. Doberstein: Heat Transfer and Fluid Friction During Flow Across Banks of Tubes, IV, A Study of Transition Zone Between Viscous and Turbulent Flow, *Trans. ASME*, **74**:953–960 (1952).
20 Muller, A. C.: Thermal Design of Shell-and-Tube Heat Exchangers for Liquid-to-Liquid Heat Transfer, *Purdue Univ., Eng. Exp. Sta., Eng. Bull., Res. Ser.*, 121, 1954.
21 Tinker, T.: Shell-Side Characteristics of Shell-and-Tube Heat Exchangers, *Trans. ASME*, **80**:36–52 (1958).
22 London, A. L., J. W. Mitchell, and W. A. Sutherland: Heat Transfer and Flow Friction Characteristics of Crossed Rod Matrices, *J. Heat Transfer*, **82C**:199–213 (1960).
23 Fraas, A. P.: Design Precepts for High Temperature Heat Exchangers, *Nucl. Sci. Eng.*, **8**:21–31 (1960).
24 Morton, D. S.: Thermal Design of Heat Exchangers, *Ind. Eng. Chem.*, **52**:474–478 (1960).
25 Bayley, F. J., and C. W. Rapley: Heat Transfer and Pressure Loss Characteristics of Matrices for Regenerative Heat Exchangers, *ASME Paper 65-HT-35*, 1965.
26 Nance, G. R.: Fundamental Relationships in the Design of Cooling Towers, *Trans. ASME*, **61**:721–725 (1939).
27 London, A. L., W. E. Mason, and L. M. K. Boelter: Performance Characteristics of

Mechanically Induced Draft Counterflow Packed Cooling Towers, *Trans. ASME*, **62**:41–50 (1940).

28 Lichtenstein, J.: Performance and Selection of Mechanical Draft Cooling Towers, *Trans. ASME*, **65**:779–787 (1943).

29 Kelly, N. W., and L. K. Swenson: Comparative Performance of Cooling Tower Packing Arrangements, *Chem. Eng. Prog.*, **52**:263–268 (1956).

30 Moore, F. K.: On the Minimum Size of Large Dry Cooling Towers with Combined Mechanical and Natural Draft, *J. Heat Transfer*, **95C**:383–389 (1973).

31 Williams, R. B., and D. L. Katz: Performance of Finned Tubes in Shell-and-Tube Heat Exchangers, *Trans. ASME*, **74**:1307–1320 (1952).

32 Hausen, H.: On the Theory of Heat Exchange in Regenerators, *Z. Angew. Math. Mech.*, **9**:193–200 (1929).

33 Karlsson, H., and S. Holm: Heat Transfer and Fluid Resistance in Ljungstrom-Type Air Preheaters, *Trans. ASME*, **65**:61–72 (1943).

34 Coppage, J. E., and A. L. London: The Periodic Flow Regenerator—A Summary of Design Theory, *Trans. ASME*, **75**:779–787 (1953).

35 Harper, D. B., and W. M. Rohsenow: Effect of Rotary Regenerator Performance on Gas-Turbine-Plant Performance, *Trans. ASME*, **75**:759–765 (1953).

36 Lambertson, T. J.: Performance Factors of a Periodic Flow Heat Exchanger, *Trans. ASME*, **80**:586–592 (1953).

37 Harper, D. B.: Seal Leakage in the Rotary Regenerator and Its Effect on Rotary Regenerator Design for Gas Turbines, *Trans. ASME*, **79**:233–245 (1957).

38 Jakob, M.: "Heat Transfer," vol. 2, John Wiley & Sons, Inc., New York, 1957.

39 Bowman, R. A., A. C. Mueller, and W. M. Nagle: Mean Temperature Difference in Design, *Trans. ASME*, **62**:283–294 (1940).

40 Gardner, K. A.: Variable Heat Transfer Rate Correction in Multipass Exchangers, Shell-Side Film Controlling, *Trans. ASME*, **67**:31–38 (1945).

41 Stevens, R. A., J. Fernandes, and J. R. Woolf: Mean Temperature Difference in One, Two and Three-Pass Cross Flow Heat Exchangers, *Trans. ASME*, **79**:287–297 (1957).

42 Butler, R. C., and W. N. McCurdy, Jr: Fouling Rates and Cleaning Methods in Refinery Heat Exchangers, *Trans. ASME*, **71**:843–847 (1949).

43 Bethon, H. E.: Fouling of Marine-Type Heat Exchangers, *Trans. ASME*, **71**:855–869 (1949).

44 Bengelin, O. P.: The Fouling and Cleaning of Surfaces of Unfired Heat Exchangers–Panel Discussion, *Trans. ASME*, **71**:871–883 (1949).

45 Taborek, J., T. Aoki, R. B. Ritter, J. W. Palen, and J. G. Knudsen: Fouling: Major Unresolved Problem in Heat Transfer, Pts. I and II, *Chem. Eng. Prog.*, **88** (2): 59–67; (7): 69–78 (1972).

Sixteen

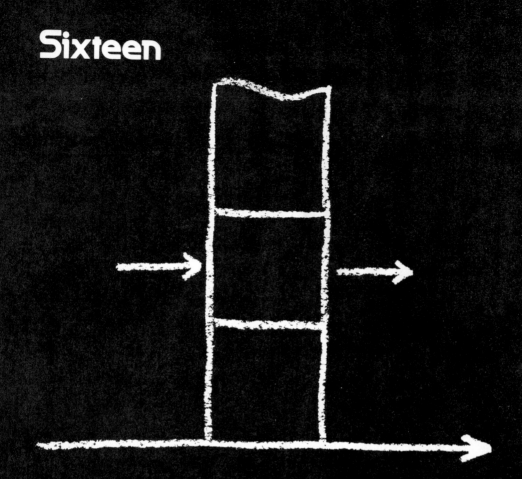

Mass Transfer

Mass-transfer processes occur in a variety of applications in the fields of mechanical, chemical, and aerospace engineering, physics, chemistry, and biology. Typical examples include the transpiration cooling of jet engines and rocket motors, the ablative cooling of space vehicles during reentry into the atmosphere, the mass transfer from laminar and turbulent streams onto the surfaces of a conduit, and evaporation or condensation on the surface of a tube or plate. Processes such as absorption, desorption, distillation, solvent extraction, drying, humidification, sublimation, and many others involve mass transfer. In absorption, a gas is brought into direct contact with a liquid solvent in order to remove the soluble components of the gas. The reverse process occurs in desorption; that is, the transfer of solute takes place from the liquid to the gas. In solvent extraction, one or more components of a liquid mixture are extracted by solution in a selective solvent. In humidification, water is transferred from the liquid to the air. The biological applications include oxygenation of blood, food and drug assimilation, respiration mechanism, and numerous others. The theory and application of mass transfer is such a widespread field that numerous books have been written on the subject [1-16]. Here, in the one chapter devoted to mass transfer we present only an introductory treatment of this subject, with emphasis on the similarity between heat- and mass-transfer processes. When mass transfer takes place in a fluid at rest, the mass is transferred by purely molecular diffusion resulting from concentration gradients; the process is analogous to heat diffusion resulting from temperature gradients. When the fluid is in motion, mass transfer takes place by both molecular diffusion and convective motion of the bulk fluid; then a knowledge of the velocity field is needed to solve the mass-transfer problem. For low concentrations of the mass in the fluid and low mass-transfer rates, the convective heat- and mass-transfer processes are analogous, and many of the results derived in connection with convective heat transfer are applicable to convective mass transfer. Therefore, the mass-transfer equations and the mass-transfer coefficients presented in this chapter are obtained by analogy directly from the corresponding heat-transfer equations. However, under high-mass-flux conditions and with chemical reactions there are significant differences between heat- and mass-transfer processes; such situations are not considered here. The reader should consult Refs. [7,8] for a discussion of mass transfer in chemically reacting systems and Ref. [1] for mass transfer at high mass flux and high concentrations.

16-1   DEFINITIONS OF FLUX

We consider a fluid mixture of two components, say A and B, the composition of which is characterized by the *molal concentration* of the components. The molal concentration c_A of component A is defined as the number of molecules of component A per unit volume of the mixture and may be given in the units lb mol/ft^3 or kg mol/m^3, and so on. Various other definitions are also in use in the literature for expressing the composition. For example, the *mole fraction* χ_A of component A is defined as $\chi_A = c_A/c$, where c is the total molal concentration of the mixture. The *mass concentration* ρ_A of component A is the mass of component A per unit volume and may be given in the units lb/ft^3 or kg/m^3. The *mass fraction* w_A of component A is defined as $w_A = \rho_A/\rho$, where ρ is the total mass density of the mixture.

Various Velocities in the Mixture

Consider a two-component mixture whose concentration varies in the x direction and the fluid undergoes a bulk motion in the same direction. Let u_A and u_B be the *statistical mean velocities* of components A and B, respectively, in the x direction with respect to the stationary coordinate. The *molal average velocity* U of the mixture in the x direction is defined by

$$U = \frac{1}{c}(u_A c_A + u_B c_B) \tag{16-1}$$

where c_A and c_B are the molal concentrations of species A and B, respectively, and c is the *total molal concentration of the mixture*.

The *diffusion velocities* of species A and B with respect to the molal average velocity U are defined as

$$u_A - U = \text{diffusion velocity of species A}$$
$$u_B - U = \text{diffusion velocity of species B} \tag{16-2}$$

Thus, the diffusion velocity indicates the motion of a species relative to the local average motion of the mixture.

Various Fluxes in the Mixture

The *molal fluxes N_A and N_B of species A and B, respectively, relative to stationary coordinates* in the x direction are given by

$$N_A = c_A u_A \quad \text{and} \quad N_B = c_B u_B \tag{16-3}$$

That is, N_A and N_B characterize the moles of species A and B, respectively, that pass through a unit area perpendicular to the x axis per unit time.

The *molal fluxes J_A and J_B of species A and B, respectively, relative to the molal average velocity U* are defined as

$$J_A = c_A(u_A - U) \quad \text{and} \quad J_B = c_B(u_B - U) \tag{16-4}$$

That is, the molal fluxes J_A and J_B are the measure of the diffusion rates of species A and B, respectively, in the mixture.

Relationship Among Various Fluxes

The relationship among various molal fluxes is now derived by combining the above results.

We substitute U from Eq. (16-1) into Eqs. (16-4) to obtain

$$J_A = c_A u_A - \frac{c_A}{c} (u_A c_A + u_A c_B) \tag{16-5a}$$

$$J_B = c_B u_B - \frac{c_B}{c} (u_A c_A + u_B c_B) \tag{16-5b}$$

When the definition of N_A and N_B as given by Eqs. (16-3) is introduced into Eqs. (16-5), we obtain the relations among various fluxes:

$$J_A = N_A - \frac{c_A}{c} (N_A + N_B) \tag{16-6a}$$

$$J_B = N_B - \frac{c_B}{c} (N_A + N_B) \tag{16-6b}$$

These results show that the *molal diffusion flux* J_i of species i is equal to the difference between the molal flux N_i and the bulk flow in the mixture of species i.

It is apparent from Eqs. (16-6) that the sum of J_A and J_B is zero, that is,

$$J_A + J_B = 0 \quad \text{or} \quad J_A = -J_B \tag{16-7}$$

since $c = c_A + c_B$. This result implies that in a binary mixture the diffusion fluxes J_A and J_B of the two components are of equal magnitude and in opposite directions.

Fick's First Law

In a binary mixture in which composition varies in the x direction, and molecular diffusion occurs within the fluid due to the nonuniformity of composition, the molal fluxes J_A and J_B in the x direction of species A and B are related to the concentration gradients by *Fick's first law*:

$$J_A = -D_{AB} \frac{dc_A}{dx} \tag{16-8a}$$

$$J_B = -D_{BA} \frac{dc_B}{dx} \tag{16-8b}$$

where D_{AB} is the *mass diffusivity* (or *the diffusion coefficient*) of A in B and D_{BA} is the *mass diffusivity* of B in A; they are equal to each other, i.e.,

$$D_{AB} = D_{BA} \equiv D \tag{16-9}$$

Because mass diffusion takes place in the direction of decreasing concentration, a minus sign is included in Eqs. (16-8) to make the mass flux in the positive x direction a positive quantity when the concentration decreases in the positive x direction.   Thus, *when J_A is positive, the mass flux of species A is in the positive x direction, and vice versa.*   To give some idea on the units of various quantities in Eqs. (16-8), we list below the units:

c_i = concentration of component i in mixture, i = A or B, lb mol/ft^3 (kg mol/m^3)

D = mass diffusivity, or diffusion coefficient, ft^2/h (m^2/s)

J_i = molal flux of component i in x direction, i = A or B, lb mol/ft$^2 \cdot$h (kg mol/m$^2 \cdot$s)

x = distance in x direction, ft (m)

We note that the mass-flux relation given above by Fick's first law is similar to the heat-flux expression given by the Fourier law as

$$q = -k \frac{dT}{dx} = -\alpha \frac{d}{dx} (\rho c_p T) \tag{16-10}$$

and to the momentum-flux expression given by

$$\tau g = -\mu \frac{du}{dy} = -v \frac{d(\rho u)}{dy} \tag{16-11}$$

Clearly, the mass diffusivity D, the heat diffusivity α, and the momentum diffusivity v have the same units, i.e., ft^2/h or m^2/s, and Eqs. (16-8), (16-10), and (16-11) are of the same form.

If the mixture is considered to be a perfect gas, the molal concentrations c_A and c_B are related to the partial pressures p_A and p_B of the species A and B in the mixture by

$$p_i = c_i \mathscr{R} T \qquad i = \text{A or B} \tag{16-12}$$

where c_i = molal concentration of component i in mixture, lb mol/ft^3 (kg mol/m^3)

p_i = partial pressure of component i in mixture, atm

$\mathscr{R}$ = gas constant = 0.730 ft$^3 \cdot$atm/lb mol$\cdot$°R = 0.08205 m$^3 \cdot$atm/kg mol$\cdot$K

Then, Eqs. (16-8) can be written as

$$J_i = -\frac{D}{\mathscr{R}T}\frac{dp_i}{dx} \qquad i = \text{A or B} \qquad \text{lb mol/ft}^2\cdot\text{h (kg mol/m}^2\cdot\text{s)} \tag{16-13}$$

Various relations given in this section for the definition of mass fluxes will be applied in the next sections in the analysis of mass-diffusion problems for binary mixtures.

16-2   STEADY-STATE EQUIMOLAL COUNTERDIFFUSION IN GASES

Consideration is now given to the application of the foregoing relations for flux in the prediction of concentration distribution for the *steady-state equimolal counterdiffusion* in a binary gas mixture composed of components A and B.   In this mass-transfer process the gases A and B diffuse simultaneously in the opposite directions through each other.   That is, component A diffuses through component B, and vice versa, and they diffuse at the same molal rate but in opposite directions.   This process is approximated in the distillation of a binary system.

Consider that two large vessels containing uniform mixtures of A and B at different concentrations are suddenly connected by a small pipe.   It is assumed that both vessels are at the same total pressure P and uniform temperature T. Component A will diffuse from the higher concentration to the lower concentration, and component B will diffuse at the same rate but in the opposite direction through the connecting pipe.   Since the vessels are sufficiently large, steady-state equimolal counterdiffusion takes place in the connecting pipe; that is, *the total molal flux with respect to stationary coordinates is zero* and we have

$$N_A + N_B = 0 \qquad \text{or} \qquad N_A = -N_B \tag{16-14}$$

For this type of mass-diffusion process the molal fluxes of species A and B relative to stationary coordinates are equal and in the opposite directions.   The substitution of Eqs. (16-14) into Eqs. (16-6) yields

$$J_A = N_A = -N_B = -J_B \tag{16-15}$$

Substitution of this result into Eq. (16-13) for $i = A$ gives

$$N_A = -\frac{D}{\mathscr{R}T}\frac{dp_A}{dx} \tag{16-16}$$

At steady state N_A and N_B are constant.   Then, Eq. (16-16) for constant D implies that the distribution of the partial pressure p_A of component A along the connecting pipe is linear with the distance.   A similar conclusion can be drawn for the distribution of p_B either by writing an equation for N_B analogous to Eq. (16-15) or by the fact that the sum of the partial pressures p_A and p_B is equal to the total pressure p which remains constant, i.e.,

$$p_A + p_B = p = \text{const} \tag{16-17}$$

Let p_{A1} and p_{A2} be the partial pressures of component A at the two ends of the connecting pipe, $x = x_1$ and $x = x_2$, respectively.   The integration of Eq. (16-16) from $x = x_1$ to $x = x_2$ gives

$$N_A = -\frac{D}{\mathscr{R}T}\frac{p_{A2} - p_{A1}}{x_2 - x_1} \qquad \text{lb mol/ft}^2\cdot\text{h (kg mol/m}^2\cdot\text{s)} \tag{16-18}$$

A similar relation can be written for N_B.   If it is assumed that $p_{A1} > p_{A2}$, then for the component B we must have $p_{B2} > p_{B1}$, where p_{B1} and p_{B2} are the partial pressures of component B at $x = x_1$ and $x = x_2$, respectively.   Figure 16-1 shows schematically the distribution of partial pressures of the two components as a function of the distance for the case $p_{A1} > p_{A2}$.   Clearly, component A diffuses in the direction from x_1 to x_2, and component B in the opposite direction.

Example 16-1   Consider two large vessels each containing uniform mixtures of nitrogen (i.e., component A) and carbon dioxide (i.e., component B) at 1-atm pressure, $T = 520°R$ but at different concentrations.   Vessel 1 contains 90 mol percent N_2 and 10 mol percent CO_2, whereas vessel 2 contains 20 mol percent N_2 and 80 mol percent CO_2.   The two vessels are connected by a duct of $d = 6$ in (0.1524 m) inside diameter and $L = 4$ ft (1.22 m) long.   Determine the rate of transfer of nitrogen between the two vessels by assuming that steady-state transfer takes place in view of the large capacity of the two

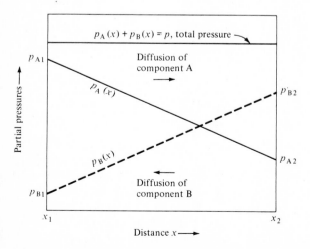

FIG. 16-1   Distribution of partial pressures p_A and p_B in equimolal counterdiffusion of a binary gas mixture.

reservoirs.   The mass diffusivity for the $N_2 - CO_2$ mixture at 1-atm pressure and $520°R$ can be taken as $D = 0.62$ ft^2/h $(0.16 \times 10^{-4}$ $m^2/s)$.

Solution   In this mass-transfer process, nitrogen (component A) is transferred from vessel 1 containing a higher concentration of nitrogen to vessel 2 containing a lower concentration of nitrogen.   In the early stages of the mass transfer, the partial pressure of nitrogen in both vessels is considered to remain constant, hence a steady-state transfer can be assumed.   Then, the mass-transfer process can be characterized as an *equimolal counterdiffusion* as described above, and the mass flux of nitrogen through the connecting duct can be determined by using Eq. (16-18):

$$N_A = \frac{D}{\mathscr{R}T} \frac{p_{A1} - p_{A2}}{x_2 - x_1} \qquad \text{lb mol/ft}^2 \cdot \text{h (kg mol/m}^2 \cdot \text{s)}$$

The total mass-transfer rate Q_A of nitrogen is given by

$$Q_A = \text{area} \times N_A = \left(\frac{\pi}{4} d^2\right) \frac{D}{\mathscr{R}T} \frac{p_{A2} - p_{A1}}{x_2 - x_1} \qquad \text{lb mol/h (kg mol/h)}$$

The numerical values of various quantities in this equation are

$d = \frac{1}{2}$ ft $= 0.1524$ m $D = 0.62$ $ft^2/h = 0.16 \times 10^{-4}$ m^2/s

$T = 520°R = 288.9$ K $x_2 - x_1 = L = 4$ ft $= 1.22$ m

$\mathscr{R} = 0.730$ $ft^2 \cdot atm/lb$ $mol \cdot °R = 0.08205$ $m^3 \cdot atm/kg$ $mol \cdot K$

$p_{A1} = 0.9 \times 1$ atm $= 0.9$ atm $p_{A2} = 0.2 \times 1$ atm $= 0.2$ atm

Then, the mass-transfer rate Q_A is determined by using the English system of units as

$$Q_A = \left(\frac{\pi}{4}\frac{1}{4}\right) \frac{0.62}{0.730 \times 520} \frac{0.9 - 0.2}{4} = 5.6 \times 10^{-5} \text{ lb mol/h } (0.71 \times 10^{-8} \text{ kg mol/s})$$

or, by using the SI system of units, as

$$Q_A = \frac{\pi}{4} \times 0.1524^2 \frac{0.16 \times 10^{-4}}{0.08205 \times 288.9} \frac{0.9 - 0.2}{1.22} = 0.71 \times 10^{-8} \text{ kg mol/s}$$

16-3   STEADY-STATE UNIDIRECTIONAL DIFFUSION IN GASES

An application of various flux relations discussed previously is now given in relation to the *steady-state unidirectional diffusion* in a binary gas mixture of components A and B.   In this process one of the components, say A, diffuses through component B which remains motionless relative to the stationary coordinates, i.e., the molal flux N_B of component B relative to the stationary coordinates is zero. The situation is better envisioned if we consider a medium containing gas components A and B, in which component A is supplied at $x = x_1$ at a steady rate and diffuses in the x direction through gas B to an interface at $x = x_2$ where gas A is absorbed but B is not. This type of diffusion process is approximated in gas absorption, desorption, and adsorption. The determination of the distribution of concentrations of gases A and B in the medium is of interest.

Since in this mass-transfer process the net molal flux of component B relative to the stationary coordinates is zero, we set

$$N_B = 0 \qquad (16\text{-}19)$$

and the molal diffusion flux J_A of component A due to molecular diffusion is given by

$$J_A = -D \frac{dc_A}{dx} \qquad (16\text{-}20)$$

where D is the mass diffusivity of A in B. The mass flux J_A is now obtained from Eq. (16-6a) as

$$J_A = N_A - \frac{c_A}{c}(N_A + N_B) \qquad (16\text{-}21)$$

When the results in Eqs. (16-19) and (16-20) are substituted into Eq. (16-21) we obtain

$$-D \frac{dc_A}{dx} = N_A - \frac{c_A}{c} N_A \qquad (16\text{-}22)$$

Assuming perfect gas, we replace $c_A = p_A/\mathscr{R}T$, and Eq. (16-22) becomes

$$-\frac{D}{\mathscr{R}T} \frac{dp_A}{dx} = N_A\left(1 - \frac{p_A}{p}\right)$$

or

$$N_A\, dx = -\frac{Dp}{\mathscr{R}T} \frac{dp_A}{p - p_A} \qquad (16\text{-}23)$$

where p is the total pressure of the mixture which is assumed to be constant and p_A is the partial pressure of A in the mixture; then we have

$$p_A + p_B = p = \text{const} \qquad (16\text{-}24)$$

The integration of Eq. (16-23) for constant D, p, T, N_A with x varying from x_1 to x_2 and p_A varying from p_{A1} to p_{A2} gives

$$N_A \int_{x_1}^{x_2} dx = -\frac{Dp}{\mathscr{R}T} \int_{p_{A1}}^{p_{A2}} \frac{dp_A}{p - p_A}$$

or

$$N_A = \frac{Dp}{\mathscr{R}T(x_2 - x_1)} \ln \frac{p - p_{A2}}{p - p_{A1}} = \frac{Dp}{\mathscr{R}T(x_2 - x_1)} \ln \frac{p_{B2}}{p_{B1}} \qquad (16\text{-}25)$$

since

$$p - p_{A1} = p_{B1} \quad \text{and} \quad p - p_{A2} = p_{B2} \qquad (16\text{-}26)$$

We note from Eqs. (16-26) that the following relation holds: $p_{B2} - p_{B1} = p_{A1} - p_{A2}$. Equation (16-25) can be arranged as

$$N_A = \frac{Dp}{\mathscr{R}T(x_2 - x_1)p_{B\,ln}} (p_{B2} - p_{B1}) = \frac{Dp}{\mathscr{R}T(x_2 - x_1)p_{B\,ln}} (p_{A1} - p_{A2}) \qquad (16\text{-}27a)$$

where the *logarithmic mean partial pressure* $p_{B\,ln}$ of component B is defined as

$$p_{B\,ln} = \frac{p_{B2} - p_{B1}}{\ln (p_{B2}/p_{B1})} \qquad (16\text{-}27b)$$

Figure 16-2 illustrates the distribution of partial pressures (or the concentration) of species A and B with distance x in the medium; clearly the distribution is not linear with the position.   It is also to be noted that there is a diffusion of species

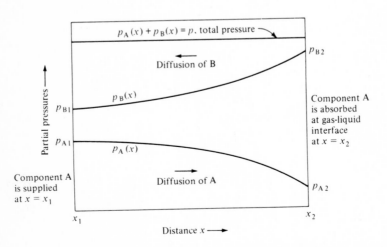

FIG. 16-2   Distribution of partial pressures p_A and p_B for unidirectional diffusion of gas A through gas B.

B in the medium due to the presence of the concentration gradient set up; but the *net flux of B remains zero* because species B is supplied by the bulk movement at the same rate as it diffuses away.

Example 16-2 A deep, narrow cylindrical vessel which is open at the top contains some toluene at the bottom. The air within the vessel is considered motionless, but there is sufficient air current at the top surface of the vessel so that any toluene vapor arriving at the top surface is immediately removed to ensure zero toluene concentration at the top surface. The entire system is at atmospheric pressure and 65°F. Under these conditions the diffusivity of air-toluene vapor is $D \cong 0.32$ ft^2/h $= 0.826 \times 10^{-5}$ m^2/s (see Table 16-3) and the saturated vapor pressure of toluene at the liquid surface in the vessel is 0.026 atm. Determine the rate of evaporation of toluene into the air per unit area of the liquid surface if the distance between the liquid toluene surface and the top of the vessel is 5 ft (1.524 m).

Solution In this problem toluene vapor is supplied to the air inside the vessel as a result of evaporation from the liquid surface. The toluene vapor diffuses through the stagnant air layer to the top of the vessel where it is immediately removed as a result of air currents; but there is no net removal of air from the vessel. Therefore, the process can be characterized as the steady-state unidirectional diffusion of toluene (component A) through the stationary air (component B) layer of thickness $x_2 - x_1 = 5$ ft. Equations (16-27) can be used to determine the rate of evaporation of toluene into the air, that is,

$$N_A = \frac{Dp}{\mathscr{R}T(x_2 - x_1)p_{B\,\text{ln}}}\,(p_{A1} - p_{A2})$$

where

$$p_{B\,\text{ln}} = \frac{p_{B2} - p_{B1}}{\text{ln}\,(p_{B2}/p_{B1})}$$

and the numerical values of various quantities are given as

$$D = 0.32 \text{ ft}^2/\text{h} = 0.826 \times 10^{-5} \text{ m}^2/\text{s} \qquad T = 460 + 65 = 525°\text{R} = 291.7 \text{ K}$$

$$x_2 - x_1 = 5 \text{ ft} = 1.524 \text{ m}$$

$$\mathscr{R} = 0.730 \text{ ft}^3 \cdot \text{atm/lb mol} \cdot °\text{R} = 0.08205 \text{ m}^3 \cdot \text{atm/kg mol} \cdot \text{K}$$

$$p = 1 \text{ atm} \qquad p_{A1} = 0.026 \text{ atm} \qquad p_{A2} = 0 \text{ atm}$$

$$p_{B1} = 1 - p_{A1} = 1 - 0.026 = 0.974 \text{ atm}$$

Then,

$$p_{B\,\text{ln}} = \frac{1 - 0.974}{\text{ln}\,(1/0.974)} = 0.987$$

The rate of evaporation of toluene into the air per unit area of the liquid surface is now determined by substituting these numerical values into the above equation. By using the English system of units, we find

$$N_A = \frac{0.32 \times 1 \times (0.026 - 0)}{0.730 \times 525 \times 5 \times 0.987} = 0.44 \times 10^{-5} \text{ lb mol/ft}^2 \cdot \text{h} \, (0.597 \times 10^{-8} \text{ kg mol/m}^2 \cdot \text{s})$$

or, by using the SI system of units, we obtain

$$N_A = \frac{(0.826 \times 10^{-5}) \times 1 \times (0.026 - 0)}{0.08205 \times 291.7 \times 1.524 \times 0.987} = 0.597 \times 10^{-8} \text{ kg mol/m}^2 \cdot \text{s}$$

16-4 STEADY-STATE DIFFUSION IN LIQUIDS

In the previous two sections we considered steady-state equimolal counterdiffusion and unidirectional diffusion in binary gas mixtures. We now examine the same diffusion processes for binary mixtures of liquids.

Steady-State Equimolal Counterdiffusion in Liquids

The analysis follows the same approach as that discussed previously for gases. Equation (16-16) is now written in terms of the molal concentration c_A as

$$N_A = -D \frac{dc_A}{dx} \tag{16-28}$$

The integration of the equation from $x = x_1$ to $x = x_2$ with c_A varying from c_{A1} to c_{A2} gives

$$N_A = D \frac{c_{A1} - c_{A2}}{x_2 - x_1} \tag{16-29}$$

where c_{A1}, c_{A2} = molal concentration of species A at x_1 and x_2, respectively, lb mol/ft^3 (kg mol/m^3)

D = mass diffusivity for liquid-to-liquid diffusion in binary liquid mixture, ft^2/h (m^2/s)

N_A = molal flux of species A, lb mol/ft$^2 \cdot$h (kg mol/m$^2 \cdot$s)

x = distance, ft (m)

Clearly, Eq. (16-29) is analogous to Eq. (16-18). In the analysis of diffusion in liquids, mole fraction χ_i is sometimes defined as

$$\chi_i = \frac{c_i}{c} = \text{mole fraction of component } i \qquad i = \text{A or B} \tag{16-30a}$$

where

$$c_A + c_B = c = \text{total molal concentration of mixture, lb mol/ft}^3 \text{ (kg mol/m}^3) \tag{16-30b}$$

Then, in terms of the mole fraction, Eq. (16-29) is written

$$N_A = Dc \frac{\chi_{A1} - \chi_{A2}}{x_2 - x_1} \quad \text{lb mol/ft}^2 \cdot \text{h (kg mol/m}^2 \cdot \text{s)} \tag{16-31}$$

Steady-State Undirectional Diffusion in Liquids

The analysis for the determination of the concentration distribution is exactly the same as that discussed previously for gases. Therefore, omitting the details, Eqs. (16-27) derived previously for gases in terms of partial pressures are now written for liquids in terms of the molal concentrations c_A and c_B:

$$N_A = D \frac{c}{x_2 - x_1} \frac{c_{B2} - c_{B1}}{c_{B \text{ ln}}} = D \frac{c}{x_2 - x_1} \frac{c_{A1} - c_{A2}}{c_{B \text{ ln}}} \tag{16-32a}$$

where the *logarithmic mean concentration* $c_{B \text{ ln}}$ of component B is defined as

$$c_{B \text{ ln}} = \frac{c_{B2} - c_{B1}}{\ln (c_{B2}/c_{B1})} \tag{16-32b}$$

and we note that $c = c_A + c_B$, $c_{B2} - c_{B1} = c_{A1} - c_{A2}$.

Equations (16-32) can also be written in terms of mole fractions:

$$N_A = D \frac{c}{x_2 - x_1} \frac{\chi_{B2} - \chi_{B1}}{\chi_{B \text{ ln}}} = D \frac{c}{x_2 - x_1} \frac{\chi_{A1} - \chi_{A2}}{\chi_{B \text{ ln}}} \tag{16-33a}$$

where the *logarithmic mean mole fraction* $\chi_{B \text{ ln}}$ of component B is defined as

$$\chi_{B \text{ ln}} = \frac{\chi_{B2} - \chi_{B1}}{\ln (\chi_{B2}/\chi_{B1})} \tag{16-33b}$$

In Eqs. (16-32) and (16-33) the units of various quantities may be taken as

c_i = molal concentration, i = A, B, or B_{ln}, lb mol/ft^3 (kg mol/m^3)

$\chi_i = c_i/c$ = mole fraction

c = total molal concentration of mixture, lb mol/ft^3 (kg mol/m^3)

D = mass diffusivity, ft^2/h (m^2/s)

N_A = molal flux of species A, lb mol/ft$^2 \cdot$ h (kg mol/m$^2 \cdot$ s)

x = distance, ft (m)

16-5 UNSTEADY DIFFUSION

In the previous sections we considered mass-transfer problems in which the concentration distribution is a function of one space variable only. There are many engineering applications in which the concentration distribution in the medium varies with both time and position. The mathematical analysis of such unsteady mass-transfer problems is complicated because they involve the solution of differential equations with more than one independent variable. However, there are a large number of unsteady mass-transfer problems which can be expressed in a form analogous to the one-dimensional time-dependent heat-conduction problems considered in a previous chapter. Therefore, the same mathematical techniques discussed for the solution of a time-dependent heat-conduction equation become applicable to the solution of one-dimensional unsteady mass-diffusion problems. In this section we present the mathematical formulation of the one-dimensional unsteady mass-diffusion equation in the rectangular coordinate system and discuss its solution for typical applications.

When there are no mass sources, sinks, or chemical reactions in the medium and the mass transfer takes place as a result of pure molecular diffusion, the mass-conservation equation for a given species may be stated as

$$
\begin{pmatrix} \text{Net rate of diffusion} \\ \text{of species A into} \\ \text{medium} \end{pmatrix} = \begin{pmatrix} \text{rate of increase of} \\ \text{species A in} \\ \text{medium} \end{pmatrix}
\tag{16-34}
$$

We now consider a one-dimensional mass diffusion in the x direction for a differential volume element of thickness Δx and area a, as illustrated in Fig. 16-3,

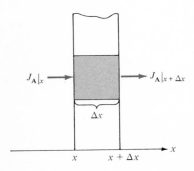

FIG. 16-3 Nomenclature for the derivation of the one-dimensional unsteady mass-diffusion equation.

and write the mathematical expressions for various terms in the mass conservation Eq. (16-34) as

$$\left(J_A\bigg|_x - J_A\bigg|_{x+\Delta x}\right)a = a\,\Delta x\,\frac{\partial c_A}{\partial t} \tag{16-35}$$

where J_A = molal diffusion flux
$\quad c_A$ = molal concentration of component A
$\quad t$ = time

Equation (16-35) is written

$$-\frac{\partial J_A}{\partial x} = \frac{\partial c_A}{\partial t} \tag{16-36}$$

By Fick's law of diffusion, the diffusion flux J_A is related to the concentration gradient by [see Eqs. (16-8)]

$$J_A = -D\frac{\partial c_A}{\partial x} \tag{16-37}$$

Substituting Eq. (16-37) into Eq. (16-36) and assuming a constant D, we obtain

$$\frac{\partial^2 c_A}{\partial x^2} = \frac{1}{D}\frac{\partial c_A}{\partial t} \tag{16-38}$$

where c_A = molal concentration of component A, lb mol/ft^3 (kg mol/m^3)
$\quad D$ = mass diffusivity for diffusion of component A in medium, ft^2/h (m^2/s)
$\quad t$ = time, h (s)
$\quad x$ = distance, ft (m)

Equation (16-38) is called *the one-dimensional, time-dependent mass-diffusion equation* in the rectangular coordinate system for constant mass diffusivity. It is analogous to the heat-conduction (or diffusion) equation given previously in the form

$$\frac{\partial^2 T}{\partial x^2} = \frac{1}{\alpha}\frac{\partial T}{\partial t} \tag{16-39}$$

We note that in the mass-diffusion equation the mass concentration c_A replaces temperature T, and the mass diffusivity D replaces the heat diffusivity α.

Equation (16-38) can readily be generalized to the three-dimensional case:

$$\nabla^2 c_A = \frac{1}{D} \frac{\partial c_A}{\partial t} \qquad (16\text{-}40a)$$

where ∇^2 is the laplacian operator and in the rectangular coordinate system is given as

$$\nabla^2 \equiv \frac{\partial^2}{\partial x^2} + \frac{\partial^2}{\partial y^2} + \frac{\partial^2}{\partial z^2} \qquad (16\text{-}40b)$$

Again Eq. (16-40a) is analogous to the three-dimensional, time-dependent heat-conduction equation, with no energy generation and constant thermal diffusivity given by Eq. (2-15).

The mass-transfer problems characterized by the mass-diffusion equation as given above are encountered in processes such as drying, vapor penetration through porous media, and numerous other applications. The analytical solution of the resulting partial differential equation subject to a given set of boundary and initial conditions can be obtained for simple geometries such as slabs, cylinders, and spheres. The reader should consult Refs. [5,6,17,18] for analytical methods of solution of the diffusion equation and for solutions under different boundary conditions. In the following example we illustrate the solution of the above one-dimensional unsteady mass-diffusion equation for the problem of drying of a slab. *The mass-conservation equation (16-39) being exactly of the same form as the one-dimensional time-dependent heat-conduction equation, the same mathematical techniques described previously for the solution of heat-conduction problems are applicable.*

Example 16-3 A large slab of clay of thickness L initially contains c_0 mass percent of water distributed uniformly over the volume. The surface of the slab at $x = 0$ is impermeable to moisture flow. The clay is subjected to drying by blowing a stream of low-humidity air over the boundary surface at $x = L$. It is assumed that the drying takes place by the process of moisture diffusion through the clay, and the mass diffusivity D for the diffusion of water vapor through the medium is uniform everywhere. The concentration of water vapor at the boundary surface $x = L$ may be taken as zero since dry air is swept over the surface. Determine a relation for the concentration of water vapor in the clay as a function of time and position. Also give a relation for the concentration of water vapor at the impermeable boundary as a function of time.

Solution The partial differential equation and appropriate boundary and initial conditions for the determination of distribution of concentration $c(x, t) \equiv c$ in mass percent of water in the slab are given as

$$\frac{\partial^2 c}{\partial x^2} = \frac{1}{D} \frac{\partial c}{\partial t} \qquad \text{in } 0 \le x \le L, t > 0 \qquad (16\text{-}41a)$$

$$\frac{\partial c}{\partial x} = 0 \qquad \text{at } x = 0, t > 0 \qquad (16\text{-}41b)$$

$$c = 0 \qquad \text{at } x = L, t > 0 \tag{16-41c}$$

$$c = c_0 \qquad \text{in } 0 \le x \le L, t = 0 \tag{16-41d}$$

Clearly, boundary condition (16-41b) implies that the boundary surface at $x = 0$ is impermeable to moisture, and boundary condition (16-41c) is a statement of vapor concentration being zero at the surface $x = L$. The mass-transfer problem described above by Eqs. (16-41) is a special case of the transient-heat-conduction problem given by Eqs. (4-59) and (4-60). That is, if we set in Eqs. (4-59) and (4-60) $\alpha \equiv D$, $h = 0$, $F(x) = c_0$, and $T = c$ we obtain the above mass-transfer problem. Therefore, the same technique described in Chap. 4 to solve transient-heat-conduction problems can be used to solve Eqs. (16-41). Omitting the details, the solution of Eqs. (16-41) is given as

$$c(x, t) = \sum_{n=1}^{\infty} \frac{1}{N} e^{-D\lambda_n^2 t} \cos \lambda_n x \int_0^L c_0 \cos \lambda_n x' \, dx' \tag{16-42a}$$

where

$$\frac{1}{N} = \frac{2}{L} \tag{16-42b}$$

and λ_n's are the roots of $\cos \lambda L = 0$, or λ_n's are given by

$$\lambda_n = \frac{2n - 1}{2L} \pi \tag{16-42c}$$

The integration in Eq. (16-42a) is readily performed since c_0 is constant, and the resulting expression for the distribution of water concentration in the slab becomes

$$\frac{c(x, t)}{c_0} = \frac{2}{L} \sum_{n=1}^{\infty} e^{-D\lambda_n^2 t} \cos \lambda_n x \frac{\sin \lambda_n L}{\lambda_n} \tag{16-43a}$$

where

$$\lambda_n = \frac{2n - 1}{2L} \pi \tag{16-43b}$$

The concentration of water at the impermeable boundary $x = 0$ becomes

$$\frac{c(0, t)}{c_0} = \frac{2}{L} \sum_{n=1}^{\infty} e^{-D\lambda_n^2 t} \frac{\sin \lambda_n L}{\lambda_n} \tag{16-44}$$

16-6   MASS DIFFUSIVITY

The prediction of mass diffusivity D has been the subject of extensive investigations, and the mass diffusivity for gases, liquids, and solids requires different considerations. Here we present a brief discussion of the methods of determination of mass diffusivities.

Mass Diffusivity in Gases

Various models have been proposed for the determination of mass diffusivity $D_{AB} = D_{BA}$ for a binary mixture of gases A and B. The reader should consult Refs. [15,19,20] for detailed discussion of various mathematical models. Here we present the pertinent result for the mass diffusivity of a binary mixture of gas, obtained from the kinetic-theory derivation by considering the gas molecules as rigid spheres experiencing elastic collisions and assuming an ideal-gas behavior. According to this model, the mass diffusivity $D_{AB} = D_{BA}$ for a mixture of two gases composed of species A and B, for pressures below 20 atm, can be predicted from the following relation:

$$D_{AB} = \frac{1.8583 \times 10^{-3}}{p\sigma_{AB}{}^2 \Omega_{D,\,AB}} T^{3/2} \sqrt{\frac{1}{M_A} + \frac{1}{M_B}} \tag{16-45}$$

where D_{AB} = mass diffusivity, cm²/s
    p = pressure, atm
    T = absolute temperature, K
    M_A, M_B = molecular weights of components A and B, respectively
    σ_{AB} = *collision diameter*, Å
    $\Omega_{D,\,AB}$ = *collision integral* or a dimensionless function depending on temperature and intermolecular forces

TABLE 16-1   The Parameter σ and $\frac{\varepsilon}{k}$ for Substances†

Substance		Molecular Weight, M	σ, Å	ε/k, K
Air	Air	28.97	3.617	97.0
Argon	A	39.94	3.418	124.0
Benzene	C_6H_6	78.11	5.270	440.0
Bromine	Br_2	159.83	4.268	520.0
Carbon dioxide	CO_2	44.01	3.996	190.0
Carbon disulfide	CS_2	76.14	4.438	488.0
Carbon monoxide	CO	28.01	3.590	110.0
Carbon tetrachloride	CCl_4	153.84	5.881	327.0
Chlorine	Cl_2	70.91	4.115	357.0
Ethane	C_2H_6	30.07	4.418	230.0
Ethylene	C_2H_4	28.05	4.232	205.0
Fluorine	F_2	38.00	3.653	112.0
Helium	He	4.003	2.576	10.2
Hydrogen	H_2	2.016	2.915	38.0
Iodine	I_2	253.82	4.982	550.0
Krypton	Kr	83.80	3.61	190.0
Methane	CH_4	16.04	3.822	137.0
Methyl chloride	CH_3Cl	50.49	3.375	855.0
Neon	Ne	20.183	2.789	35.7
Nitric oxide	NO	30.01	3.470	119.0
Nitrogen	N_2	28.02	3.681	91.5
Nitrous oxide	N_2O	44.02	3.879	220.0
Oxygen	O_2	32.00	3.433	113.0
Sulfur dioxide	SO_2	64.07	4.290	252.0
Water‡	H_2O	18.0	2.641	809.1
Xenon	Xe	131.3	4.055	229.0

† From Hirschfelder, Curtiss, and Bird [20].
‡ From Svehla [21].

TABLE 16-2   Values of Collision Integral $\Omega_{D,\,AB}$[†]

kT/ε[‡]	$\Omega_{D,\,AB}$[‡]	kT/ε	$\Omega_{D,\,AB}$	kT/ε	$\Omega_{D,\,AB}$
0.30	2.662	1.65	1.153	4.0	0.8836
0.35	2.476	1.70	1.140	4.1	0.8788
0.40	2.318	1.75	1.128	4.2	0.8740
0.45	2.184	1.80	1.116	4.3	0.8694
0.50	2.066	1.85	1.105	4.4	0.8652
0.55	1.966	1.90	1.094	4.5	0.8610
0.60	1.877	1.95	1.084	4.6	0.8568
0.65	1.798	2.00	1.075	4.7	0.8530
0.70	1.729	2.1	1.057	4.8	0.8492
0.75	1.667	2.2	1.041	4.9	0.8456
0.80	1.612	2.3	1.026	5.0	0.8422
0.85	1.562	2.4	1.012	6	0.8124
0.90	1.517	2.5	0.9996	7	0.7896
0.95	1.476	2.6	0.9878	8	0.7712
1.00	1.439	2.7	0.9770	9	0.7556
1.05	1.406	2.8	0.9672	10	0.7424
1.10	1.375	2.9	0.9576	20	0.6640
1.15	1.346	3.0	0.9490	30	0.6232
1.20	1.320	3.1	0.9406	40	0.5960
1.25	1.296	3.2	0.9328	50	0.5756
1.30	1.273	3.3	0.9256	60	0.5596
1.35	1.253	3.4	0.9186	70	0.5464
1.40	1.233	3.5	0.9120	80	0.5352
1.45	1.215	3.6	0.9058	90	0.5256
1.50	1.198	3.7	0.8998	100	0.5130
1.55	1.182	3.8	0.8942	200	0.4644
1.60	1.167	3.9	0.8888	400	0.4170

[†] From Hirschfelder, Curtiss, and Bird [20].
[‡] Reference [20] uses the symbol T^* for kT/ε and $\Omega^{(1,\,1)*}$ in place of $\Omega_{D,\,AB}$.

The parameters σ_{AB} and $\Omega_{D,\,AB}$ for a mixture of two gases composed of components A and B can be determined from Tables 16-1 and 16-2 for the species shown in Table 16-1 as now described.

We obtain from Table 16-1 the values of σ_A, $(\varepsilon/k)_A$ and σ_B, $(\varepsilon/k)_B$ for the components A and B, respectively.   Then, the parameter σ_{AB} is given by

$$\sigma_{AB} = \tfrac{1}{2}(\sigma_A + \sigma_B) \tag{16-46}$$

Table 16-2 lists the values of the parameter $\Omega_{D,\,AB}$ as a function of kT/ε which actually implies

$$\frac{kT}{\varepsilon} \equiv \frac{T}{(\varepsilon/k)_{AB}} \tag{16-47a}$$

and the value of the parameter $(\varepsilon/k)_{AB}$ for the mixture is taken as

$$\left(\frac{\varepsilon}{k}\right)_{AB} = \sqrt{\left(\frac{\varepsilon}{k}\right)_A \left(\frac{\varepsilon}{k}\right)_B} \tag{16-47b}$$

where $(\varepsilon/k)_A$ and $(\varepsilon/k)_B$ for the components A and B are readily obtainable from Table 16-1. Thus, knowing the value of kT/ε, the corresponding value of $\Omega_{D, AB}$ is obtained from Table 16-2.

We present in Table 16-3 measured values of mass diffusivities for typical binary gas systems at 1-atm pressure. Clearly, according to Eq. (16-45), if the experimental value of mass diffusivity D_1 at temperature T_1 is available, the mass diffusivity D_2 at temperature T_2 may be estimated from the relation

$$D_2 = D_1 \left(\frac{T_2}{T_1}\right)^{3/2} \frac{(\Omega_{D, AB})_{T_1}}{(\Omega_{D, AB})_{T_2}} \tag{16-48}$$

TABLE 16-3   Mass Diffusivities of Binary Gas Systems at Atmospheric Pressure (Measured Values) †

System	T, K	D, Mass Diffusivity cm^2/s	ft^2/h
Air–ammonia	273	0.198	0.769
Air–aniline	298	0.0726	0.282
Air–benzene	298	0.0962	0.374
Air–carbon dioxide	273	0.136	0.528
Air–carbon disulfide	273	0.0883	0.343
Air–chlorine	273	0.124	0.482
Air–ethyl alcohol	298	0.132	0.513
Air–iodine	298	0.0834	0.524
Air–mercury	614	0.473	1.837
Air–naphthalene	298	0.0611	0.237
Air–oxygen	273	0.175	0.680
Air–sulfur dioxide	273	0.122	0.474
Air–toluene	298	0.0844	0.328
Air–water	298	0.260	1.010
CO_2–benzene	318	0.0715	0.278
CO_2–carbon disulfide	318	0.0715	0.278
CO_2–ethyl alcohol	273	0.0693	0.269
CO_2–hydrogen	273	0.550	2.136
CO_2–nitrogen	298	0.158	0.614
CO_2–water	298	0.164	0.637
Oxygen–ammonia	293	0.253	0.983
Oxygen–benzene	296	0.039	0.365

† Compiled from Reid and Sherwood [22].

Example 16-4   Determine the mass diffusivity of the binary mixture of air-CO_2 at 273 K and 1-atm pressure from Eq. (16-45) and compare the result with the experimental value given by Table 16-3.

Solution   The parameters σ, ε/k and the molecular weights M for air and CO_2 are obtained from Table 16-1 as

$$\text{Air (A):} \quad \sigma_A = 3.617 \quad \left(\frac{\varepsilon}{k}\right)_A = 97.0 \quad M_A = 28.97$$

$$\text{CO}_2 \text{ (B):} \quad \sigma_B = 3.996 \quad \left(\frac{\varepsilon}{k}\right)_B = 190.0 \quad M_B = 44.01$$

Then,

$$\sigma_{AB} = \tfrac{1}{2}(\sigma_A + \sigma_B) = \tfrac{1}{2}(3.617 + 3.996) = 3.806 \text{ Å}$$

$$\left(\frac{\varepsilon}{k}\right)_{AB} = \sqrt{\left(\frac{\varepsilon}{k}\right)_A \left(\frac{\varepsilon}{k}\right)_B} = \sqrt{97 \times 190} = 135.76 \text{ K}$$

$$\frac{kT}{\varepsilon} = \frac{T}{(\varepsilon/k)_{AB}} = \frac{273}{135.76} = 2.01$$

and the value of the collision integral for $kT/\varepsilon = 2.01$ is obtained from Table 16-2 as

$$\Omega_{D,\,AB} = 1.073$$

Various quantities as given above, the mass diffusivity for $p = 1$ atm, and $T = 273$ K are determined from Eq. (16-45) as

$$D_{AB} = \frac{1.8583 \times 10^{-3}}{p \times \sigma_{AB}{}^2 \times \Omega_{D,\,AB}} T^{3/2} \sqrt{\frac{1}{M_A} + \frac{1}{M_B}}$$

$$D_{AB} = \frac{1.8583 \times 10^{-3}}{1 \times 3.806^2 \times 1.073} 273^{3/2} \sqrt{\frac{1}{28.97} + \frac{1}{44.01}}$$

$$D_{AB} = 0.129 \text{ cm}^2/\text{s}$$

This result should be compared with the measured value $D_{AB} = 0.136$ cm^2/s given in Table 16-3.

Mass Diffusivity in Liquids

A rigorous theory is yet to be developed for the prediction of mass diffusivity in liquids.   The existing approximate theories lack agreement with experiments; therefore a number of semiempirical relations have been proposed for the determination of mass diffusivity in liquids.   The reader should consult Refs. [1,22] for a collection of such expressions.   One of these relations developed by Wilke and Chang [23] gives the mass diffusivity D_{AB} for small concentrations of A in B (that is, A is the solute, B is the solvent) as

$$D_{AB} = 7.4 \times 10^{-8} \frac{(\xi_B M_B)^{1/2} T}{\mu \tilde{V}_A{}^{0.6}} \tag{16-49}$$

where D_{AB} = mass diffusivity (A is solute, B is solvent), cm^2/s
 M_B = molecular weight of solvent B
 T = absolute temperature, K
 $\tilde{V}_A$ = molal volume of solute A as liquid at its normal boiling point, cm^3/g mol
 μ = viscosity of solution, centipoises
 ξ_B = "association" factor for solvent B; its recommended values are 2.6 for water; 1.9 for methanol; 1.5 for ethanol; and 1.0 for benzene, ether, heptane, and other unassociating solvents

The liquid molal volumes $\tilde{V}$ at the normal boiling point are given in Table 16-4 for a number of liquids; the reader is referred to Reid and Sherwood [22] for a discussion of various methods of estimating the liquid molal volumes $\tilde{V}$. Equation (16-49) is good within ± 10 percent for the prediction of mass diffusivity of dilute solutions of nondissociating solutes. In Table 16-5 we present the measured values of mass diffusivities for dilute solutions of various liquids in water, methanol, and benzene.

Example 16-5 Determine the mass diffusivity D_{AB} for a dilute solution of CCl_4 (i.e., component A, the solute) in benzene (i.e., component B, the solvent) at 25°C and compare the predicted value with the measured value given in Table 16-5.

Solution Equation (16-49) can be used to determine the mass diffusivity, and the values of various quantities in this equation are taken as

$M_B = 78.10$ $\tilde{V}_A = 103$ (from Table 16-4) $T = 298$ K

$\xi_B = 1$ (for benzene)   and $\mu = 0.6$ centipoise

TABLE 16-4 Liquid Molal Volumes $\tilde{V}$ at the Normal Boiling Point Calculated by the Method of Benson†

Compound	Molal Volume $\tilde{V}$, cm^3/g mol	Compound	Molal Volume $\tilde{V}$, cm^3/g mol
Acetic acid	62.9	Ethyl mercaptan	76.0
Acetone	81.0	Fluorobenzene	101.0
Acetylene	41.3	Heptane	165.0
Ammonia	25.5	Hydriodic acid	45.9
Benzene	97.0	Hydrochloric acid	31.2
Bromobenzene	115.0	Iodobenzene	131.0
Carbon tetrachloride	103.0	Methane	37.1
Chlorine	44.6	Methanol	42.4
Chlorobenzene	115.0	Methyl formate	63.0
Diethylamine	113.0	Methyl chloride	49.8
Diethyl sulfide	122.0	Propane	75.3
Dimethyl ether	62.7	Sulfur dioxide	44.3
Ethyl acetate	108.0	Water	19.0
Ethylene	45.9		

† Compiled from Reid and Sherwood [22].

TABLE 16-5   Liquid-Phase Mass Diffusivities of Dilute Solutions in Water, Methanol, and Benzene (Measured Values) †

System		$T,$ °C	$D \times 10^5$ cm²/s	$D \times 10^5$ ft²/h
Bromine	in water	12	0.90	3.50
CO_2	in water	18	1.71	6.64
Chlorine	in water	12	1.40	5.44
Glucose	in water	15	0.52	2.02
Hydrogen	in water	25	3.36	3.05
Iodine	in water	25	1.25	4.85
Methanol	in water	15	1.28	4.97
Nitrogen	in water	22	2.02	7.84
Oxygen	in water	25	2.60	10.10
Aniline	in methanol	15	1.49	5.79
CCl_4	in methanol	15	1.70	6.60
Chloroform	in methanol	15	2.07	8.04
Iodoform	in methanol	15	1.33	5.17
Lactic acid	in methanol	15	1.36	5.28
Acetic acid	in benzene	15	1.92	7.46
Bromine	in benzene	12	2.00	7.77
CCl_4	in benzene	25	2.00	7.77
Chloroform	in benzene	15	2.11	8.19
Iodine	in benzene	20	1.95	7.57

† Compiled from Reid and Sherwood [22].

Then Eq. (16-49) gives

$$D_{AB} = 7.4 \times 10^{-8} \frac{(\xi_B M_B)^{1/2} T}{\mu \tilde{V}_A^{0.6}} = 7.4 \times 10^{-8} \frac{(1 \times 78.1)^{1/2} \times 298}{0.6 \times 103^{0.6}}$$

$$= 2.01 \times 10^{-5} \text{ cm}^2/\text{s}$$

which should be compared with the measured value $D_{AB} = 2.00 \times 10^{-5}$ cm²/s given in Table 16-5.

Mass Diffusivity in Solids

The theoretical prediction of mass diffusivity in solids containing pores and capillaries is an extremely complicated matter, and no fundamental theory is yet available on the subject.   When a solid contains capillaries and pores, there are situations in which the capillary forces are opposed to the concentration gradient and as a result the mass flux may not even be proportional to the concentration gradient.   To describe diffusion inside the tortuous void passages in such a medium by taking into consideration the interaction between different mechanisms of transport appears not to be possible analytically.   Therefore, an experimental approach is the only means to determine an *effective mass diffusivity* for a given solid structure and fluid combination. The effective mass diffusivity determined in this manner can be used in the equation of mass conservation (16-38) or (16-40) to predict the mass diffusion in solids. It is to be noted that, for solids which are not so porous, the mass diffusivity is several orders of magnitude smaller than that for liquids.

16-7   MASS TRANSFER IN LAMINAR AND TURBULENT FLOW

There are many engineering applications in which mass transfer takes place through a fluid stream that is in laminar or turbulent flow. For example, in transpiration cooling, a cold fluid is injected into the hot gas stream through the perforations in the plate of duct surface to protect the duct surfaces from the hot gas. In high-temperature gas-cooled nuclear reactors the fission products that are released into the coolant gas stream are transported by molecular and eddy diffusion from the gas stream to the surfaces of the conduits and the heat-exchanger tubes. When the concentration of mass in the fluid stream is very small, the mass transfer to and from the fluid stream is governed by the mass-conservation equation which is similar to the energy equation with no viscous dissipation. For example, the steady-state mass-conservation equation for incompressible laminar flow inside a circular tube with cylindrical symmetry in the distribution of mass concentration is given as

$$u \frac{\partial c_A}{\partial x} + v \frac{\partial c_A}{\partial r} = D \left(\frac{\partial^2 c_A}{\partial r^2} + \frac{1}{r} \frac{\partial c_A}{\partial r} + \frac{\partial^2 c_A}{\partial x^2} \right) \tag{16-50}$$

where $c_A = c_A(r, x)$ is the concentration of mass, D is the mass diffusivity, x and r are the axial and radial coordinates, respectively, and u and v are the velocity components in the axial and radial directions, respectively. [See Eq. (6-36) for $\Phi = 0$.] The velocity components u and v are to be determined from the solution of the continuity and momentum equations.

If the axial diffusion is neglected, Eq. (16-50) simplifies to

$$u \frac{\partial c_A}{\partial x} + v \frac{\partial c_A}{\partial r} = D \left(\frac{\partial^2 c_A}{\partial r^2} + \frac{1}{r} \frac{\partial c_A}{\partial r} \right) \tag{16-51}$$

and, in the case of fully developed flow, $v = 0$ and Eq. (16-51) is further simplified as

$$u \frac{\partial c_A}{\partial x} = D \left(\frac{\partial^2 c_A}{\partial r^2} + \frac{1}{r} \frac{\partial c_A}{\partial r} \right) \tag{16-52}$$

[See Eq. (7-32) for this relationship with axial conduction neglected.]

The solution of these equations subject to appropriate boundary conditions gives the distribution of mass concentration in the flow field. Then, the mass flux to and from the conduit surface, in the absence of fluid injection or suction at the wall, is determined from

$$N_A \bigg|_{\text{wall}} = -D \frac{\partial c_A}{\partial r} \bigg|_{\text{wall}} \tag{16-53}$$

(Note that $q = -k\,\partial T/\partial x$ for heat flux.)

Clearly, the mathematical formulation of the mass-transfer problem to and from fluid streams is analogous to heat-transfer problems to and from fluid streams. In mass transfer, the mass concentration c_A replaces the temperature T, and the mass diffusivity D replaces the heat diffusivity α. The similarity between heat and mass transfer illustrated above for laminar flow is also valid for turbulent flow. Therefore, when the concentration of mass in the fluid stream is very small, various expressions given in the previous chapters for heat transfer from laminar and turbulent streams become applicable to mass transfer by proper modification of such expressions as now described. The reader should consult Ref. [1] for mass transfer at high mass flux and high concentration and Refs. [7,8] for mass transfer in chemically reacting systems.

A *mass transfer coefficient* k_m for the transfer of a species A between the fluid stream and the wall surface is defined, in analogy with the heat-transfer coefficient, as [see Eq. (1-2) for the definition of heat transfer coefficient]

$$N_A = k_m(c_{Am} - c_{Aw}) \tag{16-54}$$

where c_{Am} = mean concentration of species A in bulk fluid, lb mol/ft^3 (kg mol/m^3)

c_{Aw} = concentration of species A at immediate vicinity of wall surface, lb mol/ft^3 (kg mol/m^3)

N_A = mass flux of species A to the wall, lb mol/ft$^2\cdot$h (kg mol/m$^2\cdot$s)

k_m = mass-transfer coefficient for species A, ft/h (m/s), i.e.,

$$\frac{\text{lb mol}}{\text{ft}^2\cdot\text{h(lb mol/ft}^3)} = \frac{\text{ft}}{\text{h}}\left(\frac{\text{m}}{\text{s}}\right)$$

In heat-transfer problems the Nusselt number is defined as

$$\text{Nu} = \frac{hd}{k} \qquad \text{for flow inside tube} \tag{16-55a}$$

$$\text{Nu}_x = \frac{hx}{k} \qquad \text{for boundary-layer flow} \tag{16-55b}$$

and in an analogous manner in mass-transfer problems the *Sherwood number,* Sh, is defined as

$$Sh = \frac{k_m d}{D} \qquad \text{for mass transfer inside tube} \qquad (16\text{-}56a)$$

$$Sh_x = \frac{k_m x}{D} \qquad \text{for mass transfer in boundary-layer flow} \qquad (16\text{-}56b)$$

where d = tube diameter
   D = mass diffusivity
   x = distance along plate

In previous chapters on convective heat transfer we have shown that the Nusselt number for heat transfer in incompressible flow is a function of Reynolds and Prandtl numbers, that is,

$$Nu = f(Re, Pr) \quad [\text{see Eq. (6-51)}] \qquad (16\text{-}57)$$

A similar dimensional analysis of the governing equations for mass transfer from an incompressible flow shows that the Sherwood number is a function of Reynolds and Schmidt numbers, that is,

$$Sh = f(Re, Sc) \qquad (16\text{-}58a)$$

Where the Schmidt number is defined as

$$Sc = \frac{v}{D} \qquad (16\text{-}58b)$$

and the Reynolds and Prandtl numbers have their usual definitions.

It is to be reiterated that *the foregoing similarity between heat- and mass-transfer processes is valid if the concentration of the species in the fluid stream is small.* For such situations, the mass-transfer coefficient k_m is obtained from the analogous heat-transfer coefficient by replacing the Nusselt number by the Sherwood number and the Prandtl number by the Schmidt number. We illustrate the procedure for typical cases.

Mass Transfer in Laminar Flow with Fully Developed Velocity and Concentration Distributions

In Chap. 7 we discussed heat transfer for hydrodynamically and thermally developed flow in circular tubes under constant wall temperature and constant-wall-heat-flux boundary conditions. The solution of the mass-transfer problem for laminar flow with fully developed velocity and concentration distribution yields analogous results except the Nusselt number is replaced by the Sherwood number.   That is,

$$\text{Sh} = \frac{k_m d}{D} = 3.66 \qquad \text{for uniform wall mass concentration} \qquad (16\text{-}59a)$$

$$= 4.36 \qquad \text{for uniform wall mass flux} \qquad (16\text{-}59b)$$

The Sherwood number for mass transfer in laminar flow inside conduits having noncircular cross section can be obtained from those given in Table 7-1 for heat transfer.

Mass Transfer in Turbulent Flow Inside Pipes

The heat-transfer relation for turbulent flow inside a pipe was given in Chap. 9 as

$$\text{Nu} = 0.023 \text{Re}^{0.8} \, \text{Pr}^{1/3} \qquad \text{for } 0.7 < \text{Pr} < 100 \qquad \text{Re} > 10{,}000 \qquad (16\text{-}60)$$

In the case of mass transfer, experiments performed with a wetted-wall column have shown that the analogous relation is in the form

$$\text{Sh} = 0.023 \text{Re}^{0.83} \, \text{Sc}^{1/3} \qquad \text{for } 0.6 < \text{Sc} < 2500 \qquad 2000 < \text{Re} < 35{,}000 \qquad (16\text{-}61)$$

The slight increase in the exponent of the Reynolds number may be justified under the experimental conditions leading to this expression.   That is, a wetted-wall column is an experimental vertical tube in which liquid flows in a thin film down the inside surface and a gas flows upward in the tube.   Mass transfer takes place from the liquid to the gas, or vice versa.   Under these conditions, the presence of rippling and wave formation on the liquid surface increases the interface available for mass transfer which is characterized by slightly stronger dependence of the mass-transfer coefficient on Reynolds number. In the case of mass-transfer experiments with liquids flowing in turbulent flow inside soluble tubes so that mass is transferred from the inside surface of the tube to the fluid stream, the surface remains smooth and the mass-transfer relation takes a form analogous to Eq. (16-60), that is,

$$\text{Sh} = 0.023 \text{Re}^{0.8} \, \text{Sc}^{1/3} \qquad (16\text{-}62)$$

Other mass-transfer relations are obtainable from the heat-transfer relations in a similar manner.   For example, the results of analogies between heat and momentum transfer can be recast as analogies between mass and momentum transfer.   That is, by replacing the Nusselt number by the Sherwood number and Prandtl number by the Schmidt number, the Reynolds, Prandtl, and von Kármán analogies become the analogies between mass and momentum transfer.

Then, once the friction factor or the drag coefficient is known, the mass-transfer coefficient k_m for turbulent flow inside a pipe or along a flat plate can be determined.

Example 16-6 Air at atmospheric pressure and 77°F (25°C), containing small quantities of iodine, flows with a velocity of 17 ft/s (5.18 m/s) inside a 1.2-in (3.048 × 10^{-2}-m) ID tube. Determine the mass-transfer coefficient for iodine transfer from the gas stream to the wall surface. If c_m is the mean concentration of iodine in lb mol/ft³ (kg mol/m³) in the air stream, determine the rate of deposition of iodine on the tube surface by assuming that the wall surface is a perfect sink for iodine deposition (i.e., the iodine concentration at the immediate vicinity of the wall is assumed to be zero).

Solution The kinematic viscosity of air is $v = 1.7 \times 10^{-4}$ ft²/s (1.58 × 10^{-5} m²/s) and the mass diffusivity for the air-iodine system at 1-atm pressure, 77°F (25°C), is obtained from Table 16-3 as $D = 0.32$ ft²/h or 0.89×10^{-4} ft²/s or 0.826×10^{-5} m²/s. The Reynolds number for the flow is

$$\text{Re} = \frac{ud}{v} = \frac{17 \times 1.2}{(1.7 \times 10^{-4}) \times 12} = 10^4$$

or, by using the SI system of units, we obtain the same result:

$$\text{Re} = \frac{5.18 \times (3.048 \times 10^{-2})}{1.58 \times 10^{-5}} \simeq 10^4$$

The flow is turbulent, and the mass-transfer coefficient can be determined by Eq. (16-62):

$$\text{Sh} = 0.023 \, \text{Re}^{0.8} \, \text{Sc}^{1/3}$$

where

$$\text{Sc} = \frac{v}{D} = \frac{1.7 \times 10^{-4} \text{ ft}^2/\text{s}}{0.89 \times 10^{-4} \text{ ft}^2/\text{s}} = 1.91$$

$$\text{Sh} = \frac{k_m d}{D} = \frac{k_m \times 1.2}{0.32 \times 12} = \frac{k_m}{3.2}$$

Then,

$$\frac{k_m}{3.2} = 0.023 \times (10^4)^{0.8} \times 1.91^{1/3}$$

$$k_m = 0.023 \times 10^{3.2} \times 1.91^{1/3} \times 3.2 = 144.6 \text{ ft/h} \ (1.224 \times 10^{-2} \text{ m/s})$$

The rate of deposition of iodine on the tube surface is determined from

$$N = k_m(c_m - c_w) \qquad \text{lb mol/ft}^2 \cdot \text{h (kg mol/m}^2 \cdot \text{s)}$$

For the perfect sink condition at the wall, $c_w = 0$; then

$$N = k_m c_m = 144.6 \times c_m \text{ lb mol/ft}^2 \cdot \text{h (kg mol/m}^2 \cdot \text{s)}$$

where c_m is the mean iodine concentration in lb mol/ft³ (kg mol/m³) in the air stream.

REFERENCES

1   Skelland, A. H. P.: "Diffusional Mass Transfer," John Wiley & Sons, Inc., New York, 1974.

2   Sawistowski, H., and W. Smith: "Mass Transfer Process Calculations," Interscience Publishers, Inc., New York, 1963.

3   Hobler, T.: "Mass Transfer and Absorbers" (translated from Polish by J. Bandrowski), Pergamon Press, New York, 1966.

4   Spalding, D. B.: "Convective Mass Transfer," Edward Arnold (Publishers) Ltd., London, 1963.

5   Jost, W.: "Diffusion in Solids, Liquids, and Gases," Academic Press, Inc., New York, 1960.

6   Crank, J.: "The Mathematics of Diffusion," Oxford University Press, London, 1956.

7   Astarita, G.: "Mass Transfer with Chemical Reaction," Elsevier Publishing Company, Amsterdam, 1967.

8   Danckwerts, P. V.: "Gas-Liquid Reactions," McGraw-Hill Book Company, New York, 1970.

9   Barrer, R. M.: "Diffusion in and through Solids," Cambridge University Press, New York, 1941.

10 Treybal, R. E.: "Mass Transfer Operations," McGraw-Hill Book Company, New York, 1955.

11 Brian, P. L. T.: "Staged Cascades in Chemical Processing," Prentice-Hall, Inc., Englewood Cliffs, N.J., 1972.

12 Henley, E. J., and H. K. Staffin: "Stagewise Process Design," John Wiley & Sons, Inc., New York, 1963.

13 Smith, Buford D.: "Design of Equilibrium Stage Processes," McGraw-Hill Book Company, New York, 1963.

14 Sherwood, T. K., and R. L. Pigford: "Absorption and Extraction," McGraw-Hill Book Company, New York, 1952.

15 Bird, R. B., W. E. Stewart, and E. N. Lightfoot: "Transport Phenomena," John Wiley & Sons, Inc., New York, 1960.

16 Bennett, C. O., and J. E. Myers: "Momentum, Heat and Mass Transfer," McGraw-Hill Book Company, New York, 1962.

17 Carslaw, H. S., and J. C. Jaeger: "Conduction of Heat in Solids," Oxford University Press, London, 1959.

18 Özışık, M. N.: "Boundary Value Problems of Heat Conduction," International Textbook Company, Scranton, Pa., 1968.

19 Present, R. D.: "Kinetic Theory of Gases," McGraw-Hill Book Company, New York, 1958.

20 Hirschfelder, J. O., C. F. Curtiss, and R. B. Bird: "Molecular Theory of Gases and Liquids," John Wiley & Sons, Inc., New York, 1954.

21 Svehla, R. A.: *NACA Tech. Rep.* R-132, Lewis Research Center, Cleveland, Ohio, 1962.

22 Reid, R. C., and T. K. Sherwood: "The Properties of Gases and Liquids," chap. 11, McGraw-Hill Book Company, New York, 1966.

23 Wilke, C. R., and P. Chang: Correlations of Diffusion Coefficients in Dilute Solutions, *AIChE J.*, **1**:264–270 (1955).

Appendix A

Physical Properties of Gases, Liquids, Liquid Metals, Metals, and Nonmetals

TABLE A-1 Physical Properties of Gases at Atmospheric Pressure

Temperature		c_p	$c_p \times 10^{-3}$	k	k	μ	$\mu \times 10^5$	ρ	ρ	ν	$\nu \times 10^4$	α	$\alpha \times 10^4$	Pr
°F	°C	$\dfrac{\text{Btu}}{\text{lb·°F}}$	$\dfrac{\text{W·s}}{\text{kg·°C}}$	$\dfrac{\text{Btu}}{\text{h·ft·°F}}$	$\dfrac{\text{W}}{\text{m·°C}}$	$\dfrac{\text{lb}}{\text{ft·h}}$	$\dfrac{\text{kg}}{\text{m·s}}$	$\dfrac{\text{lb}}{\text{ft}^3}$	$\dfrac{\text{kg}}{\text{m}^3}$	$\dfrac{\text{ft}^2}{\text{h}}$	$\dfrac{\text{m}^2}{\text{s}}$	$\dfrac{\text{ft}^2}{\text{h}}$	$\dfrac{\text{m}^2}{\text{s}}$	
Air														
−200	−128.9	0.2392	1.001	0.0079	0.0137	0.0252	1.042	0.153	2.462	0.165	0.0426	0.216	0.0557	0.760
0	−17.8	0.2400	1.005	0.014	0.0242	0.0415	1.716	0.0864	1.390	0.480	0.1239	0.675	0.1742	0.711
200	93.3	0.2414	1.011	0.0181	0.0313	0.0519	2.146	0.0602	0.969	0.862	0.2225	1.245	0.3213	0.692
400	204.4	0.2451	1.026	0.0224	0.0388	0.0624	2.580	0.0462	0.743	1.351	0.3487	1.977	0.6103	0.683
600	315.6	0.2505	1.049	0.0263	0.0455	0.0720	2.976	0.0375	0.603	1.920	0.4956	2.800	0.7223	0.686
800	426.7	0.2567	1.075	0.0300	0.0519	0.0805	3.328	0.0316	0.508	2.548	0.6576	3.698	0.9545	0.688
1000	537.8	0.263	1.101	0.0332	0.0574	0.0884	3.654	0.0272	0.438	3.250	0.8388	4.641	1.1978	0.700
1200	648.9	0.2692	1.127	0.0363	0.0628	0.0960	3.969	0.0239	0.385	4.168	1.0758	5.642	1.4562	0.712
1400	760.0	0.2755	1.153	0.0391	0.0677	0.1035	4.279	0.0214	0.344	4.836	1.2482	6.632	1.7117	0.728
N₂														
−200	−128.9	0.252	1.055	0.0079	0.0137	0.0237	0.980	0.148	2.381	0.160	0.0413	0.212	0.0547	0.756
0	−17.8	0.2484	1.040	0.0132	0.0228	0.039	1.612	0.0835	1.344	0.467	0.1205	0.635	0.1639	0.734
200	93.3	0.249	1.042	0.0173	0.0299	0.0498	2.059	0.0582	0.936	0.856	0.2209	1.194	0.3082	0.717
400	204.4	0.2515	1.053	0.021	0.0363	0.0601	2.485	0.0448	0.721	1.342	0.3464	1.864	0.4811	0.719
600	315.6	0.2562	1.073	0.0248	0.0429	0.0696	2.877	0.0362	0.582	1.923	0.4963	2.674	0.7134	0.719
800	426.7	0.262	1.097	0.0283	0.0488	0.0775	3.204	0.0305	0.481	2.541	0.6558	3.542	0.9142	0.717
1000	537.8	0.2687	1.125	0.0317	0.0549	0.0849	3.510	0.0263	0.423	3.228	0.8331	4.501	1.1617	0.720
1200	648.9	0.2755	1.153	0.0345	0.0600	0.0918	3.795	0.0221	0.356	4.154	1.0721	5.666	1.4624	0.733
1400	760.0	0.282	1.181	0.0372	0.0644	0.0982	4.060	0.0207	0.333	4.744	1.2244	6.373	1.6449	0.745
O₂														
−200	−128.9	0.2175	0.911	0.0079	0.0137	0.0272	1.124	0.169	2.719	0.161	0.0417	0.215	0.0555	0.749
0	−17.8	0.2182	0.914	0.0135	0.0234	0.044	1.819	0.096	1.545	0.458	0.1821	0.644	0.1662	0.711
200	93.4	0.2223	0.931	0.018	0.0311	0.0583	2.410	0.0665	1.070	0.877	0.2264	1.217	0.3141	0.720
400	204.4	0.2305	0.965	0.0233	0.0403	0.0712	2.943	0.0512	0.824	1.391	0.3590	1.973	0.5092	0.704
600	315.6	0.2385	0.999	0.0278	0.0481	0.0825	3.411	0.0415	0.668	1.988	0.5131	2.807	0.7245	0.707
800	426.7	0.2463	1.031	0.0317	0.0548	0.0925	3.824	0.0350	0.563	2.643	0.6822	3.677	0.9490	0.718
1000	537.8	0.2525	1.057	0.0352	0.0609	0.1018	4.208	0.0302	0.486	3.371	0.8701	4.616	1.2375	0.730
1200	648.9	0.257	1.076	0.0385	0.0666	0.1108	4.580	0.0265	0.426	4.181	1.0781	5.653	1.4590	0.739
1400	760.0	0.2615	1.095	0.0416	0.0720	0.119	4.919	0.0236	0.380	5.042	1.3013	6.737	1.7388	0.748

Temp													
0	0.248	1.038	0.0122	0.0211	0.038	1.571	0.0835	1.344	0.455	0.1174	0.588	0.1518	0.772
200	0.2495	1.045	0.0174	0.0301	0.0507	2.096	0.0582	0.936	0.871	0.2248	1.198	0.3092	0.726
400	0.2528	1.058	0.0215	0.0372	0.0613	2.534	0.0448	0.721	1.368	0.3531	1.897	0.4896	0.721
600	0.2587	1.083	0.0254	0.0439	0.0702	2.902	0.0362	0.582	1.939	0.5005	2.710	0.6995	0.715
800	0.2655	1.112	0.0288	0.0498	0.0785	3.245	0.0305	0.491	2.574	0.664	3.554	0.9173	0.724
1000	0.272	1.139	0.0317	0.0549	0.086	3.555	0.0263	0.423	3.270	0.844	4.430	1.1434	0.738
1200	0.2782	1.165	0.0347	0.0600	0.093	3.844	0.0221	0.356	4.208	1.086	5.643	1.456	0.746
1400	0.2834	1.187	0.0377	0.0652	0.0998	4.126	0.0207	0.333	4.821	1.244	6.425	1.658	0.750
H₂													
−240.0	2.46	10.30	0.014	0.0242	0.0043	0.178	0.045	0.724	0.096	0.0248	0.124	0.0320	0.756
−128.9	2.975	12.46	0.055	0.0952	0.0131	0.542	0.0105	0.169	1.248	0.3221	1.752	0.4522	0.709
−17.8	3.385	14.17	0.092	0.1592	0.0195	0.806	0.0059	0.0949	3.306	0.8533	4.593	1.1855	0.717
93.3	3.45	14.44	0.122	0.2111	0.0248	1.025	0.00415	0.0668	6.049	1.561	8.610	2.222	0.700
204.4	3.46	14.49	0.152	0.263	0.0297	1.228	0.0032	0.0515	9.281	2.395	13.719	3.541	0.676
315.6	3.47	14.53	0.18	0.311	0.0342	1.414	0.0026	0.0418	13.154	3.395	19.923	5.142	0.659
426.7	3.48	14.57	0.207	0.358	0.0394	1.629	0.0021	0.0338	18.762	4.842	28.286	7.301	0.662
537.8	3.48	14.57	0.223	0.386	0.0421	1.740	0.00186	0.0299	23.388	5.036	35.556	9.177	0.657
648.9	3.49	14.61	0.241	0.417	0.0461	1.906	0.00165	0.0265	28.813	7.437	43.125	11.331	0.667
760.0	3.50	14.65	0.257	0.445	0.0497	2.055	0.00147	0.0237	33.133	8.552	48.933	12.630	0.676
He													
−128.9	1.25	5.23	0.052	0.090	0.0395	1.633	0.021	0.338	1.881	0.486	1.981	0.5113	0.949
−17.8	1.25	5.23	0.08	0.138	0.0434	1.794	0.012	0.193	3.617	0.934	5.333	1.3744	0.678
93.3	1.25	5.23	0.0985	0.170	0.0545	2.253	0.0083	0.1335	6.566	1.695	9.493	2.550	0.691
204.4	1.25	5.23	0.118	0.204	0.066	2.728	0.0064	0.1030	10.312	2.662	14.750	3.807	0.699
315.6	1.25	5.23	0.137	0.237	0.077	3.183	0.0051	0.0821	15.098	3.897	21.490	5.547	0.702
426.7	1.25	5.23	0.156	0.270	0.088	3.638	0.0044	0.0708	20.000	5.162	28.364	7.321	0.705
537.8	1.25	5.23	0.176	0.315	0.099	4.093	0.0037	0.0595	26.757	6.906	38.054	9.822	0.703
648.9	1.25	5.23	0.194	0.336	0.109	4.506	0.0033	0.0531	33.030	8.525	47.030	12.138	0.702
760.0	1.25	5.23	0.212	0.367	0.119	4.919	0.0029	0.0467	41.034	10.590	58.483	15.094	0.701

Table A-1 (continued)

Temperature		c_p	$c_p \times 10^{-3}$	k	k	μ	$\mu \times 10^5$	ρ	ρ	ν	$\nu \times 10^4$	α	$\alpha \times 10^4$	Pr
°F	°C	$\dfrac{\text{Btu}}{\text{lb·°F}}$	$\dfrac{\text{W·s}}{\text{kg·°C}}$	$\dfrac{\text{Btu}}{\text{h·ft·°F}}$	$\dfrac{\text{W}}{\text{m·°C}}$	$\dfrac{\text{lb}}{\text{ft·h}}$	$\dfrac{\text{kg}}{\text{m·s}}$	$\dfrac{\text{lb}}{\text{ft}^3}$	$\dfrac{\text{kg}}{\text{m}^3}$	$\dfrac{\text{ft}^2}{\text{h}}$	$\dfrac{\text{m}^2}{\text{s}}$	$\dfrac{\text{f}^2}{\text{h}}$	$\dfrac{\text{m}^2}{\text{s}}$	
Argon														
0	−17.8	0.124	0.519	0.009	0.0156	0.049	2.025	0.1135	1.826	0.432	0.1115	0.639	0.1649	0.675
200	93.3	0.124	0.519	0.012	0.0208	0.064	2.646	0.0792	1.274	0.808	0.2085	1.221	0.3151	0.661
400	204.4	0.124	0.519	0.0147	0.0254	0.078	3.224	0.0607	0.977	1.285	0.3317	1.952	0.5038	0.658
600	315.6	0.124	0.519	0.0172	0.0298	0.0905	3.741	0.0492	0.792	1.839	0.4746	2.819	0.7276	0.652
800	426.7	0.124	0.519	0.0194	0.0336	0.102	4.217	0.0415	0.668	2.458	0.6344	3.769	0.9728	0.652
1000	537.8	0.124	0.519	0.0218	0.0377	0.1125	4.651	0.0358	0.576	3.142	0.8190	4.911	1.2675	0.640
1200	648.9	0.124	0.519	0.0234	0.0405	0.1225	5.064	0.0314	0.505	3.901	1.0068	6.010	1.5512	0.649
1400	760.0	0.124	0.519	0.0252	0.0436	0.1315	5.436	0.0281	0.452	4.680	1.2079	7.231	1.8663	0.648
Neon														
200	93.3	0.246	1.030	0.0324	0.0561	0.0884	3.664	0.042	0.676	2.104	0.5430	3.136	0.8094	0.670
400	204.4	0.246	1.030	0.0384	0.0664	0.1040	4.299	0.032	0.515	3.250	0.8388	4.875	1.2582	0.668
600	315.6	0.246	1.030	0.0438	0.0758	0.1190	4.919	0.026	0.418	4.577	1.1813	6.846	1.7670	0.668
800	426.7	0.246	1.030	0.0488	0.0844	0.1325	5.476	0.022	0.354	6.023	1.5545	9.014	2.2265	0.668
1000	537.8	0.246	1.030	0.0535	0.0926	0.1450	5.994	0.019	0.306	7.632	1.9698	11.442	2.9532	0.666
1200	648.9	0.246	1.030	0.0585	0.1012	0.1585	6.552	0.0166	0.267	9.548	2.4643	14.325	3.6972	0.666
1400	760.0	0.246	1.030	0.0625	0.1081	0.1700	7.028	0.0149	0.240	14.093	3.6374	17.047	4.3998	0.666
CO_2														
0	17.8	0.19	0.795	0.0077	0.0133	0.031	1.281	0.1315	2.116	0.236	0.0609	0.308	0.0795	0.765
200	93.3	0.218	0.913	0.0127	0.0220	0.0433	1.790	0.0915	1.472	0.473	0.1221	0.636	0.1642	0.743
400	204.4	0.238	0.996	0.0177	0.0306	0.0548	2.265	0.0702	1.130	0.781	0.2016	1.058	0.2731	0.737
600	315.6	0.2554	1.069	0.0226	0.0391	0.0652	2.695	0.0570	0.917	1.144	0.2953	1.551	0.4003	0.736
800	426.7	0.2684	1.124	0.0273	0.0472	0.074	3.059	0.0480	0.772	1.542	0.3980	2.119	0.5469	0.727
1000	537.8	0.2793	1.169	0.0317	0.0549	0.0827	3.419	0.0415	0.668	1.993	0.5144	2.733	0.7054	0.728
1200	648.9	0.2898	1.213	0.0358	0.0619	0.091	3.762	0.0364	0.586	2.500	0.6453	3.393	0.8757	0.736
1400	760.0	0.2975	1.246	0.0396	0.0685	0.0988	4.084	0.0325	0.523	3.040	0.7846	4.095	1.0569	0.742

NH₃														
0	−17.8	0.522	2.185	0.0117	0.0202	0.0213	0.880	0.0441	0.710	0.483	0.1247	0.508	0.1311	0.95
200	93.3	0.532	2.227	0.0192	0.0332	0.0303	1.253	0.0307	0.494	0.999	0.2578	1.173	0.3028	0.84
400	204.4	0.574	2.403	0.0280	0.0484	0.0394	1.629	0.0236	0.380	1.669	0.4308	2.064	0.5327	0.807
600	315.6	0.625	2.617	0.0397	0.0687	0.0479	1.980	0.0192	0.309	2.495	0.6440	3.307	0.8535	0.755
800	426.7	0.675	2.826	0.0537	0.0929	0.0557	2.303	0.0161	0.259	3.460	0.8930	4.938	1.2745	0.700
CH₄														
0	−17.8	0.507	2.123	0.0157	0.0272	0.0237	0.980	0.0455	0.732	0.521	0.1345	0.679	0.1752	0.765
200	93.3	0.579	2.424	0.0255	0.0442	0.0317	1.310	0.0317	0.510	1.000	0.2581	1.388	0.3582	0.720
400	204.4	0.674	2.822	0.0358	0.0619	0.038	1.571	0.0243	0.391	1.564	0.4037	2.185	0.5639	0.715
600	315.6	0.772	3.232	0.0505	0.0874	0.044	1.819	0.0197	0.317	2.234	0.5766	3.320	0.8568	0.672
Freon-11														
0	−17.8	0.124	0.519	0.00412	0.00713	0.0232	0.959	0.0398	0.640	0.583	0.1505	0.829	0.2140	0.701
100	37.8	0.134	0.561	0.00519	0.00898	0.0274	1.133	0.0322	0.518	0.851	0.2196	1.205	0.3110	0.706
200	93.3	0.145	0.607	0.00627	0.01085	0.0312	1.290	0.0278	0.447	1.122	0.2896	1.555	0.4013	0.722
Steam														
212	100.0	0.451	1.888	0.0145	0.0251	0.0313	1.294	0.0372	0.599	0.842	0.2173	0.864	0.2230	0.96
300	148.9	0.456	1.909	0.0171	0.0296	0.0360	1.488	0.0328	0.528	1.098	0.2834	1.14	0.2942	0.95
400	204.4	0.462	1.934	0.0200	0.0346	0.0407	1.683	0.0288	0.463	1.422	0.3670	1.50	0.3872	0.94
600	315.6	0.477	1.997	0.0257	0.0445	0.0511	2.112	0.0233	0.375	2.196	0.5668	2.31	0.5962	0.94
800	426.7	0.494	2.068	0.0321	0.0555	0.0612	2.530	0.0196	0.315	3.078	0.7944	3.32	0.8569	0.91
1000	537.8	0.51	2.135	0.0388	0.0671	0.0691	2.857	0.0169	0.272	4.068	1.0500	4.50	1.1615	0.91
2000	1093.3	0.60	2.512	0.076	0.1315	0.1091	4.510	0.0100	0.161	10.908	2.8153	12.7	3.2779	0.86

TABLE A-2 Physical Properties of Liquids

Temperature		c_p		k		μ		ρ		ν		α		Pr
°F	°C	$\dfrac{Btu}{lb\cdot°F}$	$\dfrac{W\cdot s}{kg\cdot°C} \times 10^{-3}$	$\dfrac{Btu}{h\cdot ft\cdot°F}$	$\dfrac{W}{m\cdot°C}$	$\dfrac{lb}{ft\cdot h}$	$\dfrac{kg}{m\cdot s} \times 10^4$	$\dfrac{lb}{ft^3}$	$\dfrac{kg}{m^3}$	$\dfrac{ft^2}{h}$	$\dfrac{m^2}{s} \times 10^6$	$\dfrac{ft^2}{h}$	$\dfrac{m^2}{s} \alpha + 10^6$	
Water														
32	0	1.0293	4.309	0.337	0.583	4.32	17.859	62.54	1006.3	0.0690	0.781	0.0052	0.134	13.2
200	93.3	1.0039	4.203	0.393	0.680	0.738	3.051	60.2	968.6	0.0122	0.315	0.0065	0.168	1.88
400	204.4	1.075	4.501	0.382	0.661	0.32	1.323	53.62	862.7	0.0059	0.152	0.0066	0.170	0.91
600	315.6	1.525	6.385	0.293	0.507	0.215	0.889	42.37	681.7	0.0050	0.129	0.0045	0.116	1.08
Dowtherm A														
200	93.3	0.432	1.809	0.0863	0.1493	2.27	11.203	62.6	1007.2	0.0432	1.115	0.0030	0.077	13.56
400	204.4	0.600	2.512	0.105	0.1817	1.14	4.713	56.8	913.9	0.0200	0.516	0.0030	0.077	6.51
600	315.6	0.700	2.931	0.1037	0.1794	0.727	3.005	50.5	812.5	0.0143	0.369	0.0028	0.072	4.90
Methyl Alcohol														
0	17.8	0.57	2.386	0.124	0.2146	2.80	11.575	51.3	825.4	0.0545	1.407	0.0042	0.108	12.87
100	37.8	0.615	2.575	0.1205	0.2085	1.15	4.754	48.1	773.9	0.0239	0.617	0.0040	0.103	5.87
200	93.3	0.65	2.721	0.117	0.2024	0.666	2.753	43.1	693.5	0.0154	0.374	0.0041	0.106	3.70
Freon-11														
0	17.8	0.198	0.829	0.06	0.104	1.639	6.776	98.27	1581.2	0.0166	0.428	0.0030	0.077	5.40
100	37.8	0.212	0.888	0.053	0.0917	0.920	3.803	90.19	1451.2	0.0102	0.263	0.0023	0.059	3.68
200	93.3	0.225	0.942	0.046	0.0796	0.637	2.633	80.94	1302.3	0.0078	0.201	0.0022	0.057	3.12
Freon-114														
0	17.8	0.23	0.963	0.044	0.0761	1.452	6.003	98.62	1586.8	0.0147	0.379	0.0017	0.044	7.58
100	37.8	0.2412	1.010	0.0353	0.0611	0.809	3.344	88.37	1421.9	0.0091	0.235	0.0012	0.031	5.53
200	93.3	0.2627	1.100	0.027	0.0467	0.600	2.480	79.0	1271.1	0.0075	0.194	0.0011	0.028	5.84
Gasoline														
0	17.8	0.447	1.871	0.110	0.1903	2.60	10.748	49.7	799.7	0.0523	1.350	0.0049	0.126	10.58
200	93.3	0.565	2.365	0.103	0.1782	0.745	3.080	42.7	687.0	0.0174	0.449	0.0042	0.108	4.08
400	204.4	0.683	2.860	0.0967	0.1673	0.336	1.389	36.8	592.1	0.0091	0.235	0.0038	0.098	2.37

Kerosene

0	17.8	0.430	1.800	0.101	0.1748	17.1	70.69	52.5	844.7	0.3257	8.406	0.0044	0.114	72.8
200	93.3	0.545	2.282	0.095	0.1644	1.59	6.57	47.4	762.7	0.0335	0.865	0.0036	0.093	9.12
400	204.4	0.655	2.742	0.0892	0.1543	0.625	2.58	42.4	682.2	0.0147	0.379	0.0032	0.083	4.58
600	315.6	0.745	3.119	0.0829	0.1434	0.31	1.28	38.1	613.0	0.0081	0.209	0.0028	0.072	2.78

SAE 10 Petroleum Lubricating Oil

0	17.8	0.411	1.721	0.09375	0.1618	4730.	19,550	55.6	894.6	85.07	2195.	0.0038	0.0981	20,750
200	93.3	0.52	2.177	0.0884	0.1530	11.88	49.112	52.25	840.7	0.2273	5.867	0.0030	0.0774	69.7
300	148.9	0.575	2.407	0.0852	0.1474	4.503	18.615	48.75	784.4	0.0923	2.382	0.0029	0.0748	30.4

Ethylene Glycol

60	15.6	0.556	2.328	0.169	0.2924	62.1	256.72	96.4	1116.6	0.8948	23.095	0.0043	0.1110	204
100	37.8	0.581	2.432	0.1595	0.2760	25.1	103.76	68.7	1105.4	0.3653	9.428	0.0039	0.1007	91.4
200	93.3	0.644	2.696	0.135	0.2336	5.67	23.44	66.2	1065.2	0.0856	2.209	0.0031	0.0800	27.05
300	148.9	0.706	2.956	0.111	0.1921	2.295		63.3	1018.5	0.0362	0.934	0.0024	0.0619	14.6

H_2 (liquid)

430	256.7	1.91	7.997	0.0636	0.1100	0.0447	0.1848	4.67	75.14	0.0095	0.245	0.0071	0.1832	1.34
410	245.6	4.44	18.589	0.0769	0.1377	0.0204	0.0843	3.69	59.37	0.0055	0.142	0.0048	0.1239	1.135

N_2 (liquid)

210	134.4	0.500	2.093	0.041	0.0709	0.162	0.670	34.5	555.1	0.0046	0.119	0.0022	0.0568	1.97
110	78.9	0.474	1.985	0.095	0.1644	0.756	3.125	54.0	868.9	0.0140	0.361	0.0035	0.0903	3.75

NH_3 (liquid)

0	17.8	1.08	4.522	0.29	0.5018	0.567	2.344	42.0	675.8	0.0135	0.348	0.0063	0.1626	2.112
100	93.3	1.17	4.898	0.29	0.5018	0.172	0.711	35.6	572.8	0.0048	0.124	0.0069	0.1781	0.694

TABLE A-3 Physical Properties of Liquid Metals

Temperature °F	Temperature °C	c_p $\frac{\text{Btu}}{\text{lb}\cdot°\text{F}}$	$c_p \times 10^{-3}$ $\frac{\text{W}\cdot\text{s}}{\text{kg}\cdot°\text{C}}$	k $\frac{\text{Btu}}{\text{h}\cdot\text{ft}\cdot°\text{F}}$	k $\frac{\text{W}}{\text{m}\cdot°\text{C}}$	μ $\frac{\text{lb}}{\text{ft}\cdot\text{h}}$	$\mu \times 10^4$ $\frac{\text{kg}}{\text{m}\cdot\text{s}}$	ρ $\frac{\text{lb}}{\text{ft}^3}$	ρ $\frac{\text{kg}}{\text{m}^3}$	ν $\frac{\text{ft}^2}{\text{h}}$	$\nu \times 10^6$ $\frac{\text{m}^2}{\text{s}}$	α $\frac{\text{ft}^2}{\text{h}}$	$\alpha \times 10^6$ $\frac{\text{m}^2}{\text{s}}$	Pr
Sodium														
200	93.3	0.3305	1.384	49.1	84.96	1.725	7.131	57.9	931.6	0.02979	0.7689	2.566	56.29	0.0116
400	204.4	0.3199	1.339	46.7	80.81	1.095	4.521	56.4	907.5	0.01941	0.5010	2.588	66.80	0.0075
600	315.6	0.3115	1.304	43.8	75.78	0.797	3.294	54.6	878.5	0.01459	0.3766	2.575	66.47	0.00567
800	426.7	0.3049	1.277	40.1	69.39	0.61	2.522	53.0	852.8	0.01150	0.2968	2.482	64.05	0.00464
1000	537.8	0.302	1.264	37.2	64.37	0.56	2.315	51.2	823.8	0.01093	0.2821	2.406	62.09	0.00455
1200	648.9	0.3011	1.261	35.0	60.56	0.475	1.964	49.1	790.0	0.00967	0.2496	2.367	61.10	0.00408
1400	760.0	0.3033	1.270	32.7	56.58	0.415	1.716	47.7	767.5	0.00870	0.2245	2.260	58.34	0.00385
Na K (56% Na, 44% K)														
200	93.3	0.270	1.130	14.9	25.78	1.36	5.622	55.3	889.8	0.02459	0.6347	0.998	25.76	0.0246
400	204.4	0.260	1.089	15.3	26.47	0.92	3.803	53.8	865.6	0.01710	0.4414	1.093	28.23	0.0155
600	315.6	0.255	1.068	15.7	27.17	0.71	2.935	52.1	838.3	0.01362	0.3515	1.182	30.50	0.0115
800	426.7	0.251	1.051	16.0	27.68	0.52	2.150	50.6	814.2	0.01028	0.2652	1.259	32.52	0.0081
1000	537.8	0.250	1.047	16.0	27.68	0.49	2.026	49.0	788.4	0.01000	0.2581	1.306	33.71	0.0076
1200	648.9	0.251	1.051	16.0	27.68	0.41	1.695	47.2	759.5	0.00868	0.2240	1.350	34.86	0.0064
Potassium														
800	426.7	0.183	0.766	22.8	39.45	0.51	2.108	46.1	741.7	0.0140	0.2839	2.702	69.74	0.0041
1000	537.8	0.182	0.762	21.1	36.51	0.414	1.711	44.4	714.4	0.0093	0.2400	2.611	67.39	0.0036
1200	648.9	0.183	0.766	19.5	33.74	0.354	1.463	42.9	690.3	0.0082	0.2116	2.484	64.10	0.0033
1400	760.0	0.187	0.783	18.0	31.15	0.322	1.331	41.5	667.7	0.0077	0.1987	2.319	59.86	0.0033
Lithium														
400	204.4	1.0425	4.365	26.8	46.37	1.31	5.416	31.65	509.2	0.0413	1.1098	0.8121	20.96	0.051
600	315.6	1.02	4.270	24.9	43.08	1.08	4.465	31.0	498.8	0.0348	0.8982	0.7874	20.32	0.0443
800	426.7	1.0057	4.211	22.1	38.24	0.95	3.927	30.4	489.1	0.0312	0.8053	0.7227	18.65	0.0432
1000	537.8	0.9962	4.171	17.6	30.45	0.84	3.473	29.6	476.3	0.0283	0.7304	0.5967	15.40	0.0476
Mercury														
0	−17.8	0.0338	0.1415	5.64	9.76	4.435	18.334	851.9	13,707.1	0.0052	0.1342	0.1952	5.038	0.0266
200	93.3	0.0326	0.1365	6.00	10.38	2.957	12.224	833.4	13,409.4	0.0035	0.0903	0.2177	5.619	0.0161
400	204.4	0.0324	0.1356	7.3	12.63	2.43	10.046	818.4	13,168.1	0.0029	0.0748	0.2746	7.087	0.0108
600	315.6	0.0342	0.1432	7.9	13.67	2.27	9.384	802.6	12,913.8	0.0028	0.0723	0.2877	7.426	0.00983

TABLE A-4 Physical Properties of Metals and Nonmetals

Material	Temperature °F	Temperature °C	c_p Btu/(lb·°F)	$c_p \times 10^{-3}$ W·s/(kg·°C)	k Btu/(h·ft·°F)	k W/(m·°C)	ρ lb/ft³	ρ kg/m³	α ft²/h	$\alpha \times 10^6$ m²/s
Metals										
Aluminum	32	0	0.208	0.871	117	202.4	169	2719	3.33	85.9
Copper	32	0	0.091	0.381	224	387.6	558	8978	4.42	114.1
Gold	68	20	0.030	0.126	169	292.4	1204	19,372	4.68	120.8
Iron, pure	32	0	0.104	0.435	36	62.3	491	7900	0.70	18.1
Cast iron ($c \simeq 4\%$)	68	20	0.10	0.417	30	51.9	454	7304	0.66	17.0
Lead	70	21.1	0.030	0.126	20	34.6	705	11,343	0.95	25.5
Mercury	32	0	0.033	0.138	4.83	8.36	849	13,660	0.172	4.44
Nickel	32	0	0.103	0.431	34.4	59.52	555	8930	0.60	15.5
Silver	32	0	0.056	0.234	242	418.7	655	10,539	6.60	170.4
Steel, mild	32	0	0.11	0.460	26	45.0	490	7884	0.48	12.4
Tungsten	32	0	0.032	0.134	92	159.2	1204	19,372	2.39	61.7
Zinc	32	0	0.091	0.381	65	112.5	446	7176	1.60	41.3
Nonmetals										
Asbestos	32	0	0.25	1.047	0.087	0.151	36	579	0.010	0.258
Brick, fire clay	400	204.4	0.20	0.837	0.58	1.004	144	2317	0.020	0.516
Cork, ground	100	37.8	0.48	2.010	0.024	0.042	8	128.7	0.006	0.155
Glass, Pyrex			0.20	0.837	0.68	1.177	150	2413	0.023	0.594
Granite	32	0	0.19	0.796	1.6	2.768	168	2703	0.050	1.291
Ice	32	0	0.49	2.051	1.28	2.215	57	917	0.046	1.187
Oak, across grain	85	29.4	0.41	1.716	0.111	0.192	44	708	0.0062	0.160
Pine, across grain	85	29.4	0.42	1.758	0.092	0.159	37	595	0.0059	0.152
Quartz sand, dry			0.19	0.796	0.15	0.260	103	1657	0.008	0.206
Rubber, soft			0.45	1.884	0.10	0.173	69	1110	0.003	0.077

Appendix B

Roots of Transcendental Equations

TABLE B-1   First Six Roots β_n of $\beta \tan \beta = c$

c	β_1	β_2	β_3	β_4	β_5	β_6
0	0	3.1416	6.2832	9.4248	12.5664	15.7080
0.001	0.0316	3.1419	6.2833	9.4249	12.5665	15.7080
0.002	0.0447	3.1422	6.2835	9.4250	12.5665	15.7081
0.004	0.0632	3.1429	6.2838	9.4252	12.5667	15.7082
0.006	0.0774	3.1435	6.2841	9.4254	12.5668	15.7083
0.008	0.0893	3.1441	6.2845	9.4256	12.5670	15.7085
0.01	0.0998	3.1448	6.2848	9.4258	12.5672	15.7086
0.02	0.1410	3.1479	6.2864	9.4269	12.5680	15.7092
0.04	0.1987	3.1543	6.2895	9.4290	12.5696	15.7105
0.06	0.2425	3.1606	6.2927	9.4311	12.5711	15.7118
0.08	0.2791	3.1668	6.2959	9.4333	12.5727	15.7131
0.1	0.3111	3.1731	6.2991	9.4354	12.5743	15.7143
0.2	0.4328	3.2039	6.3148	9.4459	12.5823	15.7207
0.3	0.5218	3.2341	6.3305	9.4565	12.5902	15.7270
0.4	0.5932	3.2636	6.3461	9.4670	12.5981	15.7334
0.5	0.6533	3.2923	6.3616	9.4775	12.6060	15.7397
0.6	0.7051	3.3204	6.3770	9.4879	12.6139	15.7460
0.7	0.7506	3.3477	6.3923	9.4983	12.6218	15.7524
0.8	0.7910	3.3744	6.4074	9.5087	12.6296	15.7587
0.9	0.8274	3.4003	6.4224	9.5190	12.6375	15.7650
1.0	0.8603	3.4256	6.4373	9.5293	12.6453	15.7713
1.5	0.9882	3.5422	6.5097	9.5801	12.6841	15.8026
2.0	1.0769	3.6436	6.5783	9.6296	12.7223	15.8336
3.0	1.1925	3.8088	6.7040	9.7240	12.7966	15.8945
4.0	1.2646	3.9352	6.8140	9.8119	12.8678	15.9536
5.0	1.3138	4.0336	6.9096	9.8928	12.9352	16.0107
6.0	1.3496	4.1116	6.9924	9.9667	12.9988	16.0654
7.0	1.3766	4.1746	7.0640	10.0339	13.0584	16.1177
8.0	1.3978	4.2264	7.1263	10.0949	13.1141	16.1675
9.0	1.4149	4.2694	7.1806	10.1502	13.1660	16.2147
10.0	1.4289	4.3058	7.2281	10.2003	13.2142	16.2594
15.0	1.4729	4.4255	7.3959	10.3898	13.4078	16.4474
20.0	1.4961	4.4915	7.4954	10.5117	13.5420	16.5864
30.0	1.5202	4.5615	7.6057	10.6543	13.7085	16.7691
40.0	1.5325	4.5979	7.6647	10.7334	13.8048	16.8794
50.0	1.5400	4.6202	7.7012	10.7832	13.8666	16.9519
60.0	1.5451	4.6353	7.7259	10.8172	13.9094	17.0026
80.0	1.5514	4.6543	7.7573	10.8606	13.9644	17.0686
100.0	1.5552	4.6658	7.7764	10.8871	13.9981	17.1093
∞	1.5708	4.7124	7.8540	10.9956	14.1372	17.2788

Roots are all real if $c > 0$.

TABLE B-2　First Six Roots β_n of $\beta \cot \beta = -c$

c	β_1	β_2	β_3	β_4	β_5	β_6
-1.0	0	4.4934	7.7253	10.9041	14.0662	17.2208
-0.995	0.1224	4.4945	7.7259	10.9046	14.0666	17.2210
-0.99	0.1730	4.4956	7.7265	10.9050	14.0669	17.2213
-0.98	0.2445	4.4979	7.7278	10.9060	14.0676	17.2219
-0.97	0.2991	4.5001	7.7291	10.9069	14.0683	17.2225
-0.96	0.3450	4.5023	7.7304	10.9078	14.0690	17.2231
-0.95	0.3854	4.5045	7.7317	10.9087	14.0697	17.2237
-0.94	0.4217	4.5068	7.7330	10.9096	14.0705	17.2242
-0.93	0.4551	4.5090	7.7343	10.9105	14.0712	17.2248
-0.92	0.4860	4.5112	7.7356	10.9115	14.0719	17.2254
-0.91	0.5150	4.5134	7.7369	10.9124	14.0726	17.2260
-0.90	0.5423	4.5157	7.7382	10.9133	14.0733	17.2266
-0.85	0.6609	4.5268	7.7447	10.9179	14.0769	17.2295
-0.8	0.7593	4.5379	7.7511	10.9225	14.0804	17.2324
-0.7	0.9208	4.5601	7.7641	10.9316	14.0875	17.2382
-0.6	1.0528	4.5822	7.7770	10.9408	14.0946	17.2440
-0.5	1.1656	4.6042	7.7899	10.9499	14.1017	17.2498
-0.4	1.2644	4.6261	7.8028	10.9591	14.1088	17.2556
-0.3	1.3525	4.6479	7.8156	10.9682	14.1159	17.2614
-0.2	1.4320	4.6696	7.8284	10.9774	14.1230	17.2672
-0.1	1.5044	4.6911	7.8412	10.9865	14.1301	17.2730
0	1.5708	4.7124	7.8540	10.9956	14.1372	17.2788
0.1	1.6320	4.7335	7.8667	11.0047	14.1443	17.2845
0.2	1.6887	4.7544	7.8794	11.0137	14.1513	17.2903
0.3	1.7414	4.7751	7.8920	11.0228	14.1584	17.2961
0.4	1.7906	4.7956	7.9046	11.0318	14.1654	17.3019
0.5	1.8366	4.8158	7.9171	11.0409	14.1724	17.3076
0.6	1.8798	4.8358	7.9295	11.0498	14.1795	17.3134
0.7	1.9203	4.8556	7.9419	11.0588	14.1865	17.3192
0.8	1.9586	4.8751	7.9542	11.0677	14.1935	17.3249
0.9	1.9947	4.8943	7.9665	11.0767	14.2005	17.3306
1.0	2.0288	4.9132	7.9787	11.0856	14.2075	17.3364
1.5	2.1746	5.0037	8.0385	11.1296	14.2421	17.3649
2.0	2.2889	5.0870	8.0962	11.1727	14.2764	17.3932
3.0	2.4557	5.2329	8.2045	11.2560	14.3434	17.4490
4.0	2.5704	5.3540	8.3029	11.3349	14.4080	17.5034
5.0	2.6537	5.4544	8.3914	11.4086	14.4699	17.5562
6.0	2.7165	5.5378	8.4703	11.4773	14.5288	17.6072
7.0	2.7654	5.6078	8.5406	11.5408	14.5847	17.6562
8.0	2.8044	5.6669	8.6031	11.5994	14.6374	17.7032
9.0	2.8363	5.7172	8.6587	11.6532	14.6870	17.7481
10.0	2.8628	5.7606	8.7083	11.7027	14.7335	17.7908
15.0	2.9476	5.9080	8.8898	11.8959	14.9251	17.9742
20.0	2.9930	5.9921	9.0019	12.0250	15.0625	18.1136
30.0	3.0406	6.0831	9.1294	12.1807	15.2380	18.3018
40.0	3.0651	6.1311	9.1987	12.2688	15.3417	18.4180
50.0	3.0801	6.1606	9.2420	12.3247	15.4090	18.4953
60.0	3.0901	6.1805	9.2715	12.3632	15.4559	18.5497
80.0	3.1028	6.2058	9.3089	12.4124	15.5164	18.6209
100.0	3.1105	6.2211	9.3317	12.4426	15.5537	18.6650
∞	3.1416	6.2832	9.4248	12.5664	15.7080	18.8496

Roots are all real if $c > -1$.

Appendix C

Exponential Integrals $E_n(x)$

The nth exponential integral $E_n(x)$ of the argument x is defined by

$$E_n(x) = \int_1^\infty e^{-xt} t^{-n} \, dt = \int_0^1 e^{-(x/\mu)} \mu^{n-2} \, d\mu \tag{1}$$

Here we present a few useful relations for the functions $E_n(x)$ and a concise tabulation of the first four of these functions. The reader is referred to the books by Chandrasekhar [1, pp. 373–378] and Kourganoff [2, pp. 253–271] for a detailed discussion of the properties of the functions $E_n(x)$, and to Case, de Hoffmann, and Placzek [3, pp. 153–162] and "Handbook of Mathematical Functions," [4, pp. 228–231] for the more complete tables.
From Eq. (1), for $x = 0$, we write

$$E_n(0) = \int_1^\infty t^{-n} \, dt = \begin{cases} +\infty & \text{for } n = 1 \tag{2a} \\[2mm] \dfrac{1}{n-1} & \text{for } n = 2, 3, 4, \dots \end{cases} \tag{2b}$$

By direct differentiation of Eq. (1) we obtain

$$\frac{d}{dx} E_n(x) = \begin{cases} -\dfrac{1}{x} e^{-x} & \text{for } n = 1 \tag{3a} \\[2mm] -E_{n-1}(x) & \text{for } n = 2, 3, 4, \dots \end{cases} \tag{3b}$$

Conversely

$$\int E_n(x) \, dx = -E_{n+1}(x) \tag{4}$$

Series expansions of $E_1(x)$, $E_2(x)$, and $E_3(x)$ are of the form

$$E_1(x) = -(\gamma + \ln x) + x - \frac{x^2}{2!\,2} + \frac{x^3}{3!\,3} - \cdots \tag{5}$$

$$E_2(x) = 1 + x(\gamma - 1 + \ln x) - \frac{x^2}{2!\,1} + \frac{x^3}{3!\,2} - \frac{x^4}{4!\,3} + \cdots \tag{6}$$

$$E_3(x) = \frac{1}{2} - x + \frac{1}{2} x^2 \left(-\gamma + \frac{3}{2} - \ln x \right) + \frac{x^3}{3!\,1} - \frac{x^4}{4!\,2} + \cdots \tag{7}$$

and $\gamma = 0.577216 \cdots$ is Euler's constant.
For large values of x, the asymptotic expansion for $E_n(x)$ is given as

$$E_n(x) \cong \frac{e^{-x}}{x} \left[1 - \frac{n}{x} + \frac{n(n+1)}{x^2} - \frac{n(n+1)(n+2)}{x^3} + \cdots \right] \tag{8}$$

TABLE C-1   Values of Functions $E_n(x)$†

x	E_1	E_2	E_3	E_4
0.00	∞	1.0000	0.5000	0.3333
0.01	4.0379	0.9497	0.4903	0.3284
0.02	3.3547	0.9131	0.4810	0.3235
0.03	2.9591	0.8817	0.4720	0.3188
0.04	2.6813	0.8535	0.4633	0.3141
0.05	2.4679	0.8278	0.4549	0.3095
0.06	2.2953	0.8040	0.4468	0.3050
0.07	2.1508	0.7818	0.4388	0.3006
0.08	2.0269	0.7610	0.4311	0.2962
0.09	1.9187	0.7412	0.4236	0.2919
0.10	1.8229	0.7225	0.4163	0.2877
0.15	1.4645	0.6410	0.3823	0.2678
0.20	1.2227	0.5742	0.3519	0.2494
0.25	1.0443	0.5177	0.3247	0.2325
0.30	0.9057	0.4691	0.3000	0.2169
0.35	0.7942	0.4267	0.2777	0.2025
0.40	0.7024	0.3894	0.2573	0.1891
0.45	0.6253	0.3562	0.2387	0.1767
0.50	0.5598	0.3266	0.2216	0.1652
0.60	0.4544	0.2762	0.1916	0.1446
0.70	0.3738	0.2349	0.1661	0.1268
0.80	0.3106	0.2009	0.1443	0.1113
0.90	0.2602	0.1724	0.1257	0.0978
1.00	0.2194	0.1485	0.1097	0.0861
1.10	0.1860	0.1283	0.0959	0.0758
1.20	0.1584	0.1111	0.0839	0.0668
1.30	0.1355	0.0964	0.0736	0.0590
1.40	0.1162	0.0839	0.0646	0.0521
1.50	0.1000	0.0731	0.0567	0.0460
1.60	0.0863	0.0638	0.0499	0.0407
1.70	0.0747	0.0558	0.0439	0.0360
1.80	0.0647	0.0488	0.0387	0.0319
1.90	0.0562	0.0428	0.0341	0.0282
2.0	4.890×10^{-2}	3.753×10^{-2}	3.013×10^{-2}	2.502×10^{-2}
2.2	3.719	2.898	2.352	1.969
2.4	2.844	2.246	1.841	1.552
2.6	2.185	1.746	1.443	1.225
2.8	1.686	1.362	1.134	0.968
3.0	1.305	1.064	0.893	0.767
3.5	6.970×10^{-3}	5.802×10^{-3}	4.945×10^{-3}	4.296×10^{-3}
4.0	3.779	3.198	2.761	2.423
4.5	2.073	1.779	1.552	1.374
5.0	1.148	0.996	0.878	0.783

† This table is abbreviated from Case, de Hoffmann, and Placzek [3].

It is apparent from Eq. (8) that

$$E_n(x) \cong \frac{e^{-x}}{x} \text{ as } x \to \infty \tag{9}$$

The first four of the functions $E_n(x)$ for values of x from 0 to 5 are given in Table C-1.

REFERENCES

1   Chandrasekhar, S.: "Radiative Transfer," Oxford University Press, London, 1950; also Dover Publications, Inc., New York, 1960.
2   Kourganoff, V.: "Basic Methods in Transfer Problems," Dover Publications, Inc., New York, 1963.
3   Case, K. M., F. de Hoffmann, and G. Placzek: "Introduction to the Theory of Neutron Diffusion," Los Alamos Scientific Laboratory, Los Alamos, N.M., 1953.
4   Abramowitz, M., and I. A. Stegun (eds.): "Handbook of Mathematical Functions," Dover Publications, Inc., New York, 1965.

Appendix D

Dimensional Data for Tubes and Pipes

Tubes[†]

Outside Diam., in	BWG Gauge	Wall Thickness, in	Inside Diam., in
$\frac{1}{2}$	16	0.065	0.370
$\frac{1}{2}$	18	0.049	0.402
$\frac{1}{2}$	20	0.035	0.430
$\frac{5}{8}$	14	0.083	0.459
$\frac{5}{8}$	16	0.065	0.495
$\frac{5}{8}$	18	0.049	0.527
$\frac{5}{8}$	20	0.035	0.555
$\frac{3}{4}$	12	0.109	0.532
$\frac{3}{4}$	14	0.083	0.584
$\frac{3}{4}$	16	0.065	0.620
$\frac{3}{4}$	18	0.049	0.652
$\frac{3}{4}$	20	0.035	0.680
1	12	0.109	0.782
1	14	0.083	0.834
1	16	0.065	0.870
1	18	0.049	0.902
1	20	0.035	0.930
$1\frac{1}{4}$	12	0.109	1.032
$1\frac{1}{4}$	14	0.083	1.084
$1\frac{1}{4}$	16	0.065	1.120
$1\frac{1}{4}$	18	0.049	1.152
$1\frac{1}{2}$	12	0.109	1.282
$1\frac{1}{2}$	14	0.083	1.334
$1\frac{1}{2}$	16	0.065	1.370
$1\frac{1}{2}$	18	0.049	1.402
2	12	0.109	1.782
2	14	0.083	1.834
2	16	0.065	1.870

† Based on data from Tubular Exchanger Manufacturers Association.

Steel Pipes†

Nominal Pipe Size, in	Outside Diam., in	Schedule No.	Wall Thickness, in	Inside Diam., in
$\frac{1}{8}$	0.405	40	0.068	0.269
		80	0.095	0.215
$\frac{1}{4}$	0.540	40	0.088	0.364
		80	0.119	0.302
$\frac{3}{8}$	0.675	40	0.091	0.493
		80	0.126	0.423
$\frac{1}{2}$	0.840	40	0.109	0.622
		80	0.147	0.546
$\frac{3}{4}$	1.050	40	0.113	0.824
		80	0.154	0.742
1	1.315	40	0.133	1.049
		80	0.179	0.957
$1\frac{1}{2}$	1.900	40	0.145	1.610
		80	0.200	1.500
2	2.375	40	0.154	2.067
		80	0.218	1.939
3	3.500	40	0.216	3.068
		80	0.300	2.900
4	4.500	40	0.237	4.026
		80	0.337	3.826
5	5.563	40	0.258	5.047
		80	0.375	4.813
6	6.625	40	0.280	6.055
		80	0.432	5.761
10	10.75	40	0.365	10.020
		60	0.500	9.750

‡ Based on ASA Standards B36.10.

Appendix E

List of Symbols

Symbol	Quantity	Units	
		English Engineering	SI
a	Lateral surface area (Chap. 3)	ft^2	m^2
a_f	Heat-transfer area of the fin (Chap. 3)	ft^2	m^2
A	Area	ft^2	m^2
Bi	Biot number ($= hL/k$) [Eq. (2-43a)]		
c	Molal concentration (Chap. 16)	lb mol/ft^3	kg mol/m^3
c_D	Average drag coefficient (Chap. 9)		
C_{min}, C_{max}	Minimum and maximum heat capacity [Eq. (15-25)]	Btu/h·°F	W/K
c_p	Specific heat under constant pressure	Btu/lb·°F	J/kg·°C
D, d	Diameter	ft	m
D, D_{AB}	Diffusion coefficient (Chap. 16)	ft^2/h	m^2/s
D_e	Equivalent diameter	ft	m
e	Specific internal energy per unit mass (Chap. 6)	Btu/lb	J/kg
$E_n(z)$	Exponential integral of the nth kind (Chap. 3)		
E	Eckert number ($= U_\infty^2/c_p \Delta T$) (Chap. 6)		
f	Friction factor (Chaps. 6, 9)		
$f_{0-\lambda T}$	Fractional function of the first kind (Chap. 11)		
F	Drag force (Chap. 9)	lb$_f$	N
F	Fouling factor (Chap. 15)	h·ft^2·°F/Btu	m^2·°C/W
F_x, F_y	Body forces per unit volume in the x and y directions, respectively (Chap. 6)	lb$_f$/ft^3	N/m^3
F_{i-j}	View factor (Chap. 12)		
Fo	Fourier number ($= \alpha t/L^2$) [Eq. (2-43c); Eq. (4-92b)]		
g	Energy generation rate per unit volume	Btu/h·ft^3	W/m^3
g	Acceleration of gravity	ft/s^2	m/s^2
g_c	Gravitational conversion factor	$32.1739 \dfrac{lb \cdot ft}{lb_f \cdot s^2}$ or $4.18 \times 10^8 \dfrac{lb \cdot ft}{lb_f \cdot h^2}$	$1.0 \dfrac{kg \cdot m}{N \cdot s^2}$
G	Mass velocity ($= \rho u$)	lb/ft^2·h or lb/ft^2·s	kg/m^2·s
Gr	Grashoff number [$= (\beta g L^3 \Delta T)/v^2$] (Chap. 10)		
Gz	Graetz number [$= $ Re Pr (D/x)] (Chap. 7)		
h	Heat-transfer coefficient; h_{cr}, heat-transfer coefficient for combined convection and radiation; h_r, heat-transfer coefficient for radiation; h_x, local heat-transfer coefficient	Btu/h·ft^2·°F	W/m^2·°C
h_{fg}	Latent heat of vaporization or condensation	Btu/lb	J/kg

Symbol	Quantity	Units English Engineering	SI
H	$= \dfrac{h}{k} = \dfrac{\text{heat-transfer coefficient}}{\text{thermal conductivity}}$ (Chap. 4)	ft^{-1}	m^{-1}
I	Radiation intensity; I_b, blackbody radiation intensity	$\text{Btu/h} \cdot \text{ft}^2 \cdot \text{sr}$	$\text{W/m}^2 \cdot \text{sr}$
I_λ	Spectral radiation intensity; $I_{\lambda b}$, spectral blackbody radiation intensity	$\text{Btu/h} \cdot \text{ft}^2 \cdot \text{sr} \cdot \mu\text{m}$	$\text{W/m}^2 \cdot \text{sr} \cdot \mu\text{m}$
I^+, I^-	Radiation intensity in the forward, backward direction, respectively (Chap. 13)	$\text{Btu/h} \cdot \text{ft}^2 \cdot \text{sr}$	$\text{W/m}^2 \cdot \text{sr}$
j	Factor defined by Eq. (9-107)		
J	Molal flux (Chap. 16)	$\text{lb mol/ft}^2 \cdot \text{h}$	$\text{kg mol/m}^2 \cdot \text{s}$
J	Mechanical equivalent of heat	$778.16 \text{ ft} \cdot \text{lb}_f/\text{Btu}$	
J	Denotes joule		
k	Thermal conductivity	$\text{Btu/h} \cdot \text{ft} \cdot {}^\circ\text{F}$	$\text{W/m} \cdot {}^\circ\text{C}$
k_m	Mass-transfer coefficient (Chap. 16)	ft/h	m/s
l	Mixing length (Chap. 9)	ft	m
L	Length	ft	m
m	Denotes meter		
m	Mass-flow rate (Chap. 14)	lb/h	kg/s
m	$= Ah/\rho c_p V$, parameter defined by Eq. (4-88)	s^{-1} or h^{-1}	s^{-1}
M	Molecular weight	lb/lb mol	kg/kg mol
M	Parameter defined by Eq. (15-7b)		
N	Molal flux (Chap. 16)	$\text{lb mol/ft}^2 \cdot \text{h}$	$\text{kg mol/m}^2 \cdot \text{s}$
N	Normalization integral defined by Eq. (4-21)		
Nu	Nusselt number ($= hD/k$ or hx/k)		
p	Pressure	lb_f/ft^2 or lb_f/in^2	N/m^2
P	Dimensionless pressure defined by Eq. (6-4)		
P	Perimeter (Chap. 3)	ft or in	m
P	Parameter defined by Eq. (15-18)		
Pe	Peclet number ($=$ Re Pr)		
Pr	Prandtl number ($= v/\alpha = c_p \mu/k$)		
Pr_t	Turbulent Prandtl number $= \varepsilon_m/\varepsilon_h$ (Chap. 9)		
q	Heat flux; net radiative-heat flux (Chaps. 11-13)	$\text{Btu/h} \cdot \text{ft}^2$	W/m^2
q_λ	Spectral radiative-heat flux; $q_{\lambda b}$ spectral blackbody emissive flux (Chap. 11)	$\text{Btu/h} \cdot \text{ft}^2 \cdot \mu\text{m}$	$\text{W/m}^2 \cdot \mu\text{m}$
q_b	Blackbody emissive flux (Chap. 11)	$\text{Btu/h} \cdot \text{ft}^2$	W/m^2
q_c, q_r	Convective and radiative-heat fluxes, respectively (Chap. 1)	$\text{Btu/h} \cdot \text{ft}^2$	W/m^2
q^l, q^t	Diffusive and turbulent heat fluxes, respectively (Chap. 9)	$\text{Btu/h} \cdot \text{ft}^2$	W/m^2
Q	Heat flow rate	Btu/h	W
r	Radius	ft or in	m
r	Recovery factor defined by Eq. (8-87a)		
r	$\equiv \alpha \, \Delta t/(\Delta x)^2$ (Chap. 5)		
R	Thermal resistance (Chap. 3)	$\text{h} \cdot \text{F/Btu}$	C/W
$*R$	Specific thermal contact resistance (Chap. 3)	$\text{h} \cdot \text{ft}^2 \cdot \text{F Btu}$	$\text{m}^2 \cdot \text{C/W}$
R	Parameter defined by Eq. (15-19)		

Symbol	Quantity	Units	
		English Engineering	SI
R_i	Radiosity (Chap. 12)	$Btu/h \cdot ft^2$	W/m^2
$\mathcal{R}$	Gas constant (see Table 1-3)	1545	8.314
		$ft \cdot lb/lb\ mol \cdot °R$	$J/kg\ mol \cdot K$
Ra	Rayleigh number $= g\beta L^3 \Delta T/v\alpha = $ Gr Pr [Eq. (10-22)]		
Re	Reynolds number $= ux/v$ or ud/v or ud_e/v		
S	Distance measured along the direction of propagation of radiation (Chap. 13)	ft or in	m
S	Spacing between tubes (Chap. 9)	ft or in	m
S_t, S_L	Transverse and longitudinal pitch, respectively (Chap. 9)	ft or in	m
Sc	Schmidt number $(= v/D)$ (Chap. 16)		
Sh	Sherwood number $(= hd/D)$ (Chap. 16)		
St	Stanton number $(= h/\rho u c_p = $ Nu/Re Pr) (Chap. 9)		
t	Time	h	s
t	Fin thickness (Chap. 3)	ft	m
T	Temperature	°F or °R	°C or K
T_{aw}	Adiabatic wall temperature defined by Eq. (8-87a)	°F or °R	°C or K
T_b	Bulk temperature	°F or °R	°C or K
T_m	Mean temperature	°F or °R	°C or K
T'	Temperature fluctuation in turbulent flow (Chap. 9)	°F	°C
u, v, v_r	Velocity components in the x, y, and r directions, respectively	ft/s	m/s
U	Molal average velocity [Eq. (16-1)]	ft/s	m/s
U, V	Dimensionless velocity components defined by Eq. (6-41)		
U	Overall heat-transfer coefficient	$Btu/h \cdot ft^2 \cdot °F$	$W/m^2 \cdot °C$
u', v'	Velocity fluctuations in turbulent flow (Chap. 9)	ft/s	m/s
u^+	Dimensionless velocity defined by Eq. (9-31a)		
x, y, z	Space coordinates in the rectangular coordinate system	ft	m
X, Y	Dimensionless coordinates defined by Eq. (6-41)		
y^+	Dimensionless distance defined by Eq. (9-31b)		
w	Width	ft	m
W	Denotes watt		
α	Thermal diffusivity $(= k/\rho c_p)$	ft^2/h	m^2/s or cm^2/s
α	Absorptivity (Chaps. 11 and 13)		
α_λ	Spectral absorptivity (Chaps. 11 and 13)		
β	Temperature coefficient of thermal conductivity (Chap. 3)	$°R^{-1}$	K^{-1}
β	Eigenvalues (Chap. 4)		

Symbol	Quantity	Units English Engineering	SI
β	$= \dfrac{a_f}{a} = \dfrac{\text{fin area}}{\text{total heat-transfer area}}$ (Chap. 3)		
Γ	Condensate flow rate per unit width at the bottom of a plate $\left(= \dfrac{\text{mass flow rate}}{\pi D}, \text{ for a tube} \right)$ (Chap. 14)	lb/ft·h	kg/m·s
$\Gamma(t)$	Function of time (Chap. 4)		
δ	Thickness of velocity boundary layer (Chap. 8)	ft	m
δ	Phase lag (Chap. 4)	rad	rad
δ_t	Thickness of thermal boundary layer	ft	m
ε/k	Parameter in Table 16-1		K
ε	Emissivity (Chaps. 11–13)		
ε	Heat-exchanger effectiveness (Chap. 15)		
ε_h	Eddy conductivity (Chap. 9)	ft^2/h	m^2/s
ε_m	Eddy viscosity (Chap. 9)	ft^2/h	m^2/s
η	Similarity variable (Chap. 8)		
η	Fin efficiency (Chap. 3)		
η'	Area-weighted fin efficiency (Chap. 3)		
θ	Dimensionless temperature		
θ	Polar angle	degree	rad
κ	Absorption coefficient (Chap. 11)	ft^{-1}	m^{-1}
κ	Universal constant in turbulent flow ($= 0.4$) (Chap. 9)		
λ	Wavelength (Chap. 11)	μm	μm
λ	Eigenvalue (Chap. 4)		
μ	Viscosity	lb/ft·h	kg/m·s
μ	$= \cos\theta$, cosine of the polar angle between the direction of radiation intensity and the positive y axis (Chap. 13)		
μm	Denotes micrometer ($= 10^{-6}$ m)		
ν	Kinematic viscosity $= \mu/\rho$	ft^2/h	m^2/s
ν	Frequency (Chap. 11)	s^{-1}	s^{-1}
ρ	Mass density	lb/ft^3	kg/m^3
ρ	Reflectivity (Chap. 11)		
ρ_λ	Spectral reflectivity (Chap. 11)		
σ^*	Surface tension (Chap. 14)	lb$_f$/ft	N/m
σ	Stefan-Boltzmann constant (Chap. 11)	0.1714×10^{-8} Btu/h·ft^2·°R^4	5.6699×10^{-8} W/m^2·K^4
σ_{AB}	Collision diameter (Chap. 16)	Å	Å
σ_x, σ_y	Normal stresses in the x and y directions, respectively (Chap. 6)	lb$_f$/ft^2	N/m^2
$\Omega_{D,\,AB}$	Collision integral (Table 16-2)		

Symbol	Quantity	Units English Engineering	SI
τ	Shear stress; τ_{xy}, τ_{yx} shear stresses (Chap. 6)	$\mathrm{lb}_f/\mathrm{ft}^2$	$\mathrm{N/m}^2$
τ	Transmissivity (Chap. 11)		
τ	$\equiv \kappa y$, the optical variable (Chap. 13)		
τ_0	$\equiv \kappa L$, the optical thickness (Chap. 13)		
φ	Azimuthal angle	degrees	rad
Φ	Viscous-energy-dissipation function defined by Eq. (6-36b)		
χ	Mole fraction (Chap. 16)		
Ψ	Stream function (Chap. 8)		
ω	Angular velocity (Chap. 4)	s^{-1}	s^{-1}

Problems

Chapter 1

1-1 Determine the heat-transfer rate per unit area of a large vertical surface which is kept at a uniform temperature of 100°F and dissipates heat by free convection into the surrounding atmosphere at 40°F with a heat-transfer coefficient $h = 0.8$ Btu/h·ft²·°F.

1-2 Determine the radiation-heat-transfer coefficient h_r for Prob. 1-1 and repeat the heat-transfer calculation by including the heat transfer by thermal radiation.

1-3 Water flows in forced flow inside a 1-in-ID tube. Determine the heat-transfer rate between the water and the tube per linear foot of the tube for a mean temperature difference of 50°F and a forced-convection heat-transfer coefficient of $h = 150$ Btu/h·ft²·°F between the water and the tube surface.

1-4 A wall of thickness $L = \frac{1}{2}$ ft and thermal conductivity $k = 0.5$ Btu/h·ft·°F is maintained at 660°F at one face and at 60°F at the other face. Determine the heat-transfer rate across the wall per square foot of surface.

1-5 A temperature difference of 600°F is applied across an asbestos sheet $\frac{1}{4}$ in in thickness. Determine the heat-transfer rate across the sheet per square foot of surface for the thermal conductivity of the asbestos $k = 0.09$ Btu/h·ft·°F.

1-6 Fluid flows in forced convection along a flat plate 1 ft wide and 2 ft long. Determine the heat-transfer rate between the fluid and the plate for a mean temperature difference of 400°F and a mean heat-transfer coefficient of 30 Btu/h·ft²·°F between the fluid and the plate surface.

1-7 An insulating material of thermal conductivity $k = 0.2$ Btu/h·ft·°F is to be used to limit the heat losses to 800 Btu/h·ft² for a temperature difference of 1000°F across the insulating layer. Determine the thickness of the insulation needed.

1-8 Two large parallel plates, one at a uniform temperature of 1000°R and the other at 2000°R, are separated by a nonparticipating gas. Assuming that the plates are perfect emitters, calculate the net radiative-heat exchange between the plates per square foot of plate surface.

1-9 A large plate at 2500°R is exposed to an atmosphere which can be considered at 500°R. Assuming that the plate is a perfect emitter, determine the rate of heat loss by radiation per square foot of plate surface.

1-10 A packed glass wool of thermal conductivity $k = 0.03$ Btu/h·ft·°F is to be used to insulate an icebox. If the maximum heat loss from the icebox is not to exceed 15 Btu/h·ft² for a temperature difference of 80° across the walls of the icebox, determine the thickness of the glass-wool insulation.

1-11 A flat plate has one surface insulated and the other surface is exposed to the sun. If the surface facing the sun receives 300 Btu/h·ft² of solar energy from the sun (all of which is assumed to be absorbed by the plate) and dissipates heat by convection into the surrounding atmosphere at 70°F with a heat-transfer coefficient of $h = 3$ Btu/h·ft²·°F, determine the equilibrium temperature of the plate surface.

1-12 Derive the conversion factor for converting the thermal conductivity from the units of Btu/h·ft² (°F/in) into the units of W/m·°C.

1-13 Convert the heat-transfer coefficient $h = 50$ W/m²·°C into the units Btu/h·ft²·°F.

1-14 Derive the following conversion factors.

$$\frac{\text{Btu/h·ft}^2\cdot°\text{F}}{4.882 \text{ kcal/m}^2\cdot\text{h}\cdot°\text{C}} \quad \text{and} \quad \frac{\text{Btu/h·ft}\cdot°\text{F}}{0.0173 \text{ W/cm}\cdot°\text{C}}$$

Chapter 2

2-1 By writing the energy balance for a differential volume element, derive the one-dimensional, steady-state heat-conduction equation with internal energy generation and variable thermal conductivity in the rectangular coordinate system.

2-2 Repeat Prob. 2-1 in the cylindrical and spherical coordinates.

2-3 Write the mathematical formulation of the steady-state heat-conduction problem in a rectangular region $0 \le x \le a, 0 \le y \le b$ with internal energy generation and constant thermal conductivity for the following boundary conditions: The boundary surface at $x = 0$ is insulated, the boundary at $x = a$ dissipates heat by convection into a medium at temperature T_∞, the boundary at $y = 0$ is kept at a uniform temperature T_0, and the boundary at $y = b$ is exchanging heat by convection with a medium at zero temperature.

2-4 A solid cylinder of radius $r = b$ dissipates heat by convection at the boundary surface $r = b$ into a medium at temperature T_∞ with a heat-transfer coefficient h. Derive the mathematical expression for the boundary condition at this surface.

2-5 Consider the heat-conduction problem for a solid cylinder given in the form

$$\frac{1}{r}\frac{\partial}{\partial r}\left(r\frac{\partial T}{\partial r}\right) + \frac{g}{k} = \frac{1}{\alpha}\frac{\partial T}{\partial t} \qquad \text{in } 0 \le r \le b, t > 0$$

$$k\frac{\partial T}{\partial r} + hT = hT_\infty \qquad \text{at } r = b, t > 0$$

$$T = T_0 \qquad \text{for } t = 0, \text{ in } 0 \le r \le b$$

Express this problem in the dimensionless form by introducing the new variables as $\eta = r/b$ and $\theta = (T - T_\infty)/(T_0 - T_\infty)$

2-6   Repeat Prob. 2-5 for a solid sphere of radius $r = b$.

2-7   Consider a solid cylinder in the region $0 \leq r \leq b, 0 \leq z \leq H$ in which heat is generated at a constant rate of g Btu/h·ft^3 and the thermal conductivity is constant.   Write the mathematical formulation of the steady-state heat-conduction problem for the following boundary conditions: The boundary surface at $z = 0$ is insulated, the boundary at $z = H$ dissipates heat by convection into a medium at constant temperature T_∞ with a heat-transfer coefficient h, and heat is supplied into the cylinder at a constant rate of q Btu/h·ft^2 through the boundary surface at $r = b$.

2-8   Write the mathematical formulation of the steady-state heat conduction in a rectangular region $0 \leq x \leq a, 0 \leq y \leq b$ in which there is heat generation at a constant rate of g Btu/h·ft^3 and the thermal conductivity is constant for the following set of boundary conditions: The boundary surface at $x = 0$ is dissipating heat by convection into a medium at constant temperature T_∞ with a heat-transfer coefficient h_1, the boundary at $x = a$ is insulated, heat is supplied into the region at a constant rate of q Btu/h·ft^2 through the boundary surface at $y = 0$, and the boundary surface at $y = b$ dissipates heat by convection into a medium at zero temperature with a heat-transfer coefficient h_2.

2-9   Write the mathematical formulation of the one-dimensional, time-dependent heat-conduction problem for a solid cylinder of radius $r = b$ with constant thermal properties for the following conditions: The solid is initially at a uniform temperature T_0, and for times $t > 0$ there is heat generation in the medium at a constant rate of g Btu/h·ft^3 while the heat is dissipated by convection from the boundary surface at $r = b$ into a medium at temperature T_∞ with a heat-transfer coefficient h.

2-10   Write the mathematical formulation of the one-dimensional, steady-state heat conduction in a hollow sphere $a \leq r \leq b$ in which thermal conductivity is constant, heat is generated at a rate of cr^2 Btu/h·ft^3, the boundary surface at $r = a$ is kept at a uniform temperature T_1 while the boundary surface at $r = b$ dissipates heat by convection into a medium at temperature T_∞ with a heat-transfer coefficient h.

Chapter 3

3-1   Thermal conductivity of a plane wall varies with temperature according to the relation

$$k(T) = k_0(1 + \beta T^2)$$

Determine a relation for the heat flow through the slab per unit area if the surfaces at $x = 0$ and $x = L$ are maintained at uniform temperatures T_1 and T_2, respectively.

3-2 The inside and outside surfaces of a hollow sphere $a \le r \le b$ at $r = a$ and $r = b$ are maintained at uniform temperatures T_1 and T_2, respectively. The thermal conductivity varies with temperature as

$$k(T) = k_0(1 + \alpha T + \beta T^2)$$

Determine a relation for the total heat-flow rate Q through the sphere.

3-3 Determine the relation for the steady-state temperature distribution in a slab $0 \le x \le L$ in which heat is being generated at a rate of $g(x) = g_0 x^2$ Btu/h·ft^3, while the boundary surface at $x = 0$ is insulated and that at $x = L$ is maintained at zero temperature. Give the relation for the temperature of the insulated surface at $x = 0$.

3-4 A composite slab consisting of three layers in perfect thermal contact as shown in the accompanying figure is maintained at uniform temperatures T_0 and T_3 at the outer surfaces. k_1, k_2, k_3 are the thermal conductivities, and L_1, L_2, L_3 are the thicknesses of the layers. Derive the relations for the determination of the interface temperatures T_1 and T_2.

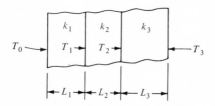

3-5 Heat is generated at a constant rate of g Btu/h·ft^3 in a thin circular rod of length L and diameter D. The two ends at $x = 0$ and $x = L$ are kept at constant temperatures T_0 and zero, respectively, while the heat is dissipated from the lateral surfaces by convection into a medium at zero temperature with a heat-transfer coefficient h.

(a) Derive the one-dimensional, steady-state energy equation for the determination of temperature distribution $T(x)$ in the rod.

(b) By solving this differential equation, determine the relation for the temperature distribution $T(x)$ in the rod.

3-6 The lateral surface of a rod is perfectly insulated while its ends at $x = 0$ and $x = L$ are kept at constant temperatures T_1 and T_2, respectively.

The cross-sectional area A of the rod is uniform, and the thermal conductivity of the material varies with temperature as

$$k(T) = k_0(1 + \alpha T)$$

Determine the relation for the heat-transfer rate Q through the rod. Calculate the heat-transfer rate per unit area by taking $k_0 = 50$ Btu/h·ft·°F, $L = 1$ ft, $\alpha = 2 \times 10^{-3}$°R^{-1}, $T_1 = 700$°F, and $T_2 = 100$°F.

3-7 A very long, thin copper rod, $k = 220$ Btu/h·ft·°F, of diameter $D = \frac{1}{2}$ in protrudes from a wall into the surrounding air. The end of the rod at the wall is kept at 500°F while the rod dissipates heat from its lateral surface by convection into the air at 100°F with a heat-transfer coefficient $h = 5$ Btu/h·ft²·°F. Determine the rate of heat loss from the rod into the surrounding air.

3-8 A pressure vessel for a nuclear reactor is approximated as a large flat plate of thickness L. The inside surface of the plate at $x = 0$ is insulated, the outside surface at $x = L$ is maintained at a uniform temperature T_2, and the gamma-ray heating of the plate can be represented as a heat-generation term in the form

$$g(x) = g_0 e^{-\gamma x}$$

where g_0 and γ are constants and x is measured from the insulated inside surface.

(a) Determine a relation for the temperature distribution in the plate.

(b) Determine the temperature at the insulated surface (that is, $x = 0$) of the plate.

(c) Determine the heat flux at the outer surface, $x = L$.

3-9 A 1-ft-long, 1-in-diameter, steel rod ($k = 20$ Btu/h·ft·°F) connects two walls, one of which is kept at 500°F and the other at 100°F. The rod dissipates heat by convection from its lateral surface into an environment at 100°F with a heat-transfer coefficient $h = 5$ Btu/h·ft²·°F. Determine (a) the temperature distribution in the rod and (b) the rate of heat transfer from the rod into the air.

3-10 A 2-ft-thick furnace wall ($k = 0.5$ Btu/h·ft·°F) is to be insulated with an insulating material ($k = 0.05$ Btu/h·ft·°F). The temperature at the inside surface of the furnace is 2600°F. If the temperature at the outer surface should not exceed 100°F for a permissible heat loss of 250 Btu/h·ft², what thickness of insulation is required?

3-11 What is the total rate of heat loss in Btu/h from a refrigerator of inside dimensions 2 by 2 by 4 ft, insulated with a 3-in insulator ($k = 0.03$ Btu/h·ft·°F) for a temperature difference of 50°F between inside and outside? Neglect the thermal resistances at the inside and outside.

3-12 Determine the heat loss per square foot of a furnace wall consisting of 1-ft thickness of fire brick ($k = 0.6$ Btu/h·ft·°F), $\frac{1}{4}$ ft of insulating brick ($k = 0.05$ Btu/h·ft·°F), and $\frac{1}{2}$-ft-thick outer layer of ordinary brick ($k = 0.4$ Btu/h·ft·°F) for a temperature difference of 2500°F between the inside and outside.   Neglect the thermal resistances at the inside and outside.

3-13 The two ends of a thin circular rod of diameter D at $x = 0$ and $x = L$ are maintained at temperatures T_0 and T_L, respectively, while heat is generated in the rod at a uniform rate of g_0 Btu/h·ft³. Determine an expression for the steady-state temperature distribution in the rod for the cases when (*a*) the lateral surface of the rod is insulated and (*b*) the lateral surface dissipates heat by convection into a medium at temperature T_L with a heat-transfer coefficient h.

3-14 The two faces of a slab at $x = 0$ and $x = L$ are kept at uniform temperatures T_1 and T_2, respectively, and the thermal conductivity of the material depends on temperature in the form

$$k(T) = k_0(T^2 - T_0{}^2)$$

where the k_0 and T_0 are constants.

   (*a*)   Find an expression for the heat-flow rate per unit area of the slab.

   (*b*)   Find an expression for the average thermal conductivity k_m for the slab.

3-15 Determine the steady-state temperature distribution $T(r)$ in a long solid cylinder of radius $r = b$, of constant thermal conductivity k, in which heat is generated at a rate

$$g(r) = g_0(1 + Ar)   \text{Btu/h·ft}^3$$

where g_0 and A are constants, while the boundary at $r = b$ is kept at a uniform temperature T_0.

3-16 Determine the steady-state temperature distribution $T(r)$ in a long hollow cylinder $a \le r \le b$ in which heat is generated at a rate of

$$g(r) = g_0(1 + Ar)   \text{Btu/h·ft}^3$$

while the boundary surfaces at $r = a$ and $r = b$ are kept at zero temperature.

3-17 Determine the steady-state temperature distribution in a long hollow cylinder $a \leq r \leq b$ in which heat is generated at a uniform rate of g_0 Btu/h·ft^3 while the boundary surface at $r = a$ is insulated and the outer boundary at $r = b$ is kept at a uniform temperature T_0.

3-18 Derive an expression for the steady-state temperature distribution $T(r)$ in a long solid cylinder of radius $r = b$, in which heat is generated at a constant rate of g_0 Btu/h·ft^3, while the heat is dissipated by convection from the outer boundary surface at $r = b$ into an environment at temperature T_∞ with a heat-transfer coefficient h.

3-19 A hollow cylinder, $a \leq r \leq b$, has its boundary surfaces at $r = a$ and $r = b$ maintained at uniform temperatures T_1 and T_2, respectively. The thermal conductivity varies with temperature in the form

$$k(T) = k_0(1 + \beta T)$$

Determine a relation for the heat flow through the cylinder per unit length of the cylinder.

3-20 The inner and outer radii of a hollow cylinder are 3 and 5 in, and they are maintained at 600 and 100°F, respectively. If the thermal conductivity of the material varies with temperature linearly in the form

$$k = 0.02 \, (1 + 5 \times 10^{-3}T) \qquad \text{Btu/h·ft·°F}$$

determine the heat-flow rate through the cylinder per foot length.

3-21 A long hollow cylinder of constant thermal conductivity has a constant heat supply at a rate of q_i Btu/h·ft^2 at the inner boundary surface at $r = r_i$ while the heat is dissipated from the outer boundary surface at $r = r_o$ by convection into a fluid at zero temperature with a heat-transfer coefficient h. Determine the steady-state temperature distribution $T(r)$ in the cylinder.

3-22 A hot fluid at a mean temperature of 480°F flows inside a 4-in-ID, $4\frac{1}{2}$-in-OD steel tube ($k = 20$ Btu/h·ft·°F). The outer surface of the tube is covered with a 1-in-thick insulating material ($k = 0.2$ Btu/h·ft·°F), and the heat is dissipated by convection from the outer surface of the insulation into the surrounding air at 80°F. The heat-transfer coefficient for the flow inside is $h_i = 50$ Btu/h·ft^2·°F and for the convection on the outside is $h_o = 1.5$ Btu/h·ft^2·°F. Determine the rate of heat loss into the air per foot length of the tube.

3-23 Determine the steady-state temperature distribution $T(r)$ in a hollow sphere, $a \leq r \leq b$, in which heat is generated at a rate of $g(r) = g_0 r$ Btu/h·ft^3, where g_0 is a constant, while the inner boundary surface at $r = a$ is insulated and the outer boundary surface at $r = b$ is kept at a uniform temperature T_0.

3-24 Determine the steady-state temperature distribution $T(r)$ in a solid sphere of radius $r = b$ in which heat is generated at a uniform rate of g_0 Btu/h·ft³ and heat is dissipated from the surface by convection into a medium at a temperature T_∞ with a heat-transfer coefficient h. Determine the expressions for the temperature at the center of the sphere and the heat flux at the boundary surface.

3-25 A hollow sphere, $r_1 \leq r \leq r_2$, is kept at a uniform temperature T_1 at the inner surface $r = r_1$ and at temperature T_2 at the outer surface $r = r_2$. Show that the total heat-transfer rate Q through the sphere is given by

$$Q = \frac{T_1 - T_2}{R}$$

where the thermal resistance of the sphere is

$$R = \frac{r_2 - r_1}{kA_m} \qquad A_m = \sqrt{(4\pi r_1{}^2)(4\pi r_2{}^2)} \equiv \sqrt{A_1 A_2}$$

and k is the thermal conductivity.

3-26 Derive an expression for the steady-state temperature distribution $T(r)$ in a solid sphere of radius $r = b$ in which heat is generated at a rate of

$$g(r) = g_0\left(1 - \frac{r}{b}\right) \qquad \text{Btu/h·ft}^3$$

where g_0 is a constant, and the boundary surface at $r = b$ is maintained at a uniform temperature T_0.

3-27 Derive an expression for the steady-state temperature distribution $T(r)$ in a hollow sphere, $a \leq r \leq b$, in which heat is generated at a uniform rate of g_0 Btu/h·ft³ and the boundary surfaces at $r = a$ and $r = b$ are kept at the same uniform temperature T_0.

3-28 Longitudinal fins of rectangular cross section are attached at the outer surface of a 1-in-OD copper tube. Fins are $\frac{1}{8}$ in thick, $\frac{1}{4}$ in long, and made of copper ($k = 200$ Btu/h·ft·°F). The ratio of the fin surface to the total heat-transfer surface is $\beta = 0.70$. The heat-transfer coefficients for flow inside and outside the tube, respectively, are $h_i = 50$ Btu/h·ft²·°F and $h_o = 5$ Btu/h·ft²·°F. Determine the heat-transfer rate per foot length of the tube for a temperature difference of 100°F between the inside and outside fluids. (Neglect tube resistance.)

3-29 Consider a tube having fins attached to both inside and outside surfaces. The thermal resistance of the tube material is negligible compared with the thermal resistances for flow on the outside and inside. Various quantities are given as follows:

	Outside	Inside
Temperatures of fluids	T_o	T_i
Heat-transfer coefficients	h_o	h_i
Fin surface area	a_{fo}	a_{fi}
Total heat-transfer area	a_{to}	a_{ti}
Tube surface areas *without fins*	a_o	a_i
Fin efficiency	η_o	η_i

Write the relations for the heat-transfer rate through the tube for the following cases:

(a) Fins are attached on both sides of the tube surface.

(b) Fins are attached on the outside surface of the tube only.

(c) There are no fins attached to the tube surfaces.

Chapter 4

4-1 Determine the eigenvalues and the eigenfunctions of the following eigenvalue problem:

$$\frac{d^2\Psi(x)}{dx^2} + \lambda^2\Psi(x) = 0 \quad \text{in } 0 \le x \le L$$

$$\frac{d\Psi}{dx} = 0 \qquad \text{at } x = 0$$

$$\Psi = 0 \qquad \text{at } x = L$$

4-2 Determine the eigenvalues and the eigenfunctions of the following eigenvalue problem:

$$\frac{d^2\Psi(x)}{dx^2} + \lambda^2\Psi(x) = 0 \quad \text{in } 0 \le x \le L$$

$$\frac{d\Psi}{dx} = 0 \qquad \text{at } x = 0$$

$$k\frac{d\Psi}{dx} + h\Psi = 0 \qquad \text{at } x = L$$

4-3 Determine the eigenvalues and eigenfunctions for cases 2 and 8 in Table 4-1.

4-4 Derive an expression for the steady-state temperature distribution $T(x, y)$ for the following heat-conduction problem in a rectangular region $0 \le x \le a, 0 \le y \le b$.

$$\frac{\partial^2 T}{\partial x^2} + \frac{\partial^2 T}{\partial y^2} = 0 \qquad \text{in } 0 \le x \le a, 0 \le y \le b$$

$$T = 0 \qquad \text{at } x = 0$$

$$T = T_0 \sin \frac{\pi y}{b} \qquad \text{at } x = a$$

$$T = 0 \qquad \text{at } y = 0$$

$$T = 0 \qquad \text{at } y = b$$

Also determine the expressions for the heat fluxes at the boundary surfaces $x = 0$ and $y = 0$.

4-5 Derive an expression for the steady-state temperature distribution $T(x, y)$ by solving the differential equation

$$\frac{\partial^2 T}{dx^2} + \frac{\partial^2 T}{\partial y^2} = 0$$

in a rectangular region $0 \le x \le a$, $0 \le y \le b$, subject to the boundary conditions shown in each of the accompanying figures.

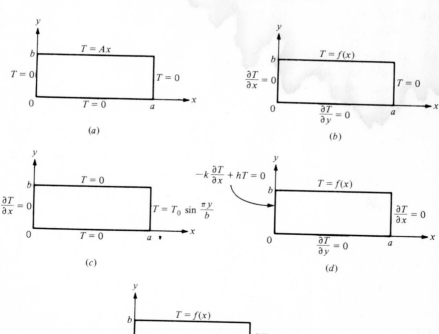

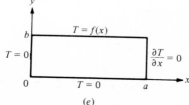

4-6 Determine an expression for the steady-state temperature distribution $T(x, y)$ by solving the differential equation

$$\frac{\partial^2 T}{\partial x^2} + \frac{\partial^2 T}{\partial y^2} = 0 \quad \text{in } 0 \le x \le \infty, 0 \le y \le b$$

for a semi-infinite strip for the boundary conditions shown in the accompanying figures.

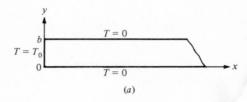

(a)

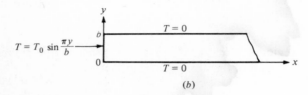

(b)

4-7 Determine an expression for the one-dimensional, time-dependent temperature distribution $T(x, t)$ in a slab, $0 \le x \le L$, by solving the heat-conduction equation

$$\frac{\partial^2 T(x, t)}{dx^2} = \frac{1}{\alpha}\frac{\partial T(x, t)}{\partial t} \quad \text{in } 0 \le x \le L, t > 0$$

subject to the boundary and initial conditions shown for each of the following cases:

(a) $T = 0$ at $x = 0, t > 0$

$T = 0$ at $x = L, t > 0$

$T = T_0 \sin \dfrac{\pi x}{L}$ for $t = 0$, in $0 \le x \le L$

(b) $T = 0$ at $x = 0, t > 0$

$T = 0$ at $x = L, t > 0$

$T = T_0$ for $t = 0$, in $0 \le x \le L$

(c) $\dfrac{\partial T}{\partial x} = 0$ at $x = 0,\ t > 0$

$\ T = 0$ at $x = L,\ t > 0$

$\ T = T_0$ for $t = 0$, in $0 \le x \le L$

(d) $\dfrac{\partial T}{\partial x} = 0$ at $x = 0,\ t > 0$

$\ k\dfrac{\partial T}{\partial x} + hT = 0$ at $x = L,\ t > 0$

$\ T = T_0$ for $t = 0$, in $0 \le x \le L$

where T_0 is constant.

Also give the expressions for the heat flux at the boundary surface $x = 0$ for cases a and b.

4-8   A 1-ft-thick brick wall ($c_p = 0.2$ Btu/lb·°F, $\rho = 150$ lb/ft^3, $\alpha = 0.018$ ft^2/h, $k = 0.6$ Btu/h·ft·°F) is initially at a uniform temperature of $T_i = 100$°F. For times $t > 0$, one of the boundary surfaces is kept insulated while the other boundary surface is exposed to hot air at 1100°F. The heat-transfer coefficient between the air and the surface is $h = 20$ Btu/h·ft^2·°F. By using the transient-temperature chart, determine the time required for the insulated surface to reach a temperature of 500°F.

4-9   A hot metal block initially at a uniform temperature T_0 is suddenly immersed in a well-stirred cold liquid bath which is maintained at a uniform temperature T_∞. The heat-transfer coefficient between the liquid and the surface of the block is h Btu/h·ft^2·°F. The metal block has a weight M lb, surface area A ft^2, and specific heat c_p Btu/lb·°F. By applying the lumped-system-analysis approach, derive an expression for the temperature $T(t)$ of the metal block as a function of time.

4-10   A 4-in-diameter copper ball ($k = 220$ Btu/h·ft·°F, $\rho = 560$ lb/ft^3, $c_p = 0.1$ Btu/lb·°F) initially at 500°F is suddenly immersed in a well-stirred fluid at temperature $T_\infty = 100$°F. The heat-transfer coefficient between the fluid and the surface is $h = 20$ Btu/h·ft^2·°F. Using the lumped-system analysis, determine the temperature of the ball at 4 min after immersion.

4-11   A long cylindrical bar ($\alpha = 1.1$ ft^2/h, $k = 70$ Btu/h·ft·°F) of 4-in diameter, initially at 1200°F, is exposed to a cooled air stream at a temperature $T_\infty = 100$°F. The heat-transfer coefficient between the air stream and the surface is $h = 15$ Btu/h·ft^2·°F. By using lumped-system analysis, determine the time required for the center of the rod to reach 800°F.

4-12 A 1-in-diameter steel ball ($k = 25$ Btu/h·ft·°F, $\rho = 490$ lb/ft³, $c_p = 0.11$ Btu/lb·°F) initially at 500°F is suddenly dropped into a well-stirred cold liquid at 100°F. The heat-transfer coefficient between the fluid and the ball surface is $h = 10$ Btu/h·ft²·°F. By using lumped-system analysis, determine the time required for the steel ball to reach a temperature of 200°F.

4-13 A 1-in-thick copper plate ($k = 220$ Btu/h·ft·°F, $\rho = 560$ lb/ft³, $c_p = 0.1$ Btu/lb·°F) is initially at a uniform temperature of 70°F. Suddenly a heat flux $q = 1500$ Btu/h·ft² is applied at one face while the other face dissipates heat by convection, with a heat-transfer coefficient $h = 10$ Btu/h·ft²·°F, into a medium at 70°F. By using lumped-system analysis, determine the temperature of the plate at 5 min after the application of the heat flux. Also determine the steady-state temperature of the plate.

4-14 A periodically oscillating heat flux $q = q_0 \sin \omega t$ is applied to one surface of a large plate of thickness L, while the other surface is dissipating heat by convection into a medium at temperature T_∞ with a heat-transfer coefficient h. Using lumped-system analysis, derive an expression for the periodic temperature variation in the slab after the temperature transients have passed.

4-15 A 2-in-thick copper plate ($k = 220$ Btu/h·ft·°F, $c_p = 0.1$ Btu/lb·°F, $\rho = 560$ lb/ft³) is subjected to a periodically varying heat flux $q = 2000 \cos \omega t$ Btu/h·ft²·°F at one of its faces and is cooled by convection from the other surface into a medium at 100°F with a heat-transfer coefficient $h = 20$ Btu/h·ft²·°F. Using lumped-system analysis, determine the temperature oscillation in the slab for a value of $\omega = 2\pi$ h⁻¹.

Chapter 5

5-1 By writing an energy balance on a differential volume element, derive the finite-difference form of the heat-conduction equation

$$\frac{\partial^2 T}{\partial x^2} + \frac{\partial^2 T}{\partial y^2} + \frac{g}{k} = 0$$

for the nodal point A in each of the accompanying figures for the boundary conditions indicated.

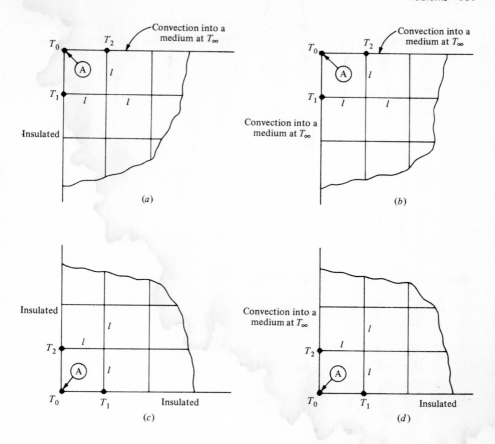

5-2   Write in the finite-difference form the heat-conduction equation

$$\frac{\partial^2 T}{\partial x^2} + \frac{\partial^2 T}{\partial y^2} + \frac{g}{k} = 0$$

for the nine nodes, $i = 1, 2, \ldots, 9$, at which the temperatures T_i are unknown, in the accompanying figure.   Here, the temperatures $f_1, f_2,$ and f_3 are considered specified.

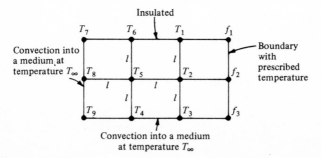

5-3 Calculate the temperatures T_1, T_2, T_3, and T_4 at the four nodes in the accompanying figure by solving the heat-conduction equation

$$\frac{\partial^2 T}{\partial x^2} + \frac{\partial^2 T}{\partial y^2} = 0$$

with finite differences.

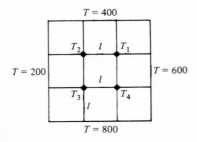

5-4 Write the finite-difference representation of the heat-conduction equation

$$\frac{\partial^2 T}{\partial x^2} + \frac{\partial^2 T}{\partial y^2} + \frac{g}{k} = 0$$

for the nodes $i = 1, 2, \ldots, 12$ at which the temperatures T_i are unknown in the accompanying figure. Here the temperatures f_1, f_2, and f_3 are considered specified.

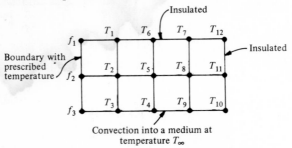

5-5 Using the *explicit* finite-differencing scheme, write the following heat-conduction problem in the finite-difference form.

$$\frac{\partial^2 T}{\partial x^2} + \frac{g}{k} = \frac{1}{\alpha}\frac{\partial T}{\partial t} \qquad \text{in } 0 \le x \le L, t > 0$$

$$\frac{\partial T}{\partial x} = 0 \qquad \text{at } x = 0, t > 0$$

$$T = T_0 \qquad \text{at } x = L, t > 0$$

$$T = 0 \qquad \text{for } t = 0, \text{ in } 0 \le x \le L$$

5-6 Using the *explicit* finite-differencing scheme, write the following heat-conduction problem in the finite-difference form for m.

$$\frac{\partial^2 T}{\partial x^2} = \frac{1}{\alpha} \frac{\partial T}{\partial t} \qquad \text{in } 0 \leq x \leq L, t > 0$$

$$-k \frac{\partial T}{\partial x} + hT = hT_\infty \qquad \text{at } x = 0, t > 0$$

$$T = T_0 \qquad \text{at } x = L, t > 0$$

$$T = 0 \qquad \text{for } t = 0 \text{ in } 0 \leq x \leq L$$

5-7 Using the *implicit* finite-differencing scheme, write in the finite-difference form the heat-conduction Prob. 5-6.

5-8 Calculate the steady-state temperatures T_1, T_2, T_3, T_4 at the four nodes in the accompanying figure by solving the heat-conduction equation

$$\frac{\partial^2 T}{\partial x^2} + \frac{\partial^2 T}{\partial y^2} = 0$$

with finite differences for these four nodes.

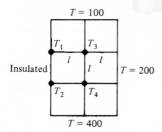

5-9 Calculate the steady-state temperatures T_1, T_2, T_3, and T_4 at the four nodes in the accompanying figure by solving the heat-conduction equation

$$\frac{\partial^2 T}{\partial x^2} + \frac{\partial^2 T}{\partial y^2} = 0$$

with finite differences for these four nodes.

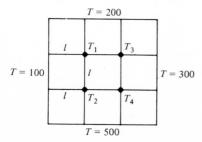

5-10 Calculate the steady-state temperatures T_1, T_2, ..., T_6 at the six nodes in the accompanying figure by solving the heat-conduction equation

$$\frac{\partial^2 T}{\partial x^2} + \frac{\partial^2 T}{\partial y^2} = 0$$

with finite differences for the four nodes.   Take the thermal conductivity as $k = 20$ Btu/h·ft·°F, the heat-transfer coefficient as $h = 10$ Btu/h·ft²·°F for the convection boundary condition, and $l = 1$ ft.

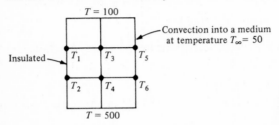

5-11 By using the finite-difference equations (5-33) given in Example 5-2, calculate the temperatures at the nodes $n = 1, 2, 3, ..., 9$ for a few consecutive time steps $i = 1, 2, 3, 4, ...$.

Chapter 6

6-1 Consider two-dimensional (x, y) steady, incompressible, constant-property laminar flow in the x direction between two parallel plates with no body forces acting on the fluid.   Assuming a *fully developed flow* (that is, $v = 0$), simplify the continuity and momentum equations given in Chap. 6 and determine the resulting momentum equation.   Discuss the appropriate boundary conditions for the solution of this equation.

6-2 Repeat Prob. 6-1 for two-dimensional (r, z), steady, incompressible, constant-property, fully developed (that is, $v_r = 0$) laminar flow in the z direction inside a circular tube.

6-3 Derive the continuity equation in the rectangular coordinate system for a three-dimensional flow having velocity components u, v, and w in the x, y, and z directions, respectively.

6-4 Consider two-dimensional (x, y), steady, incompressible, constant-property, fully developed (that is, $v = 0$) laminar flow in the x direction between two parallel plates.   The temperature gradient in the x direction is much smaller than that in the y direction so that the axial heat conduction can be neglected.   Simplify the energy equation given in Chap. 6 with these considerations and obtain the resulting energy equation.   Discuss the physical significance of various terms and the appropriate boundary conditions needed for the solution of this equation.

6-5 Repeat Prob. 6-4 for the two-dimensional (r, z), steady, incompressible, constant-property, fully developed (that is, $v_r = 0$) laminar flow in the z direction inside a circular tube.

6-6 Simplify the energy equations obtained in Probs. 6-4 and 6-5 further by the additional assumption that the flow velocities are moderate so that the viscous-energy-dissipation term can also be neglected.

6-7 Derive the continuity equation (6-33) given in the cylindrical coordinates (r, z) by considering the conservation of mass for a differential volume element chosen in the form of a ring (that is, Δr about r and Δz about z).

6-8 Verify the dimensionless continuity, momentum, and energy equations (6-42) to (6-45) by applying the dimensionless variables given by Eq. (6-41) to Eqs. (6-37) to (6-40).

6-9 Consider two-dimensional (x, y), steady, incompressible, constant-property, laminar boundary-layer flow in the x direction along a flat plate at moderate flow velocities (i.e., viscous energy dissipation can be neglected). Write the appropriate continuity, momentum, and energy equations.

6-10 Repeat Prob. 6-9 for the two-dimensional (x, y), steady, incompressible, constant-property laminar flow along a curved body with the x direction measured along the curved surface and the y direction normal to the surface.

Chapter 7

7-1 Determine the maximum temperature rise in a lubricating oil ($\mu = 0.15$ lb/ft·s, $k = 0.08$ Btu/h·ft·°F) between a bearing and its journal for a rotation velocity of $u_1 = 30$ ft/s when the journal and the bearing are maintained at the same temperature.

7-2 Lubricating oil of viscosity μ and thermal conductivity k fills the clearance L between a journal and its bearing. Determine a relation for the temperature distribution in the oil film, assuming that the bearing surface is maintained at a uniform temperature T_0, there is no heat transfer into the journal, and the velocity of rotation is u_1. Also determine an expression for the heat flux at the bearing surface.

7-3 In Prob. 7-2 determine the maximum temperature rise in the oil and the heat flux at the bearing surface for a journal 4 in in diameter, rotating at 1800 r/min, using a lubricating oil with $\mu = 0.1$ lb/ft·s and $k = 0.08$ Btu/h·ft·°F.

7-4 Oil ($v = 20$ ft²/h) fills the space between two large, horizontal plates $\frac{1}{2}$ in apart. Determine the shear stress in the oil if the upper plate moves with a velocity of 4 ft/s while the lower plate remains stationary.

7-5 Oil ($\mu = 0.2$ lb/ft·s, $\rho = 55$ lb/ft³) flows through a $\frac{1}{2}$-in-ID tube with a velocity of 3 ft/s. Determine the pressure drop over a 200-ft length of the tube.

7-6 Determine the heat-transfer coefficient under uniform wall-temperature conditions for a fully developed laminar flow of a fluid ($k = 0.08$ Btu/h·ft·°F) inside (a) a circular tube $\frac{1}{2}$ in ID, (b) a channel having a square cross section of sides $\frac{1}{2}$ in, and (c) a channel having an equilateral-triangle cross section with sides $\frac{1}{2}$ in.

7-7 For the fluid flow considered in Prob. 7-6 determine the heat-transfer rate over the 10-ft length of each of these tubes for a mean temperature difference of $\Delta T = 100°F$ between the fluid and the wall temperatures.

7-8 Oil at 70°F with a mean velocity of 2 ft/s enters a $\frac{1}{2}$-in-ID, 5-ft-long tube which is kept at a uniform temperature of 150°F. Determine the temperature rise of the fluid as it leaves the tube. Fluid properties may be taken as $\mu_b = 0.015$ lb/ft·s at 70°F, $\mu_w = 0.0055$ lb/ft·s at 150°F, $\rho = 55$ lb/ft³, $c_p = 0.45$ Btu/lb·°F, $k = 0.1$ Btu/h·ft·°F.

7-9 Verify the expressions given by Eqs. (7-23) and (7-24) for the velocity profile and the mean velocity in laminar flow inside a circular pipe by performing the details of the calculations.

7-10 Derive an expression for the velocity profile for fully developed laminar flow between two parallel plates at a distance $2L$ apart. Also determine an expression for the friction factor.

7-11 Derive an expression for the temperature distribution and the Nusselt number for laminar flow between two large parallel plates in the region of fully developed velocity and temperature profiles for a uniformly applied wall heat flux.

7-12 Water at a mean temperature of 100°F flows inside a $\frac{1}{2}$-in-ID tube whose walls are kept at a uniform temperature of 200°F. Determine the heat-transfer coefficient in the region where the velocity and temperature profiles are considered fully developed.

7-13 Air at 200°F at atmospheric pressure flows through a 1-in-ID tube whose walls are maintained at 100°F. Determine the heat-transfer coefficient for laminar flow in the region where velocity and temperature profiles are considered fully developed.

7-14 A light oil at an inlet temperature of 70°F flows with a mean velocity of .12 ft/s inside a 2-in-ID, 25-ft-long tube whose walls are kept at a uniform temperature of 220°F. Assuming that the heat-transfer coefficient for the fully developed region is applicable for the entire length of the tube, determine the outlet temperature of the oil. (Fluid properties may be taken as $v = 0.001$ ft^2/s, $\rho = 55$ lb/ft^3, $k = 0.08$ Btu/h·ft·°F, $c_p = 0.45$ Btu/lb·°F.)

Chapter 8

8-1 Assuming that the transition from laminar to turbulent flow takes place at a Reynolds number 5×10^5 in laminar boundary-layer flow along a flat plate, determine the location where the transition takes place for the flow of the following fluids with a velocity of 1 ft/s: helium (atm pressure, 100°F), air (atm pressure, 100°F), water, mercury, ethylene glycol, and Freon 114 at 100°F.

8-2 Air at 60°F and atmospheric pressure flows along a flate plate with a velocity $u_\infty = 25$ ft/s.

(*a*) Assuming that transition from laminar to turbulent flow takes place at Re $= 5 \times 10^5$, determine the length of the plate over which the boundary layer is laminar.

(*b*) Determine the boundary-layer thickness at distances 2 and 4 in from the leading edge of the plate.

(*c*) Calculate the drag coefficient at distances 2 and 4 in from the leading edge.

(*d*) Calculate the drag force exerted on the plate over the laminar-flow region per foot width of the plate.

8-3 Determine the local-drag coefficient at a distance $x = 1$ ft from the leading edge and the average-drag coefficient over the distance $x = 0$ to $x = 1$ ft for flow along a flat plate with a velocity of 2 ft/s, at a mean temperature of 100°F, of the following fluids: air at atmospheric pressure ($v = 0.18 \times 10^{-3}$ ft^2/s), water ($v = 0.74 \times 10^{-5}$ ft^2/s), and light oil ($v = 27 \times 10^{-5}$ ft^2/s).

8-4 For the flow of air and light oil in Prob. 8-3, determine the velocity and the thermal boundary-layer thicknesses at a distance 1 ft from the leading edge and compare the ratio of δ_t/δ for air and light oil (Pr $= 300$).

8-5 Determine the thickness of the velocity and thermal boundary layers at a distance 1 ft from the leading edge of a flat plate for the flow of mercury at a temperature of 200°F with a velocity of $\frac{1}{3}$ ft/s over a flat plate.

8-6 Air at 200°F, 14.7 lb/in^2 abs., flows with a velocity of $u_\infty = 50$ ft/s over a 2-ft-long flat plate maintained at a uniform temperature of 100°F. Determine the average-drag and heat-transfer coefficients, and the rate of heat transfer between the air and the plate over the entire length, per foot width of the plate.

8-7 Mercury at 70°F flows along a flat plate with a velocity of 0.5 ft/s. Assuming that transition from laminar to turbulent flow takes place at Re $= 5 \times 10^5$, determine the average heat-transfer coefficient and the Nusselt number over the region where the flow is laminar.

8-8 Air at atmospheric pressure and a temperature of 600°F flows with a velocity of 3 ft/s over a flat plate which is maintained at a uniform temperature of 400°F. Assuming that the transition from laminar to turbulent flow takes place at Re $= 5 \times 10^5$, determine the average-drag and heat-transfer coefficients over the length where the flow is laminar. Also compute the heat-transfer rate from the gas to the plate per foot width over the length where the flow is laminar.

8-9 Derive the energy integral equation by retaining the viscous-energy-dissipation term in the boundary-layer equation (6-67).

8-10 Show that the second-degree-polynomial representation of the velocity profile for flow along a flat plate subject to the conditions

$$U\bigg|_{y=0} = 0 \qquad U\bigg|_{y=\delta} = U_\infty \qquad \text{and} \qquad \frac{\partial U}{\partial y}\bigg|_{y=\delta} = 0$$

is given by

$$\frac{U}{U_\infty} = 2\frac{y}{\delta} - \left(\frac{y}{\delta}\right)^2$$

8-11 Show that the fourth-degree-polynomial representation of the velocity profile for flow along a flat plate is given by

$$\frac{U}{U_\infty} = 2\frac{y}{\delta} - 2\left(\frac{y}{\delta}\right)^3 + \left(\frac{y}{\delta}\right)^4$$

8-12 The following information is given:

Pr $\gg 1$ for oils

Pr $\cong 0.7$ for gases

Pr $\ll 1$ for liquid metals

Make a sketch of the velocity and temperature boundary-layer thicknesses for flow along a flat plate for the three cases given above, illustrating their relative thicknesses.

8-13 For laminar boundary-layer flow along a flat plate, derive the expressions for the boundary-layer thickness $\delta(x)$, the local-drag coefficient c_x, and the average-drag coefficient c_L over the length $0 \leq x \leq L$ by using a velocity profile represented by a sinusoidal expression in the form

$$\frac{U(x, y)}{U_\infty} = \sin\left(\frac{\pi}{2}\frac{y}{\delta}\right)$$

8-14 Repeat Prob. 8-13 by using a linear velocity profile given in the form

$$\frac{U(x, y)}{U_\infty} = \frac{y}{\delta}$$

8-15 Repeat Prob. 8-13 by using a second-degree-polynomial representation of the velocity profile given in the form

$$\frac{U(x, y)}{U_\infty} = 2\frac{y}{\delta} - \left(\frac{y}{\delta}\right)^2$$

8-16 Consider the laminar boundary-layer flow of a *liquid metal* with a velocity U_∞ and at a temperature T_∞ along a flat plate which is kept at a uniform temperature T_∞. Derive the expressions for the thermal boundary-layer thickness $\delta_t(x)$, and the local Nusselt number $\text{Nu}_x \equiv hx/k$ by using a linear profile for the temperature distribution given in the form

$$\frac{T(x, y) - T_w}{T_\infty - T_w} = \frac{y}{\delta_t(x)}$$

8-17 Repeat Prob. 8-16 by using a second-degree-polynomial representation for the temperature profile.

8-18 Consider laminar boundary-layer flow of a fluid having a Prandtl number $\text{Pr} \simeq 1$, with a velocity v_∞, and temperature T_∞ along a flat plate kept at a uniform temperature T_w. Derive the expressions for the thermal boundary-layer thickness $\delta_t(x)$ and the local Nusselt number $\text{Nu}_x \equiv hx/k$ by using a linear velocity profile for the velocity distribution and a second-degree-polynomial representation for the temperature distribution. Compare this result with those derived in Chap. 8 by using cubic velocity and temperature profiles.

8-19 Air at $\frac{1}{30}$ atm, at temperature $T_\infty = 400°\text{R}$, flows with a velocity $u_\infty = 3000$ ft/s over a $\frac{1}{2}$-ft-long flat plate. If the plate is to be kept at a uniform temperature $T_w = 600°\text{R}$, determine the amount of cooling needed over the $\frac{1}{2}$-ft length per foot width of the plate.

8-20 Air at 1 lb/in² abs. and $T_\infty = 500°R$ flows over a flat plate with a velocity of 2000 ft/s. The plate surface is to be maintained at a uniform temperature $T_w = 600°R$. Determine the amount of cooling needed per foot width of the plate over the length where the flow is laminar. Assume that transition from laminar to turbulent flow takes place at Re = 5×10^5.

Chapter 9

9-1 Water at 100°F flows with a velocity of 5 ft/s inside a 12-in-ID water main. The pipe is made of clean cast iron.

(a) Determine the friction factor.

(b) Determine the pressure drop over a 1-mi length of the pipe.

(c) Determine the power required (in ft·lb/s or hp) to pump the water over a length of 1 mi of the water main.

9-2 After many years of service, field tests indicate that the roughness of the pipe in Prob. 9-1 increased by 15 percent. What is the increase in the friction factor and in the pumping power required for the pumping of water over a 1-mi length of the pipe?

9-3 Compare the pressure drops for the flow of air and helium at 200°F, 20 lb/in² abs., with a velocity of 25 ft/s inside a 1-in-ID, 100-ft-long smooth tube.

9-4 Water at a mean temperature of 100°F flows through a 5-in-ID tube with a mean velocity 5 ft/s. Determine the friction factor and the heat-transfer coefficient.

9-5 Mercury at 200°F flows inside a circular tube. The Reynolds number for the flow is Re = 10^5. Compare the Nusselt number for the uniform wall heat-flux and uniform wall-temperature conditions for the following cases: (a) in the region away from the inlet; (b) at the inlet region for $x/D = 10$, 20, and 50.

9-6 Liquid sodium at 400°F flows inside a 1-in-ID, 10-ft-long tube which is kept at a uniform temperature. The Reynolds number for the flow is Re = 10^4. Determine the heat-transfer rate between the fluid and the wall for a temperature difference of 100°F from the liquid to the wall surface.

9-7 Air at 200°F and atmospheric pressure flows with a velocity 100 ft/s through a 2-in-ID, 4-ft-long tube. Determine the average heat-transfer coefficient by using the equation with appropriate L/D correction.

9-8 Using the Reynolds analogy, derive the relation between the heat-transfer coefficient and the drag coefficient for turbulent flow over a flat plate.

9-9 Explain the basic assumptions made in the Reynolds, Prandtl, and von Kármán analogies between heat and momentum transfer in turbulent flow.

9-10 Water at 100°F flows inside a 2-in-ID, long, heated tube with a velocity of 10 ft/s. By obtaining the friction factor from the chart in Fig. 9-5, determine the heat-transfer coefficient, using both the Reynolds and the Prandtl analogies. Compare the heat-transfer coefficients obtained in this manner with that calculated by the Dittus-Boelter equation (9-79).

9-11 Water at 200°F flows with a velocity of $u_\infty = 10$ ft/s across a 1-in-OD single cylinder whose surface is kept at a uniform temperature of 100°F. Determine the average heat-transfer coefficient and the heat-transfer rate per linear foot of the cylinder.

9-12 Air at 250°F and 1-atm pressure flows with a velocity of 50 ft/s across a 3-in-OD circular pipe held at 150°F. Determine (a) the average-drag coefficient, (b) the drag force exerted on the pipe per linear foot, (c) the average heat-transfer coefficient, and (d) the heat-transfer rate between the fluid and the pipe per linear foot if the tube surface is maintained at 100°F.

9-13 Oil at 70°F flows with a velocity of 5 ft/s across a 1-in-OD tube which is kept at a uniform temperature of 200°F. Determine the average heat-transfer coefficient and the heat-transfer rate between the fluid and the tube per linear foot of tube. ($v = 10$ ft²/h, $k = 0.0895$ Btu/h·ft·°F, Pr = 55.)

9-14 Water at a temperature of 80°F flows over a 1-in-diameter sphere with a velocity of 20 ft/s. Determine (a) the drag coefficient, (b) the drag force acting on the sphere, (c) the average heat-transfer coefficient, and (d) the heat-transfer rate between the fluid and the sphere if the surface of the sphere is maintained at 120°F.

9-15 Air at a temperature of 150°F and atmospheric pressure flows with a velocity of 50 ft/s over a 6-in-diameter sphere whose surface is kept at a uniform temperature of 250°F. Determine (a) the drag force acting on the sphere, and (b) the heat-transfer rate between the air and the sphere.

9-16 Air at 60°F and atmospheric pressure flows over a tube bank consisting of $\frac{3}{8}$-in-OD tubes arranged 10 rows deep. The flow velocity before the air enters the tube bank is 5 ft/s. Determine the average friction factor and the average heat-transfer coefficient for the following cases:

    (a) Tubes are in equilateral-triangular arrangement as shown in model 1 in Fig. 9-14.

    (b) Tubes are in a square arrangement as shown in model 2 in Fig. 9-14.

9-17 Air at 200°F and 100 lb/in² abs. flows across a tube bank consisting of 1-in-OD tubes in a staggered arrangement and 40 rows deep in the direction of flow. G_{max} for the flow is 4 lb/ft²·s. Determine the average heat-transfer coefficient and the pressure drop. ($S_L/D = 1.5$, $S_T/D = 2.0$.)

9-18 Air at 400°F and 20 lb/in² abs. has a velocity of 10 ft/s before entering a tube bank consisting of $\frac{1}{2}$-in-OD tubes 10 rows deep in the direction of flow and forming a stack 40 tubes high. Tubes are in a line arrangement with $S_L/D = S_T/D = 2$. Determine the heat-transfer coefficient and the pressure drop.

9-19 Repeat Prob. 9-18 for a staggered arrangement of tubes with $S_L/D = S_T/D = 2$.

9-20 Liquid sodium at 800°F flows across a tube bank consisting of $\frac{1}{2}$-in-OD tubes, 50 rows deep in the direction of flow and in an equilateral-triangular arrangement with a pitch-to-diameter ratio of 1.5. The flow velocity before the liquid enters the tube bank is 1 ft/s and the tubes are maintained at a uniform temperature of 400°F. Determine the average heat-transfer coefficient.

9-21 Mercury at a temperature of 500°F flows over a tube bank consisting of $\frac{1}{2}$-in-OD tubes in an equilateral-triangular arrangement with a pitch-to-diameter ratio 1.375 and 60 rows deep in the direction of flow. Determine the heat-transfer coefficient and the pressure drop for $T_w = 300°F$ and $U_\infty = 0.2$ ft/s.

Chapter 10

10-1 A 1-ft-long vertical plate at a uniform temperature of 200°F is in contact with air at 100°F and atmospheric pressure. Determine the average free-convection heat-transfer coefficient over the entire length of the plate and the heat-transfer rate from the plate into the air per foot width of the plate.

10-2 Compare the heat-transfer rates by free convection from a 1-ft-long vertical plate maintained at a uniform temperature of 150 to 50°F air at 1-, 2-, and 3-atm pressures.

10-3 Solve the problem in Example 10-2 by using the SI system of units.

10-4 A 1 by 1 ft electrically heated, thin, vertical plate dissipates heat by free convection from both of its surfaces into atmospheric air at 100°F. The allowable surface temperature of the plate should not exceed 300°F. What is the maximum rate of heat transfer from both surfaces of the plate if the heat-transfer coefficient h_r for radiation is $h_r = 1.5$ Btu/h·ft²·°F?

10-5 What is the heat-transfer rate by free convection from a 1 by 1 ft horizontal plate whose heated surface is maintained at 230°F and faces up into atmospheric air at 70°F?

10-6 Repeat Prob. 10-5 for a plate whose heated surface is facing down.

10-7   Compare the heat-transfer rates by free convection from a 1 by 1 ft plate whose one surface is insulated and the other surface is maintained at 200°F and exposed to atmospheric air at 100°F for the following conditions:

   (*a*)   Plate is vertical.

   (*b*)   Plate is horizontal with heated surface facing up.

   (*c*)   Plate is horizontal with heated surface facing down.

10-8   A 2-in-OD, 5-ft-long vertical tube whose outside surface is at 200°F and is exposed to atmospheric air at 70°F.   Determine the rate of heat loss from the tube by free convection into the surrounding air.

10-9   Compare the heat-transfer rates by free convection from a 1-in-OD, 4-ft-long vertical tube whose surface is kept at a uniform temperature of 140°F into a surrounding containing (*a*) air at atmospheric pressure and 60°F, (*b*) air at 2-atm pressure and 60°F, and (*c*) water at 60°F.

10-10   Repeat Prob. 10-9 for a horizontal tube.

10-11   A 1-ft-diameter sphere whose surface is kept at a uniform temperature of 150°F is submerged in water at 50°F.   Determine the rate of heat loss by free convection from the sphere into the water.

10-12   Atmospheric air is contained between two large horizontal parallel plates separated by a distance of 2 in.   The lower and upper plates are maintained at uniform temperatures of 200 and 100°F, respectively.   Determine the rate of heat transfer between the plates by free convection per square foot of plate surface.

10-13   Atmospheric air at 70°F flows with a velocity of 5 ft/s over a $1\frac{1}{2}$-ft-long vertical plate.   Determine the plate temperature for which the effect of free convection on heat transfer is less than 5 percent.

Chapter 11

11-1   The transmissivity of a plate glass for incident solar radiation at 10,000°R for various wavelength bands is given as

$\tau_1 = 0$ for $\lambda_0 = 0$ to $\lambda_1 = 0.5\ \mu m$

$\tau_2 = 0.7$ for $\lambda_1 = 0.5\ \mu m$ to $\lambda_2 = 2.8\ \mu m$

$\tau_3 = 0$ for $\lambda_2 = 2.8\ \mu m$ to $\lambda_3 \rightarrow \infty\ \mu m$

Determine the average hemispherical transmissivity of the glass over the entire wavelengths.

11-2 Fused quartz transmits 85 percent of the thermal radiation at 3000°R in the wavelength band $\lambda_1 = 0.3$ μm to $\lambda_2 = 3$ μm and is opaque to the radiation outside this range. If a blackbody radiation source at 3000°R is placed in front of this fused quartz sheet, what is the rate of energy transmitted through a 1-ft² area of the quartz sheet?

11-3 The sun radiates as a blackbody at 10,000°R. What fraction of the total energy is the ultraviolet (that is, $\lambda = 0.01$ to 0.4 μm), visible (that is, $\lambda = 0.4$ to 0.7 μm), and infrared (that is, $\lambda = 0.7$ to 1000 μm) regions?

11-4 A tungsten filament is heated to 4000°R. What is the maximum radiative-heat flux from the filament, and what fraction of this energy is in the visible range (that is, $\lambda = 0.4$ to 0.7 μm)?

11-5 A red-hot surface is at 3000°R. What fraction of the total radiation emitted is in the following wavelength bands:

$$\Delta\lambda_1 = \quad 1 \text{ to } \quad 5 \text{ } \mu m$$

$$\Delta\lambda_2 = \quad 5 \text{ to } 10 \text{ } \mu m$$

$$\Delta\lambda_3 = 10 \text{ to } 15 \text{ } \mu m$$

$$\Delta\lambda_4 = 15 \text{ to } 20 \text{ } \mu m$$

11-6 Determine the radiative energy emitted between 2- and 10-μm wavelengths by a 1 by 1 ft gray surface at 1000°R which has an emissivity $\varepsilon = 0.8$.

11-7 There is a $\frac{1}{10}$-in-diameter hole in a laboratory black enclosure which is at 1000°R. Determine the rate of emission of radiative energy through this opening.

11-8 A plain glass has transmissivity $\tau = 0.90$ in the range $\lambda = 0.2$ to 3 μm and zero transmissivity for other wavelengths. A tinted glass has transmissivity $\tau = 0.90$ in the range $\lambda = 0.5$ to 1 μm and zero transmissivity for all other wavelengths. If the solar energy is incident on both glasses, compare the energy transmitted through each of these glasses.

11-9 A tungsten filament is heated to 4000°R. What fraction of the total energy is emitted in the wavelength range $\lambda = 0.4$ to 0.8 μm?

11-10 A blackbody radiation is emitted on a quartz sheet which transmits 90 percent of the incident radiation in the wavelength range $\lambda = 0.2$ to 4 μm. Calculate the percent of the incident radiation transmitted through the quartz sheet for each of the blackbody radiation sources at 1000, 2000, and 4000°R.

11-11 A laboratory black enclosure at 1500°R has a small opening into the atmosphere. Calculate (a) the blackbody radiation intensity emerging from the opening and (b) the blackbody radiation heat flux from the enclosure.

11-12 The average inside temperature of an oven is 3000°R, and the emissivity of the inside surface is $\varepsilon = 0.8$ at this temperature. Determine the radiative energy leaving the oven through an opening 6 by 6 in.

11-13 The surface of a satellite receives solar radiation at a rate of 400 Btu/h·ft². The surface has an absorptivity of $\alpha = 0.8$ for solar radiation and an emissivity of $\varepsilon = 0.9$. Assuming no heat losses into the satellite and a heat dissipation by thermal radiation into the space at absolute zero, calculate the equilibrium temperature of the surface.

11-14 A surface receives solar radiation at a rate of 300 Btu/h·ft² while the other side is kept insulated. The absorptivity of the surface to solar radiation is $\alpha = 0.9$ while its emissivity is $\varepsilon = 0.5$. Assuming the surface loses heat by radiation into a clear sky at an effective temperature of $-50°F$, determine the equilibrium temperature of the surface.

11-15 Repeat Prob. 11-14 for an aluminum surface which has a solar absorptivity $\alpha = 0.15$ and emissivity $\varepsilon = 0.1$.

11-16 A space radiator is to dissipate 2.46×10^4 Btu/h·ft² by thermal radiation into an environment at absolute zero temperature. If the surface has an emissivity $\varepsilon = 0.9$, determine the equilibrium temperature of the radiator surface.

11-17 A space radiator is to dissipate heat by thermal radiation into an environment at absolute zero temperature. If the maximum allowable surface temperature is 2500°R, what is the maximum heat-transfer rate per square foot of surface for a surface emissivity of $\varepsilon = 0.8$?

Chapter 12

12-1 Determine the view factor F_{1-2} between an elemental surface dA_1 and the finite rectangular surface A_2 for the geometrical arrangements shown in the accompanying figure.

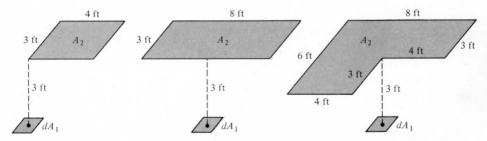

12-2 Determine the view factor F_{1-2} between two rectangular surfaces A_1 and A_2 for the geometrical arrangements shown in the accompanying figure.

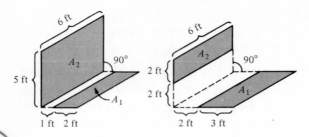

12-3 Two black square plates each 3 by 3 ft in size, 3 ft apart, as shown in the accompanying figure, are located in a large room whose walls are black and maintained at 500°R. Determine the net radiative-heat exchange between the plates if one of the plates is kept at 1500°R and the other at 1000°R.

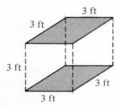

12-4 In a cubical enclosure there are 36 individual view factors between the six surfaces of the enclosure. How many different view factors are to be computed?

12-5 Two black rectangular surfaces A_1 and A_2, arranged as shown in the accompanying figure, are located in a large room whose walls are black and kept at 460°R. Determine the net radiative-heat exchange between these two surfaces when A_1 is kept at 2000°R and A_2 at 1000°R. (Neglect the radiation from the room.)

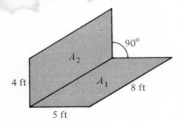

12-6 A 1-ft-diameter sphere whose surfaces are kept at 1000°R is suspended in a large room whose walls are black and kept at 500°R. Determine the radiative-heat loss from the sphere for the following cases:

(a) The sphere surface is black.

(b) The sphere surface has an emissivity $\varepsilon = 0.1$.

12-7   A furnace whose inside surface is at 2000°R and black has a 1-ft² opening into a room whose walls are kept at 500°R.   Determine the rate of radiative-heat loss from the furnace.

12-8   A 5-in-OD, 10-ft-long steam pipe whose surfaces are at 240°F passes through a room whose walls are at 40°F.   Assuming the emissivity of the pipe $\varepsilon = 0.9$, determine the rate of heat loss from the pipe by radiation.

12-9   In a 10 by 10 by 10 ft test room the ceiling is kept at 140°F while the walls and the floor are at 40°F.   Assuming that all surfaces have an emissivity $\varepsilon = 0.8$, determine the rate of heat loss from the ceiling by radiation.

12-10   In a 5 by 5 by 5 ft cubical furnace the ceiling is at 2000°R, the floor at 1000°R, and the walls are refractory (reradiating) surfaces.   Assuming that both the ceiling and the floor are black, determine the net radiative-heat exchange between the ceiling and the floor.

12-11   Repeat Prob. 12-10 for the following surface emissivities:

(*a*) $\varepsilon_{ceiling} = 1.0$ $\varepsilon_{floor} = 0.8$

(*b*) $\varepsilon_{ceiling} = 0.8$ $\varepsilon_{floor} = 0.8$

12-12   Consider a three-zone enclosure which is infinitely long in the direction perpendicular to the plane of the accompanying figure.   Surfaces 1 and 2 are kept at prescribed uniform heat fluxes q_1 and q_2, respectively, while surface 3 is kept at a uniform temperature T_3.

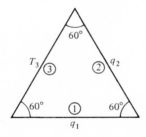

(*a*)   Write the equations for the determination of radiosities R_1, R_2, and R_3; give also the numerical values of the view factors.

(*b*)   Write the equations for the determination of temperatures T_1 and and T_2 on surfaces 1 and 2.

(*c*)   Write the equation for the determination of radiative-heat flux q_3 on surface 3.

12-13   Consider a three-zone enclosure as shown in the accompanying figure. The enclosure is of infinite extent in the direction perpendicular to the plane of the figure.   The surfaces are opaque, gray, diffuse emitters and diffuse reflectors.   Surfaces 1 and 2 are maintained at uniform temperatures T_1 and T_2, respectively, and surface 3 is a *reradiating* surface.

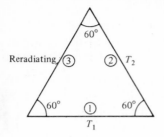

(a)   Write the appropriate equations for the determination of the radiosities R_1, R_2, and R_3 for the three zones.

(b)   Write the appropriate equations for the determination of radiative-heat fluxes at the surfaces of zones 1 and 2.

(c)   Give the numerical values of the view factors appearing in these equations.

12-14 Consider a four-zone enclosure with all sides equal as shown in the accompanying figure.   Temperatures are prescribed for surfaces 1 and 2 and the net heat fluxes are prescribed for surfaces 3 and 4.   Emissivities and reflectivities are as indicated on the surfaces.

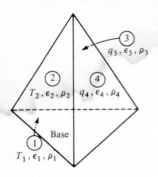

(a)   Write the appropriate equations for determining the radiosities R_1, R_2, R_3, and R_4 for these four surfaces.

(b)   Write the values of the view factors appearing in these equations.

12-15 A 1-in-OD, long pipe (surface 1) whose surfaces are kept at $T_1 = 2000°R$ is enclosed by an outer cylinder (surface 2) of 3 in diameter whose surfaces are kept at $T_2 = 500°R$.   Determine the rate of heat loss by radiation per linear foot of the inner pipe for the following conditions:

(a) $\varepsilon_1 = \varepsilon_2 = 1$

(b) $\varepsilon_1 = 1$ $\varepsilon_2 = 0.1$

(c) $\varepsilon_1 = 0.1$ $\varepsilon_2 = 0.1$

12-16 Consider two concentric spheres of 1 and 2 in diameter. The inner sphere is black and kept at a temperature $T_1 = 2000°R$, while the outer sphere has an emissivity $\varepsilon_2 = 0.8$ and is kept at a temperature $T_2 = 1000°R$. Determine the rate of radiative-heat exchange between the two spheres.

Chapter 13

13-1 Two infinitely large, parallel, black plates are separated by an absorbing, emitting gas. The optical thickness of the gas between the plates is $\tau_0 = 1$. If the gas is at $T_0 = 1500°R$ and the plates are at $T_1 = T_2 = 500°R$, determine the net radiative-heat flux at the surface of the plates.

13-2 A gas mass at a temperature $T_g = 2000°R$ and at a total pressure $P_T = 2$ atm contains 5% water vapor. Determine the emissivity of this gas mass for an equivalent path length of $L = 5$ ft.

13-3 A beam of radiation passes through an absorbing gas layer of thickness L and absorptivity κ_λ at the wavelength λ. Show that the spectral absorptivity α_λ of this gas layer is given by

$$\alpha_\lambda = 1 - e^{-\kappa_\lambda L}$$

and the spectral transmissivity τ_λ is given by

$$\tau_\lambda = e^{-\kappa_\lambda L}$$

13-4 An absorbing and emitting gas at a temperature $T_g = 2500°R$ flows between two parallel plates kept at temperatures $T_1 = 1500°R$ and $T_2 = 1000°R$. Determine the net radiative-heat flux at the surface of plate 2 for the following optical thicknesses of the gas: $\tau_0 = 0$, 0.1, and 1.

13-5 Determine the emissivity of a gas mass at a temperature $T_g = 1500°R$, at a total pressure of $P_T = 3$ atm, when the gas contains 15% water vapor and has a geometry of equivalent path length of $L = 10$ ft.

13-6 A flue gas at $T_g = 2000°R$, at a total pressure of $P_T = 2$ atm, and containing 15% water vapor flows over a tube bank with the tube centers on squares and the clearance equal to the tube diameter (i.e., case 6 in Table 13-1). Tube surfaces are black and at a temperature $T_w = 1000°R$. Determine the net radiative-heat flux between the gas and the tubes. ($D = 2$ in.)

13-7 A flue gas at a temperature $T_g = 2500°R$, at a total pressure of $P_T = 3$ atm, and containing 10% water vapor flows over a sphere of diameter

$D = 1$ ft (i.e., case 1 in Table 13-1). The sphere is black and at a temperature of $T_w = 1000°R$. Determine the net radiative-heat exchange between the gas and the sphere.

Chapter 14

14-1 Air-free steam at atmospheric pressure condenses on a 2-in-OD, 1-ft-long vertical tube maintained at a uniform temperature of 150°F. Assuming filmwise condensation, determine the average condensation heat-transfer coefficient over the entire length of the tube and calculate the total rate of condensate flow at the bottom of the tube.

14-2 Repeat Prob. 14-1 for the tube in a horizontal position.

14-3 Saturated steam at 170°F is to be condensed on a 2-in-OD vertical tube whose surface is kept at 150°F. Determine the tube length to condense 20 lb/h steam.

14-4 Air-free steam at 2.222 lb/in² abs. (i.e., saturation temperature 130°F) condenses on a 1-ft-long inclined plate making an angle $\varphi = 60°$ with the horizontal. The plate surface is kept at a uniform temperature of 100°F. Determine the plate length beyond which the condensate flow changes from laminar to turbulent. Calculate the average-condensation heat-transfer coefficient over the laminar-flow region and the rate of condensate flow per foot width of the plate at the location where the condensate flow changes from laminar to turbulent.

14-5 Compare the average-condensation heat-transfer coefficient for the filmwise condensation of air-free steam at atmospheric pressure on a 2-ft-long vertical surface and on twenty 1-in-OD horizontal tubes arranged in a vertical tier.

14-6 Air-free, saturated steam at 50 lb/in² abs. condenses inside a 3-in-ID, 2-ft-long vertical tube maintained at a uniform temperature of $T_w = 230°F$. Assuming filmwise condensation and small steam velocity, determine the average value of the condensation heat-transfer coefficient over the entire length of the tube and the rate of condensate flow at the bottom of the tube.

14-7 Air-free, saturated steam at atmospheric pressure condenses on a vertical tube whose surface is kept at 150°F. What length of tube would produce a turbulent film condensation?

14-8 A heated, *scored-copper plate* is immersed in a pool of water at saturation temperature and atmospheric pressure. If the plate surface is kept 40°F above the saturation temperature of the water, determine the rate of evaporation per square foot of surface of the plate. What would be the rate of evaporation with a heater of *clean copper plate*?

14-9 Water at saturation temperature and atmospheric pressure is boiled by an electrically heated platinum wire of 0.075 in diameter. Boiling takes place in the stable film-boiling region with a temperature difference of $T_w - T_s = 800°F$. Determine the boiling heat-transfer coefficient.

14-10 Determine the heat flux and the nucleate-boiling heat-transfer coefficient for the pool boiling of water at saturation temperature and atmospheric pressure on an emery-polished copper-surface heater plate with a temperature difference of $T_w - T_s = 25°F$.

14-11 Determine the maximum heat flux obtainable with a nucleate pool boiling of water at earth's gravitational field at 1- and 10-atm pressures.

14-12 Repeat Prob. 14-11 for boiling of water in a gravitational field equal to one-eighth that of the earth.

14-13 Determine the rate of evaporation per square foot of the heater surface for the nucleate boiling of water at saturation temperature and atmospheric pressure with a vertical stainless steel heater at 240°F.

14-14 Determine the stable film-boiling heat-transfer coefficient for the film boiling of saturated water at atmospheric pressure on an electrically heated, 0.08-in-diameter horizontal platinum wire with a temperature difference of $T_w - T_s = 1000°F$.

Chapter 15

15-1 The inlet and outlet temperature differences in a tube-and-shell-type heat exchanger are, respectively, $\Delta T_0 = 80°F$ and $\Delta T_L = 20°F$. Compare the logarithmic mean temperature difference with the arithmetic mean temperature difference.

15-2 Repeat Prob. 15-1 for the case of temperature differences $\Delta T_0 = 80°F$ and $\Delta T_L = 40°F$.

15-3 The design heat load for an oil cooler is 10^6 Btu/h. Determine the total heat-transfer area required for the inlet and outlet temperature differences $\Delta T_0 = 75°F$ and $\Delta T_L = 25°F$, respectively, while the average value of the overall heat-transfer coefficient is $U = 120$ Btu/h·ft^2·°F.

15-4 Water is to be cooled from 72 to 44°F by using brine at an inlet temperature of 29°F with a temperature rise 8°F. Determine the total heat-transfer area required for the design heat load of 30,000 Btu/h with a counterflow arrangement and an overall heat-transfer coefficient of 120 Btu/h·ft^2·°F.

15-5 A shell-and-tube type of heat exchanger is to cool 48,000 lb/h of oil $(c_p = 0.5$ Btu/lb·°F$)$ from 150 to 102°F by using 32,000 lb/h of water at an inlet temperature of 70°F. The average overall heat-transfer coefficient is 136 Btu/h·ft²·°F. By using the effectiveness method, determine the heat-transfer area required for a single-shell pass heat exchanger of the type shown in Fig. 15-17.

15-6 20,000 lb/h of water is to be heated from 80 to 180°F by the hot exhaust gases $(c_p = 0.25$ Btu/lb·°F) entering the heat exchanger at 420°F and leaving it at 200°F. Determine the surface area required for a heat exchanger of the counterflow arrangement with an overall heat-transfer coefficient of $U = 40$ Btu/h·ft²·°F.

15-7 Determine the overall heat-transfer coefficient based on the outer surface area of a 1.125-in-OD, 0.995-in-ID brass tube $(k = 60$ Btu/h·ft·°F) if the heat-transfer coefficients for flow inside and outside the tube are $h_i = 300$ Btu/h·ft²·°F and $h_o = 600$ Btu/h·ft²·°F, respectively, and the unit fouling resistances for the inside and outside surfaces are $F_i = F_o = 0.004$ h·ft²·°F/Btu.

15-8 Determine the overall heat-transfer coefficient based on the outer surface of a 1-in-OD, 0.9-in-ID heat-exchanger tube $(k = 70$ Btu/h·ft·°F) if the heat-transfer coefficients at the inside and outside of the tube are $h_i = 1000$ Btu/h·ft²·°F and $h_o = 700$ Btu/h·ft²·°F, respectively, and the unit fouling resistances are $F_o = F_i = 0.002$ h·ft²·°F/Btu.

15-9 25,000 lb/h of water is to be cooled from 180 to 140°F by passing it through a heat exchanger in which 40,000 lb/h of cold water enters at 80°F. If the overall heat-transfer coefficient is 200 Btu/h·ft²·°F, calculate the total heat-transfer area required for both the counterflow and the parallel-flow arrangements.

15-10 Determine the effectiveness of a counterflow heat exchanger in which oil $(c_p = 0.5$ Btu/lb·°F) is heated from 80 to 130°F with hot water entering at 190°F and leaving at 100°F.

15-11 10,000 lb/h of water is to be heated from 80 to 180°F by using oil $(c_p = 0.5$ Btu/lb·°F) at a rate of 18,000 lb/h at an inlet temperature of 300°F. The overall heat-transfer coefficient is $U = 100$ Btu/h·ft²·°F. Using the effectiveness method, determine the heat-transfer area required for a counterflow double-pipe heat exchanger.

15-12 Steam condensing at atmospheric pressure on the shell side of a heat exchanger will heat 25,000 lb/h of water from 80 to 180°F flowing inside the tubes. Determine the heat-transfer area required for an overall heat-transfer coefficient of $U = 500$ Btu/h·ft²·°F.

15-13 A single-pass tube-and-shell-type steam condenser is designed to condense 200 lb/h steam at 2.22 lb/in² abs. (i.e., saturation temperature 130°F) by using cooling water at 70°F with a temperature rise of 20°F. Determine the condenser surface area for an overall heat-transfer coefficient of $U = 500$ Btu/h·ft²·°F.

15-14 Steam is condensed at 220°F on the outside of a 2-in-OD tube with water flowing at a rate of 30,000 lb/h inside the tube. The overall heat-transfer coefficient based on the outer surface of the tube is $U = 150$ Btu/h·ft²·°F. Determine the tube length to raise the water temperature from 70 to 150°F.

Chapter 16

16-1 Two large vessels contain uniform mixtures of air (component A) and sulfur dioxide (component B) at 1-atm pressure, 492°R, but at different concentrations. Vessel 1 contains 80 mole percent air and 20 mole percent sulfur dioxide, while vessel 2 contains 30 mole percent air and 70 mole percent sulfur dioxide. The vessels are connected by a 4-in-ID, 6-ft-long pipe. Determine the rate of transfer of air between these two vessels by assuming a steady-state transfer takes place.

16-2 Repeat Prob. 16-1 for an air and ammonia mixture.

16-3 An open tank contains benzene at atmospheric pressure and 536°R temperature. The distance from the surface of the benzene layer to the top of the tank is 10 ft. It is assumed that the air in the vessel is motionless while there is sufficient air motion outside to remove the benzene vapor arriving at the top surface. Determine the rate of evaporation of benzene per square foot of the benzene surface. ($p_{A1} = 0.01$ atm.)

16-4 Repeat Prob. 16-3 for naphthalene inside the tank. ($p_{A1} = 0.01$ atm.)

16-5 Calculate the mass diffusivity of the binary gas mixture of air-iodine at 273 K at 1-atm pressure, and compare the result with that given in Table 16-3.

16-6 Repeat Prob. 16-5 for pressures of 5 and 10 atm.

16-7 Repeat Prob. 16-5 for a temperature of 819 K by using Eq. (16-48).

16-8 Calculate the mass diffusivity D_{AB} for a dilute solution of methanol (component A, the solute) in water (component B, the solvent) at 18°C, and compare the result with that given in Table 16-5.

16-9 Determine the mass diffusivity D_{AB} for a dilute solution of acetone (component A, the solute) and benzene (component B, the solvent) at 15°C, and compare the result with that given in Table 16-5. ($\tilde{V}_A = 77.$)

16-10 Repeat Prob. 16-9 for acetic acid (component A) in benzene (component B) at 15°C. ($\tilde{V}_A = 64.$)

16-11 A slab of clay of thickness $2L$ contains C_0 mass percent of water distributed uniformly over the entire volume. The slab is subjected to drying by blowing a stream of low-humidity air over both of its surfaces. The mass diffusivity of water vapor through the clay is D.

(a) Give the mathematical formulation of this mass-diffusion problem. (Note that this problem can be expressed in the same form as that given by Example 16-3.)

(b) Derive an expression for the concentration of water vapor as a function of time and position in the slab.

(c) Derive an expression for the rate of removal of water vapor per unit area of the surface of the slab.

16-12 Air at 50°F and atmospheric pressure flows over a plane surface covered with naphthalene. The flow velocity is such that the Reynolds number at a distance of 2 ft from the leading edge of the plate is $Re = 9 \times 10^4$. Determine the average mass-transfer coefficient for the transfer of naphthalene over the 2-ft length of the surface by making use of the analogy between heat and mass transfer in laminar flow along a flat plate.

16-13 Air at atmospheric pressure and 77°F, containing small quantities of iodine, flows along a flat plate. The flow velocity is such that the Reynolds number at a distance of $x = 4$ ft from the leading edge of the plate is $Re = 9 \times 10^4$. Determine the average mass-transfer coefficient for the transfer of iodine from the air stream into the wall surface over the 4-ft length of the plate.

16-14 In Prob. 16-13, let c_0 be the mass concentration of iodine in the main air stream. It is assumed that the plate surface is a perfect sink for iodine so that the mass concentration of iodine in the gas at the immediate vicinity of the plate surface is zero. Write the expression for the rate of transfer of iodine from the gas stream to the plate surface per foot width over the 4-ft length of the plate.

16-15 Air at 77°F and atmospheric pressure flows with a velocity of 24 ft/s inside a 1-in-ID pipe. The inside surface of the tube contains a deposit of naphthalene. Determine the mass-transfer coefficient for the transfer of naphthalene from the pipe surface into the air in regions away from the inlet.

16-16 Dry air at atmospheric pressure and 50°F flows over a flat plate with a velocity of 3 ft/s. The plate is covered with a film of water which evaporates into the air stream. Determine the average mass-transfer coefficient for the transfer of water vapor over a distance of 2 ft from the leading edge of the plate.

Index

CONVERSION FACTORS

1. Acceleration
   $1 \text{ ft/s}^2 = 0.3048 \text{ m/s}^2$
   $1 \text{ m/s}^2 = 3.2808 \text{ ft/s}^2$

2. Area
   $1 \text{ in}^2 = 6.4516 \text{ cm}^2$
   $1 \text{ in}^2 = 6.4516 \times 10^{-4} \text{ m}^2$
   $1 \text{ ft}^2 = 929 \text{ cm}^2$
   $1 \text{ ft}^2 = 0.0929 \text{ m}^2$
   $1 \text{ m}^2 = 10.764 \text{ ft}^2$

3. Density
   $1 \text{ lb/in}^3 = 27.680 \text{ g/cm}^3$
   $1 \text{ lb/in}^3 = 27.680 \times 10^3 \text{ kg/m}^3$
   $1 \text{ lb/ft}^3 = 16.019 \text{ kg/m}^3$
   $1 \text{ kg/m}^3 = 0.06243 \text{ lb/ft}^3$
   $1 \text{ slug/ft}^3 = 515.38 \text{ kg/m}^3$
   $1 \text{ lb mol/ft}^3 = 16.019 \text{ kg mol/m}^3$
   $1 \text{ kg mol/m}^3 = 0.06243 \text{ lb mol/ft}^3$

4. Diffusivity (heat, mass, momentum)
   $1 \text{ ft}^2/\text{s} = 0.0929 \text{ m}^2/\text{s}$
   $1 \text{ ft}^2/\text{h} = 0.2581 \text{ cm}^2/\text{s}$
   $1 \text{ ft}^2/\text{h} = 0.2581 \times 10^{-4} \text{ m}^2/\text{s}$
   $1 \text{ m}^2/\text{s} = 10.7639 \text{ ft}^2/\text{s}$
   $1 \text{ cm}^2/\text{s} = 3.8745 \text{ ft}^2/\text{h}$

5. Energy, heat, power
   $1 \text{ J} = 1 \text{ W} \cdot \text{s} = 1 \text{ N} \cdot \text{m}$
   $1 \text{ J} = 10^7 \text{ erg}$
   $1 \text{ Btu} = 1055.04 \text{ J}$
   $1 \text{ Btu} = 1055.04 \text{ W} \cdot \text{s}$
   $1 \text{ Btu} = 1055.04 \text{ N} \cdot \text{m}$
   $1 \text{ Btu} = 252 \text{ cal}$
   $1 \text{ Btu} = 0.252 \text{ kcal}$
   $1 \text{ Btu} = 778.161 \text{ ft} \cdot \text{lb}_f$
   $1 \text{ Btu/h} = 0.2931 \text{ W}$
   $1 \text{ Btu/h} = 0.2931 \times 10^{-3} \text{ kW}$
   $1 \text{ Btu/h} = 3.93 \times 10^{-4} \text{ hp}$
   $1 \text{ cal} = 4.1868 \text{ J (or W} \cdot \text{s or N} \cdot \text{m)}$
   $1 \text{ cal} = 3.968 \times 10^{-3} \text{ Btu}$
   $1 \text{ kcal} = 3.968 \text{ Btu}$
   $1 \text{ hp} = 550 \text{ ft} \cdot \text{lb}_f/\text{s}$
   $1 \text{ hp} = 745.7 \text{ W}$
   $1 \text{ Wh} = 3.413 \text{ Btu}$
   $1 \text{ kWh} = 3413 \text{ Btu}$

6. Heat capacity, heat per unit mass, specific heat
   $1 \text{ Btu/h} \cdot {}^\circ\text{F} = 0.5274 \text{ W/}^\circ\text{C}$
   $1 \text{ W/}^\circ\text{C} = 1.8961 \text{ Btu/h} \cdot {}^\circ\text{F}$
   $1 \text{ Btu/lb} = 2325.9 \text{ J/kg}$
   $1 \text{ Btu/lb} = 2.3259 \text{ kJ/kg}$
   $1 \text{ Btu/lb} \cdot {}^\circ\text{F} = 4186.69 \text{ J/kg} \cdot {}^\circ\text{C}$
   $1 \text{ Btu/lb} \cdot {}^\circ\text{F} = 4.18669 \text{ kJ/kg} \cdot {}^\circ\text{C}$
   (or $\text{J/g} \cdot {}^\circ\text{C}$)
   $1 \text{ Btu/lb} \cdot {}^\circ\text{F} = 1 \text{ cal/g} \cdot {}^\circ\text{C} = 1 \text{ kcal/kg} \cdot {}^\circ\text{C}$

7. Heat flux
   $1 \text{ Btu/h} \cdot \text{ft}^2 = 3.1537 \text{ W/m}^2$
   $1 \text{ Btu/h} \cdot \text{ft}^2 = 3.1537 \times 10^{-3} \text{ kW/m}^2$
   $1 \text{ W/m}^2 = 0.31709 \text{ Btu/h} \cdot \text{ft}^2$

8. Heat-generation rate
   $1 \text{ Btu/h} \cdot \text{ft}^3 = 10.35 \text{ W/m}^3$
   $1 \text{ Btu/h} \cdot \text{ft}^3 = 8.9 \text{ kcal/h} \cdot \text{m}^3$
   $1 \text{ W/m}^3 = 0.0966 \text{ Btu/h} \cdot \text{ft}^3$

9. Heat-transfer coefficient
   $1 \text{ Btu/h} \cdot \text{ft}^2 \cdot {}^\circ\text{F} = 5.677 \text{ W/m}^2 \cdot {}^\circ\text{C}$
   $1 \text{ Btu/h} \cdot \text{ft}^2 \cdot {}^\circ\text{F} = 5.677 \times 10^{-4} \text{ W/cm}^2 \cdot {}^\circ\text{C}$
   $1 \text{ W/m}^2 \cdot {}^\circ\text{C} = 0.1761 \text{ Btu/h} \cdot \text{ft}^2 \cdot {}^\circ\text{F}$
   $1 \text{ Btu/h} \cdot \text{ft}^2 \cdot {}^\circ\text{F} = 4.882 \text{ kcal/h} \cdot \text{m}^2 \cdot {}^\circ\text{C}$

10. Length
    $1 \text{ Å} = 10^{-8} \text{ cm}$
    $1 \text{ Å} = 10^{-10} \text{ m}$
    $1 \text{ μm} = 10^{-3} \text{ mm}$
    $1 \text{ μm} = 10^{-4} \text{ cm}$
    $1 \text{ μm} = 10^{-6} \text{ m}$
    $1 \text{ in} = 2.54 \text{ cm}$
    $1 \text{ in} = 2.54 \times 10^{-2} \text{ m}$
    $1 \text{ ft} = 0.3048 \text{ m}$
    $1 \text{ m} = 3.2808 \text{ ft}$
    $1 \text{ mile} = 1609.34 \text{ m}$
    $1 \text{ mile} = 5280 \text{ ft}$
    $1 \text{ light year} = 9.46 \times 10^{15} \text{ m}$

11. Mass
    $1 \text{ oz} = 28.35 \text{ g}$
    $1 \text{ lb} = 16 \text{ oz}$
    $1 \text{ lb} = 453.6 \text{ g}$
    $1 \text{ lb} = 0.4536 \text{ kg}$
    $1 \text{ kg} = 2.2046 \text{ lb}$
    $1 \text{ g} = 15.432 \text{ grains}$
    $1 \text{ slug} = 32.1739 \text{ lb}$
    $1 \text{ ton (metric)} = 1000 \text{ kg}$
    $1 \text{ ton (metric)} = 2205 \text{ lb}$
    $1 \text{ ton (short)} = 2000 \text{ lb}$
    $1 \text{ ton (long)} = 2240 \text{ lb}$

12. Mass flux
    $1 \text{ lb mol/ft}^2 \cdot \text{h}$
    $\quad = 1.3563 \times 10^{-3} \text{ kg mol/m}^2 \cdot \text{s}$
    $1 \text{ kg mol/m}^2 \cdot \text{s} = 737.3 \text{ lb mol/ft}^2 \cdot \text{h}$
    $1 \text{ lb/ft}^2 \cdot \text{h} = 1.3563 \times 10^{-3} \text{ kg/m}^2 \cdot \text{s}$
    $1 \text{ lb/ft}^2 \cdot \text{s} = 4.882 \text{ kg/m}^2 \cdot \text{s}$
    $1 \text{ kg/m}^2 \cdot \text{s} = 737.3 \text{ lb/ft}^2 \cdot \text{h}$
    $1 \text{ kg/m}^2 \cdot \text{s} = 0.2048 \text{ lb/ft}^2 \cdot \text{s}$

13. Pressure, force
    $1 \text{ N} = 1 \text{ kg} \cdot \text{m/s}^2$
    $1 \text{ N} = 0.22481 \text{ lb}_f$
    $1 \text{ N} = 7.2333 \text{ poundals}$
    $1 \text{ N} = 10^5 \text{ dyn}$